MATHEMATICS

A CALCULUS

(For 12th , Engineering)

Table of Contents

Functions 3
Limit, Continuity & Differentiability 28
Maxima and Minima 55
Tangent and Normal 69
Differentiation 89
Differential Equation 221

Functions

Important Definitions: – (1) Defined functions, Dependent and independent variables. A function f(x)of x is defined for a certain value of x in its domain provided it attains a unique and definite value for that value of x.

e.g. $y = \frac{3x}{2x-3}$ is not defined for $x = \frac{3}{2}$ (not definite)

$y = \frac{\sin x}{x}$ is not defined for $x = 0$ (Indeterminate)

$y = \sqrt{5x^2}$ is not defined for any $x \in R$ (not unique)

If $y = f(x)$, then x is called independent variable and y the dependent variable. Just as a area of a circle is a function of radius r.

i.e. $A = \pi r^2 = f(r)$, Area is dependent on r, r is independent variable and A is dependent.

(2) **Range and Domain of a function**: – Let $y = f(x)$ and if this function is defined for all values of x which lie between a and b if $a < x < b$ *then the open interval* (a, b) or]a, b[will constitute the domain. If, however, f(x) is alsodefined both for x = a and b, then the domain will consist of the closed interval [a, b].

Range of $f(x) = \{y : y = f(x), x \in \text{domain}\}$.

e.g. $y = \frac{3x-2}{x-1}$, let $y = f(x) = \frac{3x-2}{x-1}$, f(x)is not defined $x - 1 = 0$ or $x = 1$

Hence, domain consist of all real value of x except x = 1. or domain $= R - \{1\}, D = x \in R - \{1\}$, whereas domain denoted by D

Range, $y = f(x) = \frac{3x-2}{x-1}$ or $(x-1)y = 3x - 2$ or $xy - y = 3x - 2$ or $xy - 3x = y - 2$ or $x(y-3) = y - 2$ or $x = \frac{y-2}{y-3}$ or $x = f(y)$

or x is not defined for value y = 3, x is defined all real values of y except y = 3 Finally, Range $= R - \{3\}$

The following formula will be useful for finding the domain: –

Let $f_1(x)$ and $f_2(x)$ be two functions whose domain are D_1 and D_2 respectively, then (i) $\text{Dom}[f_1(x) + f_2(x)] = D_1 \cap D_2$

(ii) $\text{Dom}[f_1(x).f_2(x)] = D_1 \cap D_2$ (iii) $\text{Dom}\left[\frac{f_1(x)}{f_2(x)}\right] = D_1 \cap D_2 \cap \{x : f_2(x) \neq 0\}$ (iv) $\text{Dom}\left[\sqrt{f_1(x)}\right] = D_1 \cap \{x : f_1(x) > 0\}$

(3) **Periodic function**: – If $f(x + T) = f(x)\ \forall x, T > 0$ *then* $f(x)$is called a periodic function and T(Least) is called its fundamental period.

Since T is period of f(x), we have $f(x) = f(x + T) = f(x + 2T) = f(x + 3T) = \cdots \ldots\ldots\ldots\ldots\ldots\ldots\ldots$

We can say T, 2T, 3T, ………………are all periods of f(x) but only smallest of these number i.e T will be called the fundamental period.

we know that $\sin x = \sin(x + 2\pi) = \sin(x + 4\pi) = \cdots \ldots\ldots\ldots\ldots\ldots\ldots\ldots\ldots\ldots\ldots$

$\cos x = \cos(x + 2\pi) = \cos(x + 4\pi) = \cdots \ldots\ldots\ldots\ldots\ldots\ldots\ldots\ldots\ldots\ldots$
$\tan x = \tan(x + \pi) = \tan(x + 2\pi) = \cdots \ldots\ldots\ldots\ldots\ldots\ldots\ldots\ldots\ldots\ldots$

Hence, sinx, cosx and tanx are periodic functions and their periods are 2π, 2π and π respectively. i.e. least value of T.

Period of (ax + b): – If period of f(x) is T, then period of f(ax + b) is $\frac{T}{|a|}$.

Hence , period of $\mathbf{\sin 2x}$, $\mathbf{\cos\frac{3x}{2}}$ and $\mathbf{\tan\frac{x}{2}}$ will be $\boldsymbol{\pi}, \boldsymbol{\frac{4\pi}{3}}$ and $\boldsymbol{\frac{\pi}{2}}$ respcetively.

Period of f(ax + b) is $\frac{T}{|a|}$. $\therefore$ Period of $\sin 3x = \frac{\text{period of } \sin x}{|a|}$, $\therefore a = 3$ or Period of $\sin 3x = \frac{2\pi}{|3|} = \frac{2\pi}{3}$.

Period of f(x) + g(x): – If period of f(x)and g(x)be T and T′ respectively then the L. C. M of T and T′ will be period of f(x) + g(x).

Note: –If there exists a number $< L.C.M.$ (T, T′) such that f(x + r) = g(x) and g(x + r) = f(x)

then r itself will be the period instead of L. C. M of T, T′.

Consider $|\sin x| = \sqrt{\sin^2 x} = \sqrt{\dfrac{1-\cos 2x}{2}}$, $\therefore$ Period is $\dfrac{2\pi}{2} = \pi$ Similarly, Period of $|\cos x| = \pi$

Now if $F(x) = |\cos x| + |\sin x|$ then its period is not π because there exists $\frac{\pi}{2} < \pi$ *such that* $\left|\cos\left(x+\frac{\pi}{2}\right)\right| = |\sin x|$

and $\left|\sin\left(x+\frac{\pi}{2}\right)\right| = |\cos x|$ Hence, for F(x) the period will be $\frac{\pi}{2}$ and not π

(4) Odd and Even functions: – A function f(x) is said to be odd if it changes sign when the sign of independent variable x is changed

i. e. if $f(-x) = -f(x)$.

For Example: – $f(x) = \sin x.\cos x$, $f(-x) = \sin(-x).\cos(-x)$ or $f(-x) = -\sin x.\cos x$ or $f(-x) = -f(x)$

so, $f(x) = \sin x.\cos x$ is odd function.

A function f(x) is said to be even if its sign does not change when the sign of the independent variable x is changed.

i. e. $f(-x) = f(x)$ For example, $f(x) = ax^4 + bx^2 + c$ and $f(-x) = a(-x)^4 + b(-x)^2 + c = ax^4 + bx^2 + c$

or $f(-x) = f(x)$ so, $f(x) = ax^4 + bx^2 + c$ is an even function.

(5) Increasing and Decreasing function: – Let y = f(x) is an increasing or decreasing function a certain interval. if $\frac{dy}{dx} = +ve$ is an increasing and $\frac{dy}{dx} = -ve$ is an decreasing in that interval. Because the tangent will make an acute angle for increasing function and will make an obtuse angle for decreasing function.

Ex. (1)Determine the intervals of monotonocity of the function $f(x) = y = 3x^4 - 24x^3 + 66x^2 - 72x + 36 = 0$

or $\frac{dy}{dx} = 12x^3 - 72x^2 + 132x - 72 = 12(x^3 - 6x^2 + 11x - 6) = 12(x-1)(x^2 - 5x + 6)$ or $\frac{dy}{dx} = 12(x-1)(x-2)(x-3) = 0$

or x = 1,2,3 Ans.

Very important rule: – I stands for increasing and D stands for decreasing.

+ve		– ve		+ ve		– ve		+ ve
I	2	D	3	I	4	D	5	I

Let function $f(x) = (x-2)(x-3)(x-4)(x-5)$. Hence, for $x > 5$, $f(x)$ is increasing, $4 < x < 5, f(x)$ is decreasing,

$3 < x < 4, f(x)$ is increasing, $2 < x < 3, f(x)$ is decreasing, $x < 2, f(x)$ is increasing

or f(x) is increasing in (5, ∞) ∪ (3, 4) ∪ (2, −∞)and f(x) is decreasing in (4, 5) ∪ (2, 3)

(6) Composite functions: – **(fog)x = f(g(x)), (gof)x = g(f(x))**.

Properties of functions: – **(a.)**Linear property for algebraic function: – If y = f(x) is a linear algebraic function of the form

$y = mx + c = f(x)$ then $f(ax + b) = af(x) + f(b)$, $f(x + y) = f(x) + f(y)$, $f(x - z) = f(x) - f(z)$ $\therefore f(2x + 3) = 2f(x) + f(3)$

(b) Logarithmic and Exponential functions: – We know that, $\log(pq) = \log p + \log q$ and $a^{p+q} = a^p.a^q$

Hence for logarithmic functions, $f(xy) = f(x) + f(y)$ For exponential functions, $f(x + y) = f(x).f(y)$

Closed and open intervals: – If $a \le x \le b$ then we say $x \in [a, b]$ is closed interval if $a < x < b$ *then we say* $x \in (a, b)$ *is open interval*

Similarly, $x \in [a, b)$ or $[a, b[$ or $a \le x < b$ *is semi – closed or open and* $x \in (a, b]$ *or* $]a, b]$ *or* $a < x \le b$ *is semi – open or closed.*

if $x > a$, *or* $x \in (a, \infty)$, $x \ge a$, or $x \in [a, \infty)$ or $x \in [a, \infty[$, $x < b$, *or* $x \in (-\infty, b)$ or $x \le b$, or $x \in (-\infty, b]$ or $x \in]-\infty, b]$

Example: – $x > 2$, $-\infty$ 2 ∞ or $x \in (2, \infty)$

$x \ge 5$, $-\infty$ 5 ∞ or $x \in [5, \infty)$

[∴ • → closed, o → open]

$x < 3$, $-\infty$ 3 ∞ or $x \in (-\infty, 3)$

$x \le 4$, $-\infty$ 4 ∞ or $x \in (-\infty, 4]$

$2 < x < 3$, $-\infty$ 2 3 ∞ or $x \in (2,3)$

(8) Modulus: – Modulus of x i.e $|x|$ {$|a|$ = modulus of a.}

$|x| = x$, if x is + ve i.e. $|5| = 5$, $|x| = -x$, if x is – ve i.e $|-3| = -(-3) = 3$

$|x - a| = x - a$, if $x - a$ is + ve, $x > a$, $|x - a| = -(x - a)$, if $x - a$ is – ve, $x < a$

or $|X| = 3$ or $x = \pm 3$

Properties of Modulus function: – (a) $|x^n| = |x|^n$, $|2^3| = |2|^3$, $|8| = (|2|)^3$ (b) $|xy| = |x||y|$ (c) $\left|\frac{x}{y}\right| = \frac{|x|}{|y|}$

(d) modulus of $(|x| - |y|) \le |x + y| \le |x| + |y|$

Example: – $f(x) = |x - 3| = x - 3$, if $x - 3$ is +ve or $x \ge 3$, $x - 3 = +ve$ and $f(x) = -(x - 3)$, if $x - 3 = -ve$ or $x < 3$

Consider, $f(x) = |x - 3| + |x + 1|$

The critical points are $-1, 3$

	−ve		+ ve		+ ve	
$-\infty$	− ve	−1	− ve	3	+ ve	∞

case I: – $x < -1$, *so that* $x - 3 < 0$, $x + 1 < 0$ ∴ $|x - 3| = -(x - 3)$, $|x + 1| = -(x + 1)$

$$f(x) = -(x - 3) - (x + 1) = -x + 3 - x - 1 = -2x + 2 = -2(x - 1)$$

case II: – $-1 \le x < 3$, *so that* $x - 3 < 0$, $x + 1 \ge 0$ *or* $f(x) = -(x - 3) + (x + 1) = -x + 3 + x + 1 = 4$

case III: – $x \ge 3$, then $x - 3 > 0, x + 1 > 0$ *or* $f(x) = x - 3 + x + 1 = 2x - 2$

Hence, Combined function $f(x) = \begin{cases} -(x+1), & \text{if } x < -1 \\ 4, & \text{if } -1 \le x < 3 \\ 2x - 2 & \text{if } x \ge 3 \end{cases}$

The graph of modulus function $y = |x|$ The critical point $x = 0$

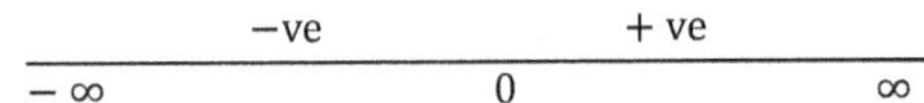

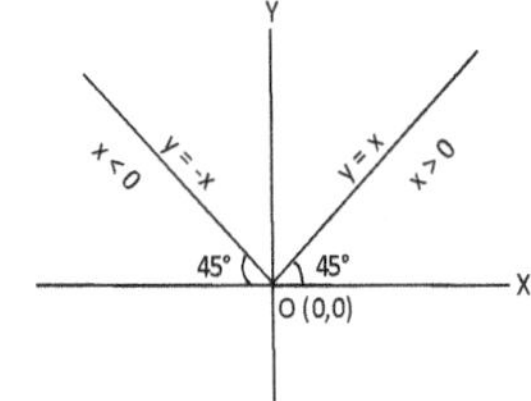

If $y = x,\ x > 0$ *and if* $y = -x,\ x < 0$ *or* $y = \begin{cases} x, & \text{if } x > 0 \\ -x, & \text{if } x < 0 \end{cases}$

(9) Greatest Integer Function [x]: − [x] means the greatest integer not exceeding x,

if, $x = 3\frac{1}{2}, 2\frac{1}{4}, \frac{1}{3}, 5$ then $[x] = 3,2,0,5$ respectively.

If $x = -3\frac{1}{2}, -2\frac{1}{4}, -\frac{1}{3}, -5$ then $[x] = -4, -3, -1, -5$ respectivley. Domain and Range of [x].

The domain of [x] is set of all real numbers its Range $y = [x]$ is set of all integers.

Properties of [x]

(a) $[x] = a, \Leftrightarrow a \le x < a + 1$ *i.e.* $[x] = 3, \Longrightarrow 3 \le x < 3 + 1 \Rightarrow 3 \le x < 4$

(b) $[x] > a, \Rightarrow [x] = a + 1 \Rightarrow x \ge a + 1$ (c) $[x] < a, \Rightarrow [x] = a - 1 \Rightarrow x < a$ (d) $a \le [x] \le b$, then $x \in [a, b + 1]$

e.g. $-2 \le [x] \le 5, \Rightarrow x \in [-2{,}5 + 1], \Rightarrow x \in [-2{,}6]$ Ans.

Example: − (1) If $[x]^2 - 5[x] + 6 \le 0$ then $[x]^2 - 3[x] - 2[x] + 6 \le 0$ or $([x] - 2)([x] - 3) \le 0, \Rightarrow x \in [2{,}3)$ Ans.

Example: − (2) $[x]^2 - 8[x] + 15 > 0$ *then* $[x]^2 - 5[x] - 3[x] + 15 > 0$

or $[x]\{[x] - 5\} - 3\{[x] - 5\} > 0$ *or* $([x] - 3)([x] - 5) > 0, \Rightarrow [x] < 3$ *or* $[x] > 5$

$\Rightarrow [x] < 3$ *or* $x \ge 5 + 1, \Rightarrow x < 3$ *or* $x \ge 6$ $\therefore x \in (-\infty, 3) \cup [6, \infty)$ Ans.

(10) Sign of $(x - a)(x - b), x < b$ *(Important)*

$(x - a)(x - b)$ is + ve, if $x < a$ *or* $x > b$ *i.e. x does not lie between a and* b

$(x - a)(x - b)$ is − ve, if $a < x < b$ *i.e. x lies between a and* b

(a) $|x| < a \Rightarrow x^2 < a^2 \Rightarrow x^2 - a^2 < 0$ *is − ve* *or* $x^2 < a^2$, or $x < \pm a$, *or* $-a < x < a$

(b) $|x| \ge a$, or $x^2 \ge a^2$, or $x \ge \pm a$ $\therefore x \le -a$ or $x \ge a$

(11) Inequalities: − Triangular inequalities: − (a) $|x + y| \le |x| + |y|$ (b) $|x - y| \ge |x| - |y|$

Example: − $|x - 3| > 5$, *Let* $y = x - 3$, $|y| > 5$, $y < -5$ *or* $y > 5$ *or* $x - 3 < -5$ *or* $x - 3 > 5$

or $x < -5 + 3$ *or* $x > 8$, $\therefore x < -2$ *or* $x > 8$

− ∞ − 2 8 ∞

$x \in (-\infty, 2) \cup (8, \infty)$

Remember:− (i) $|x| = a$ or $x = \pm a$ (ii) $|x| > a$, $\therefore x < -a$ or $x > a$ (iii) $|x| \leq a$, $\therefore -a \leq x \leq a$

(iv) $a < |x| \leq b$, $\therefore |x| > a$ or $|x| \leq b$ (v) $a \geq |x| > b$, $\therefore |x| \leq a$ or $|x| > b$

(vi) $a \leq |x| \leq b$, $\therefore |x| \geq a$ or $|x| \leq b$ (vii) $a \geq |x| \geq b$, $\therefore |x| \leq a$ or $|x| \geq b$

(12) $\log_a x$ is defined if both x and a are + ve and $x \neq 0$ and $a \neq 1$, Also $\log_a x = y$ then $x = a^y$ exponential form

∗ Exponential function is always positive

∗ $\log_a x > \log_a y$ or $x > y$, if $a > 1$ or $x < y$, if $a < 1$

Even and odd Extension

Let f(x) be a function defined on $A = [0, a]$ and $B = [-a, a]$ is a super set of A then an extension of f(x) on $B = [-a, a]$ will be even or odd extension if f(x) becomes an even or odd function on B.

Solved Example

(1) If $f(x) = ax^2 + bx + c$ then find f(0), f(a), f(b)and f(1).

Solution:− $f(x) = ax^2 + bx + c$ At $x = 0$, then $f(0) = a \times 0 + b \times 0 + c$ $\therefore f(0) = c$ Ans.

At $x = a$, then $f(a) = a \times a^2 + b \times a + c$ $\therefore f(a) = a^3 + ab + c$ Ans.

At $x = b$, then $f(b) = a \times b^2 + b \times b + c$ $\therefore f(b) = ab^2 + b^2 + c$ Ans.

At $x = 1$, then $f(1) = a \times 1 + b \times 1 + c$ $\therefore f(1) = a + b + c$ Ans.

(2) If $f(x) = x - \frac{1}{x}$, prove that $f(x) = -f\left(\frac{1}{x}\right)$.

Solution: − $f(x) = x - \frac{1}{x}$, $f\left(\frac{1}{x}\right) = \frac{1}{x} - \frac{1}{\frac{1}{x}} = \frac{1}{x} - x = -\left(x - \frac{1}{x}\right) = -f(x)$ or $f\left(\frac{1}{x}\right) = -f(x)$, $\therefore f(x) = -f\left(\frac{1}{x}\right)$ Proved.

Type of functions

(i) Rational function:− This function is defined as the ratio of two polynomials

$$y = \frac{a_0x^n + a_1x^{n-1} + \cdots \ldots\ldots\ldots\ldots + a_n}{b_0x^m + b_1x^{m-1} + \cdots \ldots\ldots\ldots\ldots + b_m}$$

For Example, $y = \frac{x^2 + 1}{x^3 + 3}$ is rational function

(ii) Irrational function: − If in the $y = f(x)$, the operations of addition, subtraction, multiplication, division and raising to a power with rational non − integral exponents are performed on the right − hand side the function $y = f(x)$is said to be irrational.

Example:− $y = \frac{3x^2 + \sqrt{x}}{\sqrt{1 + 2x}}$ and $y = \sqrt{x}$.

(iii) Explicit function: − If the dependent variable, say y, is expressible explicitly in terms of the independent variable, the function $y = f(x)$ is called an explicit function. otherwise it is said to be an implicit function.

Example:− $y = \cos^3x + x^3$ is an explicit function whereas $y = x^3 + y^3 - 3yx = 0$ is an explicit function

(iv) Onto function (or Surjective function): – If a fucntion $f: A \to B$ is such that each element in B is the f – image of at least one element in A, then we say that f is a function of A "onto" B equivalently a function f is an onto function if co – domain of f = Range of f.

Example: – $f: R \to [-1,1]$ defined by $f(x) = \sin x$ is an onto function but $f: R \to R$ defined by $f(x) = \sin x$ is not onto since Range of $f = [-1,1]$ and co – domain of $f = R$.

(v) One – to – one function (or injective function): – A function f is said to be one – to – one if it does not take the same values at two distinct points in its domain.

Example: – $f(x) = x^3$ is one – to – one. whereas $f(x) = x^2$ is not, as $f(1) = 1$ and $f(-1) = 1, f(-1) = f(1)$

so, $f(x) = x^2$ is not a one – to – one function.

(vi) Bijective function (or one – to – one and onto): – If a function f is both one – to – one and onto, then f is said to be a bijective function.

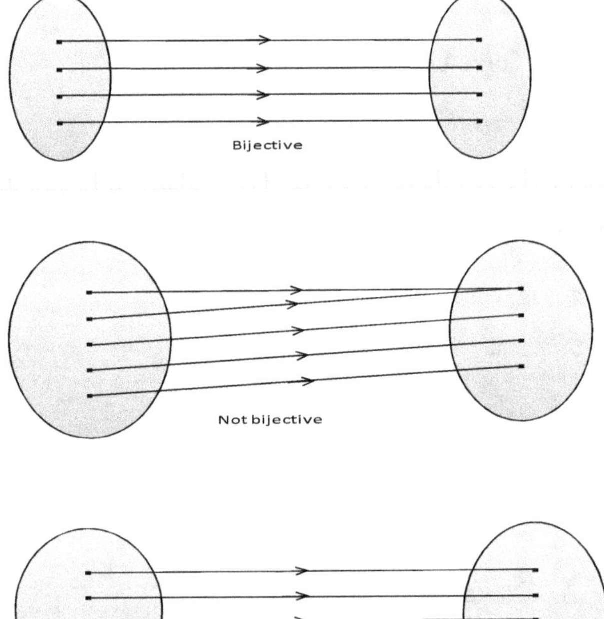

Important Function

#. The domain and Range of trigonometric function are as follows: –

(i) The domain of $f(x) = \sin x$ is R and its Range is $[-1,1] = \{x \in R;\ -1 \le x \le 1\}$

(ii) The domain of $f(x) = \cos x$ is R and its Range is $[-1,1]$.

(iii) The domain of $f(x) = \tan x$ is $R - \left\{(2m+1)\frac{\pi}{2}, m \in I\right\}$ or $f(x) = \left\{x \in R: x \neq (2m+1)\frac{\pi}{2},\ m \text{ is an integer}\right\}$ and its Range is R.

(iv) The domain of $f(x) = \cot x$ is $R - \{m\pi: m \in I\}$ or $f(x) = \{x \in R, x \neq m\pi, m \text{ is an integer}\}$ and its Range is R.

(v) The domain of $f(x) = \sec x$ is $R - \left\{(2m+1)\frac{\pi}{2}, m \in I\right\}$ and its Range is $(\infty, -1] \cup [1, \infty) = \{x \in R, x \notin (-1,1)\}$ so, $|\sec x| \ge 1$

(vi) The domain of $f(x) = \text{cosec}x$ is $R - \{m\pi, m \in I\}$ and its Range is the same as that of secx.

#. The domain and Ranges of the inverse trigonometric functions are as follows: –

(i) The domain of $f(x) = \sin^{-1} x$ is $[-1,1]$ and its Range is $\left[-\frac{\pi}{2}, \frac{\pi}{2}\right]$.

(ii) The domain of $f(x) = \cos^{-1} x$ is $[-1,1]$ and its Range is $[0, \pi]$.

(iii) The domain of $f(x) = \tan^{-1} x$ is R and its Range is $\left(-\frac{\pi}{2}, \frac{\pi}{2}\right) = \{x \in R, -\frac{\pi}{2} < x < \frac{\pi}{2}\}$.

(iv) The domain of $f(x) = \cot^{-1} x$ is R and its Range is $(0, \pi)$.

(v) The domain of $f(x) = \sec^{-1} x$ is $R - (-1,1)$ and its Range is $\left[0, \frac{\pi}{2}\right) \cup \left(\frac{\pi}{2}, \pi\right] = \left\{x \in R, 0 \le x \le \pi, x \ne \frac{\pi}{2}\right\}$.

(vi) The domain of $f(x) = \csc^{-1} x$ is $R - (-1,1)$ and its Range is $\left[-\frac{\pi}{2}, 0\right) \cup \left(0, \frac{\pi}{2}\right] = \left\{x \in R, -\frac{\pi}{2} \le x \le \frac{\pi}{2}, x \ne 0\right\}$.

Solved Example

(1) The domain of the function $f(x) = \sqrt{x-2} + \sqrt{5-x}$ is (a) $[2, \infty)$ (b) $[2,5]$ (c) $(-\infty, 5)$ (d) $(2,5)$

Solution: – $f(x) = \sqrt{x-2} + \sqrt{5-x}$, function f(x) to be defined as $f(x) = \begin{cases} x-2 \ge 0 \\ 5-x \ge 0 \end{cases} = \begin{cases} x \ge 2 \\ -x \ge -5 \end{cases} = \begin{cases} x \ge 2 \\ x \le 5 \end{cases}$

$-\infty \quad 2 \quad 5 \quad \infty$

domain of f(x) is $x \in [2,5]$, $\therefore$ domain $= [2,5]$ Ans. (b) $[2,5]$

(2) The domain of the function $f(x) = \sqrt{\frac{1-x}{1+x}}$ is (a)$(-\infty, 1]$ (b)$(-\infty, -1) - \{1\}$ (c)$(-\infty, 1) - \{-1\}$ (d)$R - (-1,1)$

Solution: – $f(x) = \sqrt{\frac{1-x}{1+x}}$ to be defined as $f(x) = \begin{cases} \frac{1-x}{1+x} > 0 \\ 1+x \ne 0 \end{cases} = \begin{cases} 1-x > 0 \\ x \ne -1 \end{cases} = \begin{cases} x < 1 \\ x \ne -1 \end{cases}$

$-\infty \quad -1 \quad 1 \quad \infty$ domain $= (-\infty, 1) - \{-1\}$, Ans. (c)

(3) If D is the domain of the function $f(x) = \sqrt{3-4x} + 2\sin^{-1}\frac{3x-2}{5}$ then D contains (a) $\left[-1, \frac{3}{4}\right)$ (b) $\left[-1, \frac{3}{4}\right]$ (c) $(-1,0)$ (d) $[-1,0]$

Solution: – $f(x) = \sqrt{3-4x} + 2\sin^{-1}\frac{3x-2}{5}$

f(x)to be defined as $f(x) = \begin{cases} 3-4x \ge 0 \\ -1 \le \frac{3x-2}{5} \le 1 \end{cases} = \begin{cases} -4x \ge -3 \\ -5 \le 3x-2 \le 5 \end{cases} = \begin{cases} 4x \le 3 \\ -3 \le 3x \le 7 \end{cases} = \begin{cases} x \le \frac{3}{4} \\ -1 \le x \le \frac{7}{3} \end{cases}$

$$\text{domain}(D) = \left(-\infty, \frac{3}{4}\right] \cap \left[-1, \frac{7}{3}\right]$$

$-\infty \quad -1 \quad 0 \quad \frac{3}{4} \quad \frac{7}{3} \quad \infty$

$D = [-1,0] \cup \left[0, \frac{3}{4}\right] = [-1,0], \left[-1, \frac{3}{4}\right]$ Ans. (b) $\left[-1, \frac{3}{4}\right]$, (d) $[-1,0]$

(4) The function $f(x) = \log\left(\frac{1-x}{1+x}\right)$ satisfies the equation (a) $f(x+2) - 2f(x+1) + f(x) = 0$

(b) $f(x)+f(x-1)=f(x(x+1))$ (c) $f(x_1).f(x_2)=f(x_1+x_2)$ (d) $f(x_1)+f(x_2)=f\left(\frac{x_1+x_2}{1+x_1.x_2}\right)$

Solution: – $f(x)=\log\frac{1-x}{1+x}$, $f(x_1)=\log\frac{1-x_1}{1+x_1}$, $f(x_2)=\log\frac{1-x_2}{1+x_2}$

$$f(x_1)+f(x_2)=\log\frac{1-x_1}{1+x_1}+\log\frac{1-x_2}{1+x_2}=\log\left(\frac{1-x_1}{1+x_1}\times\frac{1-x_2}{1+x_2}\right)=\log\left(\frac{1-x_2-x_1+x_1x_2}{1+x_1+x_2+x_1x_2}\right)$$

$$f(x_1)+f(x_2)=\log\left(\frac{1-\frac{x_1+x_2}{1+x_1x_2}}{1+\frac{x_1+x_2}{1+x_1x_2}}\right)\qquad\therefore\ f(x_1)+f(x_2)=f\left(\frac{x_1+x_2}{1+x_1x_2}\right)\quad \text{Ans. (d)}$$

(5) Which of the following functions is an even function: – (a) $f(x)=x\log\left(\frac{1-x}{1+x}\right)$ (b) $f(x)=\frac{a^x-1}{a^x+1}$

(c) $f(x)=x\frac{a^x+1}{a^x-1}$ (d) $f(x)=\frac{a^{2x}+1}{a^{2x}-1}$

Solution: – (a) $f(x)=x\log\left(\frac{1-x}{1+x}\right)$, $f(-x)=-x\log\left(\frac{1+x}{1-x}\right)=x\log\left(\frac{1+x}{1-x}\right)^{-1}=x\log\left(\frac{1-x}{1+x}\right)=f(x)$

$f(-x)=f(x)$ is an even function.

(b) $f(x)=\frac{a^x-1}{a^x+1}$, $f(-x)=\frac{a^{-x}-1}{a^{-x}+1}=\frac{1-a^x}{1+a^x}=-\left(\frac{a^x-1}{a^x+1}\right)=-f(x)$ $\therefore f(-x)=-f(x)$is an odd function.

(c) $f(x)=x\frac{a^x+1}{a^x-1}$, $f(-x)=-x\frac{a^{-x}+1}{a^{-x}-1}=-x\frac{1+a^x}{1-a^x}=-x\frac{1+a^x}{-(a^x-1)}=x\frac{a^x+1}{a^x-1}=f(x)$

$f(-x)=f(x)$ is an even funciton.

(d) $f(x)=\frac{a^{2x}+1}{a^{2x}-1}$, $f(-x)=\frac{a^{-2x}+1}{a^{-2x}-1}=\frac{1+a^{2x}}{1-a^{2x}}=-\frac{a^{2x}+1}{a^{2x}-1}=-f(x)$ or $f(-x)=-f(x)$ is an odd funciton. Ans: – (a) and (c)

(6) The domain of definition of the function $y=\frac{1}{\log_2(3-x)}+\sqrt{1-x}$ is (a) $(-\infty,1)-\{2\}$ (b) $(-\infty,1)$ (c) $(-\infty,2)$ (d) $(3,\infty)$

Solution: – $y=f(x)=\frac{1}{\log_2(3-x)}+\sqrt{1-x}$

f(x) to be defined as $f(x)=\begin{cases}\log_2(3-x)\neq 0\\ 3-x>0\\ 1-x\geq 0\end{cases}=\begin{cases}3-x\neq 1\\ x<3\\ x\leq 1\end{cases}=\begin{cases}x\neq 2\\ x<3\\ x\leq 1\end{cases}$

Remember
$f(x)=\log_a x$
f(x)to be defined as
$f(x)=\begin{cases}a>0, a\neq 1\\ x>0, x\neq 0\end{cases}$

$-\infty$ 1 2 3 ∞ or $D=(-\infty,1)$ (b) Ans.

(7) Find domain and Range of the function: – (a) $f(x)=\frac{4x}{2x-3}$ (b) $f(x)=\frac{3x-2}{2x-3}$ (c) $f(x)=\frac{1}{\sqrt{1-x^2}}$

Solution: – (a) $f(x)=\frac{4x}{2x-3}$ function f(x)to be defined $2x-3\neq 0$ $\therefore x\neq\frac{3}{2}$

$-\infty$ $\frac{3}{2}$ ∞ or $\boxed{\text{Domain (D)}=R-\left\{\frac{3}{2}\right\}}$ Ans.

Range: – Let $y=f(x)=\frac{4x}{2x-3}$, $\therefore y(2x-3)=4x$

or $2xy - 3y = 4x$ or $2xy - 4x = 3y$ or $x(2y - 4) = 3y$ or $x = \frac{3y}{2y-4}$, $x = f(y)$

function f(y)to be defined $2y - 4 \neq 0, y \neq 2$

$-\infty$ 2 ∞

Range = R − {2} Ans.

(b) $f(x) = \frac{3x-2}{2x-3}$ function f(x)to be not defined is $2x - 3 = 0$ or $x = \frac{3}{2}$ Domain (D) = $R - \left(\frac{3}{2}\right)$

Range: − $y = f(x) = \frac{3x-2}{2x-3}$ or $2xy - 3y = 3x - 2 = x(2y - 3) = 3y - 2$ $\therefore$ $x = \frac{3y-2}{2y-3}$

function f(y)to be not defined as $2y - 3 = 0$ or $y = \frac{3}{2}$

$-\infty$ $\frac{3}{2}$ ∞

Range = $R - \left\{\frac{3}{2}\right\}$ Ans.

(c) $f(x) = \frac{1}{\sqrt{1-x^2}}$ function f(x)to be defined $f(x) = \begin{cases}\sqrt{1-x^2} > 0 \\ 1 - x^2 \neq 0\end{cases}$ or $\begin{cases}1 - x^2 > 0 \\ x^2 \neq 1\end{cases}$

$-\infty$ −1 1 ∞

Domain (D) = $-1 < x < 1$

Range: − $y = f(x) = \frac{1}{\sqrt{1-x^2}}$ or $y^2 = \frac{1}{1-x^2}$ or $1 - x^2 = \frac{1}{y^2}$ or $x^2 = 1 - \frac{1}{y^2}$ or $x^2 = \frac{y^2-1}{y^2}$ or $x = \sqrt{\frac{y^2-1}{y^2}}$

function f(y)to be defind $\begin{cases}\frac{y^2-1}{y^2} \geq 0 \\ y^2 \neq 0\end{cases} = \begin{cases}y^2 - 1 \geq 0 \\ y \neq 0\end{cases} = \begin{cases}y^2 \geq 1 \\ y \neq 0\end{cases} = \begin{cases}y \geq \pm 1 \\ y \neq 0\end{cases}$

$-\infty$ −1 0 1 ∞

Range = $(-\infty, -1) \cup (1, \infty)$ Ans.

(8) find the range of each of the following function: − (a) $y = \frac{x-3}{2x-5}$ (b) $y = 2 - \frac{1}{1+x}$ (c) $y = \sin^{-1} x$ (d) $y = x - |x|$

Solution : −(a) $y = \frac{x-3}{2x-5}$ or $2xy - 5y = x - 3$ or $x(2y - 1) = 5y - 3$ $\therefore$ $x = \frac{5y-3}{2y-1} = f(x)$

Hence, function f(x)to be defined $2y - 1 \neq 0$ or $y \neq \frac{1}{2}$

$-\infty$ $\frac{1}{2}$ ∞

Range = $R - \left\{\frac{1}{2}\right\}$ Ans.

(b) $y = 2 - \frac{1}{1+x} = \frac{2+2x-1}{1+x} = \frac{2x+1}{1+x}$

Hence y to be defined $1 + x \neq 0$ or $x \neq -1$

$-\infty$ −1 ∞

Domain (D) = R − {−1}

Now, $y = \frac{2x+1}{1+x}$ or $y + xy = 2x + 1$ or $x(y - 2) = 1 - y$ $\therefore$ $x = \frac{1-y}{y-2}$

Hence x to be defined $y - 2 \neq 0$ or $y \neq 2$

$-\infty \quad 2 \quad \infty$ or Range = R − {2}

(c) $y = \sin^{-1} x$, Here y to be defined $-1 \le x \le 1$ $\boxed{\text{Domain (D)} = [-1,1]}$

Now, $y = \sin^{-1} x \quad \therefore x = \sin y = f(x)$ Hence f(x) to be defined of all real value.

$\boxed{\text{Range} = R}$

(d) $y = x - |x|$ Ans: − Range = R

(9) If f: R → R is defined by $f(x) = x^3 + 2$, the value of $f^{-1}(29)$ and $f^{-1}(10)$ are respectively.

Solution: − $f(x) = x^3 + 2 \quad \therefore f(x) = y = x^3 + 2$ or $x^3 = y - 2$ or $x = \sqrt[3]{y-2} \quad \therefore y = f(x)$

or $x = f^{-1}(y)$ or $f^{-1}(y) = \sqrt[3]{y-2} \quad \therefore f^{-1}(29) = \sqrt[3]{29-2} = \sqrt[3]{27} = 3$ Ans. $\therefore f^{-1}(10) = \sqrt[3]{10-2} = \sqrt[3]{8} = 2$ Ans.

(10)find the period of the following function: − (a) $y = \sin\left(\frac{2t-5}{4\pi}\right)$ (b) $y = \sin|x| + \cos|x|$ (c) $y = \frac{\sin x + \sin 2x}{\cos x + \cos 2x}$ (d) $y = \frac{\tan x - \cot x}{\tan x + \cot x}$

(e) $y = \cos x + \cot\frac{x}{2} + \cos\frac{x}{2^2} + \cot\frac{x}{2^3}$ (f) $y = \sin\left|2x + \frac{\pi}{2}\right|$ (g) $y = \tan\left(\frac{3x-2}{5}\right)$

Solution: − (a) $y = \sin\left(\frac{2t-5}{4\pi}\right) \quad \therefore y = \sin(ax+b)$ then period $= \frac{|T|}{|a|}$ $\boxed{\text{Time Period of } y = \frac{2\pi}{\frac{1}{2\pi}} = 4\pi^2 \text{ Ans.}}$

(b) $y = \sin|x| + \cos|x|$, Period of $y = \pi$

(c) $y = \frac{\sin x + \sin 2x}{\cos x + \cos 2x}$ Formula, $\operatorname{Sin}C + \operatorname{Sin}D = 2\sin\frac{C+D}{2}.\cos\frac{C-D}{2}$ and $\operatorname{Cos}C + \operatorname{Cos}D = 2\cos\frac{C+D}{2}.\cos\frac{C-D}{2}$

$$y = \tan\frac{3x}{2}, \text{ Period} = \frac{T}{|a|} = \frac{\pi}{\frac{3}{2}} = \frac{2\pi}{3} \quad \text{Ans.}$$

(d) $y = \frac{\tan x - \cot x}{\tan x + \cot x} = \sin^2 x - \cos^2 x = -\cos 2x = \cos(\pi - 2x)$

$$y = \cos[\pi + (-2x)] \therefore a = -2, \text{ Period (T)} = \frac{T}{|a|} \quad \boxed{\therefore \text{Period} = \frac{2\pi}{|-2|} = \pi \quad \text{Ans.}}$$

(e) $y = \cos x + \cot\frac{x}{2} + \cos\frac{x}{2^2} + \cot\frac{x}{2^3}$

$\boxed{\text{Period of } f(x) + g(x) \text{ are T and T' respectively. then, period of } f(x) + g(x) = \text{L.C.M of T and T'}}$

$$\text{Period of } y = 2\pi + \frac{\pi}{\frac{1}{2}} + \frac{2\pi}{\frac{1}{4}} + \frac{\pi}{\frac{1}{8}} = 2\pi + 2\pi + 8\pi + 8\pi = \text{L.C.M of all} = 4\pi \quad \text{Ans.}$$

(f) $y = \sin\left|2x + \frac{\pi}{2}\right|$ period of $\sin|x| = \pi$, $a = 2$ Period of $y = \frac{T}{|a|} = \frac{\pi}{2}$ Ans.

(g) $y = \tan\left(\frac{3x-2}{5}\right) = \tan\left(\frac{3x}{5} + \frac{-2}{5}\right), a = \frac{3}{5}$ Period of $y = \frac{T}{|a|} = \frac{\pi}{\frac{3}{5}} = \frac{5\pi}{3}$ Ans.

(11) If f is an even function defined on the interval (−3,3), then the one value of x satisfying the equation $f(x) = f\left(\frac{x+3}{x-1}\right)$ are ... − 1 ... ...

Solution: − $f(x) = f\left(\frac{x+3}{x-1}\right)$ or $x = \frac{x+3}{x-1}$ or $x^2 - x = x + 3$ or $x^2 - 2x - 3 = 0$ or $x = \frac{2 \pm \sqrt{4+12}}{2} = \frac{2 \pm 4}{2} \quad \therefore x = 3, -1$

Again $f(-x) = f(x)$ as f is an even function. $f(-x) = f(x) = f\left(\frac{x+3}{x-1}\right)$ or $-x = \frac{x+3}{x-1}$ or $-x^2 + x = x + 3$

or $x^2 = -3$ it is not defined, satisfying all real value. only one value is satisfying the equation or $x = -1$ Ans.

(12) Let f and g be functions defined by $f(x) = \frac{3x}{5-2x}$ and $g(x) = \frac{1-2x}{x+1}$ then $(fog)^{-1}(x) = \frac{1-x}{3x+2}$.

Solution: $-$ $f(x) = \frac{3x}{5-2x}$ and $g(x) = \frac{1-2x}{x+1}$ then $(fog)^{-1}(x)$ equal to ……

$$(fog)^{-1}(x) = f(g(x))^{-1}(x) \therefore f(g(x)) = f\left(\frac{1-2x}{x+1}\right) = \frac{\frac{3(1-2x)}{x+1}}{5-2\frac{1-2x}{x+1}} = \frac{1-2x}{3x+1} = h \text{ (say)}$$

$$\therefore h(x) = \frac{1-2x}{3x+1} = y \text{ or } 3xy + y = 1 - 2x \text{ or } x(3y+2) = 1 - y$$

or $x = \frac{1-y}{3y+1}$ let $h(x) = y$ or $x = h^{-1}(y)$ or $h^{-1}(y) = \frac{1-y}{3y+2}$ $\therefore h^{-1}(x) = \frac{1-x}{3x+2}$ or $f(g(x))^{-1}x = \frac{1-x}{3x+2}$ Ans.

(13) (i) find fofof if $f(x) = \frac{1-x}{2+3x}$ (ii) find fofog if $f(x) = \frac{\sqrt{1-x^2}}{x}$ and $g(x) = \frac{x^2}{1+x^2}$

(iii)find fofofog if $f(x) = \sqrt{1-x}$ and $g(x) = \frac{2}{1+x}$

Solution: $-$(i) $f(x) = \frac{1-x}{2+3x}$ then find fofof i.e $f(f(f(x)))$. or $f(f(x)) = f\left(\frac{1-x}{2+3x}\right) = \frac{1-\frac{1-x}{2+3x}}{2+3\left(\frac{1-x}{2+3x}\right)} = \frac{1+4x}{7+3x}$

or $f(f(f(x))) = f\left(\frac{1+4x}{7+3x}\right) = \frac{1+4\left(\frac{1+4x}{7+3x}\right)}{2+3\left(\frac{1+4x}{7+3x}\right)}$ $\therefore$ $f(f(f(x))) = \frac{19x+11}{18x+17}$ Ans.

(ii) $f(x) = \frac{\sqrt{1-x^2}}{x}$ and $g(x) = \frac{x^2}{1+x^2}$ then find fofog i.e $f(f(g(x)))$.

$$f(f(g(x))) = f\left(f\left(\frac{x^2}{1+x^2}\right)\right) \quad \text{or } f\left(\frac{x^2}{1+x^2}\right) = \frac{\sqrt{1-\frac{(x^2)^2}{(1+x^2)^2}}}{\frac{x^2}{1+x^2}} = \frac{\sqrt{(1+x^2)^2-(x^2)^2}}{x^2}$$

$$\text{or } f\left(\frac{x^2}{1+x^2}\right) = \frac{\sqrt{1+2x^2}}{x^2} \quad \text{or } f\left(\frac{\sqrt{1+2x^2}}{x^2}\right) = \frac{\sqrt{1-\frac{(\sqrt{1+2x^2})^2}{x^2}}}{\frac{\sqrt{1+2x^2}}{x^2}} = \frac{\sqrt{x^4-2x^2-1}}{\sqrt{1+2x^2}} \quad \boxed{\therefore f(f(g(x))) = \sqrt{\frac{x^4-2x^2-1}{1+2x^2}} \text{ Ans.}}$$

(iii) $f(x) = \sqrt{1-x}$ and $g(x) = \frac{2}{1+x}$ then find fofofog, i.e $f(f(f(g(x))))$.

$$\text{or } f(g(x)) = f\left(\frac{2}{1+x}\right) = \sqrt{1-\frac{2}{1+x}} = \sqrt{\frac{x-1}{x+1}} \quad \therefore f\left(\sqrt{\frac{x-1}{x+1}}\right) = \sqrt{1-\sqrt{\frac{x-1}{x+1}}} = \sqrt{\frac{\sqrt{x+1}-\sqrt{x-1}}{\sqrt{x+1}}}$$

$$\therefore f\left(\sqrt{\frac{\sqrt{x+1}-\sqrt{x-1}}{\sqrt{x+1}}}\right) = \sqrt{1-\sqrt{\frac{\sqrt{x+1}-\sqrt{x-1}}{\sqrt{x+1}}}} \quad \boxed{\therefore f(f(f(g(x)))) = \frac{(\sqrt{x+1}+\sqrt{x^2-1}-2)^{\frac{1}{2}}}{(\sqrt{x+1}+\sqrt{x^2-1})^{\frac{1}{2}}} \text{ Ans.}}$$

(14)(i)find inverse of the function $y = \frac{5^x-5^{-x}}{5^x+5^{-x}}$ is ______ (ii)If $f(x) = \frac{\sqrt{x}}{\sqrt{x+1}}$, the domain of $f^{-1}(x)$contains ______

(iii)The inverse of the function $y = \frac{1}{\sqrt{1-2\sin^2 x}}$ is ______ (iv)The inverse of the function $y = \log\left(\frac{x+2}{x-1}\right)$

(v)The inverse of the function $y = \tan^{-1}x + \cot^{-1}x$

Solution: $-$(i) $y = \frac{5^x - 5^{-x}}{5^x + 5^{-x}} = f(x)$ $\therefore x = f^{-1}(y)$ or $x = \frac{1}{2}\log_5\left(\frac{y+1}{1-y}\right)$ $\quad\therefore\ y = \frac{1}{2}\log_5\left(\frac{y+1}{1-y}\right)$ Ans.

(ii) $f(x) = \frac{\sqrt{x}}{\sqrt{x+1}} = y \Rightarrow y^2 = \frac{x}{x+1} \Rightarrow xy^2 + y^2 = x \ \therefore\ x(y^2 - 1) = -y^2$

$\Rightarrow x = \frac{-y^2}{y^2 - 1} = \frac{y^2}{1 - y^2}$ $\therefore y = f(x)$ $\therefore x = f^{-1}(y) = \frac{y^2}{1-y^2}$ $\therefore f^{-1}(x) = \frac{x^2}{1-x^2}$

Hence $f^{-1}(x)$to be defined or $1 - x^2 \neq 0 \Rightarrow x^2 \neq 1$ or $x \neq \pm 1$ $\quad\therefore$ Domain (D) $= R - \{-1,1\}$ Ans.

(iii) $y = \frac{1}{\sqrt{1 - 2\sin^2 x}} = \frac{1}{\sqrt{\cos 2x}} \Rightarrow y^2 = \frac{1}{\cos 2x}$ or $x = \frac{1}{2}\cos^{-1}\left(\frac{1}{y^2}\right)$ or $y = f(x)$

or $x = f^{-1}(y) = \frac{1}{2}\cos^{-1}\left(\frac{1}{y^2}\right)$ $\quad\boxed{\therefore\ y = \frac{1}{2}\cos^{-1}\left(\frac{1}{x^2}\right) \text{ Ans.}}$

(iv) $y = \log\left(\frac{x+2}{x-1}\right) = f(x)$ or $e^y = \frac{x+2}{x-1} \Rightarrow x = \frac{e^y + 2}{e^y - 1} \Rightarrow x = \frac{e^y + 2}{e^y - 1}$

or $y = f(x) \Rightarrow x = f^{-1}(y) = \frac{e^y + 2}{e^y - 1}$ $\therefore y = \frac{e^y + 2}{e^y - 1}$ $\quad\boxed{\therefore y = \frac{e^{x+2}}{e^x - 1} \text{ Ans.}}$

(v) $y = \tan^{-1}x + \cot^{-1}x = \frac{(\tan^{-1}x)^2 + 1}{\tan^{-1}x}$,let $\tan^{-1}x = z$ or $x = \tan z$

$y = \frac{z^2 + 1}{z} \Rightarrow z^2 - zy + 1 = 0 \Rightarrow z = \frac{y \pm \sqrt{y^2 - 4}}{2} = \tan^{-1}x$ $\therefore x = \tan\left(\frac{y \pm \sqrt{y^2-4}}{2}\right)$ $\quad\boxed{\therefore y = \tan\left(\frac{x \pm \sqrt{x^2 - 4}}{2}\right) \text{ Ans.}}$

(15)Determine the function $y = f(x)$satisfying the condition $f\left(x - \frac{1}{x}\right) = -\frac{1}{x^2} + x^2\ (x \neq 0)$.

Solution: $-$ $f\left(x - \frac{1}{x}\right) = -\frac{1}{x^2} + x^2$ $\therefore t = x - \frac{1}{x} \Rightarrow t^2 = x^2 - \frac{1}{x^2} + 2 \Rightarrow t^2 - 2 = -\frac{1}{x^2} + x^2$

$\therefore f(t) = t^2 - 2$, $\quad f(x) = x^2 - 2$ is required function which will satisfy the given condition. Ans.

(16) Let $f(x) = \begin{cases} x^2 - 7x + 10, & x < 5 \\ x - 7, & x \geq 5 \end{cases}$ and $g(x) = \begin{cases} x - 5, & x < 7 \\ x^2 + 3x + 3, & x \geq 7 \end{cases}$ Describe the function $\frac{f}{g}$ and find its domain.

Solution: $-$ The critical points are 5,7 and so we redefine the function for $x < 5, 5 \leq x < 7, x \geq 7$

$$f(x) = \begin{cases} x^2 - 7x + 10, & x < 5 \\ x - 7, & 5 \leq x < 7 \\ x - 7, & x \geq 7 \end{cases}, \quad g(x) = \begin{cases} x - 5, & x < 5 \\ x - 5, & 5 \leq x < 7 \\ x^2 + 3x + 3, & x \geq 7 \end{cases}$$

$$\frac{f}{g} = \begin{cases} \frac{x^2 - 7x + 10}{x - 5}, & x < 5 \\ \frac{x - 7}{x - 5}, & 5 \leq x < 7 \\ \frac{x - 7}{x^2 + 3x + 3}, & x \geq 7 \end{cases} = \begin{cases} x - 2, & x < 5 \\ \frac{x - 7}{x - 5}, & 5 \leq x < 7 \\ \frac{x - 7}{x^2 + 3x + 3}, & x \geq 7 \end{cases} \quad \text{Domain(D)} = R - \{5\} \text{ Ans.}$$

$$f + g = \begin{cases} x^2 - 7x + 10 + x - 5, & x < 5 \\ x - 7 + x - 5, & 5 \leq x < 7 \\ x - 7 + x^2 + 3x + 3, & x \geq 7 \end{cases} = \begin{cases} x^2 - 6x + 5, & x < 5 \\ 2x - 12, & 5 \leq x < 7 \\ x^2 + 4x - 4, & x \geq 7 \end{cases} \quad \text{Ans.}$$

(17) Let $f(x) = \frac{9^x}{9^x + 3}$. Prove that $f(x) + f(1 - x) = 1$, hence prove that $f\left(\frac{1}{1987}\right) + f\left(\frac{2}{1987}\right) + \cdots \ldots \ldots + f\left(\frac{1986}{1987}\right) = 993$

Solution: $-\ f(x)=\dfrac{9^x}{9^x+3}$ and $f(1-x)=\dfrac{9^{1-x}}{9^{1-x}+3}=\dfrac{\frac{9}{9^x}}{\frac{9}{9^x}+3}=\dfrac{\frac{9}{9^x}}{\frac{9+3.9^x}{9^x}}=\dfrac{9}{3(3+9^x)}=\dfrac{3}{3+9^x}$

$$f(x)+f(1-x)=\frac{9^x}{9^x+3}+\frac{3}{3+9^x}=\frac{3+9^x}{3+9^x}=1 \text{ Proved.}$$

Solved Example

(1) solve the following inequalities: $-$ (a) $\left|3x-\frac{1}{2}\right|<1$ $\quad \therefore |x|<a \quad or -a<x<a$

$$\therefore a=1, x=3x-\frac{1}{2} \ \text{ or } -1<3x-\frac{1}{2}<1 \ or -\frac{1}{2}<3x<\frac{3}{2} \quad \boxed{or -\frac{1}{6}<x<\frac{1}{2} \ \text{ Ans.}}$$

(b) $\left|\frac{x}{2}-1\right|\geq 2 \quad \therefore |x|\geq a \Rightarrow x\geq a \text{ or } x\leq -a \quad \Rightarrow \frac{x}{2}-1\geq 2 \text{ or } \frac{x}{2}-1\leq -2 \Rightarrow \frac{x}{2}\geq 3 \text{ or } \frac{x}{2}\leq -1 \quad \therefore x\geq 6 \text{ or } x\leq -2$ Ans.

(c) $\left|\frac{3x}{5}+\frac{3}{2}\right|\leq\frac{1}{2} \Rightarrow -\frac{1}{2}\leq\frac{3x}{5}+\frac{3}{2}\leq\frac{1}{2} \text{ or } -\frac{1}{2}-\frac{3}{2}\leq\frac{3x}{5}\leq\frac{1}{2}-\frac{3}{2} \text{ or } -2\leq\frac{3x}{5}\leq -1 \text{ or } -10\leq 3x\leq -5 \quad \boxed{\therefore -\frac{10}{3}\leq x\leq -\frac{5}{3} \ \text{ Ans.}}$

(d) $|-x-1|>\frac{3}{2} \Rightarrow -x-1>\frac{3}{2} \text{ or } -x-1<-\frac{3}{2} \Rightarrow -x>\frac{3}{2}+1 \text{ or } -x<-\frac{3}{2}+1$

$$\therefore -x>\frac{5}{2} \text{ or } -x<-\frac{1}{2} \quad \boxed{\therefore x<\frac{5}{2} \text{ or } x>\frac{1}{2} \ \text{ Ans.}}$$

(e) $(2x-1)^2\geq 9 \ \therefore (2x-1)^2-3^2\geq 0 \ \therefore (2x-1+3)(2x-1-3)\geq 0 \ \therefore (2x+2)(2x-4)\geq 0$

$$\therefore f(x)\geq 0 \ \therefore f(x)=0 \Rightarrow (2x+2)(2x-4)=0 \text{ or } x=-1,2$$

f(x) ≥ 0

$-\infty$	+ ve	-1	− ve	2	+ ve	∞

$$\boxed{\therefore x\geq 2 \text{ or } x\leq -1 \ \text{ Ans.}}$$

(f) $4x^2+x-3\leq 0 \ \therefore f(x)=4x^2+x-3 \ \therefore f(x)\leq 0$

$\therefore f(x)=0$ and find x. $\therefore 4x^2+x-3=0 \Rightarrow 4x^2+4x-3x-3=0 \Rightarrow 4x(x+1)-3(x+1)=0$

$$\Rightarrow (4x-3)(x+1)=0 \ \therefore x=-1,\frac{3}{4}$$

f(x) ≤ 0

$-\infty$	+ ve	-1	− ve	$\frac{3}{4}$	+ ve	∞

$$\therefore f(x)\leq 0$$

$$\boxed{\therefore -1\leq x\leq\frac{3}{4} \ \text{ Ans.}}$$

(2)(a) Given the function $f(x)=\dfrac{2x-1}{2x+3}$, find $f\left(\frac{x}{2}\right), 2f\left(\frac{x}{2}\right), f\left(\frac{x^2}{4}\right), \left[f\left(\frac{x}{2}\right)\right]^2$.

Solution: $-\ f(x)=\dfrac{2x-1}{2x+3}, f\left(\frac{x}{2}\right)=\dfrac{2.\frac{x}{2}-1}{2.\frac{x}{2}+3}=\dfrac{x-1}{x+3}$ Ans. $\quad 2f\left(\frac{x}{2}\right)=2.\dfrac{x-1}{x+3}$ Ans.

$f\left(\frac{x^2}{4}\right)=\dfrac{2\frac{x^2}{4}-1}{2.\frac{x^2}{4}+3}=\dfrac{x^2-2}{x^2+6}$ Ans. $\quad \left[f\left(\frac{x}{2}\right)\right]^2=\left(\dfrac{x-1}{x+3}\right)^2=\dfrac{(x-1)^2}{(x+3)^2}$ Ans.

(b) Given the function $f(x) = \log\left(\frac{2-x}{2+x}\right)^2$ then prove that $f(x_1) + f(x_2) = f\left(\frac{4 + x_1.x_2}{x_1 + x_2}\right)$.

Solution: – $f(x) = \log\left(\frac{2-x}{2+x}\right)^2 = 2\log\left(\frac{2-x}{2+x}\right), f(x_1) = 2\log\left(\frac{2-x_1}{2+x_1}\right), f(x_2) = 2\log\left(\frac{2-x_2}{2+x_2}\right)$

$$L.H.S = f(x_1) + f(x_2) = 2\log\left(\frac{2-x_1}{2+x_1}\right) + 2\log\left(\frac{2-x_2}{2+x_2}\right) = 2\left[\log\frac{(2-x_1)(2-x_2)}{(2+x_1)(2+x_2)}\right] = 2\log\left(\frac{4-2x_1-2x_2+x_1x_2}{4+2x_1+2x_2+x_1x_2}\right)$$

$$R.H.S = f\left(\frac{4+x_1x_2}{x_1+x_2}\right) = \log\left(\frac{2-\frac{4+x_1x_2}{x_1+x_2}}{2+\frac{4+x_1x_2}{x_1+x_2}}\right)^2 = \log\left(\frac{2x_1+2x_2-4-x_1x_2}{2x_1+2x_2+4+x_1x_2}\right)^2 = \log\left[\frac{-(4-2x_1-2x_2+x_1x_2)}{4+2x_1+2x_2+x_1x_2}\right]^2$$

$$R.H.S = \log(-1)^2.\left(\frac{4-2x_1-2x_2+x_1x_2}{4+2x_1+2x_2+x_1x_2}\right)^2 = 2\log\left(\frac{4-2x_1-2x_2+x_1x_2}{4+2x_1+2x_2+x_1x_2}\right)^2$$

∴ L.H.S = R.H.S Proved.

(3) (a)Given the function $f(x) = \frac{3^x + 3^{-x}}{2}$. show that $f(x+y) + f(x-y) = 2f(x)f(y)$.

Solution: – $f(x) = \frac{3^x+3^{-x}}{2}, f(x+y) = \frac{3^{x+y}+3^{-(x+y)}}{2}, f(x-y) = \frac{3^{x-y}+3^{-(x-y)}}{2}$

$$L.H.S = f(x+y) + f(x-y) = \frac{3^{x+y}+3^{-(x+y)}}{2} + \frac{3^{x-y}+3^{-(x-y)}}{2} = \frac{3^{x+y}+3^{-(x+y)}+3^{x-y}+3^{-(x-y)}}{2} = \frac{3^{x^2-y^2}+3^{x^2-y^2}}{2} = \frac{2.3^{x^2-y^2}}{2} = 3^{x^2-y^2}$$

$$R.H.S = 2f(x)f(y) = 2\left(\frac{3^x+3^{-x}}{2}.\frac{3^y+3^{-y}}{2}\right) = \frac{3^x3^y+3^x3^{-y}+3^{-x}3^y+3^{-x}3^{-y}}{2} = \frac{3^{x+y}+3^{x-y}+3^{-(x-y)}+3^{-(x+y)}}{2} = \frac{3^{x^2-y^2}+3^{x^2-y^2}}{2}$$

$$= \frac{2.3^{x^2-y^2}}{2} = 3^{x^2-y^2}$$

L.H.S = R.H.S Proved

(b) Given the function $f(x) = \frac{(x+2)^2}{x^2-4}$.

find $f(1), f(-1), f(2), f(-2), f(a+1), f(a)+1, f(a+2), f(a)+2, f(a-1), f(a)-1, f(a-2), f(a)-2, f(2a)$and $f(2a+1), f(2a)+1$.

Solution: – $f(x) = \frac{(x+2)^2}{x^2-4} = \frac{(x+2)^2}{x^2-2^2} = \frac{(x+2)^2}{(x+2)(x-2)} = \frac{x+2}{x-2}$

$\therefore f(1) = \frac{1+2}{1-2} = -3$ Ans. $\therefore f(-1) = \frac{-1+2}{-1-2} = -\frac{1}{3}$ Ans. $\therefore f(2) = \frac{2+2}{2-2} = \infty$ Ans.

$\therefore f(-2) = \frac{-2+2}{-2-2} = \frac{0}{-4} = 0$ Ans. $\therefore f(a+1) = \frac{a+1+2}{a+1-2} = \frac{a+3}{a-1}$ Ans.

$\therefore f(a)+1 = \frac{a+2}{a-2}+1 = \frac{a+2+a-2}{a-2} = \frac{2a}{a-2}$ Ans. $\therefore f(a+2) = \frac{a+2+2}{a+2-2} = \frac{a+4}{a}$ Ans.

$\therefore f(a)+2 = \frac{a+2}{a-2}+2 = \frac{a+2+2a-4}{a-2} = \frac{3a-2}{a-2}$ Ans. $\therefore f(a-1) = \frac{a-1+2}{a-1-2} = \frac{a+1}{a-3}$ Ans.

$\therefore f(a)-1 = \frac{a+2}{a-2}-1 = \frac{a+2-a+2}{a-2} = \frac{4}{a-2}$ Ans. $\therefore f(a-2) = \frac{a-2+2}{a-2-2} = \frac{a}{a-4}$ Ans.

$\therefore f(a)-2 = \frac{a+2}{a-2}-2 = \frac{a+2-2a+4}{a-2} = \frac{-a+6}{a-2}$ Ans. $\therefore f(2a) = \frac{2a+2}{2a-2} = \frac{a+1}{a-1}$ Ans.

$\therefore f(2a+1) = \frac{2a+1+2}{2a+1-2} = \frac{2a+3}{2a-1}$ Ans. $\therefore f(2a)+1 = \frac{2a+2}{2a-2}+1 = \frac{2a+2+2a-2}{2a-2} = \frac{4a}{2a-2} = \frac{4a}{2(a-1)} = \frac{2a}{a-1}$ Ans.

(4)(a) Given the function $f(x) = \frac{x^3+1}{x+1}$. find $\frac{f(b)-1}{b-1}$ and $f\left(\frac{b+1}{2}\right)$.

(b) Given the function $f(x) = \frac{x^3-8}{x^2+2x+4}$. find $\frac{f(a+1)-f(b+1)}{a-b}$ and $f\left(\frac{\sqrt{3}a+b}{b}\right), f\left(\frac{\sqrt{2}b-a}{a}\right)$.

Solution: – given $f(x) = \frac{x^3+1}{x+1}$ $\therefore f(x) = \frac{(x+1)(x^2-x+1)}{(x+1)} = x^2-x+1$, $f(b) = b^2-b+1$

$$\therefore \frac{f(b)-1}{b-1} = \frac{b^2-b+1-1}{b-1} = \frac{b(b-1)}{(b-1)} = b \quad \text{Ans.}$$

or $f\left(\frac{b+1}{2}\right) = \left(\frac{b+1}{2}\right)^2 - \left(\frac{b+1}{2}\right) + 1 = \frac{b^2+2b+1}{4} - \frac{b+1}{2} + 1 = \frac{b^2+2b+1-2b-2+4}{4} = \frac{b^2+3}{4}$ Ans.

(b) given $f(x) = \frac{x^3-8}{x^2+2x+4}$ $\therefore f(x) = \frac{(x-2)(x^2+2x+4)}{(x^2+2x+4)} = x-2$ $\therefore f(a+1) = a+1-2 = a-1, f(b+1) = b+1-2 = b-1$

or $\frac{f(a+1)-f(b+1)}{a-b} = \frac{a-1-b+1}{a-b} = \frac{a-b}{a-b} = 1$ Ans. or $f\left(\frac{\sqrt{3}a+b}{b}\right) = \frac{\sqrt{3}a+b}{b} - 2 = \frac{\sqrt{3}a+b-2b}{b} = \frac{\sqrt{3}a-b}{b}$ Ans.

$$\therefore f\left(\frac{\sqrt{2}b-a}{a}\right) = \frac{\sqrt{2}b-a}{a} - 2 = \frac{\sqrt{2}b-a-2a}{a} = \frac{\sqrt{2}b-3a}{a} \quad \text{Ans.}$$

(5)(a) Given the function $f(x) = \begin{cases} 3^x - 2^{-x} + 1, & -2 \le x < 0 \\ \sin\frac{x}{2} + \cos\frac{x}{2}, & 0 \le x < \pi \\ \frac{2x}{x^2-1}, & \pi \le x \le 4 \end{cases}$ find $f(-1), f(-2), f\left(\frac{\pi}{3}\right), f\left(\frac{2\pi}{3}\right), f\left(\frac{\pi}{2}\right)$, $f(\pi), f(0), f(2), f(4)$.

(b) Given the function $f(x) = \begin{cases} 3x^2+2, & x \le 1 \\ \frac{3}{2x-3}, & 1 < x \le 2 \\ \frac{x}{2} - 3, & x > 2 \end{cases}$ find $f(\sqrt{3}), f(\sqrt{2}), f(\sqrt{2}-1), f(\sqrt{3}+1), f(\sqrt{2}+2), f(\sqrt{3}-2)$.

Solution: – (2)(a) $f(x) = \begin{cases} 3^x - 2^{-x} + 1, & -2 \le x < 0 \\ \sin\frac{x}{2} + \cos\frac{x}{2}, & 0 \le x < \pi \\ \frac{2x}{x^2-1}, & \pi \le x \le 4 \end{cases}$

or $f(-1) = 3^x - 2^{-x} + 1 = 3^{-1} - 2^1 + 1$ $\therefore f(-1) = \frac{1}{3} - 2 + 1 = \frac{1-6+3}{3} = \frac{-2}{3}$ Ans.

or $f(-2) = 3^x - 2^{-x} + 1 = 3^{-2} - 2^2 + 1 = \frac{1}{9} - 4 + 1 = \frac{1-36+9}{9} = -\frac{26}{9}$ Ans.

or $f\left(\frac{\pi}{3}\right) = \sin\frac{x}{2} + \cos\frac{x}{2} = \sin\frac{\pi}{6} + \cos\frac{\pi}{6} = \frac{\sqrt{3}+1}{2}$ Ans. or $f\left(\frac{2\pi}{3}\right) = \sin\frac{x}{2} + \cos\frac{x}{2} = \sin\frac{2\pi}{6} + \cos\frac{2\pi}{6} = \frac{\sqrt{3}}{2} + \frac{1}{2} = \frac{\sqrt{3}+1}{2}$ Ans.

or $f\left(\frac{\pi}{2}\right) = \sin\frac{x}{2} + \cos\frac{x}{2} = \sin\frac{\pi}{4} + \cos\frac{\pi}{4} = \frac{1}{\sqrt{2}} + \frac{1}{\sqrt{2}} = \frac{1+1}{\sqrt{2}} = \frac{2}{\sqrt{2}} = \frac{\sqrt{2}\times\sqrt{2}}{\sqrt{2}} = \sqrt{2}$ Ans.

or $f(\pi) = \frac{2x}{x^2-1} = \frac{2\pi}{\pi^2-1}$ Ans. $\therefore f(0) = \sin\frac{x}{2} + \cos\frac{x}{2} = \sin\frac{0}{2} + \cos\frac{0}{2} = 1$ Ans.

or $f(2) = \sin\frac{x}{2} + \cos\frac{x}{2} = \sin\frac{2}{2} + \cos\frac{2}{2} = \sin 1 + \cos 1$ Ans. $\therefore f(4) = \frac{2x}{x^2-1} = \frac{2\times4}{4^2-1} = \frac{8}{15}$ Ans.

Solution: –(2)(b) same as above, Ans: $-f(\sqrt{3}) = 2\sqrt{3}+3$, $f(\sqrt{2}) = -(6\sqrt{2}+9)$, $f(\sqrt{2}-1) = 11-6\sqrt{2}$

$f(\sqrt{3}+1) = -\frac{11}{\sqrt{3}+5}$, $f(\sqrt{2}+2) = -\frac{7}{\sqrt{2}+4}$, $f(\sqrt{3}-2) = \frac{97}{23+12\sqrt{3}}$.

(6)(a) Calculate $f(x) = \frac{16}{x^2} + x^2$ at the points for which $\frac{4}{x} + x = 2$. (b) Given the function $f(x) = \frac{7x^3+3}{x-1}$. find $f(2x), f(x^2), 5f(x), [f(x)]^2$.

Solution: −(6)(a) $f(x) = \frac{16}{x^2} + x^2$ at the point $\frac{4}{x} + x = 2$ $\therefore \frac{4}{x} + x = 2$ squaring both sides then, $\left(\frac{4}{x} + x\right)^2 = 4 \Rightarrow \frac{16}{x^2} + x^2 + 2.x.\frac{4}{x} = 4$

or $\frac{16}{x^2} + x^2 = 4 - 8 = -4$ $\therefore f(x) = \frac{16}{x^2} + x^2 = -4$ Ans.

(b) $f(x) = \frac{7x^3+3}{x-1}$ $\therefore f(2x) = \frac{7(2x)^3+3}{2x-1} = \frac{56x^3+3}{2x-1}$ Ans. $\therefore f(x^2) = \frac{7(x^2)^3+3}{x^2-1} = \frac{7x^6+3}{x^2-1}$ Ans.

$\therefore 5f(x) = 5\left(\frac{7x^3+3}{x-1}\right) = \frac{35x^3+15}{x-1}$ Ans. $\therefore [f(x)]^2 = \left(\frac{7x^3+3}{x-1}\right)^3 = \frac{49x^6+42x^3+9}{x^2-2x+1}$ Ans.

(7)(a) Let $f(x) = \begin{cases} 5^x, & -2 < x < 0 \\ 3, & 0 \le x < 1 \\ 5x-2, & 1 \le x \le 4 \end{cases}$ find $f(-1), f(0), f(0.5), f(-0.5), f(2), f(3), f(4), f\left(\frac{3}{2}\right)$.

(b) $f(x) = x^3 + 5, g(x) = 3x$ solve the equation $g(x) = |f(x)|$ and $f(x) + 2 = |g(x)| + 1$.

Solution: − (7) (a) Do yourself.

Ans: − $f(-1) = \frac{1}{5}$, $f(0) = 3$, $f(0.5) = 3$, $f(-0.5) = \sqrt{5}$, $f(2) = 8$, $f(3) = 13$, $f(4) = 18$, $f\left(\frac{3}{2}\right) = \frac{11}{2}$.

(b) $f(x) = x^3 + 5, g(x) = 3x$ $\therefore g(x) = |f(x)| \Rightarrow 3x = |x^3+5| \Rightarrow 3x = -(x^3+5)$ and $3x = x^3 + 5$

$$\boxed{\therefore x^3 + 3x + 5 = 0 \text{ or } x^3 - 3x + 5 = 0 \text{ Ans.}}$$

Also find $f(x) + 2 = |g(x)| + 1 \Rightarrow x^3 + 5 + 2 = 3x + 1$ or $x^3 + 5 + 2 = -3x + 1$

$x^3 + 7 = 3x + 1$ or $x^3 + 7 = -3x + 1$ $\therefore x^3 - 3x + 6 = 0$ or $x^3 + 3x + 6 = 0$ Ans.

(8)(a) Find f(x) if $f(x+2) = x^2 + 5x - 3$ (b) Find f(x) if $f(2x-1) = 4x^2 - 6x + 7$

Solution: − (8)(a) find f (x), $f(x+2) = x^2 + 5x - 3 = (x+2)^2 + (x+2) - 9$ put $x + 2 = x$ then $f(x) = x^2 + x - 9$ Ans.

(b) Do yourself (see above question). Ans: − $f(x) = x^2 - x + 5$

(9) which of the given functions is (are)even , odd , and which of them is (are)neither even , nor odd ?

(a) $f(x) = \log\left(x - \sqrt{1+x^2}\right)$ (b) $f(x) = \log\left(\frac{2+x}{2-x}\right)$ (c) $f(x) = 2x^2 - 3x + 1$ (d) $f(x) = \frac{a^{2x}+1}{a^{2x}-1}$ (e) $f(x) = \frac{x(a^x-2)}{a^x}$ (f) $f(x) = x\frac{e^x-1}{e^x+1}$

Solution: − (9)(a) $f(x) = \log\left(x - \sqrt{1+x^2}\right)$, $f(-x) = \log\left(-x - \sqrt{1+x^2}\right)$

$f(x) + f(-x) = \log\left(x - \sqrt{1+x^2}\right) + \log\left(-x - \sqrt{1+x^2}\right) = \log\left(x - \sqrt{1+x^2}\right)\left(-x - \sqrt{1+x^2}\right)$

$f(x) + f(-x) = \log\left(-x^2 - x\sqrt{1+x^2} + x\sqrt{1+x^2} + 1 + x^2\right) = \log 1 = 0$

$f(x) + f(-x) = 0$ $\therefore f(x) = -f(-x)$ function f(x) is odd. Ans.

(b) Ans: − odd (c) Ans: − function f(x)is neither even nor odd.

(d) $f(x) = \frac{a^{2x}+1}{a^{2x}-1}$, $f(-x) = \frac{a^{-2x}+1}{a^{-2x}-1} = \frac{1+a^{2x}}{1-a^{2x}} = -\frac{a^{2x}+1}{a^{2x}-1} = -f(x)$ $f(x) + f(-x) = 0$ function f(x) is odd. Ans.

(e) $f(x) = \frac{x(a^x-2)}{a^x}$, $f(-x) = -x\frac{a^{-x}-2}{a^{-x}} = -x(1-2a^x)$ ∴ function f(x)is neither even nor odd. Ans.

(f) $f(x) = x\frac{e^x-1}{e^x+1}$, $f(-x) = -x\frac{e^{-x}-1}{e^{-x}+1} = -x\frac{1-e^x}{1+e^x} = x\frac{e^x-1}{e^x+1} = f(x)$ $f(x) = f(-x)$ $\therefore f(x) - f(-x) = 0$ function f(x) is even. Ans.

(10) which of the following function is (are) even and which is (are) odd?

(a) $f(x) = 4 - \sin^2 x$ (b) $f(x) = \sqrt{-x^2 - 6x + 9} + \sqrt{-x^2 + 6x + 9}$ (c) $f(x) = \sin\frac{x}{2} + \cos\frac{x}{2}$ (d) $f(x) = \sqrt{1 + x + x^2} - \sqrt{1 - x + x^2}$

(e) $f(x) = 3 - x^2 + \cos^2 x$ (f) $f(x) = x\frac{1 - a^{kx}}{1 + a^{kx}}$ (g) $f(x) = \sqrt[3]{1 + x} + \sqrt[3]{1 - x}$ (h) $f(x) = 2x^2 + |x|$ (i) $f(x) = x^2 \sin x - x$

(j) $f(x) = \frac{(1 + 3^x)^2}{(1 - 3^x)}$ (k) $f(x) = \frac{1 + a^{kx}}{1 - a^{kx}}$ (l) $f(x) = \frac{(1 - 2^x)^2}{1 + 2^x}$

Solution: –(10)(a) Ans: – Even (b) Ans: – Even (c) Ans: – Neither even nor odd. (d) Ans: – odd

(e) Ans: –Even (f) Ans: – Even (g) Ans: – Even (h) Ans: – Even (i) Ans: – odd (j) Ans: – odd

(k) $f(x) = \frac{1 + a^{kx}}{1 - a^{kx}}$, $f(-x) = \frac{1 + a^{-kx}}{1 - a^{-kx}} = \frac{a^{kx} + 1}{a^{kx} - 1} = -\frac{a^{kx} + 1}{1 - a^{kx}} = -f(x)$ or $f(x) + f(-x) = 0$ ∴ function f(x)is odd

(l) $f(x) = \frac{(1 - 2^x)^2}{1 + 2^x}$, $f(-x) = \frac{(1 - 2^{-x})^2}{1 + 2^{-x}} = \frac{(2^x - 1)^2}{1 + 2^x} = \frac{[-(1 - 2^x)]^2}{1 + 2^x} = f(x)$

$f(-x) - f(x) = 0$ ∴ function f(x)is even.

(11) Indicate the amplitude $|A|$, frequency ω, initial phase φ and period T of the following harmonics.

(a) $f(x) = 4\sin\frac{3x}{2}$ (b) $f(x) = 3\sin\left(5x + \frac{\pi}{3}\right)$ (c) $f(x) = 2\sin\frac{x}{3} + 3\cos\frac{x}{3}$ (d) $f(x) = \sin 2x + \cos 2x$

Solution: – (11)(a) $f(x) = 4\sin\frac{3x}{2}$ ∴ $f(x) = A\sin(\omega x + \varphi)$

$f(x) = 4\sin\left(\frac{3x}{2} + 0\right)$, ∴ $|A| = 4, \varphi = 0, \omega = \frac{3}{2}$ ∴ Period (T) $= \frac{2\pi}{\omega} = \frac{2\pi}{\frac{3}{2}} = \frac{4\pi}{3}$ Ans.

(b) $f(x) = 3\sin\left(5x + \frac{\pi}{3}\right)$, ∴ $f(x) = A\sin(\omega x + \varphi)$ Here $|A| = 3, \omega = 5, \varphi = \frac{\pi}{3}$ Period (T) $= \frac{2\pi}{\omega} = \frac{2\pi}{5}$ Ans.

(c) $f(x) = 2\sin\frac{x}{3} + 3\cos\frac{x}{3}$, Ans: – Period(T) $= 6\pi$,

(d) $f(x) = \sin 2x + \cos 2x$, Ans: – Period(T) $= \pi$, $|A| = \sqrt{2}$ and $\varphi = \frac{\pi}{4}$

(12) Find the period for each of the following functions. (a) $f(x) = \tan\frac{3x}{2}$ (b) $f(x) = 2\cot\frac{x}{3}$ (c) $f(x) = \sin\frac{2\pi x}{3}$ (d) $f(x) = \cos 3\pi x$

Solution: – (12)(a) $f(x) = \tan\frac{3x}{2}$, Period of $\tan x = \pi$ ∴ Period of $\tan\frac{3x}{2} = \frac{\text{period of } \tan x}{|a|}$

where $a = \frac{3}{2}$ ∴ Period of $\tan\frac{3x}{2} = \frac{\pi}{\frac{3}{2}} = \frac{2\pi}{3}$ Ans.

(b) $f(x) = 2\cot\frac{x}{3}$, Period of $\cot x = \pi = (T)$ ∴ Period of $2\cot\frac{x}{3} = \frac{T}{a} = \frac{\pi}{\frac{1}{3}} = 3\pi$ where $a = \frac{1}{3}$ Ans.

(c) $f(x) = \sin\frac{2\pi x}{3}$, Period of $\sin x = 2\pi = T$ ∴ Period of $\sin\frac{2\pi x}{3} = \frac{T}{a} = \frac{2\pi}{\frac{2\pi}{3}} = 3$ where $a = \frac{2\pi}{3}$ Ans.

(d) $f(x) = \cos 3\pi x$, Period of $\cos x = 2\pi = T$ Period of $\cos 3\pi x = \frac{T}{a} = \frac{2\pi}{3\pi} = \frac{2}{3}$ where $a = 3\pi$ Ans.

Note: – $f(x) = \cos ax$ ∴ Period of $\cos x = 2\pi = T$ ∴ Period of $\cos ax = \frac{2\pi}{a} = \frac{T}{a}$

Exercise – A1

(1) find the domain of each of the follwing function: – (a) $y = \sqrt{\log_{.2}\left(\frac{x+1}{x-3}\right)} \times \frac{1}{x^2-16}$ (b) $y = \sqrt[5]{2^{2x} + 16^{\left(\frac{1}{2}\right)(x-2)} - 52 - 8^{\left(\frac{2}{3}\right)(x-2)}}$

(c) $y = \log_x\left(\frac{x-3}{x+1}\right)$ (d) $y = 3\cos^{-1}\left(\frac{2x+3}{4}\right) + 4\tan^{-1}\left(\frac{5}{x}\right) + \sin^{-1}\left(\frac{2x-1}{3}\right)$ (e) $y = \sqrt{\log_{10}\left(\frac{2x-x^2}{5}\right)} + \frac{2}{x-5}$ (f) $y = \sqrt{|x|-x}$

(g) $y = \sin\left(\frac{5-2x}{3}\right) + \sqrt{\frac{x+3}{x-5}} - \frac{1}{5-x^2+x}$ (h) $y = \log_x\left(\frac{x-5}{2}\right)$ (i) $y = 3\cos^{-1}\left(\frac{x+7}{5-2x}\right) + \sin^{-1}\left(\frac{5}{x-2}\right) - \frac{1}{\sqrt{5+x}}$

(j) $y = \frac{2x-3}{5-3x} - \frac{2}{x+\sqrt{2}} + \sqrt{x^2-16}$

(2)(a) If $f(x) = \log_e\left(\frac{1+\sqrt{x}}{1-\sqrt{x}}\right)$, then show that $f(m) + f(n) = \log_e\left[\frac{\left(1+\sqrt{m}\right)^2.\left(1+\sqrt{n}\right)^2}{(1-m)(1-n)}\right]$

(b) If $f(x) = \log\left(\frac{1-\log x}{1+\log x}\right)$, then show that $f(x) = f(y) + \log\left[\frac{(1-\log x)(1+\log y)}{(1+\log x)(1-\log y)}\right]$

(c) If $f(x) = \tan(\log x)$, then show that $f(x^2) + f(y^2) = f(x^2y^2).[1 + f(x^2)f(y^2)]$.

(3) (a) Prove that the inverse of the function $y = \frac{2^x+2^{-x}}{2^x-2^{-x}} + 2$ is $y = \frac{1}{2}\log_2\left(\frac{x-1}{x-3}\right)$.

(b) Prove that the inverse of the function $f(x) = \frac{a^x+a^{-x}}{a^x-a^{-x}} + 1$ is $f(x) = \frac{1}{2}\log_a\left(\frac{y}{y-2}\right)$.

(c) Prove that the inverse of the function $f(x) = \frac{2.5^x - 5^{-x}}{3.5^x + 5^{-x}} + 3$ is $f(x) = \frac{1}{2}\log_5\left(\frac{2-y}{3y-11}\right)$.

(4) (a) If $y = \sin(\log x)$, then prove that $x^2y'' + xy' + y = 0$. (b) If $y = \cos(\log x)$, then prove that $x^2y'' - xy' + y = 0$.

(c) If $f(x) = \frac{x-1}{x}$, then prove that $f(x) + f\left(\frac{1}{x}\right) = f(x)f\left(\frac{1}{x}\right)$.

(5) (a) Let function $f: R \to R$ be defined by $f(x) = x + \cos x$ for $x \in R$, then prove that f is one to one and onto.

(b) If $f(x) = (a + x^m)^{\frac{1}{m}}$ whose $a > 0$ *and m is a positive integer, then show that* $f(f(x)) = (2a + x^m)^{\frac{1}{m}}$.

(c) Let $f(x) = \frac{\alpha x}{x-1}$, $x \neq 1$. then for what value of α is $f(f(x)) = x$.

(6) (a) Given the function $f(x) = \begin{cases} 2^{-x} - 1 & \text{for } -2 \le x < 0 \\ \cot\frac{x}{2} & \text{for } 0 \le x < \pi \\ \frac{x}{x^3-2} & \text{for } \pi \le x \le 8 \end{cases}$ find $f(-2), f\left(\frac{\pi}{6}\right), f\left(\frac{2\pi}{3}\right), f(6), f(8), f\left(\frac{\pi}{4}\right), f\left(\frac{3\pi}{4}\right), f(1), f\left(\frac{1}{2}\right), f(-1)$.

(b) $f: R \to R$ is defined as under $f(x) = \begin{cases} x-3, & x < -8 \\ x^3+1, & x \in [-8,8] \\ 3x+5, & x > 8 \end{cases}$ Evaluate, $f(-14), f(2), f(12)$ and $(fof)4$.

(c) If $f(x) = \log_2 x$ and $g(x) = 2^x$ then show that $(fog)x = (gof)x = x$. (d) If $f(x) = \frac{x}{1-x}$ then show that $(fofof)x = \frac{x}{1-3x}$.

(e) Let $f(x) = \cot x, x \in \left(-\frac{\pi}{2}, \frac{\pi}{2}\right)$ and $g(x) = \sqrt{1+x^2}$. Determine $(fog)x$ and $(gof)x$.

(7) (a) If $g(f(x)) = |\cos x|$ and $f(g(x)) = \left(\cos\sqrt{x}\right)^2$ then find $f(x)$and $g(x)$.

(b) If $f(x)=\begin{cases}x^2-2x+3, & x<3\\ 2x-1, & x\geq 3\end{cases}$ and $g(x)=\begin{cases}x-4, & x<4\\ 2x^2+4x+5, & x\geq 4\end{cases}$ Determine $f+g$ and $\frac{f}{g}$.

(c) If $f(x)=|x-3|$ and $g(x)=f(f(x))$, then for $x>3$, $g'(x)$ is equal to ………

(8) Find the domain of definition of the functions: – (a) $f(x)=\log_x 2$ (b) $f(x)=\sqrt{\log\left(\frac{9x-4x^2}{5}\right)}$ (c) $f(x)=\log|3-x^2|$

(d) $f(x)=\sqrt{2x-5}-\frac{1}{\log_3(2+x)}$ (e) $f(x)=\sqrt{\sin^{-1}x}+\frac{1}{\log x}$ (f) $f(x)=\sqrt{2+\log_2 x}$ (g) $f(x)=\log_3\left\{\log_{\frac{1}{3}}(x^2-4x+4)\right\}$

(h) $f(x)=\log_4\log_3\log_2(1-6x-9x^2)$ (i) $f(x)=\log\{(2\log x)^2-12\log x+9\}$ (j) $f(x)=\sqrt{\log_{0.2}(4x-2x^2)}$

(k) $f(x)=\sqrt{|x|-x}$ (l) $f(x)=\frac{2}{2x+3|x|}$

(9) Find the domain of definition of the functions: – (a) $f(x)=\cos^{-1}\left(\frac{2x}{1+x^2}\right)+\sqrt{\sin(\cos x)}$ (b) $f(x)=\cos^{-1}\left[\log_3\left(\frac{1}{3}x^3\right)\right]$

(c) $f(x)=\frac{\sqrt{3-x^2}}{\cos^{-1}(1+x)}$ (d) $f(x)=\log(\sin x)$ (e) $f(x)=\log(\cos x)^2$ (f) $f(x)=\sin^{-1}\left(\frac{3-|x|}{2}\right)+\log(2-x)$

(g) $f(x)=\sqrt{4+2^x+2^{2-x}}+\sqrt{\cos^{-1}(x+2)}$

(10) (a) $y=\frac{\sqrt{x-3}}{x^2-1}$, $x\geq 3$ and $x\neq\pm1$ (b) $y=\frac{\cos^{-1}(3-x)}{\sqrt{x^2-1}}$ (c) $y=\left\{\frac{3-x^2}{x+2}\right\}^{\frac{1}{2}}$ (d) $y=\frac{\sqrt{5+3x-x^2}}{\sqrt{\sin x-\frac{1}{2}}}$

(e) $y=\log_4(x^2+1)+\sqrt{64x-x^7}$ (f) $y=\sqrt{\frac{2-|x|}{3-|x|}}$ (g) $f(x)=\sin^{-1}\sqrt{1-x^2}+\cos^{-1}\sqrt{1+x^2}$

(11) Find the range of the functions: – (a) $y=\frac{2x}{1-x^2}$ (b) $y=\frac{2}{1-\sin 2x}$ (c) $y=\frac{2}{3+x^2}$ (d) $y=\frac{x+2}{x^2-x-6}$ (e) $y=2-\frac{3}{x^2-3x+4}$

(f) $y=\log\sqrt{x^2+4x+10}$ (g) $y=3\cos x+2\sin\left(x+\frac{\pi}{6}\right)+5$ (h) $y=\cos^{-1}\left[\frac{3}{2}+x^2\right]$ where $[\,.\,]$denotes the greatest integer function.

(12) Find the domain and range of the functions: – (a) $y=\frac{x+1}{x^2-6x-3}$ (b) $y=\frac{2}{\sqrt{1+2\cos x}}$ (c) $y=\cos\left[\log\left(\frac{\sqrt{1-x^2}}{2-x}\right)\right]$

(d) If $f(x)=5-x_{P_{x-1}}$, find domain and range of f(x).

(e) If $f(\sin 2x)=\frac{(2\tan x+\sec^2 x)(1+\sin 2x)}{4\sin^2 x}$ then determine the domain and range of f(t).

(13) (a) If the function f: $[2,\infty]\rightarrow[2,\infty[$ is defined by f: $(2,\infty)$ $f^{-1}(x)$ is the $f(x)=3^{x(x-2)}$, then $f^{-1}(x)$ is $1\pm\sqrt{1+\log_3 x}$.

$f^{-1}(x)$ is positive $f^{-1}(x)=1+\sqrt{1+\log_3 x}$

(b) If f: $[2,\infty[\rightarrow[3,\infty[$ is given by $f(x)=x-\frac{2}{x}$ then $f^{-1}(x)$ is equal to $\frac{x\pm\sqrt{x^2+8}}{2}$, $f^{-1}(x)$ is positive $f^{-1}(x)=\frac{x+\sqrt{x^2+8}}{2}$

(14) Find the identical function are following: – (a) $f(x)=\frac{x+1}{x+1}$ and $g(x)=1$ (b) $f(x)=1$ and $g(x)=\sec^2 x-\tan^2 x$

(c) $f(x)=\log x^4$ and $g(x)=4\log x$ (d) $f(x)=\log(x+1)+\log x$ and $g(x)=\log\{x(x+1)\}$

(e) $f(x)=2x$ and $g(x)=\sqrt{4x^2}$ (f) $f(x)=\log x-\log(x-2)$ and $g(x)=\log\left(\frac{x}{x+2}\right)$

(15) Classify the functions are even or odd: – (a) $f(x)=\log\left(x-\sqrt{1+x^2}\right)$ (b) $f(x)=\cos^3 x+2\tan^2 x$ (c) $f(x)=x^4+|x|$

(d) $f(x) = x\dfrac{e^x - 1}{e^x + 1}$ (e) $f(x) = \dfrac{\tan x . \cot x}{\sin^2 x + \cos x}$ (f) $f(x) = \sec^2 x + \csc^2 x$

(16) (a) If f is an even function defined on the interval $(-3,3)$, then real values of x satisfying the equation

$f(x) = f\left(\dfrac{x-2}{x-3}\right)$ are $1-\sqrt{3}, 2-\sqrt{2}, 1+\sqrt{3}$.

(b) show that the function $f(x) = \int_0^x \log_e\left(\dfrac{2+x}{2-x}\right)dx$ is an even function.

(17) Find the fundamental time period of the function: –

(a) $\sin\left(\dfrac{2x}{3}\right)$ (b) $\cos(\pi - x)$ (c) $\tan 2x$ (d) $\sin\dfrac{x}{2}$ (e) $|\sin x|$ (f) $\sin^3 x + \cos^3 x$ (g) $2\sin\dfrac{2}{3}(x+\pi)$

Solution

(1) (a) Domain (D) $= R - \{[-1,3]\}$ (b) Domain (D) $= [52, \infty[$ (c) Domain (D) $= (3, \infty)$

(d) Domain (D) $= \left[-1, \dfrac{1}{2}\right]$ (e) Domain (D) $= (-\infty, -1] \cup [5, \infty)$ (f) Domain (D) $=]-\infty, 0]$

(g) $D = [1,4] - \left\{\dfrac{1+\sqrt{21}}{2}\right\}$ (h) $D = (5, \infty)$ (i) $D = \left[-\dfrac{2}{3}, \dfrac{2}{3}\right]$ (j) Domain (D) $= (-\infty, -4) \cup [4, \infty[$

(2) (a) $f(x) = \log_e\left(\dfrac{1+\sqrt{x}}{1-\sqrt{x}}\right)$, then show that $f(m) + f(n) = \log_e\left[\dfrac{(1+\sqrt{m})^2.(1+\sqrt{n})^2}{(1-m)(1-n)}\right]$

$f(x) = \log_e\left(\dfrac{1+\sqrt{x}}{1-\sqrt{x}}\right)$, $f(m) = \log_e\left(\dfrac{1+\sqrt{m}}{1-\sqrt{m}}\right)$, $f(n) = \log_e\left(\dfrac{1+\sqrt{n}}{1-\sqrt{n}}\right)$

$$f(m) + f(n) = \log_e\left(\frac{1+\sqrt{m}}{1-\sqrt{m}}\right) + \log_e\left(\frac{1+\sqrt{n}}{1-\sqrt{n}}\right) = \log_e\left\{\frac{(1+\sqrt{m})(1+\sqrt{n})}{(1-\sqrt{m})(1-\sqrt{n})}\right\} = \log_e(1+\sqrt{m})(1+\sqrt{n}) - \log_e(1-\sqrt{m})(1-\sqrt{n})$$

$$\text{or } f(x) = \log_e\left(\frac{1+\sqrt{x}}{1-\sqrt{x}}\right) = \log_e\left\{\frac{(1+\sqrt{x})(1+\sqrt{x})}{(1-\sqrt{x})(1+\sqrt{x})}\right\} = \log_e\frac{(1+\sqrt{x})^2}{(1-x)}$$

$f(m) = \log_e\dfrac{(1+\sqrt{m})^2}{(1-m)}$, $f(n) = \log_e\dfrac{(1+\sqrt{n})^2}{(1-n)}$

$$f(m) + f(n) = \log_e\frac{(1+\sqrt{m})^2}{(1-m)} + \log_e\frac{(1+\sqrt{n})^2}{(1-n)} = \log_e\left\{\frac{(1+\sqrt{m})^2(1+\sqrt{n})^2}{(1-m)(1-n)}\right\}$$ Proved.

$$f(m) + f(n) = \log_e(1+\sqrt{m})^2.(1+\sqrt{n})^2 - \log_e(1-m)(1-n) = \log_e(1+\sqrt{m})^2 + \log_e(1+n)^2 - \log_e(1-m) - \log_e(1-n)$$
$$= 2\log_e(1+\sqrt{m}) + 2\log_e(1+\sqrt{n}) - \log_e(1-m) - \log_e(1-n)$$ Proved.

(b) same as above question. (c) same as above question.

(3) (a) $y = \dfrac{2^x + 2^{-x}}{2^x - 2^{-x}} + 2$ is $y = \dfrac{1}{2}\log_2\left(\dfrac{x-1}{x-3}\right)$

$\Rightarrow y - 2 = \dfrac{2^{2x}+1}{2^{2x}-1}$ $\Rightarrow 2^{2x}.y - y - 2.2^{2x} + 2 = 2^{2x} + 1$ $\Rightarrow 2^{2x}(y-3) = y - 1$ $\Rightarrow 2^{2x} = \dfrac{y-1}{y-3}$

$\log_2 2^{2x} = \log_2\left(\dfrac{y-1}{y-3}\right)$ $\Rightarrow 2x = \log_2\left(\dfrac{y-1}{y-3}\right)$ $\therefore x = \dfrac{1}{2}\log_2\left(\dfrac{y-1}{y-3}\right)$ is the inverse function. $\therefore y = \dfrac{1}{2}\log_2\left(\dfrac{x-1}{x-3}\right)$ Proved.

(b) same as above question (c) same as above question

(4) (a) $y=\sin(\log x)$ $\therefore$ $y'=\cos(\log x).\frac{1}{x}$ $\therefore y''=\frac{-x\sin(\log x).\frac{1}{x}-\cos(\log x).1}{x^2}=\frac{-[\sin(\log x)+\cos(\log x)]}{x^2}$

$L.H.S=x^2y''+xy'+y=x^2\left\{\frac{-[\sin(\log x)+\cos(\log x)]}{x^2}\right\}+x\left\{\cos(\log x).\frac{1}{x}\right\}+\sin(\log x)=-\sin(\log x)-\cos(\log x)+\cos(\log x)+\sin(\log x)$

$=0=R.H.S$ Proved.

(b) same as above question (c) $f(x)=\frac{x-1}{x}, f\left(\frac{1}{x}\right)=\frac{\frac{1}{x}-1}{\frac{1}{x}}=\frac{\frac{1-x}{x}}{\frac{1}{x}}=1-x$

$L.H.S=f(x)+f\left(\frac{1}{x}\right)=\frac{x-1}{x}+1-x=\frac{x-1+x-x^2}{x}=\frac{-(x^2-2x+1)}{x}=\frac{-(x-1)^2}{x}$

$R.H.S=f(x).f\left(\frac{1}{x}\right)=\frac{x-1}{x}\times 1-x=\frac{-(x-1)^2}{x}$ or $L.H.S=R.H.S$ $\therefore$ $\frac{-(x-1)^2}{x}=\frac{-(x-1)^2}{x}$ Proved.

(5) (a) $f(x)=x+\cos x$, $f(x_1)=x_1+\cos x_1$, $f(x_2)=x_2+\cos x_2$

$\therefore f(x_1)=f(x_2) \Rightarrow x_1=x_2$ function $f(x)$is one to one function.

$\Rightarrow y=x+\cos x$ it is straight line, then range(R) $=(-\infty,\infty)$ so, $y=x+\cos x$ is onto function.

$f(x)=x+\cos x$ the function $f(x)$is one to one and onto function. Proved

(b) $f(x)=(a+x^m)^{\frac{1}{m}}$ $\Rightarrow L.H.S=f(f(x))=f\left\{(a+x^m)^{\frac{1}{m}}\right\}=\left[a+(a+x^m)^{\frac{1}{m}.m}\right]^{\frac{1}{m}}=[2a+x^m]^{\frac{1}{m}}=R.H.S$ Proved

(c) Do yourself, $a=\frac{x\pm(x-2)}{2}=\frac{x+x-2}{2},\frac{x-x+2}{2}$ $\therefore$ $a=1,(x-1)$ Ans.

(6) (a) $f(x)=\begin{cases}2^{-x}-1 & \text{for } -2\le x<0\\ \cot\frac{x}{2} & \text{for } 0\le x<\pi\\ \frac{x}{x^3-2} & \text{for } \pi\le x\le 8\end{cases}$ find $f(-2)=2^{-x}-1=2^{-(-2)}-1=2^2-1=4-1=3$ Ans.

$f\left(\frac{\pi}{6}\right)=\cot\frac{x}{2}=\cot\frac{\pi}{12}=\cot 15^0=\frac{\cos 15^0}{\sin 15^0}=\frac{\cos(45^0-30^0)}{\sin(45^0-30^0)}=\frac{\cos 45^0\cos 30^0+\sin 45^0\sin 30^0}{\sin 45^0\cos 30^0-\cos 45^0\sin 30^0}$

$f\left(\frac{\pi}{6}\right)=\frac{\frac{1}{\sqrt2}\frac{\sqrt3}{2}+\frac{1}{\sqrt2}\frac{1}{2}}{\frac{1}{\sqrt2}\frac{\sqrt3}{2}-\frac{1}{\sqrt2}\frac{1}{2}}=\frac{\frac{\sqrt3}{2\sqrt2}+\frac{1}{2\sqrt2}}{\frac{\sqrt3}{2\sqrt2}-\frac{1}{2\sqrt2}}=\frac{\sqrt3+1}{\sqrt3-1}=\frac{\sqrt3+1}{\sqrt3-1}\times\frac{\sqrt3+1}{\sqrt3+1}=\frac{(\sqrt3+1)^2}{3-1}=\frac{3+2\sqrt3+1}{2}=\frac{4+2\sqrt3}{2}=2+\sqrt3$ Ans.

$f\left(\frac{2\pi}{3}\right)=\cot\frac{x}{2}=\cot\frac{2\pi}{2.3}=\cot\frac{\pi}{3}=\frac{1}{\sqrt3}$ Ans. $f(6)=\frac{x}{x^3-2}=\frac{6}{6^3-2}=\frac{6}{216-2}=\frac{6}{214}=\frac{3}{107}$ Ans.

$f(8)=\frac{x}{x^3-2}=\frac{8}{8^3-2}=\frac{8}{512-2}=\frac{8}{510}=\frac{4}{255}$ Ans. $f\left(\frac{\pi}{4}\right)=\cot\frac{x}{2}=\cot\frac{\pi}{8}$ Ans. $f\left(\frac{3\pi}{4}\right)=\cot\frac{x}{2}=\cot\frac{3\pi}{8}$ Ans.

$f(1)=\cot\frac{x}{2}=\cot\frac{1}{2}$ Ans. $f\left(\frac{1}{2}\right)=\cot\frac{x}{2}=\cot\frac{1}{4}$ Ans. $f(-1)=2^{-(-1)}-1=2^1-1=2-1=1$ Ans.

(b) $f(x)=\begin{cases}x-3, & x<-8\\ x^3+1, & x\in[-8,8]\\ 3x+5, & x>8\end{cases}$ Ans:– $f(-14)=-17, f(2)=9, f(12)=41, (fof)4=f(f(4))=200,$

(c) $f(x)=\log_2 x$ and $g(x)=2^x$ To prove , $(fog)x=(gof)x=x$ $\Rightarrow f(g(x))=g(f(x))=x$ $\Rightarrow f(2^x)=g(\log_2 x)=x$

$\log_2 2^x=2^{\log_2 x}=x$ $\Rightarrow x=x=x$ Proved.

(d) Do yourself, same as above question (e) Do yourself, same as question no. –(5)(c)

(7) Ans: –(a)is satisfy the equation $g(f(x))=|\cos x|$ and $f(g(x))=(\cos\sqrt{x})^2$.

(b) $f(x)=\begin{cases}x^2-2x+3, & x<3\\ 2x-1, & x\geq 3\end{cases}$ and $g(x)=\begin{cases}x-4, & x<4\\ 2x^2+4x+5, & x\geq 4\end{cases}$

The critical points are 3,4 and so we redefine the functions for $x<3, 3\leq x<4, x\geq 4$

$$\therefore f(x)=\begin{cases}x^2-2x+3, & x<3\\ 2x-1, & 3\leq x<4\\ 2x-1, & x\geq 4\end{cases} \text{ and } g(x)=\begin{cases}x-4, & x<3\\ x-4, & 3\leq x<4\\ 2x^2+4x+5, & x\geq 4\end{cases}$$

$$\therefore\ f+g=\begin{cases}x^2-2x+3+x-4, & x<3\\ 2x-1+x-4, & 3\leq x<4\\ 2x-1+2x^2+4x+5, & x\geq 4\end{cases}=\begin{cases}x^2-x-1, & x<3\\ 3x-5, & 3\leq x<4\\ 2x^2+6x+4, & x\geq 4\end{cases} \text{ Ans.}$$

$$\therefore\ \frac{f}{g}=\begin{cases}\dfrac{x^2-2x+3}{x-4}, & x<3\\ \dfrac{2x-1}{x-4}, & 3\leq x<4\\ \dfrac{2x-1}{2x^2+4x+5}, & x\geq 4\end{cases} \text{ Ans.}$$

(c) $f(x)=|x-3|$ and $g(x)=f(f(x))$ for $x>3$, *find* $g'(x)$is equal to ………………

$\therefore g(x)=f(f(x))=f(|x-3|)$ for $x>3 \Rightarrow g(x)=f(x-3)=x-3-3=x-6$ $\boxed{g'(x)=1 \text{ Ans.}}$

(8) (a) $f(x)=\log_x 2$ $\boxed{\text{Formula, } f(x)=\log_a x, f(x)\text{to defined } f(x)=\begin{cases}x>0, x\neq 0\\ a>0, a\neq 1\end{cases}}$

function f(x)to be defined $f(x)=\begin{cases}2>0, & 2\neq 0 \textit{ it is true.}\\ x>0, & x\neq 1\end{cases}=x>0, x\neq 1$ $\boxed{\text{Domain (D)}=(0,\infty)-\{1\} \quad \text{Ans.}}$

(b) $f(x)=\sqrt{\log\left(\dfrac{9x-4x^2}{5}\right)}$ Ans: – Domain (D) = φ (c) $f(x)=\log|3-x^2|$ Ans: – Domain (D) $=R-\{\pm\sqrt{3}\}$.

(d) $f(x)=\sqrt{2x-5}-\dfrac{1}{\log_3(2+x)}$ Ans: – Domain (D) $=\left[\dfrac{5}{2},\infty\right[$.

(e) $f(x)=\sqrt{\sin^{-1}x}+\dfrac{1}{\log x}$, function f(x)to be defined as $f(x)=\begin{cases}\sin^{-1}x\geq 0\\ -1\leq x\leq 1\end{cases}$ or $\begin{cases}\log x\neq 0\\ x>0\end{cases}=\begin{cases}x\geq 0\\ -1\leq x\leq 1\end{cases}$ or $\begin{cases}x\neq 1\\ x>0\end{cases}$

Domain (D) = (0,1) Ans.

(f) $f(x)=\sqrt{2+\log_2 x}$, function f(x)is defined as $f(x)=\begin{cases}2+\log_2 x\geq 0\\ x>0\\ x\neq 0\end{cases}=\begin{cases}\log_2 x\geq -2=\log_2 2^{-2}\\ x>0\\ x\neq 0\end{cases}=\begin{cases}x\geq \dfrac{1}{4}\\ x>0\\ x\neq 0\end{cases}$

Domain (D) $=\left[\dfrac{1}{4},\infty\right[$ Ans.

(g) $f(x)=\log_3\left\{\log_{\frac{1}{3}}(x^2-4x+4)\right\}$ $\boxed{\text{Formula, } f(x)=\log_a x \text{ is defined as} \begin{cases}x>0, & x\neq 0\\ a>0, & a\neq 1\end{cases} \text{and } \log_a x>\log_a y \text{ then,} \begin{cases}x>y, & a>1\\ x<y, & a<1\end{cases}}$

function $f(x)=\log_3\left\{\log_{\frac{1}{3}}(x^2-4x+4)\right\}$ is defined as

$f(x)=\begin{cases}\log_{\frac{1}{3}}(x^2-4x+4)>0=\log_{\frac{1}{3}}1\\ x^2-4x+3>0\end{cases}$ or $\begin{cases}x^2-4x+4>1\\ x^2-4x+3>0 \textit{ satisfy all value.}\end{cases}$ or $x^2-4x+3>0$

or $(x-1)(x-3)>0$ $\boxed{\therefore \text{ domain (D)}=(-\infty,1)\cup(3,\infty) \text{ Ans.}}$

(h) $f(x)=\log_4\log_3\log_2(1-6x-9x^2)$, Hence, f(x)to be defined

$$f(x)=\begin{cases}\log_3\log_2(1-6x-9x^2)>0=\log_3 1\\ \log_2(1-6x-9x^2)>0=\log_2 1\\ 1-6x-9x^2>0\end{cases}=\begin{cases}\log_2(1-6x-9x^2)>1=\log_2 2\\ 1-6x-9x^2>1\\ 1-6x-9x^2>0\end{cases}=\begin{cases}1-6x-9x^2>2\\ x(3x+2)<0\\ 9x^2+6x-1<0\end{cases}$$

$$f(x)=\begin{cases}9x^2+6x+1<0\\ x(3x+2)<0\\ \dfrac{-1-\sqrt{2}}{3}<x<\dfrac{-1+\sqrt{2}}{3}\end{cases}\quad \boxed{\text{Domain (D)}=\left(\frac{-2}{3},\frac{-1}{3}\right)\ \text{Ans.}}$$

(i) $f(x)=\log\{(2\log x)^2-12\log x+9\}=\log(2\log x-3)^2=2\log(2\log x-3)$

Function f(x)is defined as

$$\begin{cases}2\log x-3>0\\ 2\log x-3\neq 0\\ x>0\ and\ x\neq 0\end{cases}=\begin{cases}\log x>\frac{3}{2}=\log_e\frac{3}{2}\\ \log x\neq\frac{3}{2}=\log_e\frac{3}{2}\\ x>0\ and\ x\neq 0\end{cases}=\begin{cases}x>e^{\frac{3}{2}}\\ x\neq e^{\frac{3}{2}}\\ x>0\ and\ x\neq 0\end{cases}\quad \boxed{\therefore\ \text{Domain (D)}=\left(e^{\frac{3}{2}},\infty\right)\ \text{Ans.}}$$

(j) $f(x)=\sqrt{\log_{0.2}(4x-2x^2)}\ \Rightarrow \log_{0.2}(4x-2x^2)\geq 0$ and $4x-2x^2>0$

$\Rightarrow 4x-2x^2\geq 1$ and $2x^2-4x<0\ \Rightarrow 2x^2-4x+1\leq 0$ and $2x(x-2)<0$

$\Rightarrow D_1=\left(-\infty,\frac{1}{2}\right), D_2=(-\infty,0)\Rightarrow D=D_1\cap D_2=\left(-\infty,\frac{1}{2}\right)\cap(-\infty,0)$ $\boxed{\therefore\ \text{Domain (D)}=(-\infty,0)\ \text{Ans.}}$

(k) $f(x)=\sqrt{|x|-x}\ \Rightarrow |x|-x\geq 0$ when, $x\geq 0,\ x-x>0,\ 0>0$ *when*, $x<0,\ -x-x\geq 0,\ -2x\geq 0,\ 2x\leq 0,\ x\leq 0$

$\boxed{\therefore\ \text{Domain (D)}=]-\infty,0]\ \text{Ans.}}$

(l) $f(x)=\dfrac{2}{2x+3|x|}\ \Rightarrow 2x+3|x|\neq 0$ when, $x>0, 2x+3x\neq 0$ *or* $5x\neq 0$ *or* $x\neq 0$

when, $x<0, 2x-3x\neq 0, x\neq 0$ $\boxed{\therefore\ \text{Domain (D)}=R-\{0\}\ \text{Ans.}}$

(9) (a) $f(x)=\cos^{-1}\left(\dfrac{2x}{1+x^2}\right)+\sqrt{\sin(\cos x)}$ Ans: – Domain (D) = $[-1,1]$

(b) $f(x)=\cos^{-1}\left[\log_3\left(\frac{1}{3}x^3\right)\right]\ \Rightarrow \begin{cases}-\frac{\pi}{2}\leq\log_3\left(\frac{1}{3}x^3\right)\leq\frac{\pi}{2}\\ \frac{1}{3}x^3>0\end{cases}=\begin{cases}\frac{1}{3}\leq\frac{1}{3}x^3\leq 3\\ x^3>0\end{cases}=\begin{cases}1\leq x^3\leq 9\\ x>0\end{cases}=\begin{cases}1\leq x\leq 9^{\frac{1}{3}}\\ x>0\end{cases}$

$\boxed{\therefore\ \text{Domain (D)}=\left[1,3^{\frac{2}{3}}\right]\ \text{Ans.}}$

(c) $f(x)=\dfrac{\sqrt{3-x^2}}{\cos^{-1}(1+x)}$ function f(x) to be defined as $f(x)=\begin{cases}3-x^2\geq 0\\ -1\leq 1+x\leq 1\\ \cos^{-1}(1+x)\neq 0=\cos^{-1}0\end{cases}$

$\boxed{\therefore\ \text{Domain (D)}=\left[-\sqrt{3},0\right]-\{-1\}\ \text{Ans.}}$

(d) $f(x)=\log(\sin x)$, Ans: – Domain (D) = $n\pi-\frac{\pi}{2}<x<n\pi+\frac{\pi}{2}$

(e) $f(x)=\log(\cos x)^2$, Ans: – Domain (D) = R

(f) $f(x)=\sin^{-1}\left(\dfrac{3-|x|}{2}\right)+\log(2-x)$, Ans: – Domain (D) = $[-5,-1]\cup[1,2)$.

(g) $f(x)=\sqrt{4+2^x+2^{2-x}}+\sqrt{\cos^{-1}(x+2)}$, Ans: – Domain (D) = $\left[\frac{\pi}{2}-2,-1\right]$.

(10) (a) $y=\dfrac{\sqrt{x-3}}{x^2-1}$, $x\geq 3$ and $x\neq\pm 1$, Ans: – Domain (D) = $[3,\infty)$

(b) $y=\dfrac{\cos^{-1}(3-x)}{\sqrt{x^2-1}}$, Ans: –[2,4]. (c) $y=\left\{\dfrac{3-x^2}{x+2}\right\}^{\frac{1}{2}}$

hence y to be defined as $y = \begin{cases} \dfrac{3-x^2}{x+2} \geq 0 \\ x+2 \neq 0 \end{cases} = \begin{cases} 3-x^2 \geq 0 \\ x+2 > 0 \\ x+2 \neq 0 \end{cases} = \begin{cases} x \leq \pm\sqrt{3} \\ x > -2 \\ x \neq -2 \end{cases}$ $\boxed{\therefore \text{Domain (D)} = \left[-\sqrt{3}, \sqrt{3}\right] \text{ Ans.}}$

(d) $y = \dfrac{\sqrt{5+3x-x^2}}{\sqrt{\sin x - \frac{1}{2}}}$ hence y to be defined $y = \begin{cases} 5+3x-x^2 \geq 0 \\ \sin x - \frac{1}{2} > 0 \end{cases} = \begin{cases} x^2-3x-5 \leq 0 \\ \sin x > \frac{1}{2} \end{cases}$

or $y = \begin{cases} x \in \left[\dfrac{3-\sqrt{29}}{2}, \dfrac{3+\sqrt{29}}{2}\right] \\ x \in \left[\dfrac{\pi}{4}, \dfrac{3\pi}{4}\right] \end{cases} = \begin{cases} \dfrac{3-\sqrt{29}}{2} \leq x \leq \dfrac{3+\sqrt{29}}{2} \\ \dfrac{\pi}{4} \leq x \leq \dfrac{3\pi}{4} \end{cases}$ $\boxed{\therefore \text{Domain (D)} = \left[\frac{\pi}{4}, \frac{3\pi}{4}\right] \text{ Ans.}}$

(e) $y = \log_4(x^2+1) + \sqrt{64x - x^7}$, hence y to be defined as

$y = \begin{cases} x^2+1 > 0 \\ 64x - x^7 \geq 0 \end{cases} = \begin{cases} x^2 > -1, \textit{it is satisfying all real value.} \\ x(64-x^6) \geq 0 \end{cases} = \begin{cases} x \geq 0 \\ 64 - x^6 \geq 0 \end{cases} = \begin{cases} x \geq 0 \\ x \leq 2 \end{cases}$

$\boxed{\therefore \text{Domain (D)} = [0,2] \text{ Ans.}}$

(f) $y = \sqrt{\dfrac{2-|x|}{3-|x|}}$, hence y to be defined $y = \begin{cases} 2-|x| \geq 0 \\ 3-|x| > 0 \end{cases} = \begin{cases} -|x| \geq -2 \\ -|x| > -3 \end{cases} = \begin{cases} |x| \leq 2 \\ |x| < 3 \end{cases}$

$\boxed{\therefore \text{Domain (D)} = (-\infty, 2] \text{ Ans.}}$

(g) $f(x) = \sin^{-1}\sqrt{1-x^2} + \cos^{-1}\sqrt{1+x^2} = \dfrac{\pi}{2}$, $f(x) = \dfrac{\pi}{2}$ $\boxed{\text{Domain (D)} = \text{R Ans.}}$

(11) (a) $y = \dfrac{2x}{1-x^2}$, hence y to be defined $\Rightarrow 1-x^2 \neq 0$ or $x \neq \pm 1$ $\boxed{\therefore \text{Domain (D)} = \text{R} - \{-1,1\} \text{ Ans.}}$

Now, $y = \dfrac{2x}{1-x^2}$ or $y - yx^2 = 2x$ or $yx^2 + 2x - y = 0$ $\therefore x = \dfrac{-1 \pm \sqrt{1+y^2}}{y}$

here x to be defined as $\begin{cases} 1+y^2 \geq 0 \\ y \neq 0 \end{cases} = \begin{cases} y^2 \geq -1 \text{ satisfy all real value of y.} \\ y \neq 0 \end{cases}$ $\boxed{\therefore \text{Range} = \text{R} - \{0\} \text{ Ans.}}$

(b) Same as above question, Ans:– Domain (D) $= \text{R} - \left\{\frac{\pi}{4}\right\}$, Range $= \left[-\frac{\pi}{4}, \frac{\pi}{4}\right]$

(c) $y = \dfrac{2}{3+x^2}$, Ans:– Domain(D) = R, Range $= \left(-\infty, \frac{2}{3}\right] - \{0\}$. see question no. –(10)(a).

(d) $y = \dfrac{x+2}{x^2-x-6} = f(x)$ here function f(x)to be defined $\Rightarrow x^2 - x - 6 \neq 0$ $\therefore (x+2)(x-) \neq 0$

$\therefore x \neq 3, -2$ $\boxed{\therefore \text{Domain (D)} = \text{R} - \{-2,3\} \text{ Ans.}}$

$\Rightarrow y = \dfrac{x+2}{x^2-x-6} = \dfrac{x+2}{(x+2)(x-3)} = \dfrac{1}{(x-3)} \Rightarrow xy - 3y = 1$ $\therefore x = \dfrac{1+3y}{y} = f(y)$,

here function f(y)to be defined $\therefore y \neq 0$ $\boxed{\therefore \text{Range} = \text{R} - \{0\} \text{ Ans.}}$

(e) $y = 2 - \dfrac{3}{x^2-3x+4}$, Ans:– Range $= \left[\frac{2}{7}, \infty\right) - \{0\}$

(f) $y = \log\sqrt{x^2+4x+10}$, Ans:– Range $= \left[\frac{1}{2}\log_e 6, \infty\right)$ or $\left[\log_e \sqrt{6}, \infty\right)$.

(g) $y = 3\cos x + 2\sin\left(x + \frac{\pi}{6}\right) + 5$, Ans:– $-\sqrt{19}+5, \sqrt{19}+5$ or $5-\sqrt{19}, 5+\sqrt{19}$

(h) $y = \cos^{-1}\left[\frac{3}{2} + x^2\right]$ where [.]denotes the greatest integer function. $\boxed{\text{Range} = \left\{0, \frac{\pi}{2}\right\} \text{ Ans.}}$

(12) (a) Ans: − Domain (D) $= R - \{3 + 2\sqrt{3}, 3 - 2\sqrt{3}\}$, Range $= \left(-\infty, -\frac{1}{4}\right] \cup \left[\frac{1}{12}, \infty\right)$.

(b) Ans: − Domain (D) $= \left(\frac{2\pi}{3}, \infty\right)$, Range $= \left(-\infty, -\frac{2}{\sqrt{3}}\right] \cup \left[\frac{2}{\sqrt{3}}, \infty\right)$.

(c) Ans: − Domain (D) $= (-1,1)$, (d) Ans: − Domain (D) $= [-1,0) \cup (0,1]$ or $[-1,1] - \{0\}$,

(e) Ans: − Domain (D) $= \{1,2,3\}$ consists of integer , Range $= \{1,2,3\}$.

(13) (a) Ans: − $1 + \sqrt{1 + \log_3 x}$ (b) Ans: − $\dfrac{x + \sqrt{x^2 + 8}}{2}$

(14) (a) Ans: − Yes, $f(0) = \dfrac{0+1}{0+1} = 1$, $g(0) = 1$ $\therefore f(0) = g(0)$

(b) Ans: − Yes, $f(0) = 1$ and $g(0) = \sec^2 0 - \tan^2 0 = 1 - 0 = 1$ $\therefore f(0) = g(0)$

(c) Ans: − No, $f(0) = \log 0$, f(x)is defined for all $x \neq 0$ and g(x)is defined only for $x > 0$.

(d) Ans: − No, f(x)is defined only for $x > 0$ *but not for* $x < -1$ *and* g(x) is defined for $x(x+1) > 0$ *i.e* $x > 0$ *and* $x < -1$.

(e) No, f(x)is defined for all x and g(x)is defined only for $x \geq 0$.

(f) No, f(x)is defined only for $x > 0$ *but not for* $x < -2$ *and* g(x) is defined for $\dfrac{x}{x+2} > 0$ *and* $x + 2 \neq 0$ *i.e* $x > 0, x \neq -2$

(15) (a) $f(x) = \log\left(x - \sqrt{1+x^2}\right)$, $f(-x) = \log\left(-x - \sqrt{1+x^2}\right) = \log\left\{\dfrac{(-x - \sqrt{1+x^2})(-x + \sqrt{1+x^2})}{(-x + \sqrt{1+x^2})}\right\}$

$$f(-x) = \log\left\{\frac{x^2 - 1 + x^2}{-(x - \sqrt{1+x^2})}\right\} = \log\left(\frac{1}{x - \sqrt{1+x^2}}\right) = \log\left(x - \sqrt{1+x^2}\right)^{-1} = -\log\left(x - \sqrt{1+x^2}\right) = -f(x)$$

$\therefore f(x) = -f(-x)$ $\therefore$ f(x) is odd.

(b) $f(x) = \cos^3 x + 2\tan^2 x$, Ans: − $f(x) = f(-x)$, function f(x)is even. (c) Ans: −Even (d) Ans: −Even (e) Ans: −Even (f) Ans: −Even

(16) Ans: − (a) $1 - \sqrt{3}, 2 - \sqrt{2}, 1 + \sqrt{3}$ (b) f(x)is even function.

(17) Ans: − (a) Period of $y = 3\pi$ (b) Period of $y = 2\pi$ (c)Period of $y = \frac{\pi}{2}$

(d) Period of $y = 4\pi$ (e) Period of $y = \pi$ (f) Period of $y = \frac{2\pi}{3}$ (g) Period of $y = 3\pi$

Formula, $y = \sin(ax + b)$, Period of $y = \dfrac{\text{period of } \sin x}{|a|}$

Limit, Continuity & Differentiability

(A)Limits: – Consider the function $y = \frac{x^2 - 4}{x - 2}$. the value of the function at $x = 2$ is of the form $\frac{0}{0}$ which is meaningless. In this case we cannot divide the numerator by denominator, since $x - 2$ is zero. Now suppose x is not actually equal to 2 but very nearly equal to 2 then $x - 2$ is not equal to zero. Hence in this case we can divide the numerator by denominator.

$$\therefore \frac{x^2 - 4}{x - 2} = \frac{(x-2)(x+2)}{(x-2)} = x + 2$$

Thus we see that when x has fixed value 2, the value of y is meaningless but when x tends to 2 (i. e $x \to 2$), y tends to 4 and we say that the limit of y is 4 when x tends to 2. This we writes as

$$\lim_{x\to 2}\frac{x^2 - 4}{x - 2} = \lim_{x\to 2}\frac{(x-2)(x+2)}{(x-2)} = \lim_{x\to 2}(x+2) = 2 + 2 = 4 \quad \text{Ans.}$$

Definition of limit: –The number A is said to be the limit of f(x)at $x = a$ if for any arbitrarily chosen positive number ϵ however small but not zero. there exists a corresponding number, δ greater then zero such that

$$|f(x) - A| < \epsilon \text{ , for all values of } x \text{ for which } 0 < |x - a| < \delta$$

where $|x - a|$ means the absolute value of $x - a$ without any regard to sign.

Right hand limits: –If x approches a from the right, that is from larger value of x then a, the limit of f is called right hand limit of f(x) and written as $\lim_{x\to a+0} f(x)$ or $\lim_{x\to a+0} f(a+0)$

put, $x \to a + h$ in f(x), where $h \to 0$ we have $\boxed{f(a+0) = \lim_{h\to 0} f(a+h) \text{ Right hand limit}}$

Left hand limit: –If x approches from the left, that is from smaller value of x than a , the limit of f is called left hand limit and is written as $\lim_{x\to a-0} f(x)$ or $\lim_{x\to a-0} f(a-0)$

put, $x \to a - h$, where $h \to 0$ we have $\boxed{f(a-0) = \lim_{h\to 0} f(a-h) \text{ left hand limit}}$

Algebra of limits: – # $\lim_{x\to a}[f(x) \pm g(x)] = \lim_{x\to a} f(x) \pm \lim_{x\to a} g(x)$ # $\lim_{x\to a}[f(x).g(x)] = \lim_{x\to a} f(x) . \lim_{x\to a} g(x)$

$\lim_{x\to a}\frac{f(x)}{g(x)} = \frac{\lim_{x\to a} f(x)}{\lim_{x\to a} g(x)} = \lim_{x\to a} f(x) \div \lim_{x\to a} g(x)$ but $\lim_{x\to a} g(x) \neq 0$

Indeterminate forms and L'Hospital rule: – If a function f(x)takes the form $\frac{0}{0}$ or $\frac{\infty}{\infty}$ at $x = a_i$ then we say that f(x) is indeterminate at $x = a$ other indeterminate forms are $\infty - \infty, 0 \times \infty, 1^\infty, 0^0, \infty^0$

L'Hospital's Rule: –If $\phi(x)$, and $\psi(x)$are functions of x such that $\phi(a) = 0, \psi(a) = 0$, then $\lim_{x\to a}\frac{\phi(x)}{\psi(x)} = \lim_{x\to a}\frac{\phi'(x)}{\psi'(x)}$.

Note: –Applying L'Hospital rule differentiate $\frac{\phi(x)}{\psi(x)}$, N^r and D^r separately.

Some important expansions: –

(i) $\sin x = x - \frac{x^3}{3!} + \frac{x^5}{5!} - \cdots$ .... (odd + –)

$\sin hx = x + \frac{x^3}{3!} + \frac{x^5}{5!} + \cdots$ .... (odd + +)

(ii) $\cos x = 1 - \frac{x^2}{2!} + \frac{x^4}{4!} - \cdots$ ……………………………………………… (even + −)

$\cos hx = 1 + \frac{x^2}{2!} + \frac{x^4}{4!} + \cdots$ ………………………………………………. (even + +)

(iii) $\tan x = x + \frac{1}{3}x^3 + \frac{2}{15}x^5 + \cdots$ ……………………………………………………..

(iv) $\tan^{-1} x = x - \frac{1}{3}x^3 + \frac{1}{5}x^5 - \cdots$ ……………………………………………………….

(v) $e^x = 1 + x + \frac{x^2}{2!} + \frac{x^3}{3!} + \cdots$ ……………………………………………… (All + +)

$e^{-x} = 1 - x + \frac{x^2}{2!} - \frac{x^3}{3!} + \cdots$ ………………………………………………. (Alternate + −)

(vi) If $|x| < 1$, *then* $\log_e(1 + x) = x - \frac{1}{2}x^2 + \frac{1}{3}x^3 - \frac{1}{4}x^4 + \cdots$ ………………………………………………………

(vii) $\log_e(1 - x) = -x - \frac{1}{2}x^2 - \frac{1}{3}x^3 - \frac{1}{4}x^4 - \cdots$ …………………………… (All−)

Some important limits should be remembered: −

(1)(a) $\lim_{x\to 0} \frac{\sin x}{x} = 1$ (b) $\lim_{x\to 0} \cos x = 1$ (c) $\lim_{x\to 0} \frac{\tan x}{x} = 1$ (d) $\lim_{x\to 0} \frac{\tan^{-1} x}{x} = 1$ (e) $\lim_{x\to 0} \frac{\sin^{-1} x}{x} = 1$

(2) $\lim_{x\to 0}(1 + x)^{\frac{1}{x}} = e$ put $x = \frac{1}{x}$, $x \to \infty$ or $\lim_{x\to\infty}\left(1 + \frac{1}{x}\right)^x = e$ or $\lim_{x\to 0}(1 + mx)^{\frac{1}{x}} = e^m$

or $\lim_{x\to\infty}\left(1 + \frac{m}{x}\right)^x = e^m$ or $\lim_{x\to 0}\left(1 + \frac{x}{m}\right)^{\frac{1}{x}} = e^{\frac{1}{m}}$ or $\lim_{x\to\infty}\left(1 + \frac{1}{mx}\right)^x = e^{\frac{1}{m}}$

(3) $\lim_{x\to 0} \frac{a^x - 1}{x} = \log a$ form $\left(\frac{0}{0}\right)$

Proof: − $\lim_{x\to 0} \frac{a^x - 1}{x}$ using L'Hospital rule.

Differentiate with respect to x, $\frac{\lim_{x\to 0} a^x - 1}{\lim_{x\to 0} x} = \frac{\lim_{x\to 0} a^x \log a}{\lim_{x\to 0} 1} = \frac{\log a}{1} = \log a$ Proved. $\lim_{x\to 0} \frac{e^x - 1}{x} = \log e = 1$ (above formula)

Example: − (a) $\lim_{x\to 0} \frac{5^x - 1}{x} = \log 5$ Ans.

(b) $\lim_{x\to 0} \frac{e^{3x} - 1}{3x} = \log 2$ Ans. Hint − $x \to 0$, $3x \to 0$ limit does not change. $\lim_{3x\to 0} \frac{e^{3x} - 1}{3x} = \log 2$ Ans.

(4) $\lim_{x\to 0} \frac{(1 + x)^n - 1}{x} = n$, Hint: − $\lim_{x\to 0} \frac{1 + nx + \frac{n(n-1)}{2}x^2 + \cdots \ldots \ldots - 1}{x} = n$ proved. (using L'Hospital rule)

(5) $\lim_{x\to a} \frac{x^n - a^n}{x - a} = na^{n-1}$, Hint: − $\lim_{x\to a} \frac{nx^{n-1} - na^{n-1}.0}{1 - 0} = na^{n-1}$ proved. (using L'Hospital rule)

(6) $\lim_{x\to 0} \frac{\log_a(1 + x)}{x} = \log_a e, (a > 0, a \neq 0)$. (7) $\lim_{x\to\infty} \frac{\log x}{x^m} = 0, (m > 0)$.

(8) (a) $\lim_{x\to\infty} \frac{1 + x}{x} = 1$ (b) $\lim_{x\to 0} \frac{x(x - 1)}{(x + 1)^n - 1} = -\frac{1}{n}, n \neq 0$ (c) $\lim_{x\to 0} \frac{(1 + \sin x)^n - 1}{x} = n, n \neq 0$

Evaluation of limits: − (1) Factorisation or substitution: − $\lim_{x\to a} \frac{x^2 - a^2}{x - a} = \lim_{x\to a} \frac{(x - a)(x + a)}{(x - a)} = \lim_{x\to a}(x + a) = a + a = 2a$ Ans.

(2) L'Hospital rule: $-\ \lim\limits_{x\to 0} x\log x$ form $(0\times\infty) = \lim\limits_{x\to 0}\frac{\log x}{\frac{1}{x}}$ form $\left(\frac{\infty}{\infty}\right)$, using L'Hospital rule.

Differentiate N^r and D^r separately, $\lim\limits_{x\to 0}\frac{\frac{1}{x}}{-\frac{1}{x^2}} = \lim\limits_{x\to 0} -\frac{1}{x}\times\frac{x^2}{1} = -\lim\limits_{x\to 0} x = 0$ Ans.

(3) Expansion rule: $-\ \lim\limits_{x\to 0}\frac{\log(1+x)}{x}$, form $\left(\frac{0}{0}\right)$

$$\lim_{x\to 0}\frac{-\left\{x+\frac{x^2}{2}+\frac{x^3}{3}+\cdots\ldots\ldots\ldots\right\}}{x} = \lim_{x\to 0}\frac{-x\left\{1+\frac{x}{2}+\frac{x^2}{3}+\cdots\ldots\ldots\ldots\right\}}{x} = \lim_{x\to 0} -\left(1+\frac{x}{2}+\frac{x^2}{3}+\cdots\ldots\ldots\ldots\right) = -1 \quad \text{Ans.}$$

(4) Rationalisation: $-\ \lim\limits_{x\to 2}\frac{x-2}{\sqrt{x-1}-\sqrt{3-x}}$

$$\text{or } \lim_{x\to 2}\frac{x-2}{\sqrt{x-1}-\sqrt{3-x}}\times\frac{\sqrt{x-1}+\sqrt{3-x}}{\sqrt{x-1}+\sqrt{3-x}} = \lim_{x\to 2}\frac{(x-2)\left(\sqrt{x-1}+\sqrt{3-x}\right)}{x-1-3+x} = \lim_{x\to 2}\frac{(x-2)\left(\sqrt{x-1}+\sqrt{3-x}\right)}{2x-4} = \lim_{x\to 2}\frac{(x-2)\left(\sqrt{x-1}+\sqrt{3-x}\right)}{2(x-2)}$$

$$= \lim_{x\to 2}\frac{\left(\sqrt{x-1}+\sqrt{3-x}\right)}{2} = \frac{2}{2} = 1 \quad \text{Ans.}$$

Continuity

Continuity: $-$ f(x)is said to be continuous at $x = a$ if $\lim\limits_{x\to a_+} f(x) = \lim\limits_{x\to a_-} f(x) = f(x)$ or $\lim\limits_{h\to a} f(a+h) = \lim\limits_{h\to 0} f(a-h) = f(a)$

The function is not defined at $x = a$, f(a) does not exist or $\lim\limits_{x\to a_+} f(x) \neq \lim\limits_{x\to a_-} f(x)$ then we say that function is discontinuous at $x = a$.

Functions continuous on the open interval: $-$ (a, b) or]a, b[or $\{a < x < b\}$, f(x) will be continuous

if it is continuous at every point of the interval.

Function continuous on the closed interval: $-$ [a, b] or $a \le x \le b$,

if will satisfy the following three condition: $-$ (i) f(x) is continuous at each point of (a, b)

(ii) $\lim\limits_{x\to a_+} f(x) = f(a)$ (iii) $\lim\limits_{x\to b_-} f(x) = f(b)$

A continuous fnction in the closed interval $a \le x \le b$ has the following properties: $-$

(a) If f(x)and f(b)are of opposite signs, then there exists at least one and in general odd solutions of the equation $f(x) = 0$ for any x in the open interval (a, b).

(b) If λ is any real number between f(a)and f(b), then there exists at least one solution of the equation $f(x) = \lambda$ in the open interval (a, b).

Type of discontinuities: $-$ (i) Discontinuity of first kind: $-$ The point of $x = a$ will be a point of discontinuity.

If both right hand and left hand limit at $x = a$ exist but not equal. $\lim\limits_{x\to a_+} f(x) \neq \lim\limits_{x\to a_-} f(x)$ or $\lim\limits_{h\to 0} f(a-h) \neq \lim\limits_{h\to 0} f(a+h)$

(ii) Discontinuity of second kind: $-$ The point at $x = a$ will be a point of discontinuity.

If either or both the RHL and LHL do not exist or if either or both the limits $\lim\limits_{x\to a_+} f(x) = \lim\limits_{x\to a_-} f(x)$ are infinite.

(iii) Removable discontinuity: $-$ If $\lim\limits_{x\to a_+} f(x) = \lim\limits_{x\to a_-} f(x)$ limit exists but it is not equal to f(a).

Differentiability

f(x) is said to be differentiable at x = a if $\lim\limits_{h\to 0_+}\dfrac{f(a+h)-f(a)}{h}=\lim\limits_{h\to 0_-}\dfrac{f(a-h)-f(a)}{-h}=f(a)$

An important point: −A function which is differentiable at a point x = a must also be continuous at that point. If a function is continuous at a point x = a , it is not necessarily differentiable at x = a.

Differentiability implies continuity always: − Since the function is differentiable

$$\lim_{h\to 0_+}\frac{f(a+h)-f(a)}{h}=\lim_{h\to 0_-}\frac{f(a-h)-f(a)}{-h}=f(a)$$

$\lim\limits_{h\to 0_+}\dfrac{f(a+h)-f(a)}{h}$ exists and is finite − − − − − − − (A)

or $\lim\limits_{h\to 0}f(a+h)=\lim\limits_{h\to 0}[f(a+h)-f(a)]+f(a)=\lim\limits_{h\to 0}h\left[\dfrac{f(a+h)-f(a)}{h}\right]+f(a)$

$\lim\limits_{h\to 0}h\;.\lim\limits_{h\to 0}\dfrac{f(a+h)-f(a)}{h}+f(a)=0+f(a)=f(a)$

similarly, $\lim\limits_{h\to 0}f(a-h)=f(a)$ since, $\lim\limits_{h\to 0}f(a+h)=\lim\limits_{h\to 0}f(a-h)=f(a)$, the function is continuous.

Continuity does not necessarily imply differentiability: −

Example: − consider $f(x)=\begin{cases}x\cos x\,, & x\neq 0\\ 0\,, & x=0\end{cases}$ differentiability at x = 0.

$$R.H.L\,[R'(0)]=\lim_{h\to 0_+}\frac{f(0+h)-f(0)}{h}=\lim_{h\to 0_+}\frac{f(h)-f(0)}{h}=\lim_{h\to 0}\frac{h\cos h}{h}=\lim_{h\to 0}\cos h=1$$

$$L.H.L=\lim_{h\to 0_-}\frac{f(0-h)-f(0)}{-h}=\lim_{h\to 0_-}\frac{f(-h)-f(0)}{-h}=\lim_{h\to 0}\frac{-h\cos h}{-h}=\lim_{h\to 0}\cos h=1$$

$f(x)=f(0)=0$

$R.H.L=L.H.L\neq f(0)$ the function is not differentiable at x = 0.

Continuity: − $\lim\limits_{x\to a_+}f(x)=\lim\limits_{x\to a_-}f(x)=f(x)$

$R.H.L=\lim\limits_{x\to 0_+}f(x)=\lim\limits_{x\to 0_+}x\cos x=0$, $L.H.L=\lim\limits_{x\to 0_-}f(x)=\lim\limits_{x\to 0_-}x\cos x=0$, $v=f(x)=f(0)=0$

$R.H.L=L.H.L=vf(x)$, the function is continuous at x = 0.

Differentiability of f(x) on closed interval [a, b] : −

(i) It should be differentiable at every point on the open interval (a, b) or]a, b[.

(ii) Right hand derivative R. H. L (R′) at a and left hand derivative L. H. L (L′)at b exist finitely.

If both the above are satisfied the we say that f(x)is differentiable in the closed interval [a, b].

Solved example

(1) Evaluate: − (a) $\lim\limits_{h\to 0}\dfrac{\sqrt{x+2h}-\sqrt{x}}{2h}$ (b) $\lim\limits_{x\to 2}\dfrac{x-2}{x^3-8}$ (c) $\lim\limits_{x\to\pi}\dfrac{\sin x-\cos x}{\cos x}$ (d) $\lim\limits_{x\to 1}\dfrac{x^2-3x+2}{2x^2-7x+5}$

Solution: − (a) $\lim\limits_{h\to 0}\dfrac{\sqrt{x+2h}-\sqrt{x}}{2h}$, (Rationalize the N^r)

$$\lim_{h\to0}\frac{\sqrt{x+2h}-\sqrt{x}}{2h}\times\frac{\sqrt{x+2h}+\sqrt{x}}{\sqrt{x+2h}+\sqrt{x}}=\lim_{h\to0}\frac{x+2h-x}{2h(\sqrt{x+2h}+\sqrt{x})}=\lim_{h\to0}\frac{1}{\sqrt{x+2h}+\sqrt{x}}=\frac{1}{2\sqrt{x}}\ \text{Ans.}$$

(b) $\lim_{x\to2}\frac{x-2}{x^3-8}=\lim_{x\to2}\frac{x-2}{x^3-2^3}=\lim_{x\to2}\frac{(x-2)}{(x^2+2x+4)(x-2)}=\lim_{x\to2}\frac{1}{(x^2+2x+4)}=\frac{1}{4+4+4}=\frac{1}{12}$ Ans.

IInd method: $-\lim_{x\to2}\frac{x-2}{x^3-8}$, form $\left(\frac{0}{0}\right)$ using L'Hospital rule. $\lim_{x\to2}\frac{x-2}{x^3-8}=\lim_{x\to2}\frac{\frac{d}{dx}(x-2)}{\frac{d}{dx}(x^3-8)}=\lim_{x\to2}\frac{1}{3x^2}=\frac{1}{3\times4}=\frac{1}{12}$ Ans.

(c) $\lim_{x\to\pi}\frac{\sin x-\cos x}{\cos x}$ put value $x=\pi$ then $\lim_{x\to\pi}\frac{\sin x-\cos x}{\cos x}=\lim_{x\to\pi}\frac{\sin\pi-\cos\pi}{\cos\pi}=\frac{0-(-1)}{-1}=-1$ Ans.

(d) $\lim_{x\to1}\frac{x^2-3x+2}{2x^2-7x+5}$, form $\left(\frac{0}{0}\right)$ using L'Hospital rule, $\lim_{x\to1}\frac{x^2-3x+2}{2x^2-7x+5}=\lim_{x\to1}\frac{2x-3}{4x-7}=\frac{2-3}{4-7}=-\frac{1}{3}$ Ans.

(2) use the formula $\lim_{x\to0}\frac{a^x-1}{x}=\log a$ to find (a) $\lim_{x\to0}\frac{(1+x)^{\frac{1}{2}}-1}{2^x-1}$ (b) $\lim_{x\to0}\frac{7^x-5^x}{5^x-3^x}$ (c) $\lim_{x\to1}\frac{\frac{3^x}{3}+\frac{5^x}{5}-2}{(x-1)}$ (d) $\lim_{x\to0}\frac{81^x-27^x-3^x+1^x}{x^2}$

(e) $\lim_{x\to0}\frac{(3^x-1)(5^x-1)}{\sqrt{2}-\sqrt{1+\cos x}}$ (f) $\lim_{x\to0}\frac{5^x-1}{(1-x)^{\frac{1}{2}}-1}$ (g) $\lim_{x\to0}\frac{7^x+3^x-2}{\sin x}$ (h) $\lim_{x\to0}\frac{x^2 3^x-x^2}{(1-\cos x)x}$

Solution: $-$(a) $\lim_{x\to0}\frac{(1+x)^{\frac{1}{2}}-1}{2^x-1}=\lim_{x\to0}\frac{\sqrt{1+x}-1}{2^x-1}\times\frac{\sqrt{1+x}+1}{\sqrt{1+x}+1}=\lim_{x\to0}\frac{1+x-1}{(2^x-1)(\sqrt{1+x}+1)}=\lim_{x\to0}\frac{x}{(2^x-1)(\sqrt{1+x}+1)}$

$$=\lim_{x\to0}\frac{1}{\frac{(2^x-1)}{x}.(\sqrt{1+x}+1)}=\frac{1}{2\log2}=\frac{1}{\log4}\ \text{Ans.}$$

(b) $\lim_{x\to0}\frac{7^x-5^x}{5^x-3^x}=\lim_{x\to0}\frac{(7^x-1)-(5^x-1)}{(5^x-1)-(3^x-1)}=\lim_{x\to0}\frac{\frac{(7^x-1)}{x}-\frac{(5^x-1)}{x}}{\frac{(5^x-1)}{x}-\frac{(3^x-1)}{x}}=\frac{\log7-\log5}{\log5-\log3}=\frac{\log\left(\frac{7}{5}\right)}{\log\left(\frac{5}{3}\right)}$ Ans.

(c) $\lim_{x\to1}\frac{\frac{3^x}{3}+\frac{5^x}{5}-2}{(x-1)}=\lim_{x\to1}\frac{3^{x-1}+5^{x-1}-2}{(x-1)}$ $,x\to1,x-1\to0$

$$\lim_{(x-1)\to0}\frac{(3^{(x-1)}-1)+(5^{(x-1)}-1)}{(x-1)}=\lim_{(x-1)\to0}\left\{\frac{(3^{(x-1)}-1)}{(x-1)}+\frac{(5^{(x-1)}-1)}{(x-1)}\right\}=\log3+\log5=\log3.5=\log15\ \text{Ans.}$$

(d) $\lim_{x\to0}\frac{81^x-27^x-3^x+1^x}{x^2}=\lim_{x\to0}\frac{(27^x-1)(3^x-1)}{x^2}=\lim_{x\to0}\frac{(27^x-1)}{x}.\lim_{x\to0}\frac{(3^x-1)}{x}=\log27.\log3=\log3^3.\log3=3\log3.\log3$

$=3(\log3)^2$ Ans.

(e) $\lim_{x\to0}\frac{(3^x-1)(5^x-1)}{\sqrt{2}-\sqrt{1+\cos x}}=\lim_{x\to0}\frac{(3^x-1)(5^x-1)}{\sqrt{2}-\sqrt{2\cos^2\frac{x}{2}}}$ use formula, $1-\cos x=2\sin^2\frac{x}{2}$, $1+\cos x=2\cos^2\frac{x}{2}$

$\lim_{x\to0}\frac{(3^x-1)(5^x-1)}{\sqrt{2}\left[1-\cos\frac{x}{2}\right]}=\lim_{x\to0}\frac{(3^x-1)(5^x-1)}{\sqrt{2}.2\sin^2\frac{x}{4}}=\lim_{x\to0}\frac{\frac{(3^x-1)}{x}\frac{(5^x-1)}{x}}{2\sqrt{2}\left(\frac{\sin\frac{x}{4}}{\frac{x}{4}}\right)^2\times\frac{1}{16}}$ formula, $1-\cos\frac{x}{2}=2\sin^2\frac{x}{4}$

$$\text{or}\ \frac{\log3.\log5}{\frac{\sqrt{2}}{8}}=8\frac{\log3.\log5}{\sqrt{2}}=4.\sqrt{2}.\sqrt{2}\frac{\log3.\log5}{\sqrt{2}}=4\sqrt{2}.\log3.\log5\ \text{Ans.}$$

(f) $\lim_{x\to0}\frac{5^x-1}{(1-x)^{\frac{1}{2}}-1}=\lim_{x\to0}\frac{5^x-1}{(1-x)^{\frac{1}{2}}-1}\times\frac{(1-x)^{\frac{1}{2}}+1}{(1-x)^{\frac{1}{2}}+1}=\lim_{x\to0}\frac{(5^x-1)\left[(1-x)^{\frac{1}{2}}+1\right]}{1-x-1}=-\lim_{x\to0}\frac{(5^x-1)}{x}.\left[(1-x)^{\frac{1}{2}}+1\right]=-\log5.2=-2\log5$

$=\log5^{-2}=\log\left(\frac{1}{25}\right)$ Ans.

(g) $\lim\limits_{x\to 0}\dfrac{7^x+3^x-2}{\sin x}=\lim\limits_{x\to 0}\dfrac{(7^x-1)+(3^x-1)}{\sin x}=\lim\limits_{x\to 0}\dfrac{\dfrac{(7^x-1)}{x}+\dfrac{(3^x-1)}{x}}{\dfrac{\sin x}{x}}=\dfrac{\log 7+\log 3}{1}=\log 7.3=\log 21$ Ans.

(h) $\lim\limits_{x\to 0}\dfrac{x^2 3^x-x^2}{(1-\cos x)x}=\lim\limits_{x\to 0}\dfrac{x^2(3^x-1)}{\left(2\sin^2\frac{x}{2}\right)x}=\lim\limits_{x\to 0}\dfrac{x(3^x-1)}{2\left(\dfrac{\sin\frac{x}{2}}{\frac{x}{2}}\right)^2\times\dfrac{x^2}{4}}=\lim\limits_{x\to 0}\dfrac{(3^x-1)}{x}\times\dfrac{2}{\left(\dfrac{\sin\frac{x}{2}}{\frac{x}{2}}\right)^2}=2\log 3=\log 3^2=\log 9$ or $2\log 3$ Ans.

(3) using the formula $\lim\limits_{x\to 0}(1+x)^{\frac{1}{x}}=e=\lim\limits_{x\to\infty}\left(1+\frac{1}{x}\right)^x$, to find the limit (a) $\lim\limits_{x\to 0}\left(\dfrac{1+5x}{1+7x}\right)^{\frac{1}{x}}$

(b) $\lim\limits_{x\to\infty}\left(\dfrac{1+\frac{5}{x^2}}{1+\frac{3}{x^2}}\right)^{x^2}$ (c) $\lim\limits_{x\to\infty}\left(\dfrac{x^2+4x+5}{x^2+2x+1}\right)^x$ (d) $\lim\limits_{x\to 0}\left(\dfrac{x^2+3x+7}{x^2+x+2}\right)^{\frac{1}{x}}$

Solution: – (a) $\lim\limits_{x\to 0}\left(\dfrac{1+5x}{1+7x}\right)^{\frac{1}{x}}=\lim\limits_{x\to 0}\dfrac{(1+5x)^{\frac{1}{5x}\times 5}}{(1+7x)^{\frac{1}{7x}\times 7}}$ put $5x=y$ and $7x=z$ formula $\lim\limits_{x\to 0}(1+mx)^{\frac{1}{x}}=e^m$

or $\lim\limits_{x\to 0}\dfrac{(1+y)^{\frac{1}{y}\times 5}}{(1+z)^{\frac{1}{z}\times 7}}$ use formula, $\lim\limits_{x\to 0}(1+x)^{\frac{1}{x}}=e$ or $\lim\limits_{x\to 0}\left(\dfrac{1+5x}{1+7x}\right)^{\frac{1}{x}}=\dfrac{e^5}{e^7}=e^{5-7}=e^{-2}=\dfrac{1}{e^2}$ Ans.

(b) $\lim\limits_{x\to\infty}\left(\dfrac{1+\frac{5}{x^2}}{1+\frac{3}{x^2}}\right)^{x^2}$ use formula, $e=\lim\limits_{x\to\infty}\left(1+\frac{1}{x}\right)^x$ or $\lim\limits_{x\to\infty}\left(1+\frac{m}{x}\right)^x=e^m$ or $\lim\limits_{x\to\infty}\dfrac{\left(1+\frac{5}{x^2}\right)^{x^2}}{\left(1+\frac{3}{x^2}\right)^{x^2}}=\dfrac{e^5}{e^3}=e^{5-3}=e^2$ Ans.

(c) $\lim\limits_{x\to\infty}\left(\dfrac{x^2+4x+5}{x^2+2x+1}\right)^x=\lim\limits_{x\to\infty}\left(1+\dfrac{2x+4}{x^2+2x+1}\right)^x=\lim\limits_{x\to\infty}\left(1+\dfrac{2x+4}{x^2+2x+1}\right)^{\frac{2x+4}{x^2+2x+1}\times\frac{x^2+2x+1}{2x+4}\times x}=\lim\limits_{x\to\infty}\left(1+\dfrac{2x+4}{x^2+2x+1}\right)^{\frac{x^2+2x+1}{2x+4}\cdot\frac{2x^2+4x}{x^2+2x+1}}$

$=\lim\limits_{x\to\infty}\left(1+\dfrac{2x+4}{x^2+2x+1}\right)^{\frac{x^2+2x+1}{2x+4}\cdot\lim\limits_{x\to\infty}\frac{2x^2+4x}{x^2+2x+1}}=e^{\lim\limits_{x\to\infty}\frac{2x^2\left(1+\frac{2}{x}\right)}{x^2\left(1+\frac{2}{x}+\frac{1}{x^2}\right)}}=e^2$ Ans.

(d) $\lim\limits_{x\to 0}\left(\dfrac{x^2+3x+7}{x^2+x+2}\right)^{\frac{1}{x}}=\lim\limits_{x\to 0}\left(1+\dfrac{2x+5}{x^2+x+2}\right)^{\frac{1}{x}}=e^{\infty}$ Ans. (Do yourself)

IInd Method: –(3) (a) $\lim\limits_{x\to 0}\left(\dfrac{1+5x}{1+7x}\right)^{\frac{1}{x}}$ Let $y=\left(\dfrac{1+5x}{1+7x}\right)^{\frac{1}{x}}$ or $\log y=\log\left(\dfrac{1+5x}{1+7x}\right)^{\frac{1}{x}}=\dfrac{1}{x}\log\left(\dfrac{1+5x}{1+7x}\right)$

We Take a limit both of side, $\lim\limits_{x\to 0}\log y=\lim\limits_{x\to 0}\dfrac{1}{x}\log\left(\dfrac{1+5x}{1+7x}\right)=\dfrac{\lim\limits_{x\to 0}\log\left(\frac{1+5x}{1+7x}\right)}{\lim\limits_{x\to 0}x}$, using L'Hospital rule

or $\log y=\lim\limits_{x\to 0}\dfrac{1}{\frac{1+5x}{1+7x}}\times\dfrac{(1+7x)5-(1+5x)7}{(1+7x)^2}=\lim\limits_{x\to 0}\dfrac{5+35x-7-35x}{(1+5x)(1+7x)}=\lim\limits_{x\to 0}\dfrac{5-7}{(1+5x)(1+7x)}=\lim\limits_{x\to 0}\dfrac{-2}{(1+5x)(1+7x)}=-2$

or $\log y=-2$ or $y=e^{-2}=\dfrac{1}{e^2}$ Ans.

(b) and (c) Do yourself, same as above question

(4) using the formula $\lim\limits_{x\to 0}\dfrac{e^x-1}{x}=1$ to find (a) $\lim\limits_{x\to 0}\dfrac{e^{x^2}-1+\sin^2 x}{x^2}$ (b) $\lim\limits_{x\to 1}\dfrac{e^{(x-1)}-1+\sin(x-1)}{(x-1)}$

Solution: – (a) $\lim\limits_{x\to 0}\dfrac{e^{x^2}-1+\sin^2 x}{x^2}=\lim\limits_{x\to 0}\left(\dfrac{e^{x^2}-1}{x^2}+\dfrac{\sin^2 x}{x^2}\right)=\lim\limits_{x\to 0}\left(\dfrac{e^{x^2}-1}{x^2}\right)+\lim\limits_{x\to 0}\left(\dfrac{\sin x}{x}\right)^2=1+1=2$ Ans.

(b) $\lim\limits_{x\to 1}\dfrac{e^{(x-1)}-1+\sin(x-1)}{(x-1)}=\lim\limits_{x-1\to 0}\dfrac{e^{(x-1)}-1+\sin(x-1)}{(x-1)}=\lim\limits_{x-1\to 0}\dfrac{e^{(x-1)}-1}{(x-1)}+\lim\limits_{x-1\to 0}\dfrac{\sin(x-1)}{(x-1)}=1+1=2$ Ans.

(5) using the formula $\lim\limits_{x\to 0}\dfrac{\log(1+x)}{x}=1$, to find (a) $\lim\limits_{x\to 0}\dfrac{\log(1-x)-\log(1+3x)}{x}$ (b) $\lim\limits_{x\to 0}\dfrac{\log(1-5x)}{\log(1+2x)}$ (c) $\lim\limits_{x\to 0}\dfrac{\log\left(\frac{1+x}{1-x}\right)}{x}$

(d) $\lim\limits_{x\to 0}\dfrac{\log(x^2+2x+1)}{\log(7x+1)}$ (e) $\lim\limits_{x\to -1}\dfrac{\log(x+2)}{\log(x^2+4x+4)}$

Solution: – (a) $\lim\limits_{x\to 0}\dfrac{\log(1-x)-\log(1+3x)}{x}=\lim\limits_{x\to 0}\left[\dfrac{\log(1-x)}{x}-\dfrac{\log(1+3x)}{x}\right]=\lim\limits_{x\to 0}\left[\dfrac{\log(1-x)}{x}\right]-\lim\limits_{3x\to 0}\left[\dfrac{\log(1+3x)}{3x}\right].3=-1-3=-4$ Ans.

(b) $\lim\limits_{x\to 0}\dfrac{\log(1-5x)}{\log(1+2x)}=\lim\limits_{x\to 0}\dfrac{\frac{\log[1+(-5)x]}{-5x}\times -5}{\frac{\log[1+2x]}{2x}\times 2}=\dfrac{\lim\limits_{x\to 0}\frac{\log[1+(-5)x]}{-5x}\times -5}{\lim\limits_{x\to 0}\frac{\log[1+2x]}{2x}\times 2}=\dfrac{1\times -5}{1\times 2}=-\dfrac{5}{2}$ Ans.

(c) $\lim\limits_{x\to 0}\dfrac{\log\left(\frac{1+x}{1-x}\right)}{x}=\lim\limits_{x\to 0}\dfrac{\log(1+x)-\log(1-x)}{x}=\lim\limits_{x\to 0}\dfrac{\log(1+x)}{x}-\lim\limits_{x\to 0}\dfrac{\log(1-x)}{x}=1-\lim\limits_{x\to 0}\dfrac{\log[1+(-x)]}{-x}\times -1=1-(-1)=1+1$
$=2$ Ans.

(d) $\lim\limits_{x\to 0}\dfrac{\log(x^2+2x+1)}{\log(7x+1)}=\lim\limits_{x\to 0}\dfrac{\log(1+x)^2}{\log(7x+1)}=\lim\limits_{x\to 0}\dfrac{\frac{2\log(1+x)}{x}}{\frac{\log(1+7x)}{7x}\times 7}=\dfrac{2\times 1}{1\times 7}=\dfrac{2}{7}$ Ans.

(e) $\lim\limits_{x\to -1}\dfrac{\log(x+2)}{\log(x^2+4x+4)}=\lim\limits_{x+1\to 0}\dfrac{\log(x+1+1)}{\log(x+2)^2}=\lim\limits_{x+1\to 0}\dfrac{\log[1+(x+1)]}{2\log(x+2)}=\lim\limits_{x+1\to 0}\dfrac{\frac{\log[1+(x+1)]}{(x+1)}}{2\frac{\log[1+(x+1)]}{(x+1)}}=\dfrac{1}{2}$ Ans.

IInd Method: – $\lim\limits_{x\to -1}\dfrac{\log(x+2)}{\log(x^2+4x+4)}=\lim\limits_{x+1\to 0}\dfrac{\log(x+2)}{\log(x+2)^2}=\lim\limits_{x+1\to 0}\dfrac{\log(x+2)}{2\log(x+2)}=\lim\limits_{x+1\to 0}\dfrac{1}{2}=\dfrac{1}{2}$ Ans.

(6) using the formula $\lim\limits_{x\to 0}\dfrac{(1+x)^n-1}{x}=n$, to find following limits (a) $\lim\limits_{x\to 0}\dfrac{\sqrt{1+2x}-\sqrt{1+3x}}{x}$

(b) $\lim\limits_{x\to 0}\dfrac{\sqrt{1+3x}-4x^2-4x-1}{x}$ (c) $\lim\limits_{x\to 0}\dfrac{x^2+2x+1-(1+2x)^{\frac{3}{2}}}{x}$ (d) $\lim\limits_{x\to -2}\dfrac{(x+3)^2-(x+3)^{\frac{5}{2}}}{(x+2)}$

Solution: – (a) $\lim\limits_{x\to 0}\dfrac{\sqrt{1+2x}-\sqrt{1+3x}}{x}=\lim\limits_{x\to 0}\dfrac{(\sqrt{1+2x}-1)-(\sqrt{1+3x}-1)}{x}=\lim\limits_{x\to 0}\dfrac{(1+2x)^{\frac{1}{2}}-1}{x}-\lim\limits_{x\to 0}\dfrac{(1+3x)^{\frac{1}{2}}-1}{x}$
$=\lim\limits_{x\to 0}\dfrac{(1+2x)^{\frac{1}{2}}-1}{2x}\times 2-\lim\limits_{x\to 0}\dfrac{(1+3x)^{\frac{1}{2}}-1}{3x}\times 3=2\dfrac{1}{2}-3\dfrac{1}{2}=1-\dfrac{3}{2}=-\dfrac{1}{2}$ Ans.

IInd Method: – $\lim\limits_{x\to 0}\dfrac{\sqrt{1+2x}-\sqrt{1+3x}}{x}$ without use the formula or $\lim\limits_{x\to 0}\dfrac{\sqrt{1+2x}-\sqrt{1+3x}}{x}$, $\left[\left(\frac{0}{0}\right)\text{form}\right]$ using L'Hospital rule

or $\lim\limits_{x\to 0}\dfrac{\sqrt{1+2x}-\sqrt{1+3x}}{x}=\lim\limits_{x\to 0}\dfrac{\frac{1}{2\sqrt{1+2x}}.2-\frac{1}{2\sqrt{1+3x}}.3}{1}$ take a limit $=1-\dfrac{3}{2}=-\dfrac{1}{2}$ Ans.

IIIrd Method: – $\lim\limits_{x\to 0}\dfrac{\sqrt{1+2x}-\sqrt{1+3x}}{x}=\lim\limits_{x\to 0}\dfrac{\sqrt{1+2x}-\sqrt{1+3x}}{x}\times\dfrac{\sqrt{1+2x}+\sqrt{1+3x}}{\sqrt{1+2x}+\sqrt{1+3x}}=\lim\limits_{x\to 0}\dfrac{1+2x-1-3x}{x(\sqrt{1+2x}+\sqrt{1+3x})}$
$=\lim\limits_{x\to 0}\dfrac{-x}{x(\sqrt{1+2x}+\sqrt{1+3x})}=\lim\limits_{x\to 0}\dfrac{-1}{(\sqrt{1+2x}+\sqrt{1+3x})}$ take a limit $=-\dfrac{1}{2}$ Ans.

(b) $\lim\limits_{x\to 0}\dfrac{\sqrt{1+3x}-4x^2-4x-1}{x}=\lim\limits_{x\to 0}\dfrac{\sqrt{1+3x}-(2x+1)^2}{x}=\lim\limits_{x\to 0}\dfrac{\left[(1+3x)^{\frac{1}{2}}-1\right]-[(2x+1)^2-1]}{x}$
$=\lim\limits_{x\to 0}\dfrac{\left[(1+3x)^{\frac{1}{2}}-1\right]}{3x}\times 3-\lim\limits_{x\to 0}\dfrac{[(2x+1)^2-1]}{2x}\times 2=1\times 3-1\times 2=3-2=1$ Ans.

Note: – without use the formula, check indeterminate form and use L'Hospital rule

(c) $\lim_{x\to 0}\frac{x^2+2x+1-(1+2x)^{\frac{3}{2}}}{x} = \lim_{x\to 0}\frac{[(x+1)^2-1]-[(1+2x)^{\frac{3}{2}}-1]}{x} = \lim_{x\to 0}\frac{[(x+1)^2-1]}{x} - \lim_{x\to 0}\frac{[(1+2x)^{\frac{3}{2}}-1]}{2x}\times 2 = 1-2.\frac{3}{2} = 1-3$

$= -2$ Ans.

IInd Method: – check indeterminate form and use L'Hospital rule.

(d) $\lim_{x\to -2}\frac{(x+3)^2-(x+3)^{\frac{5}{2}}}{(x+2)} = \lim_{x+2\to 0}\frac{\{[(x+2)+1]^2-1\}-\left\{[(x+2)+1]^{\frac{5}{2}}-1\right\}}{(x+2)} = \lim_{x+2\to 0}\frac{\{[(x+2)+1]^2-1\}}{(x+2)} - \lim_{x+2\to 0}\frac{\left\{[(x+2)+1]^{\frac{5}{2}}-1\right\}}{(x+2)}$

$= 2-\frac{5}{2} = -\frac{1}{2}$ Ans.

IInd Method: – check indeterminate form and use L'Hospital rule.

(7) using the formula $\lim_{x\to a}\frac{x^n-a^n}{x-a} = na^{n-1}$, to find following limits (a) $\lim_{x\to 2}\frac{\sqrt{x}-\sqrt{2}}{x-2}$ (b) $\lim_{x\to 1}\frac{x-1}{x^3-1}$ (c) $\lim_{x\to 2}\frac{x^{\frac{3}{2}}-2^{\frac{3}{2}}}{x^{\frac{1}{3}}-2^{\frac{1}{3}}}$

(d) $\lim_{x\to 1}\frac{\sqrt{2}-\sqrt{1+x}}{2^{\frac{3}{2}}-(1+x)^{\frac{3}{2}}}$ (e) $\lim_{x\to 1}\frac{x^p-1}{x^q-1}$

Solution: – (a) $\lim_{x\to 2}\frac{\sqrt{x}-\sqrt{2}}{x-2} = \lim_{x\to 2}\frac{x^{\frac{1}{2}}-2^{\frac{1}{2}}}{x-2}$ use formula $\lim_{x\to a}\frac{x^n-a^n}{x-a} = na^{n-1}$ or $\lim_{x\to 2}\frac{x^{\frac{1}{2}}-2^{\frac{1}{2}}}{x-2} = \frac{1}{2}\times 2^{\frac{1}{2}-1} = \frac{1}{2}\times 2^{-\frac{1}{2}} = \frac{1}{2}\times\frac{1}{\sqrt{2}} = \frac{1}{2\sqrt{2}}$ Ans.

(b) $\lim_{x\to 1}\frac{x-1}{x^3-1} = \lim_{x\to 1}\frac{1}{\frac{x^3-1}{x-1}} = \frac{1}{\lim_{x\to 1}\frac{x^3-1^3}{x-1}} = \frac{1}{3.1^{3-1}} = \frac{1}{3\times 1^2} = \frac{1}{3}$ Ans.

(c) $\lim_{x\to 2}\frac{x^{\frac{3}{2}}-2^{\frac{3}{2}}}{x^{\frac{1}{3}}-2^{\frac{1}{3}}} = \lim_{x\to 2}\frac{\frac{x^{\frac{3}{2}}-2^{\frac{3}{2}}}{x-2}}{\frac{x^{\frac{1}{3}}-2^{\frac{1}{3}}}{x-2}} = \frac{\frac{3}{2}\times 2^{\frac{3}{2}-1}}{\frac{1}{3}\times 2^{\frac{1}{3}-1}} = \frac{\frac{3}{2}\times 2^{\frac{1}{2}}}{\frac{1}{3}\times 2^{-\frac{2}{3}}} = \frac{\frac{3\times\sqrt{2}}{2}}{\frac{1}{3\times 2^{\frac{2}{3}}}} = \frac{3\sqrt{2}}{2}\times\frac{3\times 2^{\frac{2}{3}}}{1} = \frac{9\times 2^{\frac{1}{2}}.2^{\frac{2}{3}}}{2} = \frac{9}{2}\times 2^{\frac{7}{6}}$ Ans.

(d) $\lim_{x\to 1}\frac{\sqrt{2}-\sqrt{1+x}}{2^{\frac{3}{2}}-(1+x)^{\frac{3}{2}}} = \lim_{x\to 1}\frac{(1+x)^{\frac{1}{2}}-2^{\frac{1}{2}}}{(1+x)^{\frac{3}{2}}-2^{\frac{3}{2}}}$ ∴ Let $1+x=t$ ∴ $x\to 1$ then $t\to 2$

or $\lim_{t\to 2}\frac{t^{\frac{1}{2}}-2^{\frac{1}{2}}}{t^{\frac{3}{2}}-2^{\frac{3}{2}}} = \lim_{t\to 2}\frac{\left(\frac{t^{\frac{1}{2}}-2^{\frac{1}{2}}}{t-2}\right)}{\left(\frac{t^{\frac{3}{2}}-2^{\frac{3}{2}}}{t-2}\right)} = \frac{\frac{1}{2}.2^{\frac{1}{2}-1}}{\frac{3}{2}.2^{\frac{3}{2}-1}} = \frac{2^{-\frac{1}{2}}}{3.2^{\frac{1}{2}}} = \frac{1}{3.\sqrt{2}.\sqrt{2}} = \frac{1}{3.2} = \frac{1}{6}$ Ans.

(e) $\lim_{x\to 1}\frac{x^p-1}{x^q-1} = \lim_{x\to 1}\frac{\frac{x^p-1^p}{x-1}}{\frac{x^q-1^q}{x-1}} = \frac{p\times 1^{p-1}}{q\times 1^{q-1}} = \frac{p}{q}$ Ans.

(8) (a) A function f is defined as $f(x) = \begin{cases}\frac{x^2-3x+2}{x^2-1}, & \text{for } x\neq 1\\ 2, & \text{for } x=1\end{cases}$ Test the continuity at the function at $x = 1$.

(b) If $f(x) = \begin{cases}5-x, & x<0\\ 3a-x, & x\geq 0\end{cases}$ be continuous at $x = 0$ then show that $a = \frac{5}{3}$.

Solution: – (a) $f(x) = \begin{cases}\frac{x^2-3x+2}{x^2-1}, & x\neq 1\\ 2, & x=1\end{cases}$

$$\text{R.H.L} = \lim_{x\to 1_+} f(x) = \lim_{x\to 1_+}\frac{x^2-3x+2}{x^2-1} = \lim_{x\to 1_+}\frac{(x-1)(x-2)}{(x-1)(x+1)} = \lim_{x\to 1_+}\frac{(x-2)}{(x+1)} = -\frac{1}{2}$$

$$\text{L.H.L} = \lim_{x\to 1_-} f(x) = \lim_{x\to 1_-}\frac{x^2-3x+2}{x^2-1} = \lim_{x\to 1_-}\frac{(x-1)(x-2)}{(x-1)(x+1)} = \lim_{x\to 1_-}\frac{(x-2)}{(x+1)} = -\frac{1}{2}$$

we have $f(x) = f(1) = 2$ then, $\lim_{x\to1_+} f(x) = \lim_{x\to1_-} f(x) \neq f(x)$ function f(x)is not continuous at $x = 1$.

$\lim_{x\to1_+} f(x) \neq f(x)$ and $\lim_{x\to1_-} f(x) \neq f(x)$it is discontinuous function at $x = 1$.

Note: – If $\lim_{x\to1_+} f(x) = \lim_{x\to1_-} f(x) = f(x)$ the function continuous at $x = 1$.

but $\lim_{x\to1_+} f(x) \neq \lim_{x\to1_-} f(x) \neq f(x)$ or $\lim_{x\to1_+} f(x) = \lim_{x\to1_-} f(x) \neq f(x)$ the function discontinuous at $x = 1$.

(b) If $f(x) = \begin{cases} 5 - x, & x < 0 \\ 3a - x, & x \geq 0 \end{cases}$ be continuous at $x = 0$.

$\lim_{x\to0_+} f(x) = \lim_{x\to0_-} f(x) = f(x) \dots\dots\dots\dots\dots\dots\dots\dots (A)$

$R.H.L = \lim_{x\to0_+} f(x) = \lim_{x\to0_+} (3a - x) = 3a$, $L.H.L = \lim_{x\to0_-} f(x) = \lim_{x\to0_-} (5 - x) = 5$

$V = \lim_{x\to0} f(x) = \lim_{x\to0} (3a - x) = 3a$ put value L. H. L, R. H. L and f(x)in equation (A) we get,

$R.H.L = L.H.L = f(x)$ or $3a = 5 = 3a$ or $3a = 5$ $\therefore a = \frac{5}{3}$ Proved.

(9) (a) If $f(x) = x\left[\sqrt{1-x} - \sqrt{x}\right]$, then f(x)is continuous and differentiable at $x = 0$.

(b) Discuss the continuity and differentiability of $f(x) = \begin{cases} |x-2|, & x \geq 1 \\ \frac{x^2}{3} - \frac{3x}{4} + \frac{11}{4}, & x < 1 \end{cases}$ at $x = 1,2$.

Solution: – (a) $f(x) = x\left[\sqrt{1-x} - \sqrt{x}\right]$ for continuity, $R = L = V$ or f(x) at $x = 0$

$$\lim_{x\to0_+} f(x) = \lim_{x\to0_-} f(x) = f(x)$$

$R.H.L = \lim_{x\to0_+} f(x) = \lim_{x\to0_+} x\left[\sqrt{1-x} - \sqrt{x}\right] = 0$ and $L.H.L = \lim_{x\to0_-} f(x) = \lim_{x\to0_-} x\left[\sqrt{1-x} - \sqrt{x}\right] = 0$, $f(x) = 0$

$$\lim_{x\to0_+} f(x) = \lim_{x\to0_-} f(x) = f(x) = 0 \text{ or } R = L = V = 0$$

The function f(x)to be continuous at $x = 0$.

For differentiable, $\lim_{h\to0_+} \frac{f(0+h) - f(0)}{h} = \lim_{h\to0_-} \frac{f(0-h) - f(0)}{-h} = f(0)$ or $R' = L' = V$

$f(x) = x\left[\sqrt{1-x} - \sqrt{x}\right]$

$$R.H.L = \lim_{h\to0_+} \frac{f(0+h) - f(0)}{h} = \lim_{h\to0_+} \frac{h\left[\sqrt{1-h} - \sqrt{h}\right]}{h} = \lim_{h\to0_+} \left(\sqrt{1-h} - \sqrt{h}\right) = 1$$

$$L.H.L = \lim_{h\to0_-} \frac{f(0-h) - f(0)}{-h} = \lim_{h\to0_-} \frac{-h\left[\sqrt{1+h} - \sqrt{-h}\right]}{-h} = \lim_{h\to0_-} \left[\sqrt{1+h} - \sqrt{-h}\right] = 1$$

$V = f(0) = 0$ $\therefore$ $R.H.L = L.H.L \neq Vf(x)$, The function f(x) is not differentiable at $x = 0$.

$$\text{(b) } f(x) = \begin{cases} |x-2|, & x \geq 1 \\ \frac{x^2}{3} - \frac{3x}{4} + \frac{11}{4}, & x < 1 \end{cases} = \begin{cases} x - 2, & x \geq 2 \\ 2 - x, & 1 \leq x < 2 \\ \frac{x^2}{3} - \frac{3x}{4} + \frac{11}{4}, & x < 1 \end{cases}$$

For continuity at $x = 1$, $\lim_{x\to1_+} f(x) = \lim_{x\to1_-} f(x) = f(x)$ and at $x = 2$, $\lim_{x\to2_+} f(x) = \lim_{x\to2_-} f(x) = f(x)$

CaseI: – f(x) to be continuous at $x = 1$.

$$R.H.L = \lim_{x\to1_+} f(x) = \lim_{x\to1_+} (2 - x) = 1, \quad L.H.L = \lim_{x\to1_-} f(x) = \lim_{x\to1_-} \left(\frac{x^2}{3} - \frac{3x}{4} + \frac{11}{4}\right) = \frac{1}{3} - \frac{3}{4} + \frac{11}{4} = \frac{7}{3}$$

$\therefore$ R.H.L $\neq$ L.H.L the function f(x)is not continuous at x = 1.

Differentiability at x = 1, $\lim\limits_{h\to 0_+}\dfrac{f(1+h)-f(1)}{h}=\lim\limits_{h\to 0_-}\dfrac{f(1-h)-f(1)}{-h}=f(1)$

$$R.H.L=\lim_{h\to 0_+}\frac{f(1+h)-f(1)}{h}=\lim_{h\to 0_+}\frac{2-(1+h)-1}{h}=\lim_{h\to 0_+}\frac{1-1-h}{h}=\lim_{h\to 0_+}\frac{-h}{h}=-1$$

$$L.H.L=\lim_{h\to 0_-}\frac{f(1-h)-f(1)}{-h},f(x)=\frac{x^2}{3}-\frac{3x}{4}+\frac{11}{4},\ x<1$$

$$\text{or } f(1-h)=\frac{(1-h)^2}{3}-\frac{3(1-h)}{4}+\frac{11}{4}=\frac{1-2h+h^2}{3}-\frac{3(1-h)}{4}+\frac{11}{4}=\frac{4-8h+4h^2-9+9h+33}{12}=\frac{4h^2+h+28}{12}$$

$$L.H.L=\lim_{h\to 0_-}\frac{f(1-h)-f(1)}{-h}=\lim_{h\to 0_-}\frac{\frac{4h^2+h+28}{12}-\frac{7}{3}}{-h}=\lim_{h\to 0_-}\frac{\frac{4h^2+h+28-28}{12}}{-h}=\lim_{h\to 0_-}\frac{h(4h+1)}{-12h}=\lim_{h\to 0_-}\frac{(4h+1)}{-12}=-\frac{1}{12}$$

$\therefore$ R.H.L $\neq$ L.H.L, then the function f(x) is not differentiable at x = 1.

CaseII: $-$ f(x) to be continuous at x = 2, R.H.L = L.H.L = f(x) or f(2)

$$R.H.L=\lim_{x\to 2_+}f(x)=\lim_{x\to 2_+}(x-2)=0 \text{ and } L.H.L=\lim_{x\to 2_-}f(x)=\lim_{x\to 2_-}(2-x)=0$$

$$f(x)=f(2)=x-2=0\quad \therefore \lim_{x\to 2_+}f(x)=\lim_{x\to 2_-}f(x)=f(x)\ \text{ or } f(2)$$

$\therefore$ R.H.L = L.H.L = f(x) or f(2) $\therefore$ function f(x) to be continuous at x = 2.

Differentiable at x = 2

$$\lim_{h\to 0_+}\frac{f(2+h)-f(2)}{h}=\lim_{h\to 0_-}\frac{f(2-h)-f(2)}{-h}=f(2)\ \text{ or } R.H.L=L.H.L=f(2)$$

$$R.H.L=\lim_{h\to 0_+}\frac{f(2+h)-f(2)}{h},\quad f(x)=x-2,x\geq 2\ \text{ or } f(2+h)=2+h-2=h,f(2)=0$$

$$R.H.L=\lim_{h\to 0_+}\frac{f(2+h)-f(2)}{h}=\lim_{h\to 0_+}\frac{h-0}{h}=\lim_{h\to 0_+}\frac{h}{h}=1$$

$$L.H.L=\lim_{h\to 0_-}\frac{f(2-h)-f(2)}{-h},\quad f(x)=2-x,1\leq x<2\ \ or\ f(2-h)=2-2+h=h,f(2)=0$$

$$L.H.L=\lim_{h\to 0_-}\frac{f(2-h)-f(2)}{-h}=\lim_{h\to 0_-}\frac{h-0}{-h}=\lim_{h\to 0_-}\frac{h}{-h}=-1$$

$\therefore$ R.H.L $\neq$ L.H.L, the function f(x) is not differentiable at x = 2.

(10) (a) Let $f(x)=\begin{cases}x^4+x^3-16x^2+20x+15, & x\neq 2\\ m, & x=2\end{cases}$ If f(x) is continuous for all x, then m is equal to. ...?

(b) Find the values of a and b so that the function $f(x)=\begin{cases}x-a\sqrt{2}\cos x\ ,0\leq x<\frac{\pi}{4}\\ 3x\tan x+b\ ,\frac{\pi}{4}\leq x\leq\frac{\pi}{2}\\ a\sin x-b\cos 2x\ ,\frac{\pi}{2}<x\leq\pi\end{cases}$ is continuous for $0\leq x\leq\pi$.

Solution: $-$ (a) $f(x)=\begin{cases}x^4+x^3-16x^2+20x+15, & x\neq 2\\ m, & x=2\end{cases}$ to be continuous at x = 2.

For continuity, $\lim\limits_{x\to 2_+}f(x)=\lim\limits_{x\to 2_-}f(x)=f(x)$

$$R.H.L=\lim_{x\to 2_+}f(x)=\lim_{x\to 2_+}(x^4+x^3-16x^2+20x+15)=16+8-64+40+15=79-64=15$$

similarly, $L.H.L=\lim\limits_{x\to 2_-}f(x)=15$, $f(2)=m\ \therefore\ \lim\limits_{x\to 2_+}f(x)=\lim\limits_{x\to 2_-}f(x)=f(x)$ or $15=15=m$ Ans.

(b) $f(x) = \begin{cases} x - a\sqrt{2}\cos x\ , 0 \le x < \frac{\pi}{4} \\ 3x\tan x + b\ , \frac{\pi}{4} \le x \le \frac{\pi}{2} \\ a\sin x - b\cos 2x, \frac{\pi}{2} < x \le \pi \end{cases}$ is continuous for $0 \le x \le \pi$.

At $x = \frac{\pi}{4}$ we have, $\lim_{x\to\frac{\pi}{4}+} f(x) = \lim_{x\to\frac{\pi}{4}-} f(x) = f\left(\frac{\pi}{4}\right)$

$\text{R.H.L} = \lim_{x\to\frac{\pi}{4}+} f(x) = \lim_{x\to\frac{\pi}{4}+} (3x\tan x + b) = 3\frac{\pi}{4} \times \tan\frac{\pi}{4} + b = 3\frac{\pi}{4} \times 1 + b = \frac{3\pi}{4} + b$

$\text{L.H.L} = \lim_{x\to\frac{\pi}{4}-} f(x) = \lim_{x\to\frac{\pi}{4}-} \left(x - a\sqrt{2}\cos x\right) = \frac{\pi}{4} - a.\sqrt{2}\cos\frac{\pi}{4} = \frac{\pi}{4} - a.\sqrt{2} \times \frac{1}{\sqrt{2}} = \frac{\pi}{4} - a$

$f(x) = f\left(\frac{\pi}{4}\right) = 3x\tan x + b = \frac{3\pi}{4} + b$

For continuity, $\lim_{x\to\frac{\pi}{4}+} f(x) = \lim_{x\to\frac{\pi}{4}-} f(x) = f\left(\frac{\pi}{4}\right)$ or $\frac{3\pi}{4} + b = \frac{\pi}{4} - a = \frac{3\pi}{4} + b$

$\therefore \frac{3\pi}{4} + b = \frac{\pi}{4} - a$ or $a + b = \frac{\pi}{4} - \frac{3\pi}{4} = \frac{\pi - 3\pi}{4} = \frac{-2\pi}{4} = -\frac{\pi}{2}$ $\therefore a + b = -\frac{\pi}{2}$ ……………… (A)

At $x = \frac{\pi}{2}$ we have, $\lim_{x\to\frac{\pi}{2}+} f(x) = \lim_{x\to\frac{\pi}{2}-} f(x) = f\left(\frac{\pi}{2}\right)$

$\text{R.H.L} = \lim_{x\to\frac{\pi}{2}+} f(x) = \lim_{x\to\frac{\pi}{2}+} (a\sin x - b\cos x) = a\sin\frac{\pi}{2} - b\cos\frac{\pi}{2} = a - 0 = a$

$\text{L.H.L} = \lim_{x\to\frac{\pi}{2}-} f(x) = \lim_{x\to\frac{\pi}{2}-} (3x\tan x + b) = 3\frac{\pi}{2} \times \tan\frac{\pi}{2} + b = \infty + b = \infty$

$f(x) = f\left(\frac{\pi}{2}\right) = 3x\tan x + b = \infty$

$\therefore \lim_{x\to\frac{\pi}{2}+} f(x) = \lim_{x\to\frac{\pi}{2}-} f(x) = f\left(\frac{\pi}{2}\right)$ or $a = \infty = \infty$ $\therefore a = \infty$ ……………. (B)

solving equation (A)and (B), we get $\therefore$ $a + b = -\frac{\pi}{2}$, $a = \infty$ $\therefore$ $\infty + b = -\frac{\pi}{2}$ or $b = \infty$

$\boxed{\therefore\ a = \infty, b = \infty \quad \text{Ans.}}$

(11) (a) If $f(x) = \frac{(2^x - 1)^2}{\sin x.\log(1 - x)}$, $x \neq 0$ is continuous at $x = 0$ then find $f(0)$.

(b) If $f(x) = \begin{cases} 5x - 7\ , 0 \le x \le 1 \\ 3x + a\ , 1 < x \le 2 \end{cases}$ to be continuous at $x = 1$, then prove that $a = -5$.

Solution: – (a) $f(x) = \frac{(2^x - 1)^2}{\sin x.\log(1 - x)}$, $x \neq 0$ For continuity, $\lim_{x\to 0_+} f(x) = \lim_{x\to 0_-} f(x) = f(0)$

$$\text{R.H.L} = \lim_{x\to 0_+} f(x) = \lim_{x\to 0_+} \frac{(2^x - 1)^2}{\sin x.\log(1 - x)} = \lim_{x\to 0_+} \frac{\left(\frac{2^x - 1}{x}\right)^2 \times x^2}{\left(\frac{\sin x}{x}\right) \times x.\left[-\left(x + \frac{x^2}{2} + \frac{x^3}{3} + \cdots \ldots..\right)\right]} = (\log 2)^2 \lim_{x\to 0_+} \frac{x}{-x\left[1 + \frac{x}{2} + \frac{x^2}{3} \ldots \ldots ..\right]}$$

$$= (\log 2)^2 \times (-1) = -(\log 2)^2$$

similarly, $\text{L.H.L} = \lim_{x\to 0_-} f(x) = -(\log 2)^2$, $f(0) = 0$ $\therefore$ $\lim_{x\to 0_+} f(x) = \lim_{x\to 0_-} f(x) = f(0)$ or $-(\log 2)^2 = -(\log 2)^2 = f(0)$

$\boxed{\therefore\ f(0) = -(\log 2)^2 \quad \text{Ans.}}$

(b) $f(x) = \begin{cases} 5x - 7\ , 0 \le x \le 1 \\ 3x + a\ , 1 < x \le 2 \end{cases}$ to be continuous at $x = 1$.

for continuity, $\lim_{x\to 1_+} f(x) = \lim_{x\to 1_-} f(x) = f(x) = f(1)$ or $\text{R.H.L} = \text{L.H.L} = f(1)$

$\text{R.H.L} = \lim_{x\to1_+} f(x) = \lim_{x\to1_+} 3x + a = a + 3$, $\text{L.H.L} = \lim_{x\to1_-} f(x) = \lim_{x\to1_-} 5x - 7 = -2$, $f(1) = 5x - 7 = -2$

for continuity, $\lim_{x\to1_+} f(x) = \lim_{x\to1_-} f(x) = f(x) = f(1)$ $\therefore$ $a + 3 = -2 = -2$ or $a + 3 = -2$ or $a = -2 - 3 = -5$ $\therefore$ $a = -5$ Proved.

(12) (a) Test the continuity and differentiability of the function defined as under at $x = 2$ and $x = 3$. function

$$f(x) = \begin{cases} -x, & x < 2 \\ 3 + x, & 2 \le x \le 3 \\ -5 + 2x - x^2, & x > 3 \end{cases}$$

(b) Find the derivative of $f(x) = \begin{cases} \dfrac{x-2}{x^2 + 3x - 10}, & x \ne 2 \\ \dfrac{1}{7}, & x = 2 \end{cases}$ at $x = 2$.

Solution: – (a) $f(x) = \begin{cases} -x, & x < 2 \\ 3 + x, & 2 \le x \le 3 \\ -5 + 2x - x^2, & x > 3 \end{cases}$

For continuity at $x = 2$, $\lim_{x\to2_+} f(x) = \lim_{x\to2_-} f(x) = f(2)$

$\text{R.H.L} = \lim_{x\to2_+} f(x) = \lim_{x\to2_+} (3 + x) = 5$, $\text{L.H.L} = \lim_{x\to2_-} f(x) = \lim_{x\to2_-} (-x) = -2$ or $f(2) = 3 + x = 5$ $\therefore$ $\text{R.H.L} \ne \text{L.H.L}$

$f(x)$ is not continuous at $x = 2$, $f(x)$ is discontinuous at $x = 2$.

For continuity at $x = 3$, $\lim_{x\to3_+} f(x) = \lim_{x\to3_-} f(x) = f(3)$

$\text{R.H.L} = \lim_{x\to3_+} f(x) = \lim_{x\to3_+} (-5 + 2x - x^2) = -5 + 6 - 9 = -8$

$\text{L.H.L} = \lim_{x\to3_-} f(x) = \lim_{x\to3_-} (3 + x) = 3 + 3 = 6$, $f(3) = 3 + x = 6$

$\therefore$ $\text{R.H.L} \ne \text{L.H.L}$, $f(x)$is not continuous at $x = 3$. $\therefore$ $f(x)$is discontinuous at $x = 3$.

Differentiability at $x = 2$, Ist method: – $\lim_{h\to0_+} \dfrac{f(2+h) - f(2)}{h} = \lim_{h\to0_-} \dfrac{f(2-h) - f(2)}{-h} = f(2)$

IInd method: – $\lim_{x\to2_+} f'(x) = \lim_{x\to2_-} f'(x) = f'(2)$

using Ist method formula: – $\lim_{h\to0_+} \dfrac{f(2+h) - f(2)}{h} = \lim_{h\to0_-} \dfrac{f(2-h) - f(2)}{-h} = f(2)$

$\text{R.H.L} = \lim_{h\to0_+} \dfrac{f(2+h) - f(2)}{h} = \lim_{h\to0_+} \dfrac{3 + 2 + h - 5}{h} = \lim_{h\to0_+} \dfrac{h}{h} = 1$ where $f(2 + h) = 3 + 2 + h = 5 + h$

$\text{L.H.L} = \lim_{h\to0_-} \dfrac{f(2-h) - f(2)}{-h} = \lim_{h\to0_-} \dfrac{-2 + h + 2}{-h} = \lim_{h\to0_-} \dfrac{h}{-h} = -1$, $f(2) = 3 + x = 3 + 2 = 5$

$\therefore$ $\text{R.H.L} \ne \text{L.H.L}$. $f(x)$ is not differentiable at $x = 2$.

Differentiable at $x = 3$, $\lim_{h\to0_+} \dfrac{f(3+h) - f(3)}{h} = \lim_{h\to0_-} \dfrac{f(3-h) - f(3)}{-h} = f(3)$

$\text{R.H.L} = \lim_{h\to0_+} \dfrac{f(3+h) - f(3)}{h}$ $\therefore$ $f(3 + h) = -5 + 2(3 + h) - (3 + h)^2 = -5 + 6 + 2h - 9 - 6h - h^2 = -4h - h^2 - 8$

$\therefore$ $f(3) = -5 + 2x - x^2 = -5 + 6 - 9 = -8$

$\text{R.H.L} = \lim_{h\to0_+} \dfrac{f(3+h) - f(3)}{h} = \lim_{h\to0_+} \dfrac{-4h - h^2 - 8 + 8}{h} = \lim_{h\to0_+} \dfrac{-h(4 - h)}{h} = \lim_{h\to0_+} -(4 - h) = -4$

$\text{L.H.L} = \lim_{h\to0_-} \dfrac{f(3-h) - f(3)}{-h} = \lim_{h\to0_-} \dfrac{3 + 3 - h - 6}{-h} = \lim_{h\to0_-} \dfrac{-h}{-h} = 1$

$\therefore$ R.H.L $\neq$ L.H.L the function f(x) is not differentiable at x = 3.

using IInd method formula, $f(x) = \begin{cases} -x, & x < 2 \\ 3 + x, & 2 \leq x \leq 3 \\ -5 + 2x - x^2, & x > 3 \end{cases}$ Differentiate, $f'(x) = \begin{cases} -1, & x < 2 \\ 1, & 2 \leq x \leq 3 \\ 2 - 2x, & x > 3 \end{cases}$

IInd method formula, $\lim_{x \to 2_+} f'(x) = \lim_{x \to 2_-} f'(x) = f'(2)$ and $\lim_{x \to 3_+} f'(x) = \lim_{x \to 3_-} f'(x) = f'(3)$

(b) $f(x) = \begin{cases} \dfrac{x-2}{x^2 + 3x - 10}, & x \neq 2 \\ \dfrac{1}{7}, & x = 2 \end{cases} = \begin{cases} \dfrac{(x-2)}{(x-2)(x+5)}, & x \neq 2 \\ \dfrac{1}{7}, & x = 2 \end{cases}$

Derivative at x = 2, $\lim_{h \to 0_+} \dfrac{f(2+h) - f(2)}{h} = \lim_{h \to 0_-} \dfrac{f(2-h) - f(2)}{-h} = \text{finite}$

$$\text{R.H.L} = \lim_{h \to 0_+} \frac{f(2+h) - f(2)}{h} = \lim_{h \to 0_+} \frac{\frac{1}{2+h+5} - \frac{1}{7}}{h} = \lim_{h \to 0_+} \frac{\frac{7-h-7}{7(h+7)}}{h} = \lim_{h \to 0_+} \frac{-h}{7h(h+7)} = -\frac{1}{49}$$

similarly, R.H.L = L.H.L = finite $\therefore$ finite $= -\dfrac{1}{49}$ Ans.

(13) (a) If $G(x) = \sqrt{16 - x^2}$, then $\lim_{x \to 2} \dfrac{G(x) - G(2)}{x - 2}$ equal to ……………??

(b) If $f(16) = 16, f'(16) = 8$ then $\lim_{x \to 16} \dfrac{\sqrt{f(x)} - 4}{\sqrt{x} - 4}$ equal to ……………??

Solution: – (a) $G(x) = \sqrt{16 - x^2}$, $G(2) = \sqrt{16 - 4} = \sqrt{12}$

or $\lim_{x \to 2} \dfrac{G(x) - G(2)}{x - 2} = \lim_{x \to 2} \dfrac{\sqrt{16 - x^2} - \sqrt{12}}{x - 2}$ $\left(\dfrac{0}{0} \text{ form}\right)$ using L'Hospital rule.

or $\lim_{x \to 2} \dfrac{\frac{1}{2\sqrt{16 - x^2}} \times (-2x) - 0}{1} = \lim_{x \to 2} \dfrac{-x}{\sqrt{16 - x^2}} = \dfrac{-2}{\sqrt{16 - 4}} = \dfrac{-2}{\sqrt{12}} = \dfrac{-2}{2\sqrt{3}} = -\dfrac{1}{\sqrt{3}}$ Ans.

(b) Given, $f(16) = 16, f'(16) = 8$

$\Rightarrow \lim_{x \to 16} \dfrac{\sqrt{f(x)} - 4}{\sqrt{x} - 4}$ $\left(\dfrac{0}{0} \text{ form}\right)$ using L'Hospital rule.

or $\lim_{x \to 16} \dfrac{\sqrt{f(x)} - 4}{\sqrt{x} - 4} = \lim_{x \to 16} \dfrac{\frac{1}{2\sqrt{f(x)}} \times f'(x)}{\frac{1}{2\sqrt{x}}} = \dfrac{\frac{f'(16)}{2\sqrt{f(16)}}}{\frac{1}{2\sqrt{16}}} = \dfrac{8 \times f'(16)}{2\sqrt{f(16)}} = \dfrac{8 \times 8}{2\sqrt{16}} = \dfrac{64}{8} = 8$ Ans.

Exercise – A2

(1) Evaluate the following limits: – (a) $\lim_{x \to 0} \dfrac{\sin x}{x}$ (b) $\lim_{x \to 0} \dfrac{\tan x}{x}$

(2) Evaluate: – (a) $\lim_{x \to \infty} \dfrac{(3+x)^{30}.(6+x)^5}{(3-x)^{35}} = -1$ (b) $\lim_{x \to \infty} \dfrac{3.\sqrt{x} + 5.\sqrt[3]{x} + 7.\sqrt[5]{x}}{\sqrt{3x - 1} + \sqrt[3]{5x - 2}} = \cdots\cdots\cdots$

(3) Evaluate: – (a) $\lim_{x \to \frac{\pi}{4}} \dfrac{1 - \cot x}{1 - \sqrt{2}\cos x}$ (b) $\lim_{x \to 0} \dfrac{\sqrt{1 - \sin x} + \sqrt{1 + \sin x}}{x}$

(4) Evaluate: – (a) $\lim_{x \to \infty} \left(\dfrac{x+7}{x+3}\right)^{x+2}$ (b) $\lim_{x \to \infty} \left(\dfrac{x+2}{x-5}\right)^{2x+3}$

(5) Evaluate: – (a) $\lim_{x\to 0}\left(\frac{1+7x^2}{1+3x^2}\right)^{\frac{1}{x^2}}$ (b) $\lim_{x\to 0}\left(\frac{4x^2+4x+1}{9x^2+6x+1}\right)^{\frac{1}{x}}$

(6) Evaluate: – (a) $\lim_{x\to 0}\frac{x-\sin x}{x\cos x}$ (b) $\lim_{x\to\infty}\frac{\sqrt{x^3+1}-\sqrt[3]{x^6+1}}{\sqrt[5]{x^5+1}+\sqrt[6]{x^6+1}}$

(7) (a) Let $f(x)=\begin{cases}x-2, & x\le 2\\ 5+ax^2, & x>2\end{cases}$ to be continuous at $x=2$, then find the value of a.

(b) The value of f(0) for which $f(x)=\frac{3-\sqrt{x-4}}{\sin 2x}$ is continuous is … … … … … . . ??

(8) (a) $\lim_{x\to 2}(2-x)\tan\frac{\pi x}{4}$ (b) $\lim_{x\to\frac{\pi}{4}}\frac{\sin\left(\frac{\pi}{4}-x\right)}{\sqrt{2}\cos x-1}$

(9) (a) $\lim_{x\to 0}\frac{x\sin x+\log(1+x^2)}{x^2}$ (b) $\lim_{x\to\frac{\pi}{4}}\frac{\tan x-\cot x}{x-\frac{x}{4}}$

(10) (a) $\lim_{x\to 0}\frac{\log\left(1+\frac{x}{m}\right)+\log(1-nx)-\log\left(1+\frac{x}{p}\right)}{x}$ (b) $\lim_{x\to 0}\frac{(5^x-1)^2}{\sin\frac{x}{2}.\log\left(1-\frac{x^2}{4}\right)}$

(11) (a) $\lim_{x\to\tan^{-1}5}\left(\frac{\tan^2 x-3\tan x+10}{\tan^2 x-7\tan x+10}\right)$ (b) $\lim_{x\to 0}\frac{e^{2x}-e^{-2x}}{x}$

(12) (a) Discuss the limits and continuity of the function $f(x)=\begin{cases}\frac{a^{[x]-x}}{[x]-x}, & x\ne 0\\ \log a, & x<0\end{cases}$ at $x=0$.

(b) Discuss the continuity of the function $f(x)=\begin{cases}\frac{x-3}{1+e^{\frac{1}{(x-3)}}}, & x\ne 3\\ 1, & x=3\end{cases}$.

(13) (a) Find the value of A so that the function $f(x)=\begin{cases}\frac{2^{x+3}-64}{4^x-64}, & x\ne 3\\ A, & x=3\end{cases}$.

(b) Find the value of A , B such that $\lim_{x\to 0}\frac{A\cos x-B\sin x}{x^2}=1$.

(14) (a) Determine the value of x for which the following function fails to be continuous or differentiable.

$$f(x)=\begin{cases}2-x, & x<2\\ (2-x)(3-x), & 2\le x\le 3\\ 4-x, & x>3\end{cases}.$$

(b) The function $f(x)=\begin{cases}|x-2| & \text{for } x\ge 1\\ \frac{x^2}{5}+\frac{3x}{2}+\frac{13}{10} & \text{for } x<1\end{cases}$ Discuss the continuity and differentiability.

(15) (a) $f(x)=\begin{cases}ax^2+b, & |x|<2\\ \frac{1}{|x|}, & |x|\ge 2\end{cases}$ The function is continuous and differentiable , then prove that

(b) Discuss the continuity and differentiability of the function $f(x)=\begin{cases}\frac{x}{1-|x|}, & |x|\ge 1\\ \frac{x}{1+|x|}, & |x|<1\end{cases}$.

(16) (a) Prove that the function $f(x)=\cos[\pi|x|]$ is continuous at $x=0$ but is also differentiable them.

(b) $f(x) = \begin{cases} \dfrac{\left(e^{-\frac{1}{x}} + e^{\frac{1}{x}}\right)}{\left(e^{-\frac{1}{x}} - e^{\frac{1}{x}}\right)}, & x \neq 0 \\ 0, & x = 0 \end{cases}$ prove that f(x) is not differentiable at $x = 0$.

(17) (a) Find the value of a and b so that the function $f(x) = \begin{cases} x - b\sqrt{2}\sin x, & 0 \le x < \frac{\pi}{4} \\ -3x\cot x - a, & \frac{\pi}{4} \le x \le \frac{\pi}{2} \\ b\cos 2x + a\sin x, & \frac{\pi}{2} < x \le \pi \end{cases}$ is continuous for $0 \le x \le \pi$.

(b) $f(x) = \begin{cases} A\tan^{-1}\dfrac{1}{x-3}, & 0 \le x < 3 \\ \dfrac{\pi}{2}, & x = 3 \\ B\tan^{-1}\dfrac{2}{x-3}, & 3 < x < 5 \\ \sin^{-1}(6-x) + A\dfrac{\pi}{3}, & 5 \le x \le 7 \end{cases}$, Determine the value of A and B if f(x) is continuous in the interval [0,7].

(18) (a) Discuss the continuity and differentiability of the function f(x) in (0,5) where $f(x) = \begin{cases} |3x-5|[x], & 0 \le x \le 3 \\ \dfrac{x^2}{3}, & 3 < x \le 5 \end{cases}$

(b) $f(x) = \begin{cases} (x-3).3^{-\left[\frac{1}{|x-3|} + \frac{1}{(x-3)}\right]}, & x \neq 3 \\ 0, & x = 3 \end{cases}$ prove that the function is not differentiable at x = 3.

(19) (a) If $f(3) = 5$ and $f'(3) = 2$ then find $\lim_{x\to 3} \dfrac{xf(3) - 2f(x)}{x-2}$.

(b) If $f(a) = 3, f'(a) = 1, g(a) = -2, g'(a) = -1$ then the value of $\lim_{x\to a} \dfrac{g(x)f(x) - g(a)f(x)}{x-a} = -1$

(20) (a) If $P(x) = -\sqrt{36 - x^2}$ then $\lim_{x\to 2} \dfrac{P(x) - P(2)}{x-2} = \dfrac{1}{2\sqrt{2}}$. (b) If $f(25) = 25, f'(25) = 5$ then $\lim_{x\to 25} \dfrac{\sqrt{f(x)} - 5}{\sqrt{x} - 5} = 5$.

(c) If $P(x) = \sqrt{64 - x^2}$ then $\lim_{x\to 3} \dfrac{P(x) - P(3)}{x-3} = -\dfrac{3}{\sqrt{55}}$.

Answer

(1) (a) Ans: 1 (b) Ans: 1

(2) (a) $\lim_{x\to\infty} \dfrac{(3+x)^{30}.(6+x)^5}{(3-x)^{35}} = \lim_{x\to\infty} \dfrac{x^{30}\left(\frac{3}{x}+1\right)^{30}.x^5\left(\frac{6}{x}+1\right)^5}{x^{35}\left(\frac{3}{x}-1\right)^{35}} = \lim_{x\to\infty} \dfrac{\left(\frac{3}{x}+1\right)^{30}.\left(\frac{6}{x}+1\right)^5}{\left(\frac{3}{x}-1\right)^{35}} = -1$ Ans.

(b) Ans: $\sqrt{3}$. Hint, put $x = \frac{1}{y}$, $x \to \infty, y \to 0$ (3) (a) Ans: 2 (b) Ans: -1 (4) (a) Ans: e^4 (b) Ans: e^{14}

(5) (a) Ans: $\dfrac{e^7}{e^3} = e^{7-3} = e^4$. (b) Ans: e^{-2} (6) (a) Ans: 0 (b) Ans: ∞

(7) (a) Ans: $a = -\dfrac{5}{4}$ (b) Ans: (8) (a) Ans: $\dfrac{4}{\pi}$ (b) Ans: 1 (9) (a) Ans: 2 (b) Ans: 1

(10) (a) Ans: $\dfrac{p-n-m}{mp}$ (b) Ans: $(2\log 5)^3$ (11) (a) Ans: $\dfrac{7}{3}$ (b) Ans: 4 (13) (a) Ans: A = 0 (b) Ans:

(17) (a) Ans: $a = \pi, b = 2\pi$ (b) Ans: $A = 1, \frac{9}{4}$ and $B = 1, -\frac{2}{3}$

(19) (a) Ans: 1 (b) Ans: -1 (20) (a) Ans: $\dfrac{1}{2\sqrt{2}}$ (b) Ans: 5 (c) Ans: $\dfrac{-3}{\sqrt{55}}$

Exercise – A3

(1) Find $\lim\limits_{x\to\infty} f(x)$ if, (a) $f(x)=\dfrac{4x^2+3x+2}{1+x^2}$ (b) $f(x)=\dfrac{3-x^2}{x^2-2x+3}$

(c) $f(x)=\dfrac{5x^3+2x^2-x+3}{3x^3-4x+10}$ (d) $f(x)=\dfrac{1^2+2^2+\cdots\ldots\ldots+x^2}{4x^3-x+1}$ (e) $f(x)=\dfrac{1+2+3\ldots\ldots\ldots+x}{x^2+1}$

(2) Find the following limit: – (a) $\lim\limits_{n\to\infty}\left(\dfrac{2n^3-n+3}{3n^3+2n+1}\right)^2$ (b) $\lim\limits_{n\to\infty}\sqrt[n]{1+n}$ (c) $\lim\limits_{x\to\infty}\sqrt[x]{x^5+2}$ (d) $\lim\limits_{x\to\infty}\sqrt[x]{x^3}$

(3) Find the following limit: – (a) $\lim\limits_{n\to0}\dfrac{1+2+3+\cdots\ldots\ldots+n}{n}$ (b) $\lim\limits_{n\to0}\dfrac{1+2+3+\cdots\ldots\ldots+n}{n}-1$

(c) $\lim\limits_{n\to0}\dfrac{1^2+2^2+3^2+\cdots\ldots\ldots+n^2}{n}$ (d) $\lim\limits_{n\to0}\dfrac{1^2+2^2+3^2+\cdots\ldots\ldots+n^2}{n^2}$

(4) Find the following limit: – (a) $\lim\limits_{x\to0}\dfrac{(x^3-1)^5+1}{x}$ (b) $\lim\limits_{x\to1}(x^7+x^5+x^3+x)$ (c) $\lim\limits_{x\to0}\dfrac{2^x+3^x-2}{x}$ (d) $\lim\limits_{x\to0}\left(\dfrac{2\sin 2x+\cos 2x-1}{x}\right)$

(e) $\lim\limits_{x\to\frac{\pi}{4}}\dfrac{1-\tan x}{1-\sqrt{2}\sin x}$ (f) $\lim\limits_{x\to2}\dfrac{\log(x-1)}{x-2}$ (solve without use L'Hospital rule) (g) $\lim\limits_{x\to\infty}\left(\dfrac{1-x^2}{x^2}\right)$

(h) $\lim\limits_{x\to\infty} x\left(\dfrac{\pi}{2}-\tan^{-1}x\right)$ (i) $\lim\limits_{x\to0}\dfrac{5^x-3^x}{4^x-2^x}$ (j) $\lim\limits_{x\to2}\dfrac{e^{\sqrt{x}}-e^{\sqrt{2}}}{x-2}$ (k) $\lim\limits_{x\to0}\dfrac{(x.3^x-x)}{\frac{1-\cos 2x}{2}}$

(5) Find the following limit: – (a) $\lim\limits_{x\to0}\dfrac{27^x+9^x+3^x-3}{x}$ (b) $\lim\limits_{x\to0}\dfrac{81^x-27^x+9^x-3^x}{(\sqrt{24+\cos x}-3)\sin x}$ (c) $\lim\limits_{x\to\frac{\pi}{4}}\dfrac{\sqrt{2}\left(x-\frac{\pi}{4}\right)}{(\sin x-\cos x)}$ (d) $\lim\limits_{x\to0}\dfrac{x}{\sin x+\tan x}$

(e) $\lim\limits_{x\to1}\dfrac{\sqrt{x+2}-\sqrt{2x+1}}{x-1}$ (f) $\lim\limits_{x\to0}(1+\sin x)^{\frac{1}{\sin x}}$ (g) $\lim\limits_{x\to2}\dfrac{x-2}{2x^2-5x+2}$ (h) $\lim\limits_{x\to0}\dfrac{\sqrt{1+\tan x}-\sqrt{1-\sin x}}{x}$ (i) $\lim\limits_{x\to\frac{\pi}{2}}\dfrac{\sqrt{1-\sin x}}{\cos x}$

(j) $\lim\limits_{x\to2}\dfrac{(1+x)^3-27}{x-2}$ (k) $\lim\limits_{x\to3}\dfrac{\left(1+\sqrt{1+x}\right)^2-9}{x-3}$

(6) Find the following limit: – (a) $\lim\limits_{x\to\frac{\pi}{4}}\dfrac{1-\sin 2x}{x-\frac{\pi}{4}}$ (b) $\lim\limits_{x\to a}\dfrac{x^{\frac{1}{3}}-a^{\frac{1}{3}}}{x^{\frac{1}{2}}-a^{\frac{1}{2}}}$ (c) $\lim\limits_{x\to\frac{\pi}{2}}\dfrac{\log(\cos x+1)}{\sin 2x}$ (d) $\lim\limits_{x\to0}\dfrac{1+\sin x-\cos x}{\cos x-\sin x-1}$

(e) $\lim\limits_{x\to0}\dfrac{\sqrt{9+\tan 2x}-3}{\log(1+\sin 2x)}$ (f) $\lim\limits_{x\to1}\dfrac{\sqrt[4]{15+x}-2}{3(x-1)}$ (g) $\lim\limits_{x\to0}\dfrac{\sqrt[3]{1+x}+\sqrt[4]{1+x}-2}{3^x+2^x-2}$ (h) $\lim\limits_{x\to\frac{\pi}{2}}\dfrac{\sqrt{1+\cos x}-\sqrt{1-\cos x}}{\cos x}$

(i) $\lim\limits_{x\to a}\dfrac{\sqrt{x+3a}-2.\sqrt{a}}{3.\sqrt{x+a}-3.\sqrt{2a}},(a\neq0)$ (j) $\lim\limits_{x\to0}\dfrac{x\sin x+\log(1+x^2)}{x^2}$

(7) Find the following limit: – (a) $\lim\limits_{x\to\infty}\dfrac{\sqrt{x^2+1}+\sqrt[4]{x^4+1}}{\sqrt[3]{x^3+1}+\sqrt[5]{x^5+1}}$ (b) $\lim\limits_{x\to\infty}\dfrac{\sqrt[3]{x^2+1}-\sqrt[4]{x^4+1}}{\sqrt[3]{x^3+1}-\sqrt[5]{x^4+1}}$ (c) $\lim\limits_{x\to\infty}\dfrac{x+\cos x}{x-\sin^2 x}$ (d) $\lim\limits_{x\to3}\dfrac{x^2-4x+3}{x^2-5x+6}$

(e) $\lim\limits_{x\to1}\dfrac{(3x-1)\left(\sqrt{x^2}-1\right)}{(3x^3+x^2-3x-1)}$ (f) $\lim\limits_{x\to0}\dfrac{(x.3^x-x)}{\sin^2 x}$ (g) $\lim\limits_{x\to\pi}\dfrac{(x-\pi)\sin x}{1+\cos x}$ (h) $\lim\limits_{x\to0}\dfrac{\tan 2x.(1-\tan^2 x)}{x+\sin x}$ (i) $\lim\limits_{x\to0}\dfrac{x\cos x+\tan x}{\tan x}$

(j) $\lim\limits_{x\to0}\dfrac{2x-\sin x}{3x-\tan x}$ (k) $\lim\limits_{x\to0}\dfrac{\tan x+x}{x^2}$

(8) Evaluate: – (a) $\lim\limits_{x\to0}\dfrac{\cot(\pi x)}{\frac{\pi}{2x}}$ (b) $\lim\limits_{\theta\to\pi}\dfrac{\sqrt{1+\cos\theta}-\sin\theta}{\cos\left(\frac{\theta}{2}\right)}$ (c) $\lim\limits_{\theta\to\frac{\pi}{2}}\dfrac{(\sin\theta)^{\sin\theta}-1}{1-\sin\theta}$

(d) If $f'(a)$ is exists then show that $\lim\limits_{x\to a}\dfrac{(2x-a)f(x)-af(2x-a)}{x-a}=2f(a)-af'(a)$.

(e) If $f(a) = 3, f'(a) = 2, g(2a) = -1$ then the value of $\lim_{x\to a}\frac{f(a).g(x+a) - f(2a-x).g(2a)}{x-a} = -1$.

(9) (a) If $F(x) = \sqrt{24 - x^3}$, then $\lim_{x\to 2}\frac{F(x) - F(2)}{x-2}$ equal to … … … … … ?

(b) If $F(5) = 5, F'(5) = 3$, then $\lim_{x\to 5}\frac{\sqrt{F(x)+3} - \sqrt{8}}{\sqrt{x} - \sqrt{5}}$ equal to … … … … … ?

(10) Evaluate: – (a) $\lim_{x\to 0}\frac{\tan x - x - \frac{x^3}{3}}{x^5}$ (b) $\lim_{x\to\frac{\pi}{2}} \sec x.\log_e(\operatorname{cosec} x)$ (c) $\lim_{x\to 1}\frac{(1+x)-2}{(1-x)}$ (d) $\lim_{x\to 0}\frac{\log\sin\left(\frac{\pi}{2}+ax\right)}{\tan bx}$ (e) $\lim_{x\to 0}\frac{\frac{\pi}{2x}}{\operatorname{cosec}\left(\frac{\pi}{2}-x\right)}$

(11) (a) Find the value of constant a and b such that $\lim_{x\to 0}\frac{a\tan x + x(b+\cos x)}{x^3} = 1$

(b) Find the value of constant a, b and c such that $\lim_{x\to 0}\frac{axe^x - b\log(1+x) + cxe^{-x}}{x^2\sin x} = 2$

(c) $\lim_{x\to 0}\left(\frac{1-x}{1+x} + ax - b\right) = 0$, then prove that $a = 0, b = -1$ (d) $\lim_{x\to 0}\left(\frac{ax^2}{x+1} - bx + 1\right) = 1$, then prove that $a = 0, b = 0$

(12) (a) $\lim_{x\to 0}\frac{\cos 2x + a\cos x}{x^2}$ = finite, find a and the limit. (b) $\lim_{x\to 0}\frac{a\tan x + b\sin x - 2}{x^3}$ = finite, find a and b and the limit.

(c) $\lim_{x\to 0}\frac{ae^x + b\log(1+x) - ce^{-x}}{x\tan x}$ = finite, find a, b and c. (d) $\lim_{x\to 0}\frac{ax\cos x + b\log(1+x) - cxe^x}{x^2\sin x} = 2$, find a, b and c.

(13) (a) $\lim_{x\to 0}\left\{\sin\left(\frac{\pi}{2}+x\right)\right\}^{\frac{1}{x}} = 1$ (b) $\lim_{x\to\infty}\left(\frac{x+3}{x+2}\right)^{x+4} = e$ (c) $\lim_{x\to\infty}\left(\frac{x+1}{x-2}\right)^{x+3} = e^3$ (d) $\lim_{x\to 0}\left(\frac{1+2x^2}{1+3x^2}\right)^{\frac{1}{x^2}} = \frac{1}{e}$

(e) $\lim_{x\to 0}\left\{\frac{2}{3}\left(\frac{3+x}{2+x}\right)\right\}^{\frac{1}{x}} = e^{-\frac{1}{6}} = \frac{1}{e^{\frac{1}{6}}}$ (f) $\lim_{x\to 0}\frac{\log(1+x)}{x} - 2$ (g) $\lim_{x\to\infty}\left(\frac{x-5}{x+3}\right)^x$ (h) $\lim_{x\to 0}\left(\frac{\sin 2x}{2x}\right)^{\frac{1}{x^2}}$ (i) $\lim_{x\to 0}\left(\frac{\tan 3x}{3x}\right)^{\frac{1}{x^2}}$

(j) $\lim_{x\to 0}\left(\frac{\tan x}{x}\right)^{\frac{1}{x}}$ (k) $\lim_{x\to 0}\left(\frac{\sin x}{x}\right)^{\frac{1}{x}}$ (l) $\lim_{x\to 0}\left(\frac{\tan x}{x}\right)^{\frac{1}{x^3}}$ (m) $\lim_{x\to 0}\left(\frac{\sin 3x}{x}\right)^{\frac{1}{x^3}}$

(14) Evaluate: – (a) $\lim_{x\to 0}\frac{\sin 2x}{3x}$ (b) $\lim_{x\to\infty}\frac{3x^2-2}{\sqrt{x^4+2x^3-x^2}}$ (c) $\lim_{x\to 0}\frac{\log(1+\sin x)}{\sin x}$ (solve without L'Hospital rule)

(d) $\lim_{x\to\infty}\frac{\sqrt{2x^2-3}}{\sqrt{3x^2-1}}$ (e) $\lim_{x\to 2}\frac{\frac{e^x}{e^2}-1}{(x-2)}$ (f) $\lim_{x\to 1}\frac{\sqrt{3}-\sqrt{4-x}}{2-\sqrt{3+x}}$

(15) Find the following limits (without L'Hospital rule): – (a) $\lim_{x\to\frac{\pi}{2}}\frac{\sqrt{\sin x}+\sqrt{\cos x}-1}{\sqrt{\cos x}-\sqrt{\sin x}+1}$ (b) $\lim_{x\to\frac{\pi}{4}}\frac{\tan x - 1}{x-\frac{\pi}{4}}$

(c) $\lim_{x\to 3}\frac{x-\sqrt{6+x}}{x-3}$ (d) $\lim_{x\to 0}\frac{\frac{3^x}{5^x}-1}{x}$ (e) $\lim_{x\to\infty}\frac{3x^3+2x^2-5x+1}{2x^3+x^2+2x+1}$ (f) $\lim_{x\to 0}\frac{\sqrt{1-\cos 2x}}{\sqrt{2}x}$

(16) Evaluate: – (a) $\lim_{x\to 0}\frac{\log(1-e^x)-\log 2}{x}$ (b) $\lim_{x\to 0}\frac{2\left[e^{(1-\cos x)}-\cos x\right]}{\sin x}$ (c) $\lim_{x\to 1}\frac{5^{(x^2-2x+1)}-3^{(x-1)}}{(x-1)}$

(d) $\lim_{x\to\frac{\pi}{4}}\frac{(\sin x-\cos x)}{4x-\pi}$ (without L'Hospital rule). (e) $\lim_{x\to\infty}\frac{x^{32}(1-x)^8}{(2+x)^{40}}$

(f) $\lim_{x\to 0}\frac{x\sin 2x + 2x\sin x}{(1-\cos 2x)}$ (without L'Hospital rul). (g) $\lim_{x\to 0}\frac{e^{\tan x}-e^{\sin x}}{5^{\sin x}-4^{\tan x}}$ (h) $\lim_{x\to\frac{\pi}{2}}\frac{\sqrt{2}\sin\left(\frac{x}{2}\right)-1}{\sqrt{3}-\tan\left(\frac{2x}{3}\right)}$

(17) Evaluate: – (a) $\lim\limits_{x\to\frac{\pi}{4}}(\sec x - 2\cos x)(1-\tan x)$ (b) $\lim\limits_{x\to\frac{\pi}{6}}\frac{\operatorname{cosec} x - 2\sin x - 1}{\sqrt{3}\sec x - 2}$ (c) $\lim\limits_{x\to\infty}\frac{\sqrt{x^2-3}-\sqrt[3]{2+3x^3}}{x+\sqrt{1+x^2}}$

(d) $\lim\limits_{x\to3}\frac{\sin(x^2-2x-3)-\cos(2x-6)+1}{(x-3)(x+1)}$ (e) $\lim\limits_{x\to1}\sqrt{x-1}.(x^2-1)$ (f) $\lim\limits_{x\to0}\frac{(1-\sin x)^2-1}{\sin x}$ (g) $\lim\limits_{x\to2}|x-2|$

(18) Evaluate: – (a) $\lim\limits_{x\to1} x-[x]$ (b) $\lim\limits_{x\to2} e^{|x-2|}$ (c) $\lim\limits_{x\to0}\frac{x-|x|}{x}$ (d) $\lim\limits_{x\to0}\frac{\sin|x|}{x}$ (e) $\lim\limits_{x\to0}\frac{\tan|x|}{x}$

(f) $\lim\limits_{\theta\to0}\tan 2\theta.\cot\theta$ (g) $\lim\limits_{x\to0}\frac{\sqrt{1+\sin x}-\sqrt{\cos x}}{\sqrt{1-\cos x}}$ (h) $\lim\limits_{x\to0}\frac{\sec x - 1}{\tan x}$ (solve without use L'Hospital rule)

(19) Evaluate: – (a) $\lim\limits_{x\to-1}\frac{\sqrt{x+2}-1}{x+1}$ (b) $\lim\limits_{x\to-2}\frac{\sqrt[3]{x+3}-1}{x+2}$ [use formula and solve question no. (a)and (b)] (c) $\lim\limits_{x\to-\pi}\frac{\sin(x+\pi)}{1+\cos x}$

(d) $\lim\limits_{x\to-\frac{\pi}{2}}\frac{\cos x}{\left(x+\frac{\pi}{2}\right)}$ (e) $\lim\limits_{x\to-\frac{\pi}{4}}\frac{\left(x+\frac{\pi}{4}\right)(1-\tan x)}{1+\tan x}$ (f) $\lim\limits_{x\to0}(1-\sin x)^{\frac{1}{\sin x}}$ (g) $\lim\limits_{x\to\frac{\pi}{2}}\left(1+\frac{2}{\tan x}\right)^{\tan x}$ (h) $\lim\limits_{x\to0}\left(1+\frac{\sin x}{3}\right)^{\operatorname{cosec} x}$

(20) Evaluate: – (a) $\lim\limits_{x\to\infty}\left(1-\frac{3}{\sin x}\right)^{\sin x}$ (b) $\lim\limits_{x\to0}\frac{\sin(x+\pi)}{x\cos x}$ (c) $\lim\limits_{x\to-\frac{\pi}{2}}\frac{e^{2x}-e^{-\pi}}{e^{\left(x+\frac{\pi}{2}\right)}-1}$ (d) $\lim\limits_{x\to-5}\frac{\sqrt{26-x^2}-1}{1-\sqrt{x+6}}$

(e) $\lim\limits_{x\to\log 2}\frac{\log x-\log 2}{2x-\log 4}$ (f) $\lim\limits_{x\to\infty}\frac{\sqrt{x+\tan x}}{\sqrt{x-\sin x}}$ (g) $\lim\limits_{x\to\infty}\frac{\tan x-\frac{\pi}{2}}{\cot x}$ (h) $\lim\limits_{x\to e}\frac{\log(x^2+2x+1)-\log(x^2+x)}{\log(2x+3)}$

(21) Evaluate: – (a) $\lim\limits_{x\to0}\frac{9^x+7^x-5^x-3^x}{8^x+6^x-4^x-2^x}$ (b) $\lim\limits_{x\to a-1}\frac{x^2+a^2+2x+1}{x-a+1}$ $\left(\text{use formula}, \lim\limits_{x\to a}\frac{x^n-a^n}{x-a}=na^{n-1}\right)$

(c) $\lim\limits_{x\to\frac{\pi}{2}}\frac{\cos^2 x}{1-\sin x}$ $\left(\text{use formula } \lim\limits_{x\to a}\frac{x^n-a^n}{x-a}=na^{n-1}\right)$ (d) $\lim\limits_{x\to0} x^{\frac{1}{|x|}}$ (e) $\lim\limits_{x\to0}|x|^x$ (f) $\lim\limits_{x\to0} e^{\frac{1}{|x|}}$ (g) $\lim\limits_{x\to0} xe^{|x|}$ (h) $\lim\limits_{x\to1} e^{\frac{1}{|x-1|}}$

(i) $\lim\limits_{x\to2}|x-2|^x$ (j) $\lim\limits_{x\to3}\sin|x-3|$

(22) Evaluate: – (a) $\lim\limits_{x\to\frac{\pi}{4}}\left(x-\frac{\pi}{4}\right)\cot\left(x-\frac{\pi}{4}\right)$ (b) $\lim\limits_{x\to0}\frac{e^{ax}-e^{-2ax}}{\log(1+x)}$ (c) $\lim\limits_{x\to1}\left(\frac{1}{x^2-1}-\frac{2}{x^4-1}\right)$ (d) $\lim\limits_{x\to2}\left(\frac{1}{x-1}-\frac{5}{x^2+1}\right)$

(e) $\lim\limits_{x\to2}\left(\frac{x^3+x^2+x+1}{x^2+2x+1}\right)^{\frac{1-\cos[2(x-2)]}{(x-2)^2}}=L$, prove that $L=\frac{25}{9}$ (f) $\lim\limits_{x\to-1}\frac{(3x+2)\sqrt{1+x}}{3x^2+x-2}$ (g) $\lim\limits_{x\to\frac{\pi}{2}}\frac{1}{\cos x}.\log_e(\sin x)$

(h) $\lim\limits_{x\to\infty} x\left[\tan^{-1}\frac{x}{x+1}-\frac{\pi}{4}\right]$

(23) Evaluate: – (a) $\lim\limits_{x\to0}\frac{\tan(\pi\cos^2 x)}{x^2}$ (b) $\lim\limits_{x\to\infty}\left(\frac{x+5}{x+1}\right)^x$ (c) $\lim\limits_{x\to\infty}\left(\frac{x-1}{x+2}\right)^{3x}$ (d) $\lim\limits_{x\to0}(1-2\tan x)^{\frac{1}{\tan x}}$

(e) $\lim\limits_{x\to\frac{\pi}{2}}\left(1+\frac{\cos x}{3}\right)^{\frac{1}{\cos x}}$ (f) $\lim\limits_{x\to1}\frac{\sin\pi x}{x-1}+\lim\limits_{x\to\infty}\left(\frac{x+1}{x-3}\right)^{2x}$ (g) $\lim\limits_{x\to a}\frac{x^{\frac{3}{2}}-a^{\frac{3}{2}}}{x^{\frac{1}{2}}-a^{\frac{1}{2}}}$ (h) $\lim\limits_{x\to3a}\frac{\sqrt{x}-\sqrt{3a}+\sqrt{x-3a}}{\sqrt{2x^2-18a^2}}$ (i) $\lim\limits_{x\to0}\frac{\sqrt{9+\tan 3x}-3}{\log(1+\sin 2x)}$

(24) (a) $\lim\limits_{x\to\infty}\left(\frac{x^2+x+1}{x+1}-ax-b\right)=4$ (b) If $\lim\limits_{x\to0}[1+x\log(1+b^2)]^{\frac{1}{x}}=2b\sin^2\theta, b>0$ *and* $\theta\in(-\pi,\pi]$, then the value of θ.

(c) Let $L=\lim\limits_{x\to0}\frac{a-\sqrt{a^2-x^2}-\frac{x^2}{4}}{x^4}, a>0$. *If L is finite*.

Answer

(1) (a) $f(x)=\frac{4x^2+3x+2}{1+x^2}$, $\lim\limits_{x\to\infty}\frac{4x^2+3x+2}{1+x^2}=\lim\limits_{x\to\infty}\frac{x^2\left(4+\frac{3}{x}+\frac{2}{x^2}\right)}{x^2\left(1+\frac{1}{x^2}\right)}=4$ Ans. (b) Ans: 1 (c) Ans: $\frac{5}{3}$

(d) $f(x) = \dfrac{1^2 + 2^2 + \cdots \ldots \ldots + x^2}{4x^3 - x + 1}$ or $\lim\limits_{x\to\infty} \dfrac{1^2 + 2^2 + \cdots \ldots \ldots + x^2}{4x^3 - x + 1}$ $\left[\text{formula}, 1^2 + 2^2 + \cdots + x^2 = \dfrac{x(x+1)(2x+1)}{6}\right]$

$$\lim_{x\to\infty} \frac{1^2 + 2^2 + \cdots \ldots \ldots + x^2}{4x^3 - x + 1} = \lim_{x\to\infty} \frac{x(x+1)(2x+1)}{(4x^3 - x + 1)6} = \frac{1}{6}\lim_{x\to\infty} \frac{x^3\left(1+\frac{1}{x}\right)\left(2+\frac{1}{x}\right)}{x^3\left(4-\frac{1}{x^2}+\frac{1}{x^3}\right)} = \frac{1}{6}\times\frac{2}{4} = \frac{1}{12} \text{ Ans.}$$

(e) $f(x) = \dfrac{1 + 2 + 3 \ldots \ldots \ldots + x}{x^2 + 1}$ or $\lim\limits_{x\to\infty} \dfrac{1 + 2 + 3 \ldots \ldots \ldots + x}{x^2 + 1}$ $\left[\text{formula}, 1 + 2 + 3 \ldots \ldots \ldots + x = \dfrac{x(x+1)}{2}\right]$

$$\lim_{x\to\infty} \frac{1 + 2 + 3 \ldots \ldots \ldots + x}{x^2 + 1} = \lim_{x\to\infty} \frac{x(x+1)}{x^2\left(1+\frac{1}{x^2}\right).2} = \lim_{x\to\infty} \frac{x^2\left(1+\frac{1}{x}\right)}{x^2\left(1+\frac{1}{x^2}\right).2} = \frac{1}{2} \text{ Ans.}$$

(2) (a) $\lim\limits_{n\to\infty}\left(\dfrac{2n^3 - n + 3}{3n^3 + 2n + 1}\right)^2 = \lim\limits_{n\to\infty}\left[\dfrac{n^3\left(2-\frac{1}{n^2}+\frac{3}{n^3}\right)}{n^3\left(3+\frac{2}{n^2}+\frac{1}{n^3}\right)}\right]^2 = \left(\dfrac{2}{3}\right)^2 = \dfrac{4}{9}$ Ans. (b) Ans: 1 (c) Ans: 1 (d) Ans: 1

(3) (a) $\lim\limits_{n\to 0}\dfrac{1 + 2 + 3 + \cdots \ldots \ldots + n}{n} = \lim\limits_{n\to 0}\dfrac{n(n+1)}{2n} = \dfrac{1}{2}$ Ans. (b) Ans: $-\dfrac{1}{2}$ (c) Ans: $\dfrac{1}{6}$ (d)

(4) (a) $\lim\limits_{x\to 0}\dfrac{(x^3-1)^5 + 1}{x}$ use L'Hospital rule. Ans: 0 (b) Ans: 4

(c) $\lim\limits_{x\to 0}\dfrac{2^x + 3^x - 2}{x} = \lim\limits_{x\to 0}\dfrac{(2^x - 1) + (3^x - 1)}{x} = \lim\limits_{x\to 0}\left[\dfrac{(2^x - 1)}{x} + \dfrac{(3^x - 1)}{x}\right] = \lim\limits_{x\to 0}\dfrac{(2^x - 1)}{x} + \lim\limits_{x\to 0}\dfrac{(3^x - 1)}{x} = \log 2 + \log 3 = \log 6$ Ans.

(d) $\lim\limits_{x\to 0}\left(\dfrac{2\sin 2x + \cos 2x - 1}{x}\right) = \lim\limits_{x\to 0}\dfrac{2\sin 2x - 2\sin^2 x}{x} = \lim\limits_{x\to 0}\left(\dfrac{2\sin 2x}{x} - \dfrac{2\sin^2 x}{x}\right) = \lim\limits_{x\to 0} 4\left(\dfrac{\sin 2x}{2x}\right) - \lim\limits_{x\to 0}\dfrac{2\sin^2 x}{x} = 4 - 0 = 4$ Ans.

(e) $\lim\limits_{x\to\frac{\pi}{4}}\dfrac{1 - \tan x}{1 - \sqrt{2}\sin x} = 2$ Ans. (f) $\lim\limits_{x\to 2}\dfrac{\log(x-1)}{x-2} = 1$ Ans. $\left(\text{hint: —use formula } \lim\limits_{x\to 0}\dfrac{\log(1+x)}{x} = 1\right)$

(g) Ans: -1 (h) Ans: $-\left[1+\dfrac{\pi}{2}\right]$

(i) $\lim\limits_{x\to 0}\dfrac{5^x - 3^x}{4^x - 2^x} = \lim\limits_{x\to 0}\dfrac{(5^x - 1) - (3^x - 1)}{(4^x - 1) - (2^x - 1)} = \lim\limits_{x\to 0}\dfrac{\frac{(5^x - 1) - (3^x - 1)}{x}}{\frac{(4^x - 1) - (2^x - 1)}{x}} = \lim\limits_{x\to 0}\dfrac{\left[\frac{5^x - 1}{x} - \frac{3^x - 1}{x}\right]}{\left[\frac{4^x - 1}{x} - \frac{2^x - 1}{x}\right]}$,

use formula $\lim\limits_{x\to 0}\dfrac{a^x - 1}{x} = \log a$ then limit $= \dfrac{\log 5 - \log 3}{\log 4 - \log 2} = \dfrac{\log\frac{5}{3}}{\log\frac{4}{2}} = \dfrac{\log\frac{5}{3}}{\log 2} = \log_2\left(\dfrac{5}{3}\right)$ Ans.

Note: – Direct formula, $\lim\limits_{x\to 0}\dfrac{a^x - b^x}{c^x - d^x} = \dfrac{\log\left(\frac{a}{b}\right)}{\log\left(\frac{c}{d}\right)}$

(j) $\lim\limits_{x\to 2}\dfrac{e^{\sqrt{x}} - e^{\sqrt{2}}}{x - 2} = \dfrac{1}{2\sqrt{2}}e^2$ Ans. (use L'Hospital rule)

(k) $\lim\limits_{x\to 0}\dfrac{(x.3^x - x)}{\frac{1-\cos 2x}{2}} = \lim\limits_{x\to 0}\dfrac{x(3^x - 1)}{\sin^2 x} = \lim\limits_{x\to 0}\dfrac{\frac{(3^x - 1)}{x}}{\left(\frac{\sin x}{x}\right)^2} = \log 3$ Ans.

$\left(\lim\limits_{x\to 0}\dfrac{a^x - 1}{x} = \log a\,,\ \dfrac{1-\cos 2x}{2} = \dfrac{1 - (\cos^2 x - \sin^2 x)}{2} = \dfrac{1 - \cos^2 x + \sin^2 x}{2} = \dfrac{2\sin^2 x}{2} = \sin^2 x\right)$

(5) (a) $\lim\limits_{x\to 0}\dfrac{27^x + 9^x + 3^x - 3}{x} = \lim\limits_{x\to 0}\left\{\dfrac{(27^x - 1)}{x} + \dfrac{(9^x - 1)}{x} + \dfrac{(3^x - 1)}{x}\right\} = \lim\limits_{x\to 0}\dfrac{(27^x - 1)}{x} + \lim\limits_{x\to 0}\dfrac{(9^x - 1)}{x} + \lim\limits_{x\to 0}\dfrac{(3^x - 1)}{x} = \log 27 + \log 9 + \log 3$

$= \log 3^3 + \log 3^2 + \log 3 = 3\log 3 + 2\log 3 + \log 3 = 6\log 3 = \log 3^6 = \log 729$ Ans.

(b) $\lim\limits_{x\to 0}\dfrac{81^x - 27^x + 9^x - 3^x}{(\sqrt{24+\cos x} - 3)\sin x} = \log 3$ Ans.

(c) $\lim\limits_{x\to\frac{\pi}{4}}\dfrac{\sqrt{2}\left(x-\frac{\pi}{4}\right)}{(\sin x - \cos x)} = \lim\limits_{x\to\frac{\pi}{4}}\dfrac{\sqrt{2}\left(x-\frac{\pi}{4}\right)}{\sqrt{2}\left(\sin x.\cos\frac{\pi}{4} - \cos x.\sin\frac{\pi}{4}\right)} = \lim\limits_{x\to\frac{\pi}{4}}\dfrac{\left(x-\frac{\pi}{4}\right)}{\sin\left(x-\frac{\pi}{4}\right)} = \lim\limits_{x-\frac{\pi}{4}\to 0}\dfrac{1}{\dfrac{\sin\left(x-\frac{\pi}{4}\right)}{\left(x-\frac{\pi}{4}\right)}} = 1$ Ans.

(d) $\lim\limits_{x\to 0}\dfrac{x}{\sin x + \tan x} = \lim\limits_{x\to 0}\dfrac{1}{\dfrac{\sin x + \tan x}{x}} = \lim\limits_{x\to 0}\dfrac{1}{\left(\dfrac{\sin x}{x} + \dfrac{\tan x}{x}\right)} = \dfrac{1}{1+1} = \dfrac{1}{2}$ Ans. (e) $\lim\limits_{x\to 1}\dfrac{\sqrt{x+2} - \sqrt{2x+1}}{x-1} = -\dfrac{1}{2\sqrt{3}}$ Ans.

(f) $\lim\limits_{x\to 0}(1+\sin x)^{\frac{1}{\sin x}} = y$ (say)

$\Rightarrow \log y = \lim\limits_{x\to 0}\log(1+\sin x)^{\frac{1}{\sin x}} = \lim\limits_{x\to 0}\dfrac{\log(1+\sin x)}{\sin x} = \lim\limits_{\sin x\to 0}\dfrac{\log(1+\sin x)}{\sin x}$

use formula, $\lim\limits_{x\to 0}\dfrac{\log(1+x)}{x} = 1$ $\quad\therefore \log y = \lim\limits_{\sin x\to 0}\dfrac{\log(1+\sin x)}{\sin x} = 1 = \log_e e \quad \therefore y = e$

$\therefore \lim\limits_{x\to 0}(1+\sin x)^{\frac{1}{\sin x}} = y = e$ Ans.

(g) $\lim\limits_{x\to 2}\dfrac{x-2}{2x^2 - 5x + 2} = \dfrac{1}{3}$ Ans. (h) $\lim\limits_{x\to 0}\dfrac{\sqrt{1+\tan x} - \sqrt{1-\sin x}}{x} = 1$ Ans.

(i) $\lim\limits_{x\to\frac{\pi}{2}}\dfrac{\sqrt{1-\sin x}}{\cos x} = \lim\limits_{x\to\frac{\pi}{2}}\dfrac{\sqrt{1-\sin x}}{\cos x}\times\dfrac{\sqrt{1+\sin x}}{\sqrt{1+\sin x}} = \lim\limits_{x\to\frac{\pi}{2}}\dfrac{\sqrt{(1-\sin x)(1+\sin x)}}{\cos x\sqrt{1+\sin x}} = \lim\limits_{x\to\frac{\pi}{2}}\dfrac{\sqrt{\cos^2 x}}{\cos x\sqrt{1+\sin x}} = \lim\limits_{x\to\frac{\pi}{2}}\dfrac{\cos x}{\cos x\sqrt{1+\sin x}} = \lim\limits_{x\to\frac{\pi}{2}}\dfrac{1}{\sqrt{1+\sin x}}$
$= \dfrac{1}{\sqrt{2}}$ Ans.

IInd method:– $\lim\limits_{x\to\frac{\pi}{2}}\dfrac{\sqrt{1-\sin x}}{\cos x} = \lim\limits_{x\to\frac{\pi}{2}}\dfrac{\sqrt{1-\sin x}}{\sqrt{1-\sin^2 x}} = \lim\limits_{x\to\frac{\pi}{2}}\dfrac{\sqrt{1-\sin x}}{\sqrt{(1-\sin x)(1+\sin x)}} = \lim\limits_{x\to\frac{\pi}{2}}\dfrac{1}{\sqrt{(1+\sin x)}} = \dfrac{1}{\sqrt{2}}$ Ans.

(j) $\lim\limits_{x\to 2}\dfrac{(1+x)^3 - 27}{x-2} = 27$ Ans. (without use L'Hospital rule) (k) $\lim\limits_{x\to 3}\dfrac{\left(1+\sqrt{1+x}\right)^2 - 9}{x-3} = \dfrac{3}{2}$ Ans.

(6) (a) $\lim\limits_{x\to\frac{\pi}{4}}\dfrac{1-\sin 2x}{x-\frac{\pi}{4}} = 0$ Ans. (b) $\lim\limits_{x\to a}\dfrac{x^{\frac{1}{3}} - a^{\frac{1}{3}}}{x^{\frac{1}{2}} - a^{\frac{1}{2}}} = \lim\limits_{x\to a}\dfrac{\dfrac{x^{\frac{1}{3}} - a^{\frac{1}{3}}}{x-a}}{\dfrac{x^{\frac{1}{2}} - a^{\frac{1}{2}}}{x-a}} = \dfrac{\frac{1}{3}a^{\frac{1}{3}-1}}{\frac{1}{2}a^{\frac{1}{2}-1}} = \dfrac{2}{3}a^{-\frac{1}{6}} = \dfrac{2}{3a^{\frac{1}{6}}}$ Ans. $\left(\text{use formula}, \lim\limits_{x\to a}\dfrac{x^n - a^n}{x-a} = na^{n-1}\right)$

(c) $\lim\limits_{x\to\frac{\pi}{2}}\dfrac{\log(\cos x + 1)}{\sin 2x} = \dfrac{1}{2}$ Ans. (solve without L'Hospital rule) (d) $\lim\limits_{x\to 0}\dfrac{1+\sin x - \cos x}{\cos x - \sin x - 1} = -1$ Ans.

(e) $\lim\limits_{x\to 0}\dfrac{\sqrt{9+\tan 2x} - 3}{\log(1+\sin 2x)} = \dfrac{1}{3}$ Ans. (without use L'Hospital rule) (f) $\lim\limits_{x\to 1}\dfrac{\sqrt[4]{15+x} - 2}{3(x-1)} = \dfrac{1}{12}$ Ans.

(g) $\lim\limits_{x\to 0}\dfrac{\sqrt[3]{1+x} + \sqrt[4]{1+x} - 2}{3^x + 2^x - 2} = \dfrac{7}{12\log 6}$ Ans. $\left(\text{formula}, \lim\limits_{x\to 0}\dfrac{(1+x)^n - 1}{x} = n,\ \lim\limits_{x\to 0}\dfrac{a^x - 1}{x} = \log a, \log m + \log n = \log m.n\right)$

(h) $\lim\limits_{x\to\frac{\pi}{2}}\dfrac{\sqrt{1+\cos x} - \sqrt{1-\cos x}}{\cos x} = 1$ Ans. (i) $\lim\limits_{x\to a}\dfrac{\sqrt{x+3a} - 2.\sqrt{a}}{3.\sqrt{x+a} - 3.\sqrt{2a}} = \dfrac{1}{3\sqrt{2}}$ $(a \neq 0)$

(j) $\lim\limits_{x\to 0}\dfrac{x\sin x + \log(1+x^2)}{x^2} = 2$ Ans. (solve the question without use L'Hospital rule)

(7) (a) $\lim\limits_{x\to\infty}\dfrac{\sqrt{x^2+1} + \sqrt[4]{x^4+1}}{\sqrt[3]{x^3+1} + \sqrt[5]{x^5+1}} = 1$ Ans. (b) $\lim\limits_{x\to\infty}\dfrac{\sqrt[3]{x^2+1} - \sqrt[4]{x^4+1}}{\sqrt[3]{x^3+1} - \sqrt[5]{x^4+1}} = -1$ Ans.

(c) $\lim\limits_{x\to\infty}\dfrac{x+\cos x}{x - \sin^2 x} = 1$ Ans. (d) $\lim\limits_{x\to 3}\dfrac{x^2 - 4x + 3}{x^2 - 5x + 6} = 2$ Ans. (e) $\lim\limits_{x\to 1}\dfrac{(3x-1)\left(\sqrt{x^2} - 1\right)}{(3x^3 + x^2 - 3x - 1)} = \lim\limits_{x\to 1}\dfrac{(3x-1)(x-1)}{(x-1)(x+1)(3x+1)} = \dfrac{1}{4}$ Ans.

(f) $\lim\limits_{x\to 0}\dfrac{(x.3^x - x)}{\sin^2 x} = \log 3$ Ans. (g) $\lim\limits_{x\to\pi}\dfrac{(x-\pi)\sin x}{1+\cos x} = -2$ Ans.

(h) $\lim\limits_{x\to 0}\dfrac{\tan 2x.(1-\tan^2 x)}{x+\sin x} = \lim\limits_{x\to 0}\dfrac{2\tan x.(1-\tan^2 x)}{(1-\tan^2 x)(x+\sin x)} = \lim\limits_{x\to 0}\dfrac{2\tan x}{(x+\sin x)} = \lim\limits_{x\to 0}\dfrac{2\tan x}{x\left(1+\frac{\sin x}{x}\right)} = \lim\limits_{x\to 0}\left(\dfrac{\tan x}{x}\right)\dfrac{2}{\left(1+\frac{\sin x}{x}\right)} = \dfrac{2}{1+1} = \dfrac{2}{2}$

$= 1$ Ans.

(i) $\lim\limits_{x\to 0}\dfrac{x\cos x+\tan x}{\tan x} = 2$ Ans. (j) $\lim\limits_{x\to 0}\dfrac{2x-\sin x}{3x-\tan x} = \dfrac{1}{2}$ Ans. (k) $\lim\limits_{x\to 0}\dfrac{\tan x + x}{x^2} = \infty$ Ans.

(8) (a) $\lim\limits_{x\to 0}\dfrac{\cot(\pi x)}{\frac{\pi}{2x}} = \dfrac{2}{\pi^2}$ Ans.

(b) $\lim\limits_{\theta\to\pi}\dfrac{\sqrt{1+\cos\theta}-\sin\theta}{\cos\left(\frac{\theta}{2}\right)} = \lim\limits_{\theta\to\pi}\dfrac{\sqrt{2\cos^2\frac{\theta}{2}}-2\sin\frac{\theta}{2}\cos\frac{\theta}{2}}{\cos\left(\frac{\theta}{2}\right)} = \lim\limits_{\theta\to\pi}\dfrac{\cos\frac{\theta}{2}\left(\sqrt{2}-2\sin\frac{\theta}{2}\right)}{\cos\left(\frac{\theta}{2}\right)} = \lim\limits_{\theta\to\pi}\left(\sqrt{2}-2\sin\frac{\theta}{2}\right) = \left(\sqrt{2}-2\right)$ Ans.

(c) $\lim\limits_{\theta\to\frac{\pi}{2}}\dfrac{(\sin\theta)^{\sin\theta}-1}{1-\sin\theta} = -1$ Ans. $\left(\text{hint:}-\ \text{put } \sin\theta = t,\ \theta\to\frac{\pi}{2} \text{ then } t\to 1 \text{ and use L'Hospital rule}\right)$

(d) If f′(a) is exists then show that $\lim\limits_{x\to a}\dfrac{(2x-a)f(x)-af(2x-a)}{x-a} = 2f(a) - af'(a)$.

$\text{L.H.S} = \lim\limits_{x\to a}\dfrac{(2x-a)f(x)-af(2x-a)}{x-a} = \lim\limits_{x\to a}\dfrac{(2x-a)f'(x)+f(x).2-af'(2x-a).2}{1} = af'(x)+2f(a)-2af'(a)$

$= 2f(a) - af'(a)$ proved. (use L'Hospital rule)

(e) Do yourself. (same as above question, use L'Hospital rule)

(9) (a) $F(x) = \sqrt{24-x^3}$, $F'(x) = \dfrac{1}{2.\sqrt{24-x^3}}\times -3x^2$, $\lim\limits_{x\to 2}\dfrac{F(x)-F(2)}{x-2}$

use L'Hospital rule, $\lim\limits_{x\to 2}\dfrac{F(x)-F(2)}{x-2} = \lim\limits_{x\to 2}\dfrac{F'(x)}{1} = F'(2) = \dfrac{1}{2.\sqrt{24-8}}\times -3(2)^2 = \dfrac{-12}{8} = -\dfrac{3}{2}$ Ans.

(b) $F(5) = 5, F'(5) = 3$ then find $\lim\limits_{x\to 5}\dfrac{\sqrt{F(x)+3}-\sqrt{8}}{\sqrt{x}-\sqrt{5}} = \lim\limits_{x\to 5}\dfrac{\left(\sqrt{F(x)+3}-\sqrt{8}\right)\left(\sqrt{F(x)+3}+\sqrt{8}\right)\left(\sqrt{x}+\sqrt{5}\right)}{\left(\sqrt{x}-\sqrt{5}\right)\left(\sqrt{x}+\sqrt{5}\right)\left(\sqrt{F(x)+3}+\sqrt{8}\right)} = \lim\limits_{x\to 5}\dfrac{[F(x)+3-8]\left(\sqrt{x}+\sqrt{5}\right)}{(x-5)\left(\sqrt{F(x)+3}+\sqrt{8}\right)}$

$= \lim\limits_{x\to 5}\dfrac{F(x)-5}{x-5}\times\lim\limits_{x\to 5}\dfrac{\sqrt{x}+\sqrt{5}}{\sqrt{F(x)+3}+\sqrt{8}} = \dfrac{2\sqrt{5}}{2\sqrt{8}}\times\lim\limits_{x\to 5}F'(x) = F'(5).\dfrac{\sqrt{5}}{\sqrt{8}} = \dfrac{3\sqrt{5}}{2\sqrt{2}} = \dfrac{3\sqrt{5}}{2\sqrt{2}}\times\dfrac{\sqrt{2}}{\sqrt{2}} = \dfrac{3}{4}\sqrt{10}$ Ans.

(10) (a) $\lim\limits_{x\to 0}\dfrac{\tan x - x - \frac{x^3}{3}}{x^5} = \lim\limits_{x\to 0}\dfrac{\left[x+\frac{1}{3}x^3+\frac{2}{15}x^5+\cdots\ldots\ldots\right]-x-\frac{x^3}{3}}{x^5} = \lim\limits_{x\to 0}\dfrac{\frac{2}{15}x^5}{x^5} = \dfrac{2}{15}$ Ans.

(b) $\lim\limits_{x\to\frac{\pi}{2}}\sec x.\log_e(\text{cosec}x) = \lim\limits_{x\to\frac{\pi}{2}}\dfrac{\log_e(\text{cosec}x)}{\cos x} = \lim\limits_{x\to\frac{\pi}{2}}\dfrac{-\cot x.\text{cosec}x}{\text{cosec}x.-\sin x} = \lim\limits_{x\to\frac{\pi}{2}}\dfrac{\cot x}{\sin x} = 0$ Ans.

(c) $\lim\limits_{x\to 1}\dfrac{(1+x)-2}{(1-x)} = -1$ Ans. (Do yourself) (d) $\lim\limits_{x\to 0}\dfrac{\log\sin\left(\frac{\pi}{2}+ax\right)}{\tan bx} = 0$ (use L'Hospital rule) (e) $\lim\limits_{x\to 0}\dfrac{\frac{\pi}{2x}}{\text{cosec}\left(\frac{\pi}{2}-x\right)} = \dfrac{\pi^2}{4}$ Ans.

(11) (a) $\lim\limits_{x\to 0}\dfrac{a\tan x + x(b+\cos x)}{x^3} = 1$

$\Rightarrow \lim\limits_{x\to 0}\dfrac{a\left(x+\frac{1}{3}x^3+\frac{2}{15}x^5+\cdots\ldots\right)+x\left(b+1-\frac{x^2}{2!}+\frac{x^4}{4!}-\cdots\ldots\right)}{x^3} = 1$

$\Rightarrow \lim\limits_{x\to 0} ax+\frac{1}{3}ax^3+\frac{2}{15}ax^5+\cdots\ldots\ldots\ldots\ldots\ldots\ldots\ldots\ldots+bx+x-\frac{x^3}{2!}+\frac{x^5}{4!}-\cdots\ldots\ldots\ldots = x^3$

$\Rightarrow x(a+b+1)+x^3\left(\frac{1}{3}a-\frac{1}{2}\right)+\cdots\ldots\ldots\ldots\ldots\ldots = x^3+0.x$

$\therefore\ a+b+1=0$ or $\frac{1}{3}a-\frac{1}{2}=1\ \therefore\ \frac{1}{3}a=1+\frac{1}{2}=\frac{3}{2}\ \therefore a=\frac{9}{2}$

$\therefore\ a+b+1=0\ \therefore\ b=-1-a=-1-\frac{9}{2}=-\frac{11}{2}\qquad \therefore\ a=\frac{9}{2}, b=-\frac{11}{2}$ Ans.

(b) $\lim\limits_{x\to0}\frac{axe^x-b\log(1+x)+cxe^{-x}}{x^2\sin x}=2,\ a=3, b=12$ and $c=9$ Ans. (same as above question)

(c) $\lim\limits_{x\to0}\left(\frac{1-x}{1+x}+ax-b\right)=0$, then prove that $a=0, b=-1\ \therefore\ \lim\limits_{x\to0}\left(\frac{1-x}{1+x}+ax-b\right)=0$ or $\lim\limits_{x\to0}\frac{1-x+ax+ax^2-b-bx}{1+x}=0$

or $\lim\limits_{x\to0}ax^2+x(a-b-1)+(1-b)=0.x^2+0.x\ \Rightarrow\ a=0$ or $a-b-1=0\ \therefore b=-1$

$\therefore\ a=0$ and $b=-1$ proved.

(d) $\lim\limits_{x\to0}\left(\frac{ax^2}{x+1}-bx+1\right)=1$, then prove that $a=0, b=0$. (solve same as above question)

(12) (a) $\lim\limits_{x\to0}\frac{\cos 2x+a\cos x}{x^2}=$ finite, $a=-6$, limit $=\infty$ Ans. (b) $\lim\limits_{x\to0}\frac{a\tan x+b\sin x-2}{x^3}=$ finite, $a=6, b=-6$ and limit $=\infty$ Ans.

(c) $\lim\limits_{x\to0}\frac{ae^x+b\log(1+x)-ce^{-x}}{x\tan x}=$ finite, $a=1, b=-\frac{6}{5}$ and $c=\frac{1}{5}$ Ans.

(d) $\lim\limits_{x\to0}\frac{ax\cos x+b\log(1+x)-cxe^x}{x^2\sin x}=2,\ a=-\frac{34}{15}, b=-\frac{8}{5}$ and $c=-\frac{2}{3}$ Ans.

(13) (a) $\lim\limits_{x\to0}\left\{\sin\left(\frac{\pi}{2}+x\right)\right\}^{\frac{1}{x}}=\lim\limits_{x\to0}(\cos x)^{\frac{1}{x}}=y$ (say)

$\Rightarrow\ \log y=\lim\limits_{x\to0}\frac{1}{x}.\log(\cos x)$, use L'Hospital rule. or $\log y=\lim\limits_{x\to0}\frac{-\sin x}{\cos x}=0=\log 1\ \therefore\ y=1$ Ans.

(b) $\lim\limits_{x\to\infty}\left(\frac{x+3}{x+2}\right)^{x+4}=\lim\limits_{x\to\infty}\left(1+\frac{1}{x+2}\right)^{(x+2)+2}$, $\left(\text{use formula, } \lim\limits_{x\to\infty}\left(1+\frac{p}{x}\right)^x=e^p\right)$

$\Rightarrow\ \lim\limits_{x\to\infty}\left(1+\frac{1}{x+2}\right)^{(x+2)}.\lim\limits_{x\to\infty}\left(1+\frac{1}{x+2}\right)^2=e^1.1=e$ proved.

(c) $\lim\limits_{x\to\infty}\left(\frac{x+1}{x-2}\right)^{x+3}=e^3\ \left(\text{Hint:}-\ \lim\limits_{x\to\infty}\left(1+\frac{3}{x-2}\right)^{(x-2)+5}=\lim\limits_{x\to\infty}\left(1+\frac{3}{x-2}\right)^{(x-2)}.\lim\limits_{x\to\infty}\left(1+\frac{3}{x-2}\right)^5=e^3\right)$

(d) $\lim\limits_{x\to0}\left(\frac{1+2x^2}{1+3x^2}\right)^{\frac{1}{x^2}}=\lim\limits_{x\to0}\frac{(1+2x^2)^{\frac{1}{x^2}}}{(1+3x^2)^{\frac{1}{x^2}}}=\frac{e^2}{e^3}=e^{2-3}=e^{-1}=\frac{1}{e}$ proved. $\left(\text{use formula, } \lim\limits_{x\to0}(1+px)^{\frac{1}{x}}=e^p\right)$

IInd method:$-$ Let $y=\lim\limits_{x\to0}\left(\frac{1+2x^2}{1+3x^2}\right)^{\frac{1}{x^2}}\ \Rightarrow\ \log y=\lim\limits_{x\to0}\frac{1}{x^2}.\left(\frac{1+2x^2}{1+3x^2}-1\right)=\lim\limits_{x\to0}\frac{1}{x^2}.\left(\frac{-x^2}{1+3x^2}\right)=-1=\log e^{-1}$

or $\log y=\log e^{-1}\ \therefore\ y=e^{-1}=\frac{1}{e}$ proved.

(e) $\lim\limits_{x\to0}\left\{\frac{2}{3}\left(\frac{3+x}{2+x}\right)\right\}^{\frac{1}{x}}=\lim\limits_{x\to0}\left\{\frac{2}{3}.\frac{3}{2}\left(\frac{1+\frac{x}{3}}{1+\frac{x}{2}}\right)\right\}^{\frac{1}{x}}=\lim\limits_{x\to0}\left(\frac{1+\frac{x}{3}}{1+\frac{x}{2}}\right)^{\frac{1}{x}}=\lim\limits_{x\to0}\frac{\left(1+\frac{x}{3}\right)^{\frac{1}{x}}}{\left(1+\frac{x}{2}\right)^{\frac{1}{x}}}\ \left(\text{formula } \lim\limits_{x\to0}\left(1+\frac{x}{p}\right)^{\frac{1}{x}}=e^{\frac{1}{p}}\right)$

$\therefore\ \lim\limits_{x\to0}\frac{\left(1+\frac{x}{3}\right)^{\frac{1}{x}}}{\left(1+\frac{x}{2}\right)^{\frac{1}{x}}}=\frac{e^{\frac{1}{3}}}{e^{\frac{1}{2}}}=e^{\frac{1}{3}-\frac{1}{2}}=e^{\frac{2-3}{6}}=e^{-\frac{1}{6}}$ or $\frac{1}{e^{\frac{1}{6}}}$ proved.

(f) $\lim\limits_{x\to0}\frac{\log(1+x)}{x}-2=1-2=-1$ Ans. (g) $\lim\limits_{x\to\infty}\left(\frac{x-5}{x+3}\right)^x=e^{-8}$ or $\frac{1}{e^8}$ Ans. $\left(\text{use formula, } \lim\limits_{x\to\infty}\left(1+\frac{p}{x}\right)^x=e^p\right)$

(h) $\lim_{x\to 0}\left(\frac{\sin 2x}{2x}\right)^{\frac{1}{x^2}} = e^{-\frac{2}{3}}$ Ans. (i) e^3 (j) 1 (k) 1 (l) ∞ (m) ∞

(14) Ans: – (a) $\frac{2}{3}$ (b) 3 (c) 1 (d) $\sqrt{\frac{2}{3}}$ (e) 1 (f) $\frac{\sqrt{3}}{2}$ (15) Ans: – (a) 1 (b) 2 (c) $\frac{5}{6}$ (d) $\log\left(\frac{3}{5}\right)$ (e) $\frac{3}{2}$ (f) 1

(16) Ans: – (a) $\frac{1}{2}$ (b) 0 (c) $\log\left(\frac{1}{3}\right)$ (d) $\frac{1}{2\sqrt{2}}$ (e) 1 (f) 2 (g) 0 (h) $-\frac{3}{16}$

(17) Ans: – (a) $-4\sqrt{2}$ (b) $-\frac{9}{2}$ (c) $\frac{1-\sqrt{3}}{2}$ (d) $-\frac{1}{2}$ (e) 0 (f) 2 (g) 0

(18) Ans: – (a) limit does not exist. (b) 1 (c) limit does not exist. (d) limit does not exist. (e) 1 (f) 2 (g) ∞ (h) 0

(19) Ans: – (a) $\frac{1}{2}$ (b) $\frac{1}{3}$ (c) ∞ (d) 1 (e) 1 (without use L'Hospital rule) (f) e^{-1} or $\frac{1}{e}$ (g) e^2 (h) $e^{\frac{1}{3}}$

(20) Ans: – (a) e^{-3} (b) -1 (c) $2e^{-\pi}$ (d) -10 (e) $\frac{1}{\log 4}$ (f) 1 (g) 0 (h) $\frac{\log(e+1)-1}{\log(2e+3)}$

(21) Ans: – (a) $\frac{\log\left(\frac{21}{5}\right)}{\log 6}$ or $\log_6\left(\frac{21}{5}\right)$ (b) 2a (c) 2 (d)

(22) Ans: – (a) 1 (b) 3a (c) $\frac{1}{2}$ (d) 0

(24) (a) (b) $\theta = \pm\frac{\pi}{2}$ (c) $a = 2, L = \frac{1}{64}$

Exercise – A4

(1) (a) A function f is defined as $f(x) = \begin{cases} \frac{x^2-5x+6}{x^2-4} & \text{for } x \neq 2 \\ \frac{1}{4} & \text{for } x = 2 \end{cases}$ function f(x) is continuous at $x = 2$.

(b) A function f is defined as $f(x) = \begin{cases} \frac{2x^2+x-3}{3x^2+5x-2}, & x \neq 1 \\ 0, & x = 1 \end{cases}$ show that f(x) is differentiable at $x = 1$ and find its value.

(c) $f(x) = \begin{cases} \frac{1-\cos 2x}{x^2}, & x < 0 \\ a, & x = 0 \\ \frac{\sqrt{x}}{\sqrt{25+\sqrt{x}}-5}, & x > 0 \end{cases}$ find the value of a so that the function may be continuous at $x = 0$.

(2) (a) Discuss the continuity and differentiability of the function $f(x) = \begin{cases} \frac{x}{2+|x|}, & |x| \geq 2 \\ \frac{x}{2-|x|}, & |x| < 2 \end{cases}$

(b) Discuss the continuity and differentiability of the function $f(x) = \begin{cases} 3+\sqrt{4-x^2}, & |x| \leq 2 \\ 3e^{(2-x)^2}, & |x| > 2 \end{cases}$

(c) Discuss the differentiability of $\tan\{\pi(x-[x])\}$ in $(0,\pi)$.

(3) (a) Show that the value of the derivative of $|x-2|+|x-4|$ at $x = 3$ is 0.

(b) The function $f(x) = (x^2-1)|x^2+5x+6| + \sin|x|$ is not differentiable at (i) 0 (ii) -2 (iii) -3 (iv) 1.

(c) If $f(x) = \begin{cases} 2x-3, & 0 \leq x \leq 1 \\ 3x-a, & 1 < x \leq 2 \end{cases}$ and f(x) be continuous at $x = 1$ and find the value of a.

(4) (a) Prove that $f(x) = |\log(x+1)|$ is continuous at $x = 0$ but is not differentiable at $x = 0$.

(b) If $f(x) = \dfrac{2}{x+1}$, then determine the points of discontinuity of $f\left(f(f(x))\right)$. (c) Find the points of discontinuity of $f(x) = \dfrac{1}{x+|x|}$.

(5) (a) $f(x) = \begin{cases} -\cos x\,, & -\frac{\pi}{2} \le x < 0 \\ a\cos x + b\,, & 0 < x < \frac{\pi}{2} \\ 2\sin x\,, & \frac{\pi}{2} \le x \le \pi \end{cases}$ If f(x) is continuous on $\left[0, \frac{\pi}{2}\right]$ then show that $a = -3, b = 2$.

(b) $f(x) = \begin{cases} \dfrac{9^x - 3^x - 5^x + 1}{x}\,, & x > 0 \\ e^x \cos x - \pi x + a\log 3\,, & x \le 0 \end{cases}$ If f(x) is continuous at $x = 0$. then show that $a = \dfrac{\log 3 - \log 5 - 1}{\log 3}$.

(c) Let $f(x) = \begin{cases} x^2 \left|\cos\left(\frac{\pi}{x}\right)\right|\,, & x \ne 0 \\ 0\,, & x = 0 \end{cases}$ $x \in R$, then f is ………….?

(6) (a) If $f(x) = x\left[\sqrt{x+1} - \sqrt{x+3}\,\right]$, then f(x) is continuous and differentiable at $x = 0$.

(b) Discuss the continuity and differentiability of $f(x) = \begin{cases} |x-3|\,, & x \ge 2 \\ \frac{x^2}{3} - \frac{4x}{2} + \frac{9}{3}\,, & x < 2 \end{cases}$ at $x = 2,3$.

(c) Discuss the continuity and differentiability of $f(x) = \begin{cases} 2x\,, & x < 2 \\ 3 - 2x\,, & 2 \le x \le 4 \\ -3 + 2x - x^2\,, & x > 4 \end{cases}$ at $x = 2$ and $x = 4$.

(7) (a) Let $f(x) = \begin{cases} 2xe^{\left(\frac{1}{|x|} - \frac{1}{x}\right)}\,, & x \ne 0 \\ 0\,, & x = 0 \end{cases}$ The function f(x) is continuous and differentiable at $x = 0$.

(b) If $f(x) = \begin{cases} \left(x - \frac{\pi}{4}\right) \cdot \dfrac{1 - \tan x}{1 + \tan x}\,, & x \ne \frac{\pi}{4} \\ 0\,, & x = \frac{\pi}{4} \end{cases}$ then f(x) is continuous and differentiable at $x = \frac{\pi}{4}$.

(c) If $f(x) = \begin{cases} ax(x+1) - b\,, & x < -1 \\ 2x - 1\,, & -1 \le x \le 0 \\ 2cx^2 + 3dx - 2\,, & x > 0 \end{cases}$ then f(x) is continuous at $x = -1$ and $x = 0$. Determine the constant a, b, c and d.

(8) (a) If $f(x) = \cos x$ and $g(x) = \sqrt{|x| - 1}$ then calculate (fog)(x) and (gof)(x). discuss the differentiability of (gof)(x)at $x = 0$

and discuss the continuous of (fog)(x)at $x = 0$.

(b) If $f(x) = -2 + |x+2|\,, -2 \le x \le 0$ and $g(x) = 3 - |x-2|\,, -3 \le x \le 3$. then calculate (fog)(x) and (gof)(x).

Discuss the continuity of (fog)(x) at $x = -2$ and differentiability of (gof)(x) at $x = 2$.

(c) Discuss the continuity and differentiability of the function $f(x) = x + |x-3|$ at $x = 3$.

(9) (a) $f(x) = \dfrac{1 + \cos x}{\left(\frac{\pi}{2} - \frac{x}{2}\right)^2} \cdot \dfrac{\log(\cos x)}{\log(1 + 2\pi^2 - 3\pi x + x^2)}$, $x \ne \pi$. Determine $f(\pi)$, if f(x)is continuous at $x = \pi$.

(b) $f(x) = \dfrac{1 + \cos x}{(\pi - x)^4} \cdot \sin x\,(4x^2 - \pi^2)$, $x \ne \pi$. Determine $f(\pi)$, if f(x) is continuous at $x = \pi$.

(c) If $f(x) = 3x - \dfrac{x-1}{|x+3|}$ then determine the point of discontinuity of the function f(x).

(10) (a) If $f(x) = \begin{cases} \dfrac{\log(1 + |x|)}{x}\,, & x \ne 0 \\ 0\,, & x = 0 \end{cases}$ then f(x) is discontinuous at $x = 0$.

(b) Let $f(x) = \begin{cases} x^2|x|\,, & x \ne 0 \\ 0\,, & x = 0 \end{cases}$ Discuss the function f(x) is continuity and differentiability at $x = 0$.

(c) Let $f(x) = \begin{cases} ae^{\cos x} - b, & x < \frac{\pi}{2} \\ ax + b, & x > \frac{\pi}{2} \\ 1, & x = \frac{\pi}{2} \end{cases}$ then the function f(x) is continuous at $x = \frac{\pi}{2}$ find the value of a and b.

(11) (a) Discuss the continuity and differentiability of $f(x) = \begin{cases} \sec x - \tan x, & 0 \le x \le \frac{\pi}{4} \\ \sqrt{2}\cot x - 1, & \frac{\pi}{4} < x \le \frac{\pi}{2} \end{cases}$ at $x = \frac{\pi}{4}$.

(b) If $f(x) = \begin{cases} e^{|1+x|} + 2, & x \ne -1 \\ 2, & x = -1 \end{cases}$ then the function f(x) is discontinuous at $x = -1$.

(c) If $f(x) = \begin{cases} \log(ax + b) + 1, & -1 \le x < 2 \\ \log x, & 2 < x \le 3 \\ -1, & x = 2 \end{cases}$ then the function f(x) is continuous at $x = 2$.

and the function f(x) is not differentiable at $x = 2$.

(12) (a) If $f(x) = \begin{cases} \frac{1 + e^{-x}}{1 - e^{x}}, & x \ne 0 \\ 0, & x = 0 \end{cases}$ prove that f(x) is not differentiable at $x = 0$.

(b) If $f(x) = \begin{cases} \frac{|x|}{x}, & x \ne 0 \\ 0, & x = 0 \end{cases}$ Discuss its continuity at $x = 0$.

(c) Discuss the continuity and differentiable of the following function $f(x) = \begin{cases} x^3, & x < -3 \\ 9, & -3 \le x \le 3 \\ x^3, & x > 3 \end{cases}$

(13) (a) which of the following function is not differentiable at $x = 0$, $f(x) = 2\sin|x| - |x|$.

(b) If $f(x) = \cos|x| - |x|$, Discuss its continuity at $x = 0$.

(c) Discuss the continuity and differentiability at $x = 2$, if $f(x) = \begin{cases} 2^x, & -2 \le x \le 2 \\ 6 - x, & 2 < x < 5 \end{cases}$.

(14) (a) Discuss the continuity and differentiability at $x = \frac{\pi}{4}$ if $f(x) = \begin{cases} e^{a\tan x} - 1, & 0 < x \le \frac{\pi}{4} \\ \log_2(\sin x), & \frac{\pi}{4} < x < \frac{\pi}{2} \end{cases}$ find the value of a.

(b) Let $f(x) = \begin{cases} x^2 + ax + b, & 0 < x < 1 \\ 2ax - 3, & 1 \le x \le 2 \\ ax + b, & 2 < x < 3 \end{cases}$ then f(x) is continuous at $x = 1,2$ and differentiable at $x = 1,2$ also find the value of a and b.

(c) Discuss the continuity and differentiable at $x = 0$, if $f(x) = \begin{cases} \frac{2}{x^2 - 2x + 1}, & x \ne 0 \\ 2, & x = 0 \end{cases}$

(15) (a) If $f(x) = \frac{(5^x - 1)^2}{\cos x . [\log(1 + x)]^2}$, $x \ne 0$ is continuous at $x = 0$ then find f(0).

(b) If the derivative of the function $f(x) = \begin{cases} 2ax^2 + 4b, & x < -2 \\ bx^2 + 2ax + 3, & x \ge -2 \end{cases}$ is everywhere continuous. then show that $a = \frac{1}{4}$ and $b = \frac{5}{8}$.

(c) Examine the continuity and differentiability in $-\infty < x < \infty$ *of the following function* $f(x) = \begin{cases} 2, & -\infty < x < 0 \\ 1 + \cos x, & 0 \le x \le \frac{\pi}{2} \\ 1 + \left(x - \frac{\pi}{2}\right), & \frac{\pi}{2} \le x < \infty \end{cases}$

(16) Test the following functions for continuity:– (a) $f(x) = x\tan\left(\frac{1}{x}\right)$, $x \ne 0$. $f(0) = 0$ at $x = 0$

(b) $f(x) = \dfrac{e^{-\frac{1}{x}}}{1 + e^{\frac{1}{x}}}, x \neq 0.\ f(0) = 0$ at $x = 0$ (c) $f(x) = \dfrac{|x|}{x}, x \neq 0. f(0) = 0$ at $x = 0$

(17) (a) If $f(x) = \begin{cases} \dfrac{\cot x + \tan x}{\sec x}, & -\dfrac{\pi}{4} < x \leq \dfrac{\pi}{4} \\ \operatorname{cosec}x, & \dfrac{\pi}{4} < x \leq \dfrac{\pi}{2} \\ \sqrt{2}\sin x + \cos x, & \dfrac{\pi}{2} < x < \pi \end{cases}$ then the function f(x) is continuous at $x = \dfrac{\pi}{4}$ and $x = \dfrac{\pi}{2}$.

(b) If $f(x) = \begin{cases} e^{|x|}, & x \neq 0 \\ 1, & x = 0 \end{cases}$ then the function f(x) is continuous at $x = 0$ and not differentiable at $x = 0$.

(c) If $f(x) = \begin{cases} e^{-\frac{1}{|x|}}, & x \neq 0 \\ 0, & x = 0 \end{cases}$ then the function f(x) is discontinuity at $x = 0$ and not differentiable at $x = 0$

(18) (a) If $f(x) = \begin{cases} \dfrac{2^x}{1 + 2^{-x}}, & -1 < x \leq 1 \\ \dfrac{x + 3}{x + 2}, & x > 1 \end{cases}$ then the function f(x) is continuous at $x = 1$.

(b) If $f(x) = \begin{cases} \log(\sin x + \cos x), & 0 < x \leq \dfrac{\pi}{4} \\ \dfrac{a}{2}, & x > \dfrac{\pi}{4} \end{cases}$ then the function f(x) is continuous at $x = \dfrac{\pi}{4}$ and prove that $a = 2 + \log 2$.

(c) If $f(x) = \begin{cases} \dfrac{e^{|x+1|}}{1 + e^{x+1}}, & x \neq -1 \\ \dfrac{1}{2}, & x = -1 \end{cases}$ then the function f(x) is discontinuous at $x = 0$ and not differentiable at $x = 0$.

(19) (a) If $f(x) = \begin{cases} \dfrac{\sin x}{|x|}, & x \neq 0 \\ 1, & x = 0 \end{cases}$ then the function f(x) is discontinuous at $x = 0$ and not differentiable at $x = 0$.

(c) Discuss the continuity and differentiability at $x = \pi$ if $f(x) = \begin{cases} \dfrac{\sqrt{1 + x - \pi} - 1}{|x - \pi|}, & x \neq \pi \\ 0, & x = \pi \end{cases}$

(20) (a) Discuss the continuity and differentiability of the following function $f(x) = \begin{cases} \sqrt{1 + \sin x}, & -\dfrac{\pi}{2} < x \leq 0 \\ \sqrt{1 + \cos 2x}, & 0 < x \leq \dfrac{\pi}{4} \\ \dfrac{\sec x + \operatorname{cosec}x}{2}, & \dfrac{\pi}{4} < x < \dfrac{\pi}{2} \end{cases}$

(b) If $f(x) = \begin{cases} \dfrac{x^2 + 3x + 2}{x + 1}, & -2 < x \leq -1 \\ ae^{-x}, & -1 < x < 0 \end{cases}$ then the function f(x) is continuous at $x = -1$ and its value of a.

(c) If $f(x) = \begin{cases} \dfrac{e^{\sin x} - 1}{\sin x}, & -1 < x \leq 0 \\ x + 1, & 0 < x < 1 \end{cases}$ Discuss the function f(x) is continuity and differentiability at $x = 0$.

(21) (a) If $f(x) = \begin{cases} \sin x, & x < \dfrac{\pi}{2} \\ ax - b, & x \geq \dfrac{\pi}{2} \end{cases}$ then the function f(x) is continuous and differentiable at $x = \dfrac{\pi}{2}$. find the value of a and b.

(b) Discuss the continuity and differentiability at $x = 0$, if $f(x) = \begin{cases} 2\cos x - 1, & x < 0 \\ \dfrac{x}{\sin x \cos x}, & x \geq 0 \end{cases}$

(c) Discuss the continuity and differentiability at $x = -2$ and $x = -3$, if $f(x) = |x^2 + 5x + 6|$

(22) (a) If $f(x) = \begin{cases} \dfrac{(3^{x-1} - 5^{x-1}).\log x.\cos x}{(x-1)^2} & ,x \neq 1 \\ ax + \log 3 \ , & x = 1 \end{cases}$ then the function f(x) is continuous at x = 1 and prove that $a = \log\left(\frac{1}{5}\right)$.

(b) Discuss the continuity and differentiability at $x = \frac{\pi}{8}$, if $f(x) = \begin{cases} \dfrac{e^{\tan 2x - 1}}{\tan 2x + \cot 2x}, & x \neq \frac{\pi}{8} \\ \frac{1}{2}, & x = \frac{\pi}{8} \end{cases}$

(c) Discuss the continuity and differentiability at $x = \frac{\pi}{2}$, if $f(x) = \begin{cases} \dfrac{(\cos x)^{\sin x} - 1}{x}, & x \leq \frac{\pi}{2} \\ -\frac{1}{x}, & x > \frac{\pi}{2} \end{cases}$

(23) Discuss the continuity and differentiability at $x = \frac{\pi}{2}$

(a) $f(x) = \begin{cases} \dfrac{\sqrt[3]{1 + x - \frac{\pi}{2}} - 1}{x - \frac{\pi}{2}} & ,x \neq \frac{\pi}{2} \\ \frac{1}{3}, & x = \frac{\pi}{2} \end{cases}$ (b) $f(x) = \begin{cases} \dfrac{\cos^2 x}{1 - \sin x}, & x \neq \frac{\pi}{2} \\ 2, & x = \frac{\pi}{2} \end{cases}$ (c) $f(x) = \begin{cases} \dfrac{\sin\left(x - \frac{\pi}{2}\right)}{\left|x - \frac{\pi}{2}\right|}, & x \neq \frac{\pi}{2} \\ 1, & x = \frac{\pi}{2} \end{cases}$

Maxima and Minima

Rule: – It is clear that at P the function $y = f(x)$ is maximum and at Q it is minimum. At these points tangent is parallel to X – axis so that its slope is zero. (from the given below figure).

Draw graph

Criteria for maxima and minima: – Let $x = a, b, c$ be the values of x given by $\frac{dy}{dx} = 0$. consider the point $x = a$ i.e P where y is maximum. it is clear that tangent at any point

$x = a - h$ will make an acute angle with X – axis as at L, and tangent at $x = a + h$ will make an obtuse angle as at M. (from the above figure)

Thus, for $x = a$ $\therefore \frac{dy}{dx} = 0$ for x slightly $< a$, $\frac{dy}{dx} = +ve$ for x slightly $> a$, $\frac{dy}{dx} = -ve$

Hence if y is maximum at $x = a$ then $\frac{dy}{dx}$ changes sign from + ve to – ve for values of $x < a$ and $x > a$ in that order, Now consider the point $x = b$ i.e Q where y is minimum. it is clear that tangent at any point $x = b - h$ will make an obtuse angle with X – axis as at M and tangent at $x = b + h$ will make an acute angle as at N.

Thus for $x = b$ $\therefore \frac{dy}{dx} = 0$, for x slightly $< b$ $\frac{dy}{dx} = -ve$ (obtuse) for x slightly $> b$, $\frac{dy}{dx} = +ve$ (acute)

Hence if y is minimum at $x = b$ then $\frac{dy}{dx}$ changes sign from – ve to + ve for value of $x < b$ and $x > b$ in that order.

Working Rule: – Calculate $\frac{dy}{dx} = 0$ and solve for x and say $x = a, b, c$ etc.

put values of $x < a$ in $\frac{dy}{dx}$ and values of $x > a$ in $\frac{dy}{dx}$.

If $\frac{dy}{dx}$ changes sign from + ve to – ve, then maximum at $x = a$.

If $\frac{dy}{dx}$ changes sign from – ve to + ve, then minimum at $x = a$.

In case there is no change of sign, then neither a maximuma nor a minimum.

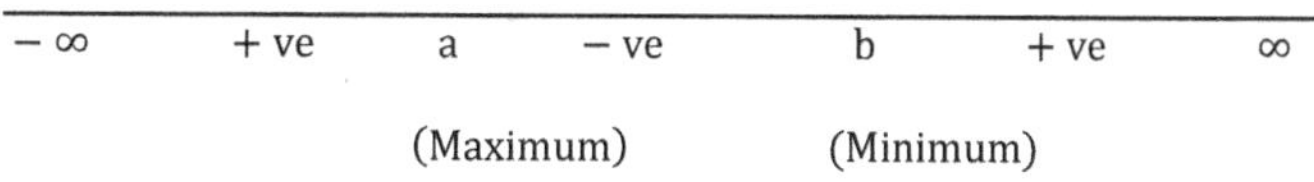

(Maximum) (Minimum)

- – ve sign to + ve sign at $x = b$ that is minimum.
- + ve sign to – ve sign at $x = a$ that is maximum.

Example: – Let $y = x^3 - 3x^2$ $\therefore \frac{dy}{dx} = 3x^2 - 6x = 3x(x - 2)$

The critical points are 0,2 $\therefore \frac{dy}{dx} = 0$ or $3x(x - 2) = 0$ then $x = 0,2$

– ∞ + ve 0 – ve 2 + ve ∞

At $x = 0$ there will be a change of sign from + ve to − ve and hence there will be Maximum at $x = 0$.

At $x = 2$ there will be a change of sign from − ve to + ve and hence there will be Minimum at $x = 2$.

Second Rule: − Calculate $\frac{dy}{dx} = 0$ and solve for x. suppose one root of $\frac{dy}{dx} = 0$ is at $x = a$.

If $\frac{d^2y}{dx^2} = -$ve for $x = a$ then Maximum at $x = a$.

If $\frac{d^2y}{dx^2} = +$ve for $x = a$ then Minimum at $x = a$.

If $\frac{d^2y}{dx^2} = 0$ at $x = a$ then find $\frac{d^3y}{dx^3}$. if $\frac{d^3y}{dx^3} \neq 0$ at $x = a$

then neither Max. nor Min. at $x = a$ if $\frac{d^3y}{dx^3} = 0$ at $x = a$ then find $\frac{d^4y}{dx^4}$. if $\frac{d^4y}{dx^4} > 0$ *i.e + ve at x = a then y is Min. at x = a* .

and if $\frac{d^4y}{dx^4} < 0$ *i.e − ve at x = a then y is Max. at x = a and so on.*

Parametric Form of a Function: − Let a function $y = f(x)$ be represented in parametric form by the equations $x = \theta(t)$, $y = \varphi(t)$

where $\theta(t)$ and $\varphi(t)$ have derivatives both of first and second orders within a certain interval of t.

Let at $t = t_0$, $\varphi'(t) = 0$, then

(a) if $\varphi''(t_0) < 0$, $f(x)$ has a Max. at $x = x_0$.

(b) if $\varphi''(t_0) > 0$, $f(x)$ has a Min. at $x = x_0$.

(c) if $\varphi''(t_0) = 0$, $f(x)$ has an extreme value at $x = x_0$.

Important Points: − (1) According to the given conditions of the problem determine the function whose Maximum and Minimum value are to be found.

(2) The above function is not of single variable but contains move than a function of the form: −

$$kf(x), k + f(x), [f(x)]^k \text{ or } [f(x)]^{\frac{1}{k}}$$

where k is a + ve constant if $f(x) > 0$ *then the function f* (x) is Maximum or Minimum.

If y is Maximum and Minimum then $\log y = z$ is also Maximum and Minimum at $y > 0$.

Also $y = f(x)$ is Max. or Min. then $z = \frac{1}{f(x)}$ is also Min. or Max.

(3) Maximum and Minimum occur alternately it may be noted that Maximum may be less than the Minimum.

Remember Geometrical Formula: −

Area of Squar $= x^2$ and Perimeter $= 4x$

Area of rectangle $= xy$, Perimeter $= 2(x + y)$

Area of equilateral $= \frac{\sqrt{3}}{4}x^2$, Perimeter $= 3x$

Area of trapezium $= \frac{1}{2}$(sum of parallel side) × distance between them.

Area of circle $= \pi r^2$, Perimeter $= 2\pi r$

Volume of a sphere $= \frac{4}{3}\pi r^3$ and surface $= 4\pi r^2$

Volume of right cone $= \frac{1}{3}\pi r^2 h$, Total surface $= \pi r(r + l)$, curved surface $= \pi rl$

Volume of a cylinder $= \pi r^2 h$, Total surface $= 2\pi r(r + h)$, curved surface $= 2\pi rh$

Volume of a cuboid $= xyz$ and surface $= 2(xy + yz + zx)$

Finding the Greatest and the Least value of a function: –

The greatest (least) value of a continuous function f(x) on an interval [a, b] is attained either at the critical points or at the end points of the interval. To find the greatest (least) value of the function we have comput its values at all the critical points on the interval [a, b] the values of f(a), f(b) of the function at the end points of the interval If a function f(x) is defined and continuous in same interval. if the interval is not closed then the function may have neither the greatest nor the least value.

Example: – (1) Find the greatest and least values of the following function on the given intervals: –

(a) $f(x) = 3x^3 - 6x^2 - 12x + 6$ on $[0,1]$ (b) $f(x) = x\log x$ on $[1, e^2]$

Solution: – (a) $f(x) = 3x^3 - 6x^2 - 12x + 6$ on the points $[0,1]$

$f'(x) = 9x^2 - 12x - 12 \quad \therefore f'(x) = 0 \quad \Rightarrow 9x^2 - 12x - 12 = 0 \quad \Rightarrow 3x^2 - 4x - 4 = 0$

$\Rightarrow 3x^2 - 6x + 2x - 4 = 0 \quad \therefore 3x(x - 2) + 2(x - 2) = 0 \quad \therefore (3x + 2)(x - 2) = 0 \quad \therefore x = -\frac{2}{3}, 2$

critical points are $-\frac{2}{3}, 0, 1, 2$

$-\infty \quad -ve \quad -\frac{2}{3} \quad +ve \quad 0 \quad -ve \quad 1 \quad +ve \quad 2 \quad -ve \quad \infty$

At $x = -\frac{2}{3}$ then f(x) is Minimum. $\therefore f(x) = 3x^3 - 6x^2 - 12x + 6$

$\therefore f\left(-\frac{2}{3}\right) = 3 \times \left(-\frac{2}{3}\right)^3 - 6 \times \left(-\frac{2}{3}\right)^2 - 12 \times \left(-\frac{2}{3}\right) + 6 = \frac{94}{9}$

At $x = 0$, then f(x) is Maximum. $\therefore f(x) = 3x^3 - 6x^2 - 12x + 6 \quad \therefore f(0) = 6$

At $x = 1$, then f(x) is Minimum. $\therefore f(x) = 3x^3 - 6x^2 - 12x + 6 \quad \therefore f(1) = 3 - 6 - 12 + 6 = -9$

At $x = 2$, then f(x) is Maximum.

$\therefore f(x) = 3x^3 - 6x^2 - 12x + 6 \quad \therefore f(2) = 3(2)^3 - 6(2)^2 - 12(2) + 6 = 24 - 24 - 24 + 6 = -18$

Hence the greatest value is $f\left(-\frac{2}{3}\right) = \frac{94}{9}$ and the least value is $f(2) = -18$ Ans.

(b) $f(x) = x\log x$ on the points $[1, e^2] \quad \therefore f'(x) = x.\frac{1}{x} + \log x.1 = 1 + \log x \quad \therefore f'(x) = 0$ or $1 + \log x = 0$

or $\log x = -1 = \log e^{-1} \quad \Rightarrow \log x = \log e^{-1} \quad \therefore x = e^{-1} = \frac{1}{e}$ critical points are $\frac{1}{e}, 1, e^2$

$-\infty \quad -ve \quad \frac{1}{e} \quad +ve \quad 1 \quad -ve \quad e^2 \quad +ve \quad \infty$

At $x = \frac{1}{e}$ then f(x) is Minimum. $\therefore f(x) = x\log x \quad \Rightarrow f\left(\frac{1}{e}\right) = \frac{1}{e}\log_e e^{-1} = -\frac{1}{e}$

At $x = 1$ then f(x) is Maximum. $\therefore\ f(x) = x\log x \Rightarrow f(1) = 1.\log 1 = 0$

At $x = e^2$ then f(x) is Minimum. $\therefore\ f(x) = x\log x \Rightarrow f(e^2) = e^2\log e^2 = 2e^2$

Hence, the greatest value is $f(e^2) = 2e^2$ and the least value is $f(1) = 0$. Ans.

Solved Example

(1) Find the Max. and Min. value of the following: – (a) Find the Maximum and Minimum value of $2x^3 - 24x + 54$ in the interval $(-3,3)$.

(b) Find the Maximum and Minimum value of the function $f(x) = \cos x\,(1 + \sin x)$ in the interval $\left[-\frac{\pi}{2}, \frac{\pi}{2}\right]$.

Solution: – (a) Let $f(x) = 2x^3 - 24x + 54$, $f'(x) = 6x^2 - 24$

For Maximum or Minimum , $f'(x) = 0$ i.e $3x^2 - 24 = 0$ or $x^2 = 4$ or $x = \pm 2$

Now , $f''(x) = 12x \Rightarrow f''(2) = 24 > 0$ *it is minimum.* $\Rightarrow f''(-2) = -24 < 0$ *it is maximum.*

The function f(x) has a maximum at $x = -2$,

Required maximum value $= 2(-2)^3 - 24 \times (-2) + 54 = 86$ Ans.

The function f(x) has a minimum at $x = 2$, Required minimum value $= 2(2)^3 - 24 \times 2 + 54 = 22$ Ans.

(b) We have $f(x) = \cos x\,(1 + \sin x)$

$f'(x) = \cos x\,(0 + \cos x) + (1 + \sin x)(-\sin x) = \cos^2 x - \sin x - \sin^2 x = (1 - \sin^2 x) - \sin x - \sin^2 x = -2\sin^2 x - \sin x + 1$

For stationary points , $f'(x) = 0 \Rightarrow -2\sin^2 x - \sin x + 1 = 0 \Rightarrow 2\sin^2 x + \sin x - 1 = 0$

or $\sin x = \frac{-1 \pm \sqrt{1+8}}{2 \times 2} = \frac{-1 \pm 3}{4} = -1$ or $\frac{1}{2}$ $\left[\text{formula, } ax^2 + bx + c = 0 \text{ then } x = \frac{-b \pm \sqrt{b^2 - 4ac}}{2a}\right]$

$$\text{or } \sin x = -1, \frac{1}{2} \quad \text{or } x = -\frac{\pi}{2}, \frac{\pi}{6}$$

Now , $f\left(-\frac{\pi}{2}\right) = 0$ and $f\left(\frac{\pi}{6}\right) = \frac{3\sqrt{3}}{4} > 0$

f(x) has maximum value $= \frac{3\sqrt{3}}{4}$ at $x = \frac{\pi}{6}$ and minimum value $= 0$ at $x = -\frac{\pi}{2}$ and also minimum value $= 0$ at $x = \frac{\pi}{2}$ Ans.

(2) Find the Maxima or Minima of the following function.

(a) $x^3 - 6x^2 + 9x + 7$ (b) $2x^3 - 54x + 108$ (c) $x^3 + 4x^2 - 3x + 2$ (d) $x^4 - 62x^2 + 120x + 9$

(e) $x^4 - 2x^2$ (f) $x^3 - 12x$ (g) $3x^2 - 4x + 5$ (h) $x^3 + 2x^2 - 4x + 1$

Solution: – (a) Let $f(x) = x^3 - 6x^2 + 9x + 7$, $f'(x) = 3x^2 - 12x + 9$

For stationary points, $f'(x) = 0 \therefore 3x^2 - 12x + 9 = 0 \Rightarrow x^2 - 4x + 3 = 0 \Rightarrow x^2 - 3x - x + 4 = 0$

$$\Rightarrow x(x-3) - 1(x-3) = 0 \Rightarrow (x-1)(x-3) = 0 \quad \therefore x = 1,3$$

$-\infty$	+ve	1	– ve	3	+ ve	∞

At $x = 1$ then $f(x) = x^3 - 6x^2 + 9x + 7$, $f(1) = 1 - 6 + 9 + 7 = 11 > 0$

and $f'(1) = 3x^2 - 12x + 9 = 3 - 12 + 9 = 0$ and $f''(1) = 6x - 12 = 6 - 12 = -6 < 0$

At $x = 3$ then $f(x) = x^3 - 6x^2 + 9x + 7$, $f(3) = 27 - 54 + 27 + 7 = 7 > 0$, $f'(3) = 27 - 36 + 9 = 0$

and $f''(3) = 6x - 12 = 18 - 12 = 6 > 0$ $f(x)$ is maximum at $x = 1$ and minimum at $x = 3$. Ans.

(b) Ans: −Hint − Let $f(x) = 2x^3 - 54x + 108$, $f'(x) = 6x^2 - 54$ (solve same as above question)

stationary points are ± 3 , f(x) is minimum at $x = 3$ and maximum at $x = -3$.

(c) Let $f(x) = x^3 + 4x^2 - 3x + 2$, $f'(x) = 3x^2 + 8x - 3$ $\therefore$ $f'(x) = 0$

For stationary points, $f'(x) = 0$ or $3x^2 + 8x - 3 = 0$ or $3x^2 + 9x - x - 3 = 0$ or $3x(x + 3) - 1(x + 3) = 0$

or $(3x - 1)(x + 3) = 0$ $\therefore x = -3, \frac{1}{3}$ or f(x) is maximum at $x = -3$ and minimum at $x = -\frac{1}{3}$. Ans.

(d) Let $f(x) = x^4 - 62x^2 + 120x + 9$, $f'(x) = 4x^3 - 124x + 120$ $\therefore f'(x) = 0$

For stationary points, $f'(x) = 0$ or $4x^3 - 124x + 120 = 0$ or $x^3 - 31x + 30 = 0$

or $x^2(x - 1) + x(x - 1) - 30(x - 1) = 0$ or $(x - 1)(x^2 + x - 30) = 0$ or $(x - 1)(x^2 + 6x - 5x - 30) = 0$

or $(x - 1)[x(x + 6) - 5(x + 6)] = 0$ or $(x - 1)(x - 5)(x + 6) = 0$ $\therefore$ $x = -6,1,5$

At $x = -6$ then $f''(x) = 12x^2 - 124$ or $f''(-6) = 12(-6)^2 - 124 = 432 - 124 = 108 > 0$ (+ve) Min.

At $x = 1$ then $f''(x) = 12x^2 - 124$ or $f''(1) = 12 \times 1 - 124 = 12 - 124 = -112 < 0$ (−ve) Max.

At $x = 5$ then $f''(x) = 12x^2 - 124$ or $f''(5) = 12(5)^2 - 124 = 300 - 124 = 176 > 0$ (+ve) Min.

f(x) is maximum at $x = 1$ and minimum at $x = -6,5$. Ans.

(e) f(x) is maximum at $x = 0$ and minimum at $x = \pm 1$. Ans. (Do yourself, same as above question)

(f) f(x) is maximum at $x = -2$ and minimum at $x = 2$. Ans. (Do yourself, same as above question)

(g) f(x) is minimum at $x = \frac{2}{3}$. Ans. (h) f(x) is maximum at $x = -2$ and minimum at $x = \frac{2}{3}$. Ans.

(3) Prove that the function $f(x) = 3x + 5$ is Monotonic for all value of $x \in R$.

Solution: − Consider two values of x (say) $x_1 , x_2 \in R$ such that $x_2 > x_1$ ……………………..(i)

Multiplying both sides of (i) by 3 , we have $3x_2 > 3x_1$ ……………………..(ii)

Adding 5 to both sides of (ii) , we get $3x_2 + 5 > 3x_1 + 5$ we have $f(x_2) > f(x_1)$

Thus , we find $f(x_2) > f(x_1)$ whenever $x_2 > x_1$

Hence the given function $f(x) = 3x + 5$ is monotonic function. Proved.

(4) Show that $f(x) = x^2$ is a strictly decreasing function for all $x < 0$.

Solution: − Consider x_1, x_2 two value of x such that $x_2 > x_1$, $x_1 x_2 < 0$ …………..(i)

it is multiplied by a negative number in equation (i) by x_2 , we have $x_2 . x_2 < x_1 . x_2$ or ${x_2}^2 < x_1 . x_2$ ……………………………(ii)

Now , Multiplying (i) by x_1 , we have $x_1 . x_2 < x_1 . x_1$ or $x_1 . x_2 < {x_1}^2$ ………………………….(iii)

from (ii) and (iii) , we have ${x_2}^2 < x_1 . x_2 < {x_1}^2$ or ${x_2}^2 < {x_1}^2$ or $f(x_2) < f(x_1)$ …………….(iv)

Thus , from (i) and (iv) , we have $x_2 > x_1$, $f(x_2) < f(x_1)$

Hence , the given function is strictly decreasing for all $x < 0$. *Proved.*

(5) Find the values of x , the function $f(x) = x^2 - 2x + 3$ is increasing and decreasing.

Solution: − $f(x) = x^2 - 2x + 3$, $f'(x) = 2x - 2$

For f(x) to be increasing $f'(x) > 0$ *i.e* $2x - 2 > 0$ *or* $2(x-1) > 0$ *or* $x - 1 > 0$ *or* $x > 1$, *The function increases for* $x > 1$.

For f(x) to be decreasing $f'(x) < 0$ *i.e* $2x - 2 < 0$ *or* $2(x-1) < 0$ *or* $x - 1 < 0$ *or* $x < 1$, *Thus, the function decreases for* $x < 1$.

(6) Find the interval in which $f(x) = x^3 + 3x^2 - 9x + 5$ is increasing or decreasing.

Solution: $-$ $f(x) = x^3 + 3x^2 - 9x + 5$, $f'(x) = 3x^2 + 6x - 9 = 3(x^2 + 2x - 3) = 3(x-1)(x+3)$

The critical points are $f'(x) = 0$ or $3(x-1)(x+3) = 0$ $\therefore$ $x = -3, 1$

	increasing		decreasing		increasing	
$-\infty$	+ ve	-3	$-$ ve	1	+ ve	∞

For f(x) to be increasing $f'(x) > 0$ *and* $f(x)$ to be decreasing $f'(x) < 0$

The function f(x) is increasing for $x > 1$ *or* $x < -3$ *and* $f(x)$ is decreasing for $-3 < x < 1$. *Ans.*

(7) Determine the intervals for which the function $f(x) = \dfrac{x}{x^2+3}$ is increasing or decreasing.

Solution: $-$ $f(x) = \dfrac{x}{x^2+3}$, $f'(x) = \dfrac{(x^2+3).1 - x.2x}{(x^2+3)^2} = \dfrac{x^2 + 3 - 2x^2}{(x^2+3)^2} = \dfrac{3 - x^2}{(x^2+3)^2}$

$$\left[\text{formula,} \quad y = \frac{f(x)}{g(x)}, \quad y' \text{ or } \frac{dy}{dx} = \frac{g(x).f'(x) - f(x).g'(x)}{[g(x)]^2}\right]$$

As $(x^2+3)^2$ is positive for all real value of x.

The stationary point are $f'(x) = 0$ or $\dfrac{3 - x^2}{(x^2+3)^2} = 0$ or $3 - x^2 = 0$ or $x^2 = 3$ $\therefore$ $x = \pm\sqrt{3}$

	decreasing		increasing		decreasing	
$-\infty$	$-$ ve	$-\sqrt{3}$	+ ve	$\sqrt{3}$	$-$ ve	∞

The function f(x) is increasing $f'(x) > 0$ *for* $-\sqrt{3} < x < \sqrt{3}$ and f(x) is decreasing $f'(x) < 0$ *for* $x > \sqrt{3}$ or $x < -\sqrt{3}$ Ans.

(8) Find the Maximum (Local Maximum) and Minimum (Local Minimum) Points of the function $f(x) = 2x^3 - 6x^2 - 18x + 3$

Solution: $-$ Here, $f(x) = 2x^3 - 6x^2 - 18x + 3$, $f'(x) = 6x^2 - 12x - 18 = 6(x^2 - 2x - 3)$

or $f'(x) = 0$, gives us $6(x^2 - 2x - 3) = 0$ $\therefore$ $x^2 - 2x - 3 = 0$ or $(x+1)(x-3) = 0$ $\therefore$ $x = 3, -1$

Critical points are $x = 3$ and $x = -1$

		Maximum		Minimum		
$-\infty$	+ ve	-1	$-$ ve	3	+ ve	∞

At $x > 3$ *then* $f'(x) = +ve > 0$ *Value + ve to − ve then the function is maximum at* $x = -1$

At $x < 3$ *then* $f'(x) = -ve < 0$ *Value − ve to + ve then the function is minimum at* $x = 3$

At $x < -1$ *then* $f'(x) = +ve > 0$

At $x > -1$ *then* $f'(x) = -ve < 0$

The function f(x)has minimum value at $x = 3$, minimum value $= f(3) = 2(3)^3 - 6(3)^2 - 18(3) + 3 = -51$

The function f(x) has maximum value at $x = -1$, maximum value $= f(-1) = 2(-1)^3 - 6(-1)^2 - 18(-1) + 3 = 13$

or $(-1, 13)$ and $(3, -51)$ are points of local maxima and local minima respectively. Ans.

(9) Find the Local Maximum and the Local Minimum of the function $f(x) = x^2 - 6x$.

Solution: $-$ $f(x) = x^2 - 6x$, $f'(x) = 2x - 6$ putting $f'(x) = 0$ or $2x - 6 = 0$ $\therefore$ $x = 3$

we have to examine whether $x = 3$ is the point of local maximum or local minimum or neither maximum nor minimum.

Let us take $x = 2.9$ which is to the left of 3 and $x = 3.1$ which is to the right of 3 and find $f'(x)$at these points.

$$f'(2.9) = 2 \times 2.9 - 6 < 0 \ \textit{and}\ f'(3.1) = 2 \times 3.1 - 6 > 0$$

Since $f'(x) < 0$ *as we approach* 3 *from the left and* $f'(x) > 0$ *as we approach* 3 *from the right.*

$\therefore$ there is a local minimum at $x = 3$

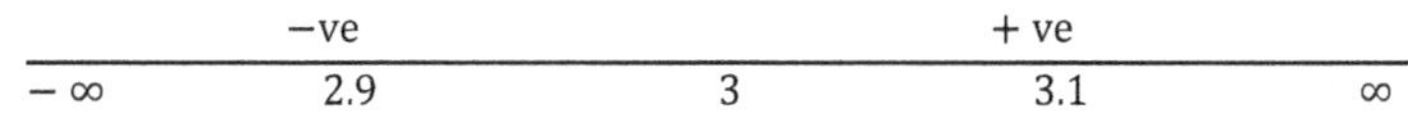

Sign of $f'(x)$ and Point $x = 3$

Left of 3	Right of 3
$f'(x) < 0$	$f'(x) > 0$

Local Minimum at $x = 3$ Ans.

(10) Find all Local Maxima and Local Minima of the function $f(x) = x^3 + 3x^2 - 9x + 7$.

Solution: − $f(x) = x^3 + 3x^2 - 9x + 7$, $f'(x) = 3x^2 + 6x - 9$

Putting $f'(x) = 0$, $3x^2 + 6x - 9 = 0$ or $x^2 + 2x - 3 = 0$ or $(x - 1)(x + 3) = 0$ $\therefore x = 1, -3$

The critical points are $x = 1$ and $x = -3$

we examine whether these points are points of local maximum or local minimum or neither of them.

solve, same as above question.

Ans: − $f'(0.9) < 0, f'(1.1) > 0$ *and* $f'(-2.9) < 0, f'(-3.1) > 0$

Then the function is a local minimum at $x = 1$ and local maximum at $x = -3$.

	+ve	$x = -3$	− ve	− ve	$x = 1$	+ ve	
$-\infty$	− 3.1	− 3	− 2.9	0.9	1	1.1	∞
		Local Maximum			Local Minimum		

(11) Find the Local Maximum and Local Minimum, if any of the following function $f(x) = \dfrac{x+1}{x^2+3}$.

Solution: − $f(x) = \dfrac{x+1}{x^2+3}$, $f'(x) = \dfrac{(x^2+3).1 - (x+1).2x}{(x^2+3)^2} = \dfrac{x^2+3-2x^2-2x}{(x^2+3)^2} = \dfrac{3-2x-x^2}{(x^2+3)^2}$

As $(x^2 + 3)^2$ is Positive for all real value of x.

For stationary Points are $f'(x) = 0$ or $\dfrac{3-2x-x^2}{(x^2+3)^2} = 0$ or $3 - 2x - x^2 = 0$

or $3(1 - x) + x(1 - x) = 0$ or $(1 - x)(x + 3) = 0$ $\therefore$ critical points are $x = 1, x = -3$

At $x > 1$ *then* $f'(x) = -ve, f'(x) < 0$

		Local Minimum		Local Maximum		
$-\infty$	− ve	− 3	+ ve	1	− ve	∞

At $x < -3$ *then* $f'(x) = -ve, f'(x) < 0$

At $x < 1$ *then* $f'(x) = +ve, f'(x) > 0$

At $x > -3$ *then* $f'(x) = +ve, f'(x) > 0$

The function f(x) is Local Maximum at $x = 1$, Local Maximum value $= \dfrac{1+1}{1+3} = \dfrac{2}{4} = \dfrac{1}{2}$ Ans.

and function f(x) is Local Minimum at $x = -3$, Local Minimum value $= \frac{-3+1}{9+3} = \frac{-2}{12} = -\frac{1}{6}$ Ans.

(12) Find the Local Maximum and Local Minimum, if any for the function $f(x) = \frac{x}{2} - \sin x$, $0 \le x \le \frac{\pi}{2}$.

Solution: – We have $f(x) = \frac{x}{2} - \sin x$, $f'(x) = \frac{1}{2} - \cos x$

For Local Maxima or Minima $f'(x) = 0$ or $\frac{1}{2} - \cos x = 0$ or $\cos x = \frac{1}{2}$ $\therefore x = \frac{\pi}{3}$ in $0 \le x \le \frac{\pi}{2}$

At $x = \frac{\pi}{3}$, for $x > \frac{\pi}{3}$ then $f'(x) = \frac{1}{2} > 0$, $\frac{1}{2} - \cos x > 0$

For $x < \frac{\pi}{3}$, $f'(x) < 0$ *or* $\frac{1}{2} - \cos x < 0$

$-\infty$	$-ve$	$\frac{\pi}{3}$	$+ve$	∞

f′(x) changes sign from negative to positive in the neithbourhood of $\frac{\pi}{3}$.

At $x = \frac{\pi}{3}$ is a point of Local Minima, then the Minimum value $= f\left(\frac{\pi}{3}\right) = \frac{\pi}{6} - \sin\left(\frac{\pi}{3}\right) = \frac{\pi}{6} - \frac{\sqrt{3}}{2} = \frac{\pi - 3\sqrt{3}}{6}$

$\therefore$ Point of Local Minima is $\left(\frac{\pi}{3}, \frac{\pi - 3\sqrt{3}}{6}\right)$ Ans.

(13) Find the Local Minimum or Maximum of the following function $f(x) = 2x^3 - 9x^2 + 12x - 8$.

Solution: – We have $f(x) = 2x^3 - 9x^2 + 12x - 8$

then $f'(x) = 6x^2 - 18x + 12 = 6(x^2 - 3x + 2) = 6(x-1)(x-2)$

For Local Maximum or Minimum $f'(x) = 0$ $\therefore 6(x-1)(x-2) = 0$ or $x = 1,2$

Now, $f''(x) = \frac{d[f'(x)]}{dx} = \frac{d[6(x^2 - 3x + 2)]}{dx} = 12x - 18 = 6(2x - 3)$

For $x = 1$, $f''(1) = 6(2-3) = -6 < 0$ *then* $f(x)$ is a Local Maximum at $x = 1$.

and $f(1) = 2(1)^3 - 9(1)^2 + 12(1) - 8 = 2 - 9 + 12 - 8 = -3$ is a Local Maximum value.

For $x = 2$, $f''(2) = 6(4-3) = 6 > 0$ *then* $f(x)$ is a Local Minimum at $x = 2$.

and $f(2) = 2(2)^3 - 9(2)^2 + 12(2) - 8 = 16 - 36 + 24 - 8 = -4$ is a Local Minimum value.

$\therefore$ $(1, -3)$ and $(2, -4)$ are Points of Local Maxima and Local Minima respectively.

		Local Maximum		Local Minimum		
$-\infty$	$+ve$	1	$-ve$	2	$+ve$	∞

(14) Find Local Maxima and Minima (if any) for the function $f(x) = \sin 3x$, $0 < x < \frac{\pi}{2}$.

Solution: – $f(x) = \sin 3x$ then $f'(x) = 3\cos 3x$ $\therefore$ putting $f'(x) = 0$ $\therefore 3\cos 3x = 0$ or $\cos 3x = 0$

or $3x = \frac{\pi}{2}, \frac{3\pi}{2}, \frac{5\pi}{2}$ or $x = \frac{\pi}{6}, \frac{\pi}{2}, \frac{5\pi}{6}$ $\therefore$ only $x = \frac{\pi}{6}$ lie in the interval $0 < x < \frac{\pi}{2}$

Now, $f''(x) = -9\sin 3x$

At $x = \frac{\pi}{6}$, $f''(x) = -9\sin 3x$ then $f''\left(\frac{\pi}{6}\right) = -9\sin\left(\frac{3\pi}{6}\right) = -9\sin\frac{\pi}{2} = -9 < 0$

The function f(x) is Maximum at $x = \frac{\pi}{6}$ and Maximum value $= f\left(\frac{\pi}{6}\right) = \sin\left(\frac{3\pi}{6}\right) = \sin\left(\frac{\pi}{2}\right) = 1$ Ans.

(15) Find the Minimum and Maximum value of the function $f(x) = 4x^3 - 48x + 105$ in the interval [1,3]and [−3,0] respectively.

Solution: − $f(x) = 4x^3 - 48x + 105$ then $f'(x) = 12x^2 - 48$

For Local Maximum or Minimum $f'(x) = 0$ i.e $12x^2 - 48 = 0$ or $x^2 = 4$ $\therefore$ $x = \pm 2$

f(x) is Minimum in the interval [1,3] only $x = 2$ belong to the interval [1,3]. find Minimum if any at $x = 2$ only.

Now, $f''(x) = 24x$ $\therefore$ $f''(2) = 24 \times 2 = 48 > 0$ (+ve) which implies the function

f(x) has a Minimum at $x = 2$, $\therefore$ Required Minimum value $= 4(2)^3 - 48(2) + 105 = 32 - 96 + 105 = 41$

Thus the Point of Minimum belonging to the given interval [1,3] is 2 and the Minimum value of the function is 41.

Now, $f''(x) = 24x, f''(-2) = 24(-2) = -48 < 0$ (−ve) $[-2$ lies in $[-3,0]]$

which implies the function f(x) shall have a Maximum at $x = -2$.

Required Maximum value $= 4(-2)^3 - 48(-2) + 105 = -32 + 96 + 105 = 169$ Ans.

(16) Find the Maximum and Minimum value of the function $f(x) = \cos x\,(1 + \sin x)$ in $(0, \pi)$.

Solution: − We have $f(x) = \cos x\,(1 + \sin x)$ then $f'(x) = \cos x\,.(0 + \cos x) + (1 + \sin x).(-\sin x)$

$f'(x) = \cos^2 x - \sin x - \sin^2 x = -2\sin^2 x - \sin x + 1$

For stationary points $f'(x) = 0$ $\therefore$ $-2\sin^2 x - \sin x + 1 = 0$ $\therefore$ $\sin x = -1, \frac{1}{2}$ $\therefore x = -\frac{\pi}{2}, \frac{\pi}{6}, \frac{5\pi}{6}$

Now, $f\left(\frac{\pi}{6}\right) = \cos\frac{\pi}{6}\left(1 + \sin\frac{\pi}{6}\right) = \frac{\sqrt{3}}{2}\left(1 + \frac{1}{2}\right) = \frac{3\sqrt{3}}{4} > 0$

$f\left(\frac{5\pi}{6}\right) = \cos\frac{5\pi}{6}\left(1 + \sin\frac{5\pi}{6}\right) = -\frac{\sqrt{3}}{2}\left(1 + \frac{1}{2}\right) = -\frac{3\sqrt{3}}{4} < 0$

f(x) has Maximum value $\frac{3\sqrt{3}}{4}$ at $x = \frac{\pi}{6}$ and Minimum value $-\frac{3\sqrt{3}}{4}$ at $x = \frac{5\pi}{6}$ and also Minimum value 0 at $x = -\frac{\pi}{2}$ Ans.

(17) Find the two positive real numbers whose sum is 50 and their product is Maximum.

Solution: − Let one number be x and other number is $(50 - x)$ as two numbers are positive.

we have $x > 0, 50 - x > 0$ $\Rightarrow 50 - x > 0$ or $-x > -50$ or $x < 50$ $\therefore$ $0 < x < 50$

Let their product be $f(x) = x(50 - x) = 50x - x^2$ we have maximize the product f(x).

find f′(x) and put that equal to zero. $f'(x) = 50 - 2x$ then $f'(x) = 0$ or $50 - 2x = 0$ or $-2x = -50$ or $2x = 50$ $\therefore$ $x = 25$

Now, $f'(x) = -2$ which is negative. Hence f(x) is Maximum at $x = 35$

then the other number is $70 - x = 35$ or $-x = 35 - 70 = -35$ $\therefore x = 35$

Hence the required numbers are 35, 35 . Ans.

(18) Show that among rectangles of given area , the square has the least perimeter.

Solution: − Let x, y be the length and breadth of the rectangle respectively.

its area = length × breadth = x. y $\therefore$ Area (A) = xy $\therefore A = xy$ or $y = \frac{A}{x}$

Now, perimeter (P) of the rectangle $= 2(x + y)$ $\therefore P = 2(x + y) = 2\left(x + \frac{A}{x}\right)$

Differentiate with respect to x then $\frac{dP}{dx} = 2\left(1 - \frac{A}{x^2}\right)$

For Minimum $\frac{dP}{dx} = 0$ then $2\left(1 - \frac{A}{x^2}\right) = 0$ or $\frac{A}{x^2} = 1$ or $A = x^2$ $\therefore$ $x = \sqrt{A}$

Now, $\frac{d^2P}{dx^2} = 2\{0 - (-2)Ax^{-3}\} = \frac{4A}{x^3}$ which is positive.

Hence perimeter (P) is Minimum when $x = \sqrt{A}$ $\therefore y = \frac{A}{x} = \frac{x^2}{x} = x$ $\therefore y = x$ $(A = x^2)$

Thus the perimeter is Minimum when rectangle is a square. proved.

(19) An open box with a square base is to be mode out of a given quantity of area a. show that the Maximum volume of the box is $\frac{a\sqrt{a}}{6\sqrt{3}}$.

Solution: − Let x be the side of the square base of the box and y its height.

Total surface area of the box $= x^2 + 4xy$ $\therefore a = x^2 + 4xy$ $\therefore$ $y = \frac{a - x^2}{4x}$

Volume of the box (V) = base area × height $= x^2 \times y = x^2\left(\frac{a - x^2}{4x}\right)$ $\therefore$ $V = \frac{ax - x^3}{4}$ ……… (i)

Differentiating equation (i) with respect to x, we have $\therefore$ $\frac{dV}{dx} = \frac{a - 3x^2}{4}$

For Maxima or Minima $\frac{dV}{dx} = 0$ then $\frac{a - 3x^2}{4} = 0$ or $a = 3x^2$ or $x^2 = \frac{a}{3}$ $\therefore$ $x = \pm\sqrt{\frac{a}{3}}$ ……… (ii)

From (i) and (ii), we have $V = \frac{1}{4}(ax - x^3) = \frac{1}{4}\left[a.\sqrt{\frac{a}{3}} - \left(\sqrt{\frac{a}{3}}\right)^3\right] = \frac{1}{4}\left[\frac{a\sqrt{a}}{\sqrt{3}} - \frac{a\sqrt{a}}{3\sqrt{3}}\right] = \frac{1}{4}\left[\frac{3a\sqrt{a} - a\sqrt{a}}{3\sqrt{3}}\right]$

$$\therefore\ V = \frac{1}{4}\left[\frac{2a\sqrt{a}}{3\sqrt{3}}\right] = \frac{a\sqrt{a}}{6\sqrt{3}}$$

Again, $\frac{d^2V}{dx^2} = \frac{d\left[\frac{a - 3x^2}{4}\right]}{dx} = \frac{1}{4}(0 - 6x) = -\frac{3x}{2}$, x being the length of the side is positive $\frac{d^2V}{dx^2} < 0$

$\therefore$ The Volume is Maximum. hence maximum volume of the box $= \frac{a\sqrt{a}}{6\sqrt{3}}$ Ans.

Putting area (a) = Any real numbers Let Area (a) = 4 then maximum value of the box $= \frac{a\sqrt{a}}{6\sqrt{3}} = \frac{4\sqrt{4}}{6\sqrt{3}} = \frac{4}{3\sqrt{3}}$ Ans.

Now, put a = 3 then maximum value of the box $= \frac{a\sqrt{a}}{6\sqrt{3}} = \frac{3\sqrt{3}}{6\sqrt{3}} = \frac{1}{2}$ Ans.

(20) Show that of all rectangle inscribed in a given circle, the square has the maximum area.

Solution: − Let ABCD be a rectangle inscribed in a circle of radius r then diameter AC = 2r.

Let AB = x and BC = y then $AB^2 + BC^2 = AC^2$ or $x^2 + y^2 = (2r)^2$ or $x^2 + y^2 = 4r^2$ ……… (i)

Now, area A of the rectangle = xy

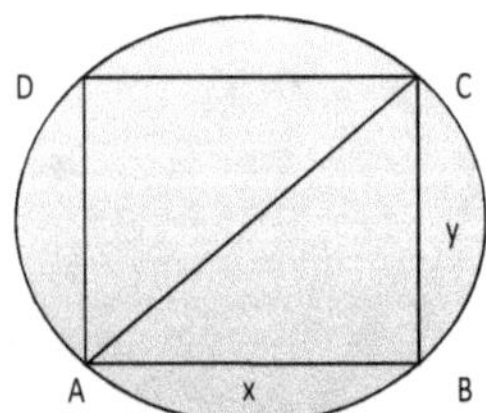

or $A = x\sqrt{4r^2 - x^2}$, $\dfrac{dA}{dx} = x.\dfrac{1}{2\sqrt{4r^2 - x^2}}.(-2x) + \sqrt{4r^2 - x^2}.1 = \dfrac{-x^2 + 4r^2 - x^2}{\sqrt{4r^2 - x^2}} = \dfrac{4r^2 - 2x^2}{\sqrt{4r^2 - x^2}}$

For Maxima or Minima, $\dfrac{dA}{dx} = 0$ or $\dfrac{4r^2 - 2x^2}{\sqrt{4r^2 - x^2}} = 0$ or $4r^2 - 2x^2 = 0$ or $x^2 = 2r^2$ $\therefore$ $x = \sqrt{2}\,r$

Now, $\dfrac{d^2A}{dx^2} = \dfrac{\sqrt{4r^2 - x^2}.(-4x) - (4r^2 - 2x^2).\dfrac{-2x}{2\sqrt{4r^2 - x^2}}}{\left(\sqrt{4r^2 - x^2}\right)^2} = \dfrac{(4r^2 - x^2)(-4x) + x(4r^2 - 2x^2)}{(4r^2 - x^2)^{\frac{3}{2}}} = \dfrac{-16xr^2 + 4x^3 + 4xr^2 - 2x^3}{(4r^2 - x^2)^{\frac{3}{2}}} = \dfrac{2x^3 - 12xr^2}{(4r^2 - x^2)^{\frac{3}{2}}}$

Putting $x = \sqrt{2}\,r$ then $\dfrac{d^2A}{dx^2} = \dfrac{4\sqrt{2}r^3 - 12\sqrt{2}r^3}{(4r^2 - 2r^2)^{\frac{3}{2}}} = \dfrac{-8\sqrt{2}\,r^3}{(2r^2)^{\frac{3}{2}}} = \dfrac{-8\sqrt{2}\,r^3}{2\sqrt{2}\,r^3} = -4 < 0$

Thus, A is Maximum when $x = \sqrt{2}\,r$.

Now from (i), $x^2 + y^2 = 4r^2$ or $y^2 = 4r^2 - x^2$ (putting $x = \sqrt{2}\,r$)

or $y^2 = 4r^2 - 2r^2 = 2r^2$ $\therefore$ $y = \sqrt{2}\,r = x$ or $x = y$ hence rectangle ABCD is a square. $\therefore$ $x = y = \sqrt{2}\,r$ Ans.

(21) Show that the height of a closed right circular cylinder of a given volume and least surface is equal to its diameter.

Solution: – Let v = volume, r = radius and h = height of the cylinder then $v = \pi r^2 h$ or $h = \dfrac{v}{\pi r^2}$ ……….(i)

Now, surface area (s) $= 2\pi rh + 2\pi r^2$ ……………… (ii)

Putting $h = \dfrac{v}{\pi r^2}$ in equation (ii), we have $s = 2\pi r.\dfrac{v}{\pi r^2} + 2\pi r^2 = \dfrac{2v}{r} + 2\pi r^2$

Now, $\dfrac{ds}{dr} = \dfrac{-2v}{r^2} + 4\pi r$

For Minimum surface area, $\dfrac{ds}{dr} = 0$ or $\dfrac{-2v}{r^2} + 4\pi r = 0$ or $-2v = -4\pi r^3$ $\therefore$ $v = 2\pi r^3$ ………….(iii)

Put value of v in equation (i), we get $h = \dfrac{v}{\pi r^2}$ or $h = \dfrac{2\pi r^3}{\pi r^2} = 2r$ or $h = 2r$ ……………………….(iv)

Aganin, $\dfrac{d^2s}{dr^2} = \dfrac{d}{dr}\left[\dfrac{-2v}{r^2} + 4\pi r\right] = \dfrac{4v}{r^3} + 4\pi$, put $v = 2\pi r^3$ then $\dfrac{d^2s}{dr^2} = \dfrac{4v}{r^3} + 4\pi = \dfrac{4.2\pi r^3}{r^3} + 4\pi = 12\pi > 0$

s is least at $h = 2r$ thus height of the cylinder = diameter of the cylinder. Ans.

(22) A square metal sheet of side 96 cm has four equal squares removed from the corners and the sides are then turned up so as to form an open box. Determine the size of the square cut so that volume of the box is maximum.

Solution: – Let the side of each of the small squares cut be x cm, so that each side of the box to be made is $(96 - 2x)$ cm and height x cm.

Now, $x > 0, 96 - 2x > 0$ *i.e* $x < 48$ $\therefore$ *x lies between* 0 *and* 48 *or* $0 < x < 48$.

Now, volume v of the box $= (96 - 2x)(96 - 2x)x$.

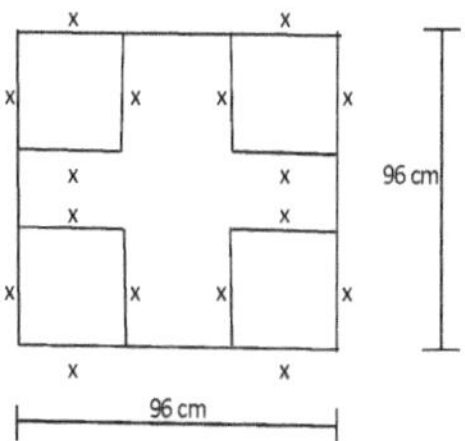

or $v = (96 - 2x)^2.x$ then $\dfrac{dv}{dx} = (96 - 2x)^2.1 + x.2(96 - 2x)(-2) = (96 - 2x)(96 - 2x - 4x)$ $\therefore$ $\dfrac{dv}{dx} = (96 - 2x)(96 - 6x)$

For Minimum or Maximum is $\frac{dv}{dx} = 0$ i.e $(96 - 2x)(96 - 6x) = 0$ $\therefore$ $96 - 2x = 0$ or $96 - 6x = 0$

or $x = 48$ and $x = 16$ $\therefore$ $0 < x < 48$ $\therefore$ *Rejecting* $x = 48$, *we have* $x = 16\ cm$

Now, $\frac{d^2v}{dx^2} = \frac{d}{dx}[(96 - 2x)(96 - 6x)] = (96 - 2x).(-6) + (96 - 6x)(-2) = 24x - 768$

$\left(\frac{d^2v}{dx^2}\right)_{x=16} = 24 \times 16 - 768 = 444 - 768 = -324 < 0,$ *Hence for* $x = 16$, *the volume is maximum.*

Hence the square of the side 16 cm should be cut from each corners. Ans.

(23) The profit function $P(x)$ of a firm, selling x items per day is given by $P(x) = (100 - x)x - 1200$.

find the number of items the firm should manufacture to get maximum profit also find the maximum profit.

Solution: $-$ It is given that x is the number of items produced and sold out by the firm every day.

In order to maximize profit, $P'(x) = 0$ i.e $\frac{dP}{dx} = 0$

or $P(x) = (100 - x)x - 1200$ then $P'(x) = 100 - 2x$, $P'(x) = 0$ i.e $100 - 2x = 0$ $\therefore 2x = 100$ $\therefore x = 50$

Now, $P''(x) = -2$ is a negative quantity hence $P(x)$ is maximum at $x = 50$.

Thus, the firm should manufacture only 50 item a day to make maximum profit.

Now, Maximum Profit $= P(x) = (100 - x)x - 1200$ or Maximum Profit at $x = 50$

then Maximum Profit $= P(50) = (100 - 50)50 - 1200 = 2500 - 1200 = 1300$ or Maximum Profit is Rs. 1300 Ans.

Exercise – A6

Find the intervals for which the following function are increasing or decreasing: –

(1) (a) $f(x) = x^2 - 7x + 12$ (b) $f(x) = 3x^2 - 7x - 6$ (c) $f(x) = x^2 - 5x + 6$

(2) (a) $f(x) = x^3 + 5x^2 + 8x - 12$ (b) $f(x) = 2x^3 - 9x^2 - 24x + 7$ (c) $f(x) = x^3 + 3x - 5$

(3) (a) $f(x) = -3x^2 - 18x + 10$ (b) $f(x) = 2 - 36x - 15x^2 - 2x^3$ (c) $f(x) = (x - 1)^2(x + 1)$

(4) (a) $f(x) = \frac{x - 3}{x + 2}, x \neq -2$ (b) $y = \frac{x}{x - 2}, x \neq 2$ (c) $y = \frac{x}{3} - \frac{3}{x}, x \neq 0$

(5) (a) Prove that the function $\sin x$ is increasing in the interval $\left[\frac{\pi}{2}, 2\pi\right]$.

(b) Prove that the function $\log(\cos x)$ is decreasing in the interval $[0, \pi]$.

(c) Find the interval in which the function $\cos\left(2x + \frac{\pi}{4}\right), 0 < x < \pi$ *is decreasing or increasing. find also the points on the graph* of the function at which the tangents are parallel to x – axis. find all points of local maxima and local minima of the following function also, find the maxima and minima of such points.

(6) (a) $x^2 - 6x + 7$ (b) $x^3 - 9x^2 + 12x + 18$ (c) $2x^3 - 24x^2 + 42x - 20$

(7) (a) $x^4 - 62x^2 + 120x + 18$ (b) $(x - 2)(x + 1)^2$ (c) $\frac{x + 2}{x^2 + x + 2}$

(8) (a) $x^3 - 7x^2 + 11x - 6$ (b) $2x^3 - 4x^2 - 8x - 3$ (c) $6x^2 - x - 2$

(9) (a) $(x - 1)(x + 2)^2$ (b) $\frac{x^2 - 3x + 2}{x^2 + 2}$ (c) $x^2 - 2x + \frac{3}{4}$

Find local maximum and local minimum for each of the following function using second order derivatives.

(10) (a) $2x^3 + 6x^2 - 18x + 12$ (b) $-x^3 + 6x^2 - 7$ (c) $(x+1)(x-2)^2$

(11) (a) $\cos 2x - x,\ -\frac{\pi}{2} \le x \le \frac{\pi}{2}$ (b) $\cos x(1 - \sin x),\ 0 < x < \frac{\pi}{2}$ (c) $\sin x - \cos x,\ 0 < x < \frac{\pi}{2}$

(12) (a) $x^5 - 10x^4 + 20x^3 - 5$ (b) $x \log x$ (c) $(x+1)e^x$ (d) $(1 + \log x)x$

(13) (a) $x^4 - 2x^3 - 3x^2 + 3$ (b) $(x-3)(x+2)^2$ (c) $\frac{2x-1}{x^2+1}$ (d) $\frac{x^3+1}{x-1}$

Find the local maxima or minima of the following function: –

(14) (a) $x^3 - 3x^2 - 9x + 1$ (b) $x^4 + 4x^2 - 12x + 9$ (c) $4x^2 - 12x + 9$

(15) (a) $\frac{1}{x^2-2}$ (b) $\frac{x}{(x-1)(x-3)}$ (c) $x\sqrt{2+x},\ x > -2$

(16) (a) $\sin x + \frac{1}{2}\cos 2x,\ 0 \le x \le \frac{\pi}{2}$ (b) $\cos 2x,\ 0 \le x \le 2\pi$ (c) $2\sin x + x,\ 0 \le x \le 2\pi$

(17) For the what value of x lying in the close interval [0,6], the slope of the tangent to $x^3 - 12x^2 + 36x + 18$ is maximum . Also find the point.

(18) Find the value of the greatest slope of a tangent to $-x^3 + 6x^2 + 3x - 15$ at a point of the other curve. find also the point.

(19) Find the two numbers whose sum is 15 and the square of one multiplied by the cube of the other is maximum.

(20) Prove that the perimeter of a right angled triangle of given hypotenuse is maximum when the triangle is isosceles.

(21) A movie theatre'smanagement is considering reducing the price of tickets from Rs. 55 in order to get more customers. After checking out various facts they decide that the average number of customers per day "P" is given by the function where x is the amount of ticket price reduced. Find the ticket price 0 that result in maximum revenue. $P = 500 + 100x$

Where x is the amount of ticket price reduced . Find the ticket price that result is maximum revenue.

(22) (a) Find the maximum and minimum value of $\frac{1-x+x^2}{1+x+x^2}$ for all real value of x.

(b) On the interval [0,1] the function $x^{25}(1-x)^{75}$ takes its maximum value at the point.

(c) Let $f: IR \to IR$ be defined as $f(x) = |x| + |x^2 - 1|$. The total number of points at which f attains either a local maximum or a local minimum.

(d) Let p(x) be a real polynomial of least degree which has a local maximum at $x = 1$ and a local Minimum at $x = 3$. If $p(1) = 6$ and $p(3) = 2$, then $p'(0)$ is

Answer – A6

(1) (a) Increasing for $x > 4, x < 3$ *and Decreasing for* $3 < x < 4$ (b) Increasing for $x > 3, x < -\frac{2}{3}$ and Decreasing for $-\frac{2}{3} < x < 3$

(c) Increasing for $x > 3, x < 2$ *and Decreasing for* $2 < x < 3$

(2) (a) Increasing for $x > -\frac{4}{3}, x < -2$ *and Decreasing for* $-2 < x < -\frac{4}{3}$ (b) Increasing for $x > 4, x < -1$ *and Decreasing for* $-1 < x < 4$

(3) (a) Increasing for $x < -3$ *and Decreasing* $x > -3$ (b) Increasing for $-3 < x < -2$ *and Decreasing for* $x > -2, x < -3$

(c) Increasing for $x > \frac{2}{3}, x < -\frac{1}{2}$ and Decreasing for $-\frac{1}{2} < x < \frac{2}{3}$

(4) (a) Increasing for $x < -2$ *and Decreasing for* $x > -2$ (b) Increasing for $x > 2$ *and Decreasing for* $x < 2$

(c) Increasing for $x > 3, x < 0$ *and Decreasing for* $0 < x < 3$

(6) (a) Local minimum at point $x = 3$, Point is $(3, -2)$

(b) Local minimum at $x = 3 + \sqrt{5}$ and maximum at $x = 3 - \sqrt{5}$, Points $(3 + \sqrt{5}, -22.36), (3 - \sqrt{5}, 22.36)$.

(c) Local minimum at $x = 7$ and maximum at $x = 1$, Points $(1,0)$ & $(7, -216)$.

(7) (a) Local minimum at $x = -6,5$ and local maximum at $x = 1$, Points are $(-6, -1638), (1,77)$ and $(5, -307)$.

(b) Local minimum at $x = 1$ and local maximum at $x = -1$, Points are $(-1,0)$ & $(1, -4)$.

(c) Local minimum at $x = -4$ and local maximum at $x = 0$, Points are $\left(-4, -\frac{1}{7}\right)$ & $(0,1)$.

(8) (a) Local minimum at $x = \frac{11}{3}$ and local maximum at $x = 1$, Points are $(1, -1)$ & $\left(\frac{11}{3}, \frac{20611}{27}\right)$.

(b) Local minimum at $x = 2$ and local maximum at $x = -\frac{2}{3}$, Points are $\left(-\frac{2}{3}, -\frac{1}{27}\right)$ and $(2, -19)$.

(c) Local minimum at $x = \frac{1}{12}$, Point is $\left(\frac{1}{12}, -\frac{49}{24}\right)$.

(9) (a) Local minimum at $x = 0$ and local maximum at $x = -2$, Points are $(0, -4)$ & $(-2,0)$.

(b) Local minimum at $x = \sqrt{2}$ and local maximum at $x = -\sqrt{2}$, Points are $\left(-\sqrt{2}, \frac{4 + 3\sqrt{2}}{4}\right)$ and $\left(\sqrt{2}, \frac{4 - 3\sqrt{2}}{4}\right)$.

(c) Local minimum at $x = 1$, Point is $\left(1, -\frac{1}{4}\right)$.

(10) (a) Local minimum at $x = -1$ (b) Local maximum at $x = 2$ (c) Local minimum at $x = 1$

(12) (a) Local minimum at $x = 0, 3 + \sqrt{3}$ and local maximum at $x = 3 - \sqrt{3}$

(13) (a) Local minimum at $x = \frac{1 + \sqrt{3}}{2}$ and local maximum at $x = \frac{1 - \sqrt{3}}{2}$ (b) Local minimum at $x = -\frac{1}{3}$ (c)

(14) (a) Local minima at $x = 3$ and local maxima at $x = -1$ (b) Local minima at $x = 1$ (c) Local minima at $x = \frac{3}{2}$

(17) Maximum at $x = 2$, Point is $(2,50)$. (22) (a) Maximum value $= 3$ and minimum value $= \frac{1}{3}$ (b) $\frac{1}{4}$ (c) 5

Tangent and Normal

(A) Tangent at (x, y): – Let $y = f(x)$ be a given curve and $P(x, y)$ and $Q(x + \delta x, y + \delta y)$ be two neighbouring point on it.

Equation of the line PQ is $Y - y = \frac{y + \delta y - y}{x + \delta x - x}(X - x)$ or $Y - y = \frac{\delta y}{\delta x}(X - x) \ldots\ldots\ldots\ldots\ldots\ldots (A)$

The line (A) will be a tangent to the given curve at P. If $Q \to P$, $\delta x \to 0$

we know that $\lim_{\delta x \to 0} \frac{\delta y}{\delta x} = \frac{dy}{dx}$

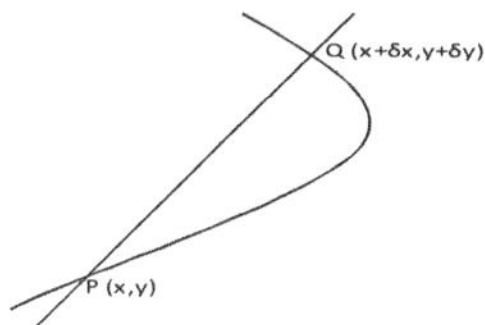

The equation of the tangent is $Y - y = \frac{dy}{dx}(X - x) \ldots\ldots\ldots\ldots (B)$ (Formula)

Normal at (x, y): – The normal at (x, y) being perpendicular to tangent will have its slope is $-1/\frac{dy}{dx}$ and

hence its equation is $Y - y = -\frac{1}{\frac{dy}{dx}}(X - x) \ldots\ldots\ldots\ldots\ldots\ldots (C)$ (Formula)

Geometrical meaning of $\frac{dy}{dx}$: – $\frac{dy}{dx}$ represent the slope of the tangent to the given curve

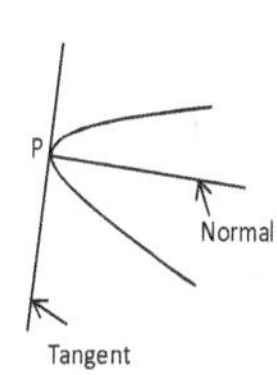

$y = f(x)$ at any point (x, y)

$$\therefore \frac{dy}{dx} = \tan \Psi$$

where Ψ is the angle which the tangent to the curve makes with + ve direction of x – axis.

find the tangent at any point (x_1, y_1) then $\left(\frac{dy}{dx}\right)_{(x_1, y_1)}$.

The value of $\frac{dy}{dx}$ at (x_1, y_1) will represent the slope of the tangent and hence its equation in this case will be

$$\boxed{y - y_1 = \left(\frac{dy}{dx}\right)_{(x_1, y_1)}(x - x_1)}$$

Normal equation is $\boxed{y - y_1 = \frac{-1}{\left(\frac{dy}{dx}\right)_{(x_1, y_1)}}(x - x_1)}$ slope of the tangent $= m_1 = \frac{dy}{dx} = \left(\frac{dy}{dx}\right)_{(x_1, y_1)}$

slope of the normal $= \frac{-1}{m_1} = -\frac{1}{\frac{dy}{dx}} = \frac{-1}{\left(\frac{dy}{dx}\right)_{(x_1, y_1)}}$

■ If a tangent is parallel to x − axis or normal is perpendicular to x − axis then m = 0 so that $\frac{dy}{dx} = 0$.

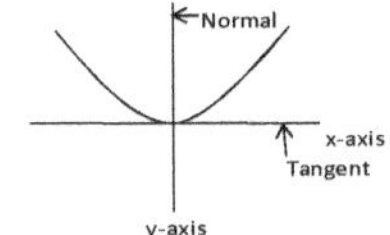

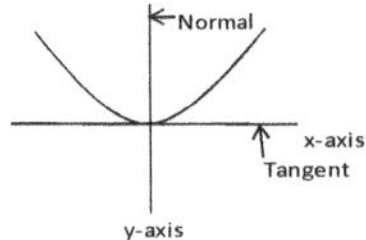

■ If a tangent is perpendicular to x − axis or normal is parallel to x − axis then m = ∞, $\frac{dy}{dx} = \infty$ or its reciprocal $\frac{dx}{dy} = 0$

Parametric form: − $\frac{dy}{dx} = \frac{dy}{dt} \div \frac{dx}{dt}$

If the equation of the curve be $x = a\sin t$, $y = b\cos t$ then $x = a\sin t \ldots\ldots\ldots (i)$

and $y = b\cos t \ldots\ldots\ldots (ii)$

Differentiate equation (i)and (ii) with respect to t, we get $\therefore \frac{dx}{dt} = a\cos t$ and $\frac{dy}{dt} = -b\sin t$

then $\frac{dy}{dx} = \frac{dy}{dt} \div \frac{dx}{dt} = -b\sin t \div a\cos t = \frac{-b\sin t}{a\cos t} = -\frac{b}{a}\tan t \ldots\ldots\ldots\ldots\ldots (iii)$

Divide equation ${}^{(i)}/_{(ii)}$ we get, $\frac{x}{y} = \frac{a\sin t}{b\cos t} = \frac{a}{b}\tan t \;\therefore\; \tan t = \frac{bx}{ay}$

put value $\tan t = \frac{bx}{ay}$ in equation (iii), we have $\frac{dy}{dx} = -\frac{b}{a}\tan t = -\frac{b}{a}.\frac{bx}{ay} = -\frac{b^2}{a^2}\left(\frac{x}{y}\right) \quad \therefore \frac{dy}{dx} = -\frac{b^2}{a^2}\left(\frac{x}{y}\right)$

Tangent is $y - b\cos t = -\frac{b^2x}{a^2y}(x - a\sin t)$ and Normal is $y - b\cos t = \frac{a^2y}{b^2x}(x - a\sin t)$

Tangent is parallel to x − axis if y = 0 and Tangent is perpendicular to x − axis if x = 0.

Partial Differentiation: − If the equation of the curve is of the form f(x, y) = c or 0 then a convenient method of finding $\frac{dy}{dx}$ is by the help of partial derivatives $\frac{dy}{dx} = -\frac{f_x}{f_y}$, where f_x is differentiation of f(x, y) with respect to x but y constant and f_y is differentiation of f(x, y) w. r. t y but x constant.

e.g. if $x^2 + y^2 - 2axy = 0 = f(x,y)$ then $f_x = 2x - 2ay$, $f_y = 2y - 2ax$ or $\frac{dy}{dx} = -\frac{f_x}{f_y} = -\frac{2x - 2ay}{2y - 2ax} = -\frac{2(x - ay)}{2(y - ax)} = -\frac{x - ay}{y - ax}$

Equation of tangent can be written as $Y - y = -\frac{f_x}{f_y}(X - x)$ or $(X - x)f_x + (Y - y)f_y = 0 \ldots\ldots\ldots\ldots\ldots (i)$

or $(X - x)f_{x(a,b)} + (Y - b)f_{y(a,b)} = 0$ at a point (a, b)

Normal will be $\frac{X - x}{f_x} = \frac{Y - y}{f_y} \ldots\ldots\ldots\ldots (ii)$ or $\frac{X - a}{f_{x(a,b)}} = \frac{Y - b}{f_{y(a,b)}}$ at the point (a, b).

(B) Angle of intersection of two curves: − By angle of intersection of two curves we mean the angle between the tangent to the two curves at their common point of intersection hence if θ be the acute angle between the tangents then $\tan\theta = \left|\frac{m_1 - m_2}{1 + m_1 m_2}\right|$

where $m_1 = \frac{dy}{dx}$ at the common point for 1st curve, $m_2 = \frac{dy}{dx}$ at the common point for 2nd curve.

■ Condition for orthogonal intersection: − Two curves are said to cut orthogonally if the angle between them is a right angle.

i.e $\theta = 90^0$ $\therefore \tan 90^0 = \infty$ or $1 + m_1 m_2 = 0$ or $m_1 m_2 = -1$ or $\left(\frac{dy}{dx}\right)_I \left(\frac{dy}{dx}\right)_{II} = -1$

■ Condition for the two curves to touch: − If the two curves touch then $\theta = 0^0$, $\tan\theta = \left|\frac{m_1 - m_2}{1 + m_1 m_2}\right|$

$$\text{or } \tan 0^0 = \left|\frac{m_1 - m_2}{1 + m_1 m_2}\right| \quad \therefore m_1 - m_2 = 0 \text{ or } m_1 = m_2 \text{ or } \left(\frac{dy}{dx}\right)_I = \left(\frac{dy}{dx}\right)_{II}$$

If the two curves be $f(x, y) = 0$ and $\phi(x, y) = 0$ then $\left(\frac{dy}{dx}\right)_I = -\frac{f_x}{f_y}$ and $\left(\frac{dy}{dx}\right)_{II} = -\frac{\phi_x}{\phi_y}$

$$\tan\theta = \left|\frac{\left(-\frac{f_x}{f_y}\right)\left(-\frac{\phi_x}{\phi_y}\right)}{1 + \left(-\frac{f_x}{f_y}\right)\left(-\frac{\phi_x}{\phi_y}\right)}\right| \quad \therefore \tan\theta = \left|\frac{f_x\phi_y - f_y\phi_x}{f_x\phi_x + f_y\phi_y}\right|$$

■ Condition for touching: − If $\theta = 0^0$ then $\tan\theta = 0$ $\therefore \left|\frac{f_x\phi_y - f_y\phi_x}{f_x\phi_x + f_y\phi_y}\right| = 0$ or $f_x\phi_y - f_y\phi_x = 0$

$$\text{or } f_x\phi_y = f_y\phi_x \text{ or } \frac{f_x}{f_y} = \frac{\phi_x}{\phi_y}$$

■ Condition to cut orthogonally: − If $\theta = 90^0$ then $\tan\theta = \infty$ $\therefore \left|\frac{f_x\phi_y - f_y\phi_x}{f_x\phi_x + f_y\phi_y}\right| = \infty$ or $f_x\phi_x + f_y\phi_y = 0$.

(C) Intercepts of tangent on the axes: − Find the equation of the tangent put $y = 0$ and find the value of x which will be intercept on axis of x. Then put $x = 0$ and find the value of y which will be intercept on y − axis.

Length of tangent and normal: − Length of tangent = PT and length of normal = PN

where P is the point of contact, T and N is the point where T (Tangent) and N (Normal) meets the axis of x.

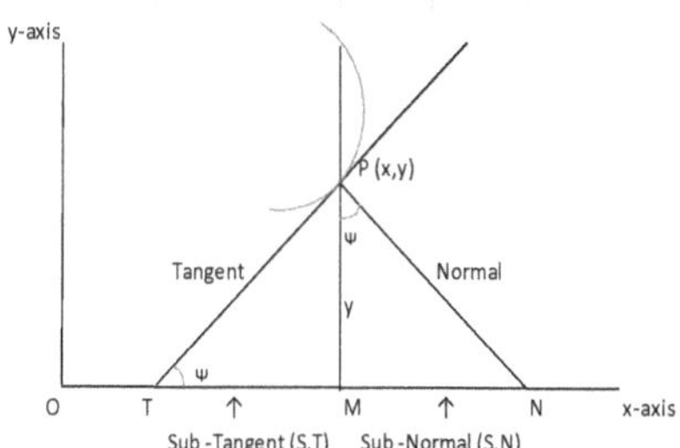

From the figure, $\frac{y}{PT} = \sin\Psi$ $\therefore PT = y\,\text{cosec}\,\Psi$ ....(i) and $\frac{y}{PN} = \cos\Psi$ $\therefore PN = y\sec\Psi$ ....(ii)

Now, $\tan\Psi = \frac{dy}{dx} = y'$ (say) $\therefore \sec\Psi = \sqrt{1 + \tan^2\Psi} = \sqrt{1 + y'^2}$ ....(iii)

$$\text{cosec}\,\Psi = \sqrt{1 + \cot^2\Psi} = \sqrt{1 + \frac{1}{\tan^2\Psi}} = \sqrt{\frac{1 + \tan^2\Psi}{\tan^2\Psi}} = \frac{\sqrt{1 + \tan^2\Psi}}{\tan\Psi} = \frac{\sqrt{1 + y'^2}}{y'} \text{ ....(iv)}$$

put value cosec Ψ in equation (i), we have $\therefore$ $PT = y\,\text{cosec}\,\Psi = y\dfrac{\sqrt{1+y'^2}}{y'}$ or $\boxed{\therefore\ PT = \frac{y}{y'}\sqrt{1+y'^2}}$

put value sec Ψ in equation (ii), we have $\therefore$ $PN = y\sec\Psi = y\sqrt{1+y'^2}$ or $\boxed{\therefore\ PN = y\sqrt{1+y'^2}}$

■ Condition for a given line to touch a given curve: − $Y - y = \dfrac{dy}{dx}(X - x)$

compare this with the given line $aX + bY + C = 0$ and eliminate x and y.

■ Length of sub − tangent (TM) $= y\cot\Psi = \dfrac{y}{\tan\Psi}$ $\therefore$ $TM = \dfrac{y}{\frac{dy}{dx}} = \dfrac{y}{y'}$ $\left\{\tan\Psi = \dfrac{dy}{dx} = y'\ (\text{say})\right\}$

■ Length of sub − normal (MN) $= y\tan\Psi = y\dfrac{dy}{dx} = yy'$ $\therefore$ $MN = \left(\dfrac{dy}{dx}\right)y = yy'$

■ Tangent to the curve $y = f(x)$, which is parallel to the line $ax + by + c = 0$ then slope of the line $= -\dfrac{a}{b} =$ slope of the tangent

If perpendicular to the line $ax + by + c = 0$ then slope of the line is $-\dfrac{a}{b}$ but slope of the tangent is $\dfrac{-1}{-\frac{a}{b}} = \dfrac{b}{a}$.

■ Normal to the curve $y = f(x)$, which is parallel to the line $ax + by + c = 0$ then slope of the line is $-\dfrac{a}{b} =$ slope of the normal

If perpendicular to the $ax + by + c = 0$ then slope of the line is $\left(-\dfrac{a}{b}\right)$ but slope of the normal is $\left(\dfrac{b}{a}\right)$.

Solved Example

(1) Find the equation of the tangent to the curve at any point (x, y), $\dfrac{x^p}{a^p} + \dfrac{y^p}{b^p} = 1 \ldots\ldots\ldots\ldots (i)$

Solution: − $\dfrac{x^p}{a^p} + \dfrac{y^p}{b^p} = 1$ Differentiating w.r.t x, we have

$$p.\frac{x^{p-1}}{a^p} + p.\frac{y^{p-1}}{b^p}.\frac{dy}{dx} = 0 \quad \therefore \quad \frac{dy}{dx} = -\frac{p.\frac{x^{p-1}}{a^p}}{p.\frac{y^{p-1}}{b^p}} = -\left(\frac{b}{a}\right)^p.\left(\frac{x}{y}\right)^{p-1} = -\left(\frac{x}{a}\right)^{p-1}.\frac{1}{a}.b\left(\frac{b}{y}\right)^{p-1}$$

Equation of tangent is $Y - y = \dfrac{dy}{dx}(X - x)$ or $Y - y = -\left(\dfrac{b}{a}\right)^p.\left(\dfrac{x}{y}\right)^{p-1}(X - x)$

or $Y - y = -\left(\dfrac{x}{a}\right)^{p-1}.\dfrac{1}{a}.b\left(\dfrac{b}{y}\right)^{p-1}(X - x)$ or $\dfrac{Y - y}{-b\left(\frac{b}{y}\right)^{p-1}} = \left(\dfrac{x}{a}\right)^{p-1}.\dfrac{1}{a}(X - x)$

or $-\dfrac{Y}{b}\left(\dfrac{y}{b}\right)^{p-1} + \dfrac{y}{b}\left(\dfrac{y}{b}\right)^{p-1} = \dfrac{X}{a}\left(\dfrac{x}{a}\right)^{p-1} - \dfrac{x}{a}\left(\dfrac{x}{a}\right)^{p-1}$ or $\dfrac{X}{a}\left(\dfrac{x}{a}\right)^{p-1} - \left(\dfrac{x}{a}\right)^p = -\dfrac{Y}{b}\left(\dfrac{y}{b}\right)^{p-1} + \left(\dfrac{y}{b}\right)^p$

or $\dfrac{X}{a}\left(\dfrac{x}{a}\right)^{p-1} + \dfrac{Y}{b}\left(\dfrac{y}{b}\right)^{p-1} = \left(\dfrac{y}{b}\right)^p + \left(\dfrac{x}{a}\right)^p$ $\therefore$ $\dfrac{X}{a}\left(\dfrac{x}{a}\right)^{p-1} + \dfrac{Y}{b}\left(\dfrac{y}{b}\right)^{p-1} = 1$ by equation (i)

Another method: − $f_x = p.\dfrac{x^{p-1}}{a^p}$, $f_y = p.\dfrac{y^{p-1}}{b^p}$

Equation of tangent is $(X - x)f_x + (Y - y)f_y = 0$ (formula) or $(X - x)p.\dfrac{x^{p-1}}{a^p} + (Y - y)p.\dfrac{y^{p-1}}{b^p} = 0$

$$\text{or } \frac{X}{a}\left(\frac{x}{a}\right)^{p-1} + \frac{Y}{b}\left(\frac{y}{b}\right)^{p-1} = \left(\frac{y}{b}\right)^p + \left(\frac{x}{a}\right)^p = 1 \text{ by equation (i) Ans.}$$

(2) Find the equation of the tangent to the curve at any point (x, y)

(a) $\left(\frac{x}{a}\right)^{\frac{1}{3}}+\left(\frac{y}{b}\right)^{\frac{1}{3}}=1$ (b) $\left(\frac{x}{a}\right)^{m/m-2}+\left(\frac{y}{b}\right)^{m/m-2}=1$ (c) $x^{\frac{1}{3}}+y^{\frac{1}{3}}=a^{\frac{1}{3}}$ (d) $\frac{x^3}{a^3}+\frac{y^3}{b^3}=1$

Solution: − (a) $\left(\frac{x}{a}\right)^{\frac{1}{3}}+\left(\frac{y}{b}\right)^{\frac{1}{3}}=1$, $\frac{x^p}{a^p}+\frac{y^p}{b^p}=1$ (above question) and equation of tangent is $\frac{X}{a}\left(\frac{x}{a}\right)^{p-1}+\frac{Y}{b}\left(\frac{y}{b}\right)^{p-1}=1$

put $p=\frac{1}{3}$ equation of tangent is $\frac{X}{a}\left(\frac{x}{a}\right)^{p-1}+\frac{Y}{b}\left(\frac{y}{b}\right)^{p-1}=1$ or $\frac{X}{a}\left(\frac{x}{a}\right)^{\frac{1}{3}-1}+\frac{Y}{b}\left(\frac{y}{b}\right)^{\frac{1}{3}-1}=1$

or $\frac{X}{a}\left(\frac{x}{a}\right)^{-\frac{2}{3}}+\frac{Y}{b}\left(\frac{y}{b}\right)^{-\frac{2}{3}}=1$ or $\frac{X}{a}\left(\frac{a}{x}\right)^{\frac{2}{3}}+\frac{Y}{b}\left(\frac{b}{y}\right)^{\frac{2}{3}}=1$ or $\frac{X}{x^{\frac{2}{3}}.a^{\frac{1}{3}}}+\frac{Y}{y^{\frac{2}{3}}.a^{\frac{1}{3}}}=1$ Ans.

(b) put $p=\frac{m}{m-2}$ $\therefore$ equation of tangent is $\frac{X}{a}\left(\frac{x}{a}\right)^{m/(m-2)-1}+\frac{Y}{b}\left(\frac{y}{b}\right)^{m/(m-2)-1}=1$ or $\frac{X}{a}\left(\frac{x}{a}\right)^{2/(m-2)}+\frac{Y}{b}\left(\frac{y}{b}\right)^{2/(m-2)}=1$ Ans.

(c) put $p=\frac{1}{3}$ $\therefore$ equation of tangent is $\frac{X}{a}\left(\frac{x}{a}\right)^{p-1}+\frac{Y}{b}\left(\frac{y}{b}\right)^{p-1}=1$ $\left[x^{\frac{1}{3}}+y^{\frac{1}{3}}=a^{\frac{1}{3}} \text{ or } \left(\frac{x}{a}\right)^{\frac{1}{3}}+\left(\frac{y}{b}\right)^{\frac{1}{3}}=1\right]$

here a = b , then $\frac{X}{a}\left(\frac{x}{a}\right)^{\frac{1}{3}-1}+\frac{Y}{a}\left(\frac{y}{a}\right)^{\frac{1}{3}-1}=1$ [put b = a] or $\frac{X}{a}\left(\frac{x}{a}\right)^{-\frac{2}{3}}+\frac{Y}{a}\left(\frac{y}{a}\right)^{-\frac{2}{3}}=1$ or $\frac{X}{x^{\frac{2}{3}}.a^{\frac{1}{3}}}+\frac{Y}{y^{\frac{2}{3}}.a^{\frac{1}{3}}}=1$ or $\frac{X}{x^{\frac{2}{3}}}+\frac{Y}{y^{\frac{2}{3}}}=a^{\frac{1}{3}}$ Ans.

(d) put p = 3 $\therefore$ equation of tangent is $\frac{X}{a}\left(\frac{x}{a}\right)^{p-1}+\frac{Y}{b}\left(\frac{y}{b}\right)^{p-1}=1$ or $\frac{X}{a}\left(\frac{x}{a}\right)^{3-1}+\frac{Y}{b}\left(\frac{y}{b}\right)^{3-1}=1$

$$\frac{X}{a}\left(\frac{x}{a}\right)^{2}+\frac{Y}{b}\left(\frac{y}{b}\right)^{2}=1 \quad \text{Ans.}$$

(3) Does the straight line $\frac{x}{a}+\frac{y}{b}=1$ touch the curve $\left(\frac{y}{b}\right)^2+\left(\frac{x}{a}\right)^2=1$?

If it touches then determine the co − ordinates of the point of contact.

Solution: − $\left(\frac{y}{b}\right)^2+\left(\frac{x}{a}\right)^2=1$ ……………….(i) Differentiating equation (i) with respect to x , we have

$$\therefore \quad \frac{1}{a^2}.2x+\frac{1}{b^2}.2y.\frac{dy}{dx}=0 \text{ or } \frac{1}{b^2}.2y.\frac{dy}{dx}=-\frac{1}{a^2}.2x \text{ or } \frac{dy}{dx}=\frac{\frac{2x}{a^2}}{\frac{2y}{b^2}}=-\frac{b^2x}{a^2y}$$

Equation of tangent is $Y-y=\frac{dy}{dx}(X-x)$ or $Y-y=-\frac{b^2x}{a^2y}(X-x)$

or $Y-y=-\frac{b^2}{y}.\frac{x}{a^2}(X-x)$ or $\frac{Yy}{b^2}-\frac{y^2}{b^2}=\frac{x^2}{a^2}-\frac{Xx}{a^2}$ or $\frac{Yy}{b^2}+\frac{Xx}{a^2}=\left(\frac{x}{a}\right)^2+\left(\frac{y}{b}\right)^2=1$ by equation (i)

or $\frac{Y}{b}\frac{y}{b}+\frac{X}{a}\frac{x}{a}=1$, comparing with straight line $\frac{x}{a}+\frac{y}{b}=1$ and $\frac{Y}{b}\frac{y}{b}+\frac{X}{a}\frac{x}{a}=1$ then $\frac{x}{a}=1, \frac{y}{b}=1$ or x = a , y = b

and equation of tangent becomes $\frac{X}{a}+\frac{Y}{b}=1$ hence the given line touches and the point of contact is (a, b).

(4) (a) $x=a\sin^2\theta$, $y=a\cos^2\theta$ or $x+y=a$ then find equation of tangent and normal.

Solution: − Parametic form , $x=a\sin^2\theta$, $y=a\cos^2\theta$ then $\frac{dx}{d\theta}=2a\cos\theta$, $\frac{dy}{d\theta}=-2a\sin\theta$ $\therefore$ $\frac{dy}{dx}=-\frac{2a\sin\theta}{2a\cos\theta}=-\tan\theta$

Eliminating θ form , $x=a\sin^2\theta$, $\left(\frac{x}{a}\right)^{\frac{1}{2}}=\sin\theta$ and $y=a\cos^2\theta$, $\left(\frac{y}{a}\right)^{\frac{1}{2}}=\cos\theta$ or $\sin^2\theta+\cos^2\theta=1$ or $\frac{x}{a}+\frac{y}{a}=1$ or $x+y=a$

Equation of the tangent is $Y-y=\frac{dy}{dx}(X-x)$ or $y-a\cos^2\theta=-\tan\theta\,(x-a\sin^2\theta)$

$$\text{or } y - a\cos^2\theta = -\frac{\sin\theta}{\cos\theta}(x - a\sin^2\theta) \text{ or } y\cos\theta - a\cos^3\theta = a\sin^3\theta - x\sin\theta$$

or $x\sin\theta + y\cos\theta = a\sin^3\theta + a\cos^3\theta = a(\sin^3\theta + \cos^3\theta)$ Ans.

or $x\sin\theta + y\cos\theta = a[(\sin\theta + \cos\theta)^3 - 3\sin\theta\cos\theta(\sin\theta + \cos\theta)]$ $[\therefore\ a^3 + b^3 = (a+b)^3 - 3ab(a+b)]$

or $x\sin\theta + y\cos\theta = a(\sin\theta + \cos\theta)[(\sin\theta + \cos\theta)^2 - 3\sin\theta\cos\theta] = a(\sin\theta + \cos\theta)[\sin^2\theta + \cos^2\theta + 2\sin\theta\cos\theta - 3\sin\theta\cos\theta]$

or $x\sin\theta + y\cos\theta = a(\sin\theta + \cos\theta)(1 - \sin\theta\cos\theta) = a(\sin^3\theta + \cos^3\theta)$

Equation of the normal is $Y - y = -\frac{1}{\frac{dy}{dx}}(X - x)$ or $y - a\cos^2\theta = \frac{1}{\tan\theta}(x - a\sin^2\theta)$

or $y - a\cos^2\theta = \frac{\cos\theta}{\sin\theta}(x - a\sin^2\theta)$ or $y\sin\theta - a\cos^2\theta\sin\theta = x\cos\theta - a\sin^2\theta\cos\theta$

or $x\cos\theta - y\sin\theta = a\sin^2\theta\cos\theta - a\cos^2\theta\sin\theta = a\sin\theta\cos\theta(\sin\theta - \cos\theta$ or $x\cos\theta - y\sin\theta = \frac{a}{2}\sin 2\theta(\sin\theta - \cos\theta)$ Ans.

(b) Normal to $x + y = a$ in the form $x\cos\theta - y\sin\theta = \frac{a}{2}\sin 2\theta(\sin\theta - \cos\theta)$ where θ is the angle which the normal makes with the axis of x.

Solution: $-$ $x + y = a$ $\therefore$ Differentiate, $1 + \frac{dy}{dx} = 0$ $\therefore$ $\frac{dy}{dx} = -1$

slope of the normal is $1 = \tan\theta$ or $\tan\theta = 1$ or $\frac{\sin\theta}{\cos\theta} = 1$ or $\sin\theta = \cos\theta$

squaring $\sin^2\theta = \cos^2\theta$ $\therefore$ $\sin^2\theta + \cos^2\theta = 1$, $x = a\sin^2\theta$, $y = a\cos^2\theta$

Hence the normal whose slope is $\tan\theta$ is given by $y - a\cos^2\theta = \frac{\sin\theta}{\cos\theta}(x - a\sin^2\theta)$

or $x\cos\theta - y\sin\theta = \frac{a}{2}\sin 2\theta(\sin\theta - \cos\theta)$ as in Q. No. $-(4)$(a)

(5) (a) The equation of the tangent to the curve $x^3 + y^2 = 3xy$ at the point (2,2) .

Solution: $-$ $x^3 + y^2 = 3xy$ $\therefore$ $3x^2 + 2y\frac{dy}{dx} = 3x\frac{dy}{dx} + 3y$ or $2y\frac{dy}{dx} - 3x\frac{dy}{dx} = 3y - 3x^2$

or $\frac{dy}{dx}(2y - 3x) = 3(y - x^2)$ $\therefore$ $\frac{dy}{dx} = \frac{3(y - x^2)}{2y - 3x}$ at a point (2,2) $\therefore$ $\left(\frac{dy}{dx}\right)_{(2,2)} = \frac{3(2-4)}{4-6} = \frac{-6}{-2} = 3$ $\therefore$ $\frac{dy}{dx} = 3$

Equation of the tangent is $y - 2 = 3(x - 2)$ or $y - 2 = 3x - 6$ $\boxed{\therefore\ 3x - y = 4 \text{ Ans.}}$

(b) The equation of the normal to the curve $y = 4 - x^2$ at the point (1,3).

Solution: $-$ $y = 4 - x^2$ $\therefore$ $\frac{dy}{dx} = 0 - 2x$ or $\frac{dy}{dx} = -2x$ at a point (1,3)

$\therefore$ $\left(\frac{dy}{dx}\right)_{(1,3)} = -2 \times 1 = -2$ slope of tangent $(m_1) = -2$ and slope of normal $(m_2) = -\frac{-1}{2} = \frac{1}{2}$

Equation of normal is $y - 3 = \frac{1}{2}(x - 1)$ or $2y - 6 = x - 1$ $\boxed{\therefore\ x - 2y = -5 \text{ Ans.}}$

(6) (a) Tangent to the parabola $x^2 = 4ay$ in the form $y = mx - 2am^2 + am^2$ or $y = m(x - am)$.

Solution: $-$ $x^2 = 4ay$ $\therefore$ $2x = 4a\frac{dy}{dx}$ or $\frac{dy}{dx} = \frac{x}{2a} = m$ (say)

slope of tangent $m = \frac{x}{2a}$ $\therefore$ $x = 2am$, $y = \frac{x^2}{4a} = \frac{4a^2m^2}{4a} = am^2$ $\therefore$ $x = 2am$, $y = am^2$

Hence, the tangent whose slope is m will be the point $(2am, am^2)$

so that its equation is $y - am^2 = m(x - 2am)$ $\therefore$ $y - am^2 = mx - 2am^2$ or $y = mx - 2am^2 + am^2$

$$\therefore\ y = mx - am^2 \text{ or } y = m(x - am) \text{ Ans.}$$

(b) Tangent to the curve $y^2 = 4x$ which passes through the point (1,2).

Solution:− $y^2 = 4x$ $\therefore$ $2y\frac{dy}{dx} = 4$ or $\frac{dy}{dx} = \frac{2}{y} = m$ or $y = \frac{2}{m}, x = \frac{y^2}{4} = \frac{4}{4m^2} = \frac{1}{m^2}$

Tangent is $y - \frac{2}{m} = m\left(x - \frac{1}{m^2}\right)$, if it passes through the point (1,2) then $2 - \frac{2}{m} = m\left(1 - \frac{1}{m^2}\right)$

or $\frac{2m-2}{m} = m\left(\frac{m^2-1}{m^2}\right)$ or $2m - 2 = m^2 - 1$ or $m^2 - 2m + 1 = 0$ or $(m-1)^2 = 0$ or $m = 1$

$\therefore$ Required tangent is $y - 2 = 1(x - 1)$ or $y - 2 = x - 1$ $\therefore\ x - y = -1$ Ans.

(7) (a) The equation of the tangent at the point P(t), where t is any parameter to the parabola $x^2 = 2ay$.

Solution:− $x^2 = 2ay$ $\therefore$ $2x = 2a\frac{dy}{dx}$ or $\frac{dy}{dx} = \frac{x}{a}$

Equation of the tangent is $Y - y = \frac{dy}{dx}(X - x)$ or $Y - y = \frac{x}{a}(X - x)$ or $aY - ay = Xx - x^2$

or $Xx = aY - ay + x^2$ or $Xx = aY - ay + 2ay$ or $Xx = aY + ay$ at a point P(t), $(at^2, 2at)$

where t is a parameter. $\therefore$ $Xat^2 = aY + a.2at$ or $aXt^2 = aY + 2a^2t$ [Put X = x and Y = y]

$$\text{or } axt^2 = ay + 2a^2t \quad \therefore\ xt^2 = y + 2at \text{ Ans.}$$

(b) The parametric equation of a curve are $x = \sqrt{2}\,a\cos 2t, y = 2a\sin t$. find the equation of tangent in the form

$mx - y = \frac{a}{2\sqrt{2}\,m}(1 + 4m^2)$ where m is the slope of the tangent. hence prove that two perpendicular tangent meet on the line $x = \frac{5a}{2\sqrt{2}}$.

Solution:− $x = \sqrt{2}\,a\cos 2t$ $\therefore$ $\frac{dx}{dt} = -2\sqrt{2}\,a\sin 2t$ and $y = 2a\sin t$ $\therefore$ $\frac{dy}{dt} = 2a\cos t$

or $\frac{dy}{dx} = \frac{\frac{dy}{dt}}{\frac{dx}{dt}} = \frac{2a\cos t}{-2\sqrt{2}\,a\sin 2t} = \frac{2a\cos t}{-4\sqrt{2}\,a\sin t\cos t} = -\frac{1}{2\sqrt{2}\sin t}$ $\therefore$ $\frac{dy}{dx} = -\frac{1}{2\sqrt{2}\sin t} = m$ or $\sin t = -\frac{1}{2\sqrt{2}\,m}$

$$\therefore\ x = \sqrt{2}\,a\cos 2t = \sqrt{2}\,a(1 - 2\sin^2 t) = \sqrt{2}\,a\left[1 - 2\left(-\frac{1}{2\sqrt{2}\,m}\right)^2\right] = \sqrt{2}\,a\left(1 - \frac{2}{8m^2}\right) = \sqrt{2}\,a\left(1 - \frac{1}{4m^2}\right)$$

$$\therefore\ x = \sqrt{2}\,a\left(1 - \frac{1}{4m^2}\right),\ y = 2a\sin t = 2a\left(-\frac{1}{2\sqrt{2}\,m}\right) = -\frac{a}{\sqrt{2}\,m}$$

Hence, equation of tangent is $Y - y = m(X - x)$ or $y + \frac{a}{\sqrt{2}m} = m\left[x - \sqrt{2}\,a\left(1 - \frac{1}{4m^2}\right)\right]$

or $y + \frac{a}{\sqrt{2}\,m} = mx - \sqrt{2}am + \frac{\sqrt{2}am}{4m^2}$ or $y + \frac{a}{\sqrt{2}\,m} = mx - \sqrt{2}am + \frac{a}{2\sqrt{2}m}$

or $mx - y = \frac{a}{\sqrt{2}\,m} + \sqrt{2}am - \frac{a}{2\sqrt{2}m}$ or $mx - y = \frac{a}{\sqrt{2}\,m}\left(1 - \frac{1}{2}\right) + \sqrt{2}am = \frac{a}{2\sqrt{2}\,m} + \sqrt{2}am = \frac{a + 2\sqrt{2}\,m.\sqrt{2}am}{2\sqrt{2}\,m} = \frac{a + 4am^2}{2\sqrt{2}\,m}$

or $mx - y = \frac{a}{2\sqrt{2}\,m}(1 + 4m^2)$ or $mx - y = \frac{a}{2\sqrt{2}\,m}(1 + 4m^2)$

Equation of tangent in terms of its slope m, if it passes through the point (x_1, y_1) then

or $mx_1 - y_1 = \frac{a}{2\sqrt{2}\,m}(1 + 4m^2)$ or $2\sqrt{2}\,x_1m^2 - 2\sqrt{2}\,my_1 = a + 4am^2$

or $2\sqrt{2}\,m^2x_1 - 4am^2 - 2\sqrt{2}\,y_1m - a = 0$ or $m^2(2\sqrt{2}\,x_1 - 4a) - 2\sqrt{2}\,y_1m - a = 0$

Above is a quadratic equation in m which shows that there will be two tangent which pass through (x_1, y_1)

if the two tangent be perpendicular, then $m_1m_2 = -1$ or $\frac{-a}{(2\sqrt{2}\,x_1 - 4a)} = -1$

or $-a = -(2\sqrt{2}\,x_1 - 4a)$ or $-a = -2\sqrt{2}\,x_1 + 4a$ or $2\sqrt{2}\,x_1 = 4a + a = 5a$ $\therefore\ x_1 = \frac{5a}{2\sqrt{2}}$

Thus, the two perpendicular tangent meet at $x = \frac{5a}{2\sqrt{2}}$ Ans.

(8) Find the equation of tangent and normal to the curve: –

(a) $y^2(2a - y) = x^2(3a + y)$ at the points where $y = a$. (b) $x^2(2a + x) = y^2(3a - x)$ at the point $x = \frac{a}{2}$.

Solution: – (a) $y^2(2a - y) = x^2(3a + y)$ at the points where $y = a$ then

or $a^2(2a - a) = x^2(3a + a)$ or $a^3 = 4ax^2$ or $x^2 = \frac{a^3}{4a} = \frac{a^2}{4}$ $\therefore\ x = \pm\frac{a}{2}$

Hence, the point are $P\left(\frac{a}{2}, a\right)$ and $Q\left(-\frac{a}{2}, a\right)$

or $y^2(2a - y) = x^2(3a + y)$ or $2ay^2 - y^3 = 3ax^2 + x^2y$ $\therefore\ 4ay\frac{dy}{dx} - 3y^2\frac{dy}{dx} = 6ax + x^2\frac{dy}{dx} + y.2x$

or $\frac{dy}{dx}(4ay - 3y^2 - x^2) = 6ax + 2xy$ or $\frac{dy}{dx} = \frac{6ax + 2xy}{4ay - 3y^2 - x^2}$ at a point $P\left(\frac{a}{2}, a\right)$

or $\left(\frac{dy}{dx}\right)_{\left(\frac{a}{2},a\right)} = \frac{6.a.\frac{a}{2} + 2.\frac{a}{2}.a}{4aa - 3a^2 - \left(\frac{a}{2}\right)^2} = \frac{4a^2}{4a^2 - 3a^2 - \frac{a^2}{4}} = \frac{4a^2}{\frac{16a^2 - 12a^2 - a^2}{4}} = \frac{16a^2}{3a^2} = \frac{16}{3} = m$

slope of the tangent is m then, equation of the tangent is $y - a = \frac{16}{3}\left(x - \frac{a}{2}\right)$

or $3y - 3a = 16x - 8a$ $\therefore\ 16x - 3y = -3a + 8a$ **or $16x - 3y = 5a$ Ans.**

slope of the normal is $-\frac{1}{m}$ then equation of normal is $y - a = -\frac{1}{\frac{16}{3}}\left(x - \frac{a}{2}\right)$

or $y - a = -\frac{3}{16}\left(x - \frac{a}{2}\right)$ or $16y - 16a = -3a + \frac{3a}{2}$ or $32y - 32a = -6x + 3a$

$\therefore\ 6x + 32y = 35a$ Ans.

Similarly, point $Q\left(-\frac{a}{2}, a\right)$ slope of the tangent is $\left(\frac{dy}{dx}\right)_{\left(-\frac{a}{2},a\right)} = \frac{6a\left(-\frac{a}{2}\right) + 2\left(-\frac{a}{2}\right)a}{4aa - 3a^2 - \left(-\frac{a}{2}\right)^2} = \frac{-4a^2}{\frac{16a^2 - 12a^2 - a^2}{4}}$

or $\left(\frac{dy}{dx}\right)_{\left(-\frac{a}{2},a\right)} = -\frac{16a^2}{3a^2} = -\frac{16}{3} = m_1$ then the equation of tangent is $y - a = -\frac{16}{3}\left(x + \frac{a}{2}\right)$

or $3y - 3a = -16x - 8a$ **$\therefore\ 16x + 3y = -5a$ Ans.**

slope of the normal $-\frac{1}{m_1}$ or $-\frac{1}{\frac{-16}{3}}$ or $\frac{3}{16}$ then the equation of the normal is $y - a = \frac{3}{16}\left(x + \frac{a}{2}\right)$

or $16y - 16a = 3x + \frac{a}{2}$ or $32y - 32a = 6x + a$ $\boxed{\therefore\ 6x - 32y = -33a \text{ Ans.}}$

(b) $x^2(2a + x) = y^2(3a - x)$ at the point $x = \frac{a}{2}$ then

or $\frac{a^2}{4}\left(2a + \frac{a}{2}\right) = y^2\left(3a - \frac{a}{2}\right)$ or $\frac{a^2}{4}\left(\frac{5a}{2}\right) = y^2\left(\frac{5a}{2}\right)$ or $y^2 = \frac{a^2}{4}$ $\therefore$ $y = \pm\frac{a}{2}$

Hence, the point are $P\left(\frac{a}{2}, \frac{a}{2}\right)$ and $Q\left(\frac{a}{2}, -\frac{a}{2}\right)$, $x^2(2a + x) = y^2(3a - x)$ or $2ax^2 + x^3 = 3ay^2 - xy^2$

or $4ax + 3x^2 = 6ay\frac{dy}{dx} - x.2y\frac{dy}{dx} - y^2.1$ $\therefore$ $\frac{dy}{dx}(6ay - 2xy) = 4ax + 3x^2 + y^2$ or $\frac{dy}{dx} = \frac{4ax + 3x^2 + y^2}{6ay - 2xy}$ at a point $P\left(\frac{a}{2}, \frac{a}{2}\right)$

then slope of the tangent (m) $= \left(\frac{dy}{dx}\right)_{\left(\frac{a}{2}, \frac{a}{2}\right)} = \frac{4a\frac{a}{2} + 3\frac{a^2}{4} + \frac{a^2}{4}}{6a\frac{a}{2} - 2\frac{a}{2}\cdot\frac{a}{2}}$ or $m = \frac{8a^2 + 3a^2 + a^2}{6a^2 - a^2} \times \frac{2}{4} = \frac{12a^2}{10a^2} = \frac{6}{5}$

Equation of the tangent is $y - \frac{a}{2} = \frac{6}{5}\left(x - \frac{a}{2}\right)$ or $\frac{2y - a}{2} = \frac{6}{5}\left(\frac{2x - a}{2}\right)$ or $10y - 5a = 12x - 6a$

or $12x - 10y = a$ $\boxed{\therefore\ 6x - 5y = \frac{a}{2} \text{ Ans.}}$

slope of the normal $-\frac{1}{m}$ or $\frac{-1}{\frac{6}{5}}$ or $-\frac{5}{6}$ then, equation of the normal is $y - \frac{a}{2} = -\frac{5}{6}\left(x - \frac{a}{2}\right)$

or $2y - a = -\frac{10}{6}\left(\frac{2x - a}{2}\right)$ or $12y - 6a = -10x + 5a$ $\boxed{\therefore\ 10x + 12y = 11a \text{ Ans.}}$

Similarly, At a point $Q\left(\frac{a}{2}, -\frac{a}{2}\right)$, slope of the tangent $(m_1) = -\frac{6}{5}$

then equation of the tangent is $y + \frac{a}{2} = -\frac{6}{5}\left(x - \frac{a}{2}\right)$ or $2y + 2a = -\frac{12}{5}\left(\frac{2x - a}{2}\right)$

or $10y + 10a = -12x + 6a$ or $12x + 10y = -4a$ $\boxed{\therefore\ 6x + 5y = -2a \text{ Ans.}}$

slope of the normal is $-\frac{1}{m_1}$ or $-\frac{1}{\frac{-6}{5}} = \frac{5}{6}$ then equation of the normal is $y + \frac{a}{2} = \frac{5}{6}\left(x - \frac{a}{2}\right)$

or $\frac{2y + a}{2} = \frac{5}{6}\left(\frac{2x - a}{2}\right)$ or $12y + 6a = 10x - 5a$ or $10x - 12y = 11a$ $\boxed{\therefore\ 5x - 6y = \frac{11a}{2} \text{ Ans.}}$

(9) (a) Normal to $x = y^2 - 2y$ which is parallel to $3x + 4y = 6$.

Solution:− $x = y^2 - 2y$ $\therefore$ $1 = 2y\frac{dy}{dx} - 2\frac{dy}{dx}$ or $\frac{dy}{dx}(2y - 2) = 1$ $\therefore$ $\frac{dy}{dx} = \frac{1}{2y - 2}$

slope of the tangent $m = \frac{dy}{dx} = \frac{1}{2y - 2}$ and slope of the normal $\left(-\frac{1}{m}\right) = -\frac{1}{\frac{dy}{dx}} = -\frac{1}{\frac{1}{2y - 2}} = -(2y - 2)$

Since, it is parallel to $3x + 4y = 6$ whose slope is $-\frac{3}{4}$ then $-(2y - 2) = -\frac{3}{4}$

or $2y = \frac{3}{4} + 2$ or $y = \frac{11}{8}$ $\therefore$ $x = y^2 - 2y$

put $y = \frac{11}{8}$ then $x = \frac{121}{64} - \frac{22}{8} = \frac{121 - 176}{64} = \frac{-55}{64}$ $\therefore$ $x = -\frac{55}{64}$, $y = \frac{11}{8}$

Hence, the point is $P\left(\frac{-55}{64}, \frac{11}{8}\right)$ then slope of the normal at which is $-\frac{3}{4}$

hence equation of the normal is

$$y-\frac{11}{8}=-\frac{3}{4}\left(x+\frac{55}{64}\right) \text{ or } \frac{8y-11}{8}=-\frac{3}{4}\left(\frac{64x+55}{64}\right) \text{ or } 32(8y-11)=-3(64x+55)$$

or $256y-352=-192x-165$ or $192x+256y=352-165$ or $192x+256y=187$

$$\therefore\ 3x+4y=\frac{187}{64} \quad \text{Ans.}$$

(b) Normal to $2y^2+3y=x-2$ which is perpendicular to $x-2y=3$.

Solution:− $2y^2+3y=x-2 \quad \therefore\ 4y\frac{dy}{dx}+3\frac{dy}{dx}=1$ or $\frac{dy}{dx}(4y+3)=1$ or $\frac{dy}{dx}=\frac{1}{4y+3}$

$$\text{slope of normal is } -\frac{1}{\frac{dy}{dx}}=-\frac{1}{\left(\frac{1}{4y+3}\right)}=-(4y+3)$$

Since , it is perpendicular to $x-2y=3$ whose slope of the line is $\frac{1}{2}$.

slope of the normal is $-\frac{1}{\frac{1}{2}}=-2$. $\quad\therefore\ -(4y+3)=-2$ or $4y+3=2$ or $4y=-1 \ \therefore\ y=-\frac{1}{4}$

Put $y=-\frac{1}{4}$ in $2y^2+3y=x-2$ then find x, $2y^2+3y=x-2$ or $2.\frac{1}{16}-\frac{3}{4}=x-2$

$$\text{or } \frac{1}{8}-\frac{3}{4}=x-2 \quad \therefore\ x=\frac{1}{8}-\frac{3}{4}+2=\frac{1-6+16}{8}=\frac{11}{8}$$

Hence the point is $P\left(\frac{11}{8},-\frac{1}{4}\right)$ slope of the normal at which is − 2

Hence equation of the normal is $y+\frac{1}{4}=-2\left(x-\frac{11}{8}\right)$ or $4y+1=-8\left(\frac{8x-11}{8}\right)$

or $4y+1=-8x+11 \quad \therefore\ 8x+4y=11-1 \ \therefore\ 8x+4y=10$ or $\therefore\ 4x+2y=5$ Ans.

(10) (a) Tangent to $2y^2+3x^2+2x+3y=0$ which are parallel to the line $2x+3y=4$.

Solution:− $2y^2+3x^2+2x+3y=0 \quad \therefore\ 4y\frac{dy}{dx}+6x+2+3\frac{dy}{dx}=0 \quad \therefore\ \frac{dy}{dx}(4y+3)=-(6x+2)$

$$\therefore\ \frac{dy}{dx}=\frac{-(6y+2)}{(4y+3)}, \text{ slope of the tangent is } \frac{-(6y+2)}{(4y+3)}.$$

Since , it is parallel to the line $2x+3y=4$ whose slope of the tangent is $-\frac{2}{3}$.

$$\therefore\ \frac{-(6y+2)}{(4y+3)}=-\frac{2}{3} \text{ or } 18x+6=8y+6 \text{ or } 18x=8y \ \therefore\ 9x=4y \text{ or } y=\frac{9}{4}x$$

put $y=\frac{9}{4}x$ in the given curve $2y^2+3x^2+2x+3y=0$ then find x,

or $2y^2+3x^2+2x+3y=0$ or $2.\frac{81}{16}x^2+3x^2+2x+3.\frac{9}{4}x=0$ or $\frac{81}{8}x^2+3x^2+2x+\frac{27}{4}x=0$

or $81x^2+24x^2+16x+54x=0$ or $105x^2+70x=0$ or $x(105x+70)=0 \quad \therefore\ x=0$ or $x=-\frac{70}{105}=-\frac{2}{3}$

$$\therefore\ x=0,\ y=\frac{9}{4}x=0 \text{ and } x=-\frac{2}{3},\ y=\frac{9}{4}x=\frac{9}{4}\left(-\frac{2}{3}\right)=-\frac{3}{2}$$

Hence the two points are $(0,0),\left(-\frac{2}{3},-\frac{3}{2}\right)$ then equation of the tangent their points

$$y - 0 = -\frac{2}{3}(x - 0) \text{ or } y + \frac{3}{2} = -\frac{2}{3}\left(x + \frac{2}{3}\right) \quad \therefore\ 3y = -2x \text{ or } 2y + 3 = -\frac{4}{3}\left(\frac{3x + 2}{3}\right) = -\frac{4}{9}(3x + 2)$$

$$\boxed{\therefore\ 2x + 3y = 0 \text{ or } 12x + 18y = -35 \quad \text{Ans.}}$$

(b) Tangent to $x^2 + y^2 + x + y = 0$ which are perpendicular to the line $-x + y = 3$.

Solution: − Given, $x^2 + y^2 + x + y = 0 \quad \therefore\ 2x + 2y\frac{dy}{dx} + 1 + \frac{dy}{dx} = 0$ or $\frac{dy}{dx}(2y + 1) = -(2x + 1) \quad \therefore\ \frac{dy}{dx} = -\frac{2x + 1}{2y + 1}$

slope of the tangent is $-\frac{2x + 1}{2y + 1}$ and slope of the line is 1 whose slope of the tangent is -1 . because tangent is perpendicular to the line .

$$\text{or } -\frac{2x + 1}{2y + 1} = -1 \text{ or } 2x + 1 = 2y + 1 \text{ or } 2x = 2y \text{ or } x = y$$

sloving with the given curve , we have $x^2 + y^2 + x + y = 0$ or $y^2 + y^2 + y + y = 0$

or $2y^2 + 2y = 0$ or $2y(y + 1) = 0 \quad \therefore 2y = 0$ or $y + 1 = 0 \quad \therefore\ y = 0, -1$

At $y = 0, x = 0$ and $y = -1, x = -1$ hence the two points are $(0,0), (-1,-1)$.

slope of the tangent is -1. hence the two points are $(0,0), (-1,-1)$

then their equation is $y - 0 = -1(x - 0)$ or $y + 1 = -1(x + 1)$

$$\therefore\ x + y = 0 \text{ or } y + 1 = -x - 1 \quad \boxed{\therefore\ x + y = 0 \text{ or } x + y = -2 \quad \text{Ans.}}$$

(11) (a) Tangent and normal to parabola $x^2 = -4y + 3$ which is parallel to $4y = 2x - 5$.

Solution: − Given, $x^2 = -4y + 3 \quad \therefore\ 2x = -4\frac{dy}{dx}$ or $\frac{dy}{dx} = -\frac{2x}{4} = -\frac{x}{2}$

$$\text{slope of the tangent } \frac{dy}{dx} \text{ or } -\frac{x}{2}, \text{slope of the normal } -\frac{1}{\frac{dy}{dx}} \text{ or } \frac{2}{x}$$

Since it is parallel to the line $4y = 2x - 5$ whose slope of tangent and normal is same $\frac{1}{2}$.

Tangent: − $-\frac{x}{2} = \frac{1}{2}$ or $x = -1 \quad \therefore\ 1 = -4y + 3$ or $y = \frac{1}{2}$

then the equation of the tangent is $y - \frac{1}{2} = \frac{1}{2}(x + 1) \quad \therefore\ 2y - 1 = x + 1 \quad \therefore\ x - 2y = -2$ Ans.

Normal: − $\frac{2}{x} = \frac{1}{2}$ or $x = 4 \quad \therefore\ 16 = -4y + 3$ or $y = -\frac{13}{4}$

then the equation of the normal is $y + \frac{13}{4} = \frac{1}{2}(x - 4) \quad \therefore\ 4y + 13 = 2x - 8$ or $2x - 4y = 21$ Ans.

(b) Tangent and normal to parabola $2y^2 + 4 = -8x$ which is perpendicular to the line $6x + 3y = 4$.

Solution: − Tangent: − $2y^2 + 4 = -8x \quad \therefore\ 4y\frac{dy}{dx} = -8$ or $\frac{dy}{dx} = \frac{-8}{4y}$ or $\frac{dy}{dx} = -\frac{2}{y} =$ slope of the tangent

Since it is perpendicular to the line $6x = -3y + 4$ whose slope of the given line is -2

then the slope of the tangent is $\frac{1}{2}$. $\quad \therefore\ -\frac{2}{y} = \frac{1}{2}$ or $y = -4 \quad \therefore\ 2 \times 16 + 4 = -8x$ or $x = -\frac{36}{8} = -\frac{9}{2}$

Equation of the tangent is $y + 4 = \frac{1}{2}\left(x + \frac{9}{2}\right)$ or $2y + 8 = \frac{2x + 9}{2}$ or $4y + 16 = 2x + 9$ or $2x - 4y = 7$ Ans.

Normal: – slope of the normal is $-\frac{1}{\frac{dy}{dx}} = -\frac{1}{\frac{-2}{y}} = \frac{y}{2}$

Since it is perpendicular to the line $6x = -3y + 4$ whose slope of the given line is -2,

slope of the normal is $\frac{1}{2}$ or $\frac{y}{2} = \frac{1}{2}$ or $y = 1$ then $2 + 4 = -8x$ or $x = -\frac{6}{8} = -\frac{3}{4}$

Equation of the normal is $y - 1 = \frac{1}{2}\left(x + \frac{3}{4}\right)$ or $2y - 2 = \frac{4x + 3}{4}$ or $8y - 8 = 4x + 3$

or $4x - 8y = -11$ Ans.

(12) (a) On the ellipse $16x^2 + 25y^2 = 1$ the points at which the tangents are perpendicular to the line $5x = 4y$.

Solution: – Given $16x^2 + 25y^2 = 1$ $\therefore 32x + 50y\frac{dy}{dx} = 0$ or $\frac{dy}{dx} = -\frac{32x}{50y}$ or $\frac{dy}{dx} = -\frac{16x}{25y}$ = slope of the tangent

Since it is perpendicular to the line $5x = 4y$ whose slope of the tangent is $-\frac{4}{5}$.

or $-\frac{16x}{25y} = -\frac{4}{5}$ or $80x = 100y$ or $4x = 5y$ $\therefore x = \frac{5}{4}y$

put $x = \frac{5}{4}y$ then find y, $16x^2 + 25y^2 = 1$ or $16 \times \frac{25}{16}y^2 + 25y^2 = 1$ or $50y^2 = 1$ or $y^2 = \frac{1}{50}$ $\therefore y = \pm\frac{1}{5\sqrt{2}}$

At $y = \frac{1}{5\sqrt{2}}$ then $x = \frac{5}{4}y = \frac{5}{4} \times \frac{1}{5\sqrt{2}} = \frac{1}{4\sqrt{2}}$ and $y = -\frac{1}{5\sqrt{2}}$ then $x = \frac{5}{4}y = \frac{5}{4}\left(-\frac{1}{5\sqrt{2}}\right) = -\frac{1}{4\sqrt{2}}$

Points are $\left(\frac{1}{4\sqrt{2}}, \frac{1}{5\sqrt{2}}\right)$ and $\left(-\frac{1}{4\sqrt{2}}, -\frac{1}{5\sqrt{2}}\right)$ Ans.

(b) Determine the equation of normal at vertex of the parabola $(y - 3)^2 = 4(x + 5)$.

Solution: – vertex is $(-5,3)$ $\therefore (y - 3)^2 = 4(x + 5)$ or $2(y - 3)\frac{dy}{dx} = 4$

or $\frac{dy}{dx} = \frac{4}{2(y - 3)} = \frac{2}{y - 3}$ at a point $(-5,3)$ $\therefore \left(\frac{dy}{dx}\right)_{(-5,3)} = \frac{2}{3 - 3} = \frac{2}{0} = \infty$

slope of normal is $-\frac{1}{\frac{dy}{dx}} = -\frac{1}{\infty} = 0$. equation of the normal at a point $(-5,3)$

or $y - 3 = 0(x + 5)$ or $y - 3 = 0$ $\therefore y = 3$ Ans.

(c) If the tangent to the curve $y = f(x + a)$ at the point $(1,3)$ makes an angle $\frac{\pi}{3}$ with the positive x – axis then find $f'(1 + a)$.

Solution: – $y = f(x + a)$ $\therefore \frac{dy}{dx} = f'(x + a)$ at a point $(1,3)$

or $\left(\frac{dy}{dx}\right)_{(1,3)} = f'(1 + a)$, $\theta = \frac{\pi}{3}$ $\therefore$ slope $(m) = \tan\theta = \tan\frac{\pi}{3} = \sqrt{3}$ $\therefore f'(1 + a) = \sqrt{3}$ Ans.

Normal: – $\left(\frac{dy}{dx}\right)_{(1,3)} = f'(1 + a)$, slope of the normal is $-\frac{1}{\frac{dy}{dx}} = -\frac{1}{f'(1 + a)}$

or $\theta = \frac{\pi}{3}$ then $\tan\theta = -\frac{1}{f'(1 + a)}$ or $\tan\frac{\pi}{3} = -\frac{1}{f'(1 + a)}$ or $\sqrt{3} = -\frac{1}{f'(1 + a)}$ or $f'(1 + a) = -\frac{1}{\sqrt{3}}$ Ans.

(13) (a) Tangent and normal to the curve $x = \frac{2at}{1 + t}, y = \frac{2at^2}{1 + t}$ at the point for which $t = \frac{3}{2}$.

Solution: $-$ Given , $x=\dfrac{2at}{1+t}, y=\dfrac{2at^2}{1+t}$ at the point $t=\dfrac{3}{2}$ $\therefore$ $x=\dfrac{6a}{5}, y=\dfrac{9a}{5}$

or $\dfrac{dx}{dt}=\dfrac{(1+t)2a-2at}{(1+t)^2}=\dfrac{2a+2at-2at}{(1+t)^2}=\dfrac{2a}{(1+t)^2}$

or $\dfrac{dy}{dt}=\dfrac{(1+t)4at-2at^2}{(1+t)^2}=\dfrac{4at+4at^2-2at^2}{(1+t)^2}=\dfrac{4at+2at^2}{(1+t)^2}$

or $\dfrac{dy}{dx}=\dfrac{\frac{dy}{dt}}{\frac{dx}{dt}}=\dfrac{\frac{4at+2at^2}{(1+t)^2}}{\frac{2a}{(1+t)^2}}=\dfrac{4at+2at^2}{2a}=\dfrac{2a(2t+t^2)}{2a}=2t+t^2$

slope of the tangent $\left(\dfrac{dy}{dx}\right)_{t=\frac{3}{2}}=2\times\dfrac{3}{2}+\dfrac{9}{4}=\dfrac{21}{4}$ and slope of the normal $-\dfrac{1}{\frac{dy}{dx}}=-\dfrac{1}{\frac{21}{4}}=-\dfrac{4}{21}$

Equation of their equation is $y-\dfrac{9a}{5}=\dfrac{21}{4}\left(x-\dfrac{6a}{5}\right)$ and $y-\dfrac{9a}{5}=-\dfrac{4}{21}\left(x-\dfrac{6a}{5}\right)$

or $20y-36a=105x-126a$ and $105y-189a=-20x+24a$ $\therefore$ $21x-4y=18a$ and $20x+105y=213a$ Ans.

(b) Normal at any point θ to the curve $x=\sin\theta-\theta\cos\theta,\ y=\cos\theta+\theta\sin\theta$ also show that it is at a constant distance from the origin.

Solution: $-$ Given, $x=\sin\theta-\theta\cos\theta,\ y=\cos\theta+\theta\sin\theta$ $\therefore$ $\dfrac{dy}{dx}=\dfrac{\frac{dy}{d\theta}}{\frac{dx}{d\theta}}=\dfrac{\theta\cos\theta}{\theta\sin\theta}=\cot\theta$

slope of the normal is $-\dfrac{1}{\frac{dy}{dx}}=-\dfrac{1}{\cot\theta}=-\tan\theta$ then the equation of the normal is $y-(\cos\theta+\theta\sin\theta)=-\dfrac{\sin\theta}{\cos\theta}[x-(\sin\theta-\theta\cos\theta)]$

solve , $x\sin\theta+y\cos\theta=\sin^2\theta+\cos^2\theta=1$ its distance from the origin is clearly $\dfrac{1}{\sqrt{\sin^2\theta+\cos^2\theta}}=1$ i.e constant Proved.

(14) (a) The rectangular co $-$ ordinates of a point on the curve are given by $x=2\tan\theta-\tan^2\theta$ and $y=2\cot\theta-\cot^2\theta$ find the equation of the tangent at any point on the curve and show that at the point P where $\theta=\dfrac{\pi}{4}$, the tangent passes through the origin.

Solution: $-$ Given, $x=2\tan\theta-\tan^2\theta,\ y=2\cot\theta-\cot^2\theta$

or $\dfrac{dx}{d\theta}=2\sec^2\theta-2\tan\theta\sec^2\theta$ and $\dfrac{dy}{d\theta}=-2\operatorname{cosec}^2\theta+2\cot\theta\operatorname{cosec}^2\theta$ then $\dfrac{dy}{dx}=\dfrac{\frac{dy}{d\theta}}{\frac{dx}{d\theta}}$

or $\dfrac{dy}{dx}=\dfrac{-2\operatorname{cosec}^2\theta+2\cot\theta\operatorname{cosec}^2\theta}{2\sec^2\theta-2\tan\theta\sec^2\theta}=\dfrac{2\operatorname{cosec}^2\theta(\cot\theta-1)}{2\sec^2\theta(1-\tan\theta)}=\dfrac{\cos^2\theta}{\sin^2\theta}\cdot\dfrac{(\cos\theta-\sin\theta)}{(\cos\theta-\sin\theta)}\cdot\dfrac{\cos\theta}{\sin\theta}=\dfrac{\cos^3\theta}{\sin^3\theta}$

slope of the tangent is $\dfrac{dy}{dx}=\dfrac{\cos^3\theta}{\sin^3\theta}$ then equation of the tangent is $y-(2\cot\theta-\cot^2\theta)=\dfrac{\cos^3\theta}{\sin^3\theta}[x-(2\tan\theta-\tan^2\theta)]$

or $y-2\dfrac{\cos\theta}{\sin\theta}+\dfrac{\cos^2\theta}{\sin^2\theta}=\dfrac{\cos^3\theta}{\sin^3\theta}\left(x-2\dfrac{\sin\theta}{\cos\theta}+\dfrac{\sin^2\theta}{\cos^2\theta}\right)$

or $y\sin^2\theta-2\sin\theta\cos\theta+\cos^2\theta=\dfrac{\cos\theta}{\sin\theta}(x\cos^2\theta-2\sin\theta\cos\theta+\sin^2\theta)$

or $y\sin^3\theta-2\sin^2\theta\cos\theta+\cos^2\theta\sin\theta=x\cos^3\theta-2\sin\theta\cos^2\theta+\sin^2\theta\cos\theta$

Divide $\sin^3\theta\cos^3\theta$ both of side , we get

or $y\sec^3\theta-2\operatorname{cosec}\theta\sec^2\theta+\operatorname{cosec}^2\theta\sec\theta=x\operatorname{cosec}^3\theta-2\operatorname{cosec}^2\theta\sec\theta+\operatorname{cosec}\theta\sec^2\theta$

or $x\operatorname{cosec}^3\theta-y\sec^3\theta=3\operatorname{cosec}^2\theta\sec\theta-3\operatorname{cosec}\theta\sec^2\theta=3\sec\theta\operatorname{cosec}\theta(\operatorname{cosec}\theta-\sec\theta)$

$$\therefore\ x\,\mathrm{cosec}^3\theta - y\sec^3\theta = 3\sec\theta\,\mathrm{cosec}\,\theta\,(\mathrm{cosec}\,\theta - \sec\theta)$$

If $\theta = \frac{\pi}{4}$ then $\mathrm{cosec}\,\theta = \mathrm{cosec}\frac{\pi}{4} = \sqrt{2} = \sec\frac{\pi}{4}$, Tangent is $2\sqrt{2}x - 2\sqrt{2}y = 0$

or $x - y = 0$ it clearly passes through the origin. Proved.

(b) The parametric equation of a curve are $x = a\sin t$, $y = a\cos 2t$. prove that the equation of tangent and normal at any point t are $x\sin 2t + y\cos 2t = a$ and $x\cos 2t = y\sin 2t$ respectively. also show that S. T (sub – tangent) $= -a\cos 2t.\cot 2t$ and sub – normal (S. N) $= -a\sin 2t$.

Solution: – Given, $x = a\sin 2t$, $y = a\cos 2t$ then $\frac{dx}{dt} = 2a\cos 2t$ and $\frac{dy}{dt} = -2a\sin 2t$

$$\therefore\ \frac{dy}{dx} = \frac{dy}{dt} \div \frac{dx}{dt} = \frac{-2a\sin 2t}{2a\cos 2t} = -\frac{\sin 2t}{\cos 2t} = -\tan 2t$$

slope of the tangent is $-\tan 2t$ and slope of the normal is $\cot 2t$ then, equation of the tangent is

$$\therefore\ y - a\cos 2t = -\frac{\sin 2t}{\cos 2t}(x - a\sin 2t) \quad \text{or } y\cos 2t - a\cos^2 2t = -x\sin 2t + a\sin^2 2t$$

or $x\sin 2t + y\cos 2t = a\sin^2 2t + a\cos^2 2t$ or $x\sin 2t + y\cos 2t = a(\sin^2 2t + \cos^2 2t)$ $\therefore$ $x\sin 2t + y\cos 2t = a$ Proved.

Equation of the normal is $y - a\cos 2t = \frac{\cos 2t}{\sin 2t}(x - a\sin 2t)$

or $y\sin 2t - a\sin 2t\cos 2t = x\cos 2t - a\sin 2t\cos 2t$ $\therefore$ $x\cos 2t = y\sin 2t$ Proved.

Also, Sub – Tangent (S. T) $= \frac{y}{y'} = \frac{a\cos 2t}{-\frac{\sin 2t}{\cos 2t}} = -\frac{a\cos^2 2t}{\sin 2t} = -a\cos 2t.\cot 2t$ Proved.

$$\text{Sub – Normal (S. N)} = yy' = a\cos 2t.\left(-\frac{\sin 2t}{\cos 2t}\right) = -a\sin 2t \quad \text{Proved.}$$

(15) (a) Tangent and normal to the curve $y = (x^2 - 4)(x - 1)$ at the points where the curve cuts the x – axis.

Solution: – Given, $y = (x^2 - 4)(x - 1)$ the curve cuts the x – axis $y = 0$

or $0 = (x^2 - 4)(x - 1)$ $\therefore$ $(x^2 - 4) = 0$ or $(x - 1) = 0$ $\therefore$ $x = \pm 2$ or $x = 1$, Points are $(1,0), (2,0), (-2,0)$

At a point $(1,0)$, $\therefore\ y = (x^2 - 4)(x - 1)$ $\therefore\ \frac{dy}{dx} = (x^2 - 4).1 + (x - 1).2x$ $\therefore\ \left(\frac{dy}{dx}\right)_{(1,0)} = (1 - 4) + 0 = -3$

slope of tangent is -3 then the equation of tangent and normal at a point $(1,0)$

$y - 0 = -3(x - 1)$ and $y - 0 = \frac{1}{3}(x - 1)$ or $y = -3x + 3$ and $3y = x - 1$ $\therefore$ $3x + y = 3$ and $x - 3y = 1$ Ans.

Similiarly, point $(2,0)$ then slope of the tangent $\left(\frac{dy}{dx}\right)_{(2,0)} = 3x^2 - 2x - 4 = 4$ and slope of the normal is $-\frac{1}{4}$.

Equation of the tangent and normal is $y - 0 = 4(x - 2)$ and $y - 0 = -\frac{1}{4}(x - 2)$

$$\boxed{\therefore\ 4x - y = 2 \text{ and } x + 4y = 2 \quad \text{Ans.}}$$

At a point $(-2,0)$, slope of the tangent $\left(\frac{dy}{dx}\right)_{(-2,0)} = 3x^2 - 2x - 4 = 12$ and slope of the normal is $-\frac{1}{12}$.

Equation of the tangent and normal is $y - 0 = 12(x + 2)$ and $y - 0 = -\frac{1}{12}(x + 2)$

$$\boxed{\therefore\ 12x - y = -24 \text{ and } x + 12y = -2 \quad \text{Ans.}}$$

(b) Find the slopes of the tangent of the curve $x = \left(y - \frac{3}{2}\right)(y+1)$ at the points where it cuts the axis of y .

Solution: − Given, $x = \left(y - \frac{3}{2}\right)(y+1)$ it cuts the axis of y i.e $x = 0$

or $x = 0,\ y = \frac{3}{2}, -1$ then the points are $(0,-1), \left(0, \frac{3}{2}\right)$ $\therefore\ x = \left(y - \frac{3}{2}\right)(y+1)$ or $x = y^2 + y - \frac{3}{2}y - \frac{3}{2}$

Differentiating, $1 = 2y\frac{dy}{dx} + \frac{dy}{dx} - \frac{3}{2}\frac{dy}{dx} - 0$ or $\frac{dy}{dx} = \frac{1}{2y + 1 - \frac{3}{2}} = \frac{1}{2y - \frac{1}{2}} = \frac{2}{4y-1}$

slope of the tangent $\left(\frac{dy}{dx}\right)_{(0,-1)} = \frac{2}{-4-1} = -\frac{2}{5}$ and $\left(\frac{dy}{dx}\right)_{\left(0,\frac{3}{2}\right)} = \frac{2}{4 \times \frac{3}{2} - 1} = \frac{2}{6-1} = \frac{2}{5}$

Then the two slope are $-\frac{2}{5}, \frac{2}{5}$ Ans.

(c) Tangent to the curve $y^2 = 2x^2 + 4x + 1$ at the point whose abscissa x is 4 .

Solution: − Given, $x = 4$ then $y^2 = 2x^2 + 4x + 1$ or $y^2 = 32 + 16 + 1 = 49$ $\therefore\ y = \pm 7$

Points are $(4,7), (4,-7)$

curve is $y^2 = 2x^2 + 4x + 1$ $\therefore$ Differentiating, $2y\frac{dy}{dx} = 4x + 4$ or $\frac{dy}{dx} = \frac{2x+2}{y}$

slopf of the tangent $= \left(\frac{dy}{dx}\right)_{(4,7)} = \frac{10}{7}$ and $\left(\frac{dy}{dx}\right)_{(4,-7)} = -\frac{10}{7}$, slope of the normal $= -\frac{7}{10}$ and $\frac{7}{10}$

Equaiton of the tangent is $y - 7 = \frac{10}{7}(x-4)$ and $y - 7 = -\frac{10}{7}(x-4)$

or $7y - 49 = 10x - 40$ and $7y - 49 = -10x + 40$ $\therefore\ 10x - 7y = -9$ and $10x + 7y = 89$ Ans.

Equation of the normal is $y - 7 = -\frac{7}{10}(x-4)$ and $y - 7 = \frac{7}{10}(x-4)$

or $10y - 70 = -7x + 28$ and $10y - 70 = 7x - 28$ $\therefore\ 7x + 10y = 98$ and $7x - 10y = -42$ Ans.

(16) (a) The points on the curve $x^4 + 3y^2 = 16x^2$ where the tangent is horizantal is (are) ... ...

Solution: − $x^4 + 3y^2 = 16x^2$

Differentiating $4x^3 + 6y\frac{dy}{dx} = 32x$ or $\frac{dy}{dx} = \frac{32x - 4x^3}{6y} = \frac{16x - 2x^3}{3y}$

If tangent is horizantal $\frac{dy}{dx} = 0$, $\frac{16x - 2x^3}{3y} = 0$ or $16x - 2x^3 = 0$ or $2x^3 = 16x$ or $2x^2 = 16$

or $x^2 = 8$ $\therefore x = \pm 2\sqrt{2}$, now $x = -2\sqrt{2}$ is a negative value so x^2 is already positive then , $x = -2\sqrt{2}$ is ruled out

Hence, $x = 2\sqrt{2}$, $x^4 + 3y^2 = 16x^2$ $\therefore$ $\left(2\sqrt{2}\right)^4 + 3y^2 = 16\left(2\sqrt{2}\right)^2$ or $3y^2 = 128 - 64$ or $y^2 = \frac{64}{3}$

or $y = \pm\frac{8}{\sqrt{3}}$, point is $\left(2\sqrt{2}, \pm\frac{8}{\sqrt{3}}\right)$ Ans.

(b) The curve $2y^2 = \log y^2 - 2x$ has a vertical tangent at the point.

Solution: − $2y^2 = \log y^2 - 2x$ $\therefore$ $2y^2 + 2x = 2\log y$ or $4y\frac{dy}{dx} + 2 = 2.\frac{1}{y}\frac{dy}{dx}$ or $2y\frac{dy}{dx} + 1 = \frac{1}{y}\frac{dy}{dx}$

or $\frac{dy}{dx}=\frac{1}{\frac{1}{y}-2y}=\frac{y}{1-2y^2}$

If tangent is vertical $\frac{dy}{dx}=\infty$ $\therefore$ $\frac{y}{1-2y^2}=\infty$ or $1-2y^2=0$ or $y^2=\frac{1}{2}$ $\therefore$ $y=\pm\frac{1}{\sqrt{2}}$

At $y=\frac{1}{\sqrt{2}}$ then $2x=\log y^2-2y^2$ $\therefore$ $2x=2\log 2^{-\frac{1}{2}}-2\times\frac{1}{2}$ or $2x=-(\log 2+1)$ $\therefore$ $x=-\frac{1}{2}(\log 2+1)$

At $y=-\frac{1}{\sqrt{2}}$ then $2x=\log y^2-2y^2$ $\therefore$ $2x=\log\left(-\frac{1}{\sqrt{2}}\right)^2-2\left(-\frac{1}{\sqrt{2}}\right)^2=\log\frac{1}{2}-1$ $\therefore$ $x=-\frac{1}{2}(\log 2+1)$

Points are $\left[-\frac{1}{2}(\log 2+1),\frac{1}{\sqrt{2}}\right]$ and $\left[-\frac{1}{2}(\log 2+1),-\frac{1}{\sqrt{2}}\right]$ Ans.

(17) (a) Normals at the points on the curve $y=x+\frac{1}{x}$ where the tangent makes an angle of $\frac{3\pi}{4}$ with $x-$ axis.

Solution: $-$ $y=x+\frac{1}{x}$ $\therefore$ $\frac{dy}{dx}=1-\frac{1}{x^2}$, $\tan\theta=\frac{3\pi}{4}=-1$

or $\frac{dy}{dx}=\tan\theta$ or $1-\frac{1}{x^2}=-1$ or $x^2-1=-x^2$ or $2x^2=1$ or $x^2=\frac{1}{2}$ $\therefore$ $x=\pm\frac{1}{\sqrt{2}}$

At $x=\pm\frac{1}{\sqrt{2}}$ then $y=x+\frac{1}{x}=\frac{1}{\sqrt{2}}+\frac{1}{\frac{1}{\sqrt{2}}}=\frac{3}{\sqrt{2}}$ and $y=x+\frac{1}{x}=-\frac{1}{\sqrt{2}}-\frac{1}{\frac{1}{\sqrt{2}}}=-\frac{3}{\sqrt{2}}$

Points are $\left(\frac{1}{\sqrt{2}},\frac{3}{\sqrt{2}}\right),\left(-\frac{1}{\sqrt{2}},-\frac{3}{\sqrt{2}}\right)$

Normal is $y-\frac{3}{\sqrt{2}}=1\left(x-\frac{1}{\sqrt{2}}\right)$ and $y+\frac{3}{\sqrt{2}}=1\left(x+\frac{1}{\sqrt{2}}\right)$ or $x-y=-\sqrt{2}$ and $x-y=\sqrt{2}$ Ans.

(b) Find the points on the curve $2x^3+y^2-24x=0$ the tangents at which are parallel to $x-$ axis.

Solution: $-$ $2x^3+y^2-24x=0$ $\therefore$ $6x^2+2y\frac{dy}{dx}-24=0$ or $\frac{dy}{dx}=\frac{24-6x^2}{2y}$

Tangent parallel to $x-$ axis $\frac{dy}{dx}=0$ or $\frac{24-6x^2}{2y}=0$ or $24-6x^2=0$ or $x^2=4$ $\therefore$ $x=\pm 2$

At $x=2$ then $y^2=24(2)-2(2)^3=48-16=32$ $\therefore$ $y=\pm 4\sqrt{2}$ point $x=2,y=\pm 4\sqrt{2}$

At $x=-2$ then $y^2=24(-2)-2(-2)^3=-48+16=-32$ $\therefore$ $y=\pm 4\sqrt{2}i$ point $x=-2,y=\pm 4\sqrt{2}i$

Points are $(2,\pm 4\sqrt{2}),(-2,\pm 4\sqrt{2}i)$ Ans.

Above Question: $-$ Tangent parallel to $y-$ axis $\frac{dy}{dx}=\infty$ or $\frac{24-6x^2}{2y}=\infty$ $\therefore$ $2y=0$ or $y=0$

At $y=0$ then $2x^3+y^2-24x=0$ or $2x^3-24x=0$ $\therefore$ $2x(x^2-12)=0$ or $x=0$ and $x=\pm 2\sqrt{3}$

Points are $(0,0),(\pm 2\sqrt{3},0)$ Ans.

(18) (a) Find the equation of the tangent and normal to the curve $y=e^{xy}+\sin^{-1}(\sin^2 x)$ at $x=0$.

Solution: $-$ Given, $y=e^{xy}+\sin^{-1}(\sin^2 x)$ at $x=0$, $y=1+0=1$

or $\frac{dy}{dx}=e^{xy}.\left(x\frac{dy}{dx}+y.1\right)+\frac{1}{\sqrt{1-\sin^4 x}}.2\sin x\cos x=\frac{ye^{xy}+\frac{2\sin x\cos x}{\sqrt{1-\sin^4 x}}}{1-xe^{xy}}$ at a point $(0,1)$.

or $\left(\frac{dy}{dx}\right)_{(0,1)} = \frac{1+0}{1-0} = 1$ slope of the tangent is 1 and slope of normal is -1.

Equation of the tangent and normal is $y - 1 = 1(x - 0)$ and $y - 1 = -1(x - 0)$ or $y - x = 1$ and $x + y = 1$ Ans.

(b) Find the equation of the tangent and normal to the curve $y = e^{\sin x} + a^{\cos x} + \log(1 + x)$.

Solution: – Given, $y = e^{\sin x} + a^{\cos x} + \log(1 + x)$ at $x = 0$, $y = e^0 + a^1 + \log 1 = 1 + a$

$\frac{dy}{dx} = e^{\sin x}.\cos x + a^{\cos x}.\log a\,(-\sin x) + \frac{1}{1+x}$ at a point $(0,1+a)$ or $\left(\frac{dy}{dx}\right)_{(0,1+a)} = 1 + a\log a.0 + 1 = 2$

slope of the tangent is 2 and slope of the normal is $-\frac{1}{2}$

then equation of the tangent and normal are $\therefore y - (1 + a) = 2(x - 0)$ and $y - (1 + a) = -\frac{1}{2}(x - 0)$

or $y - (1 + a) = 2x$ and $2y - 2(1 + a) = -x$ or $y - 2x = 1 + a$ and $x + 2y = 2(1 + a)$ Ans.

(19) Find the angle of intersection of the curve: – (a) $x - y = \sqrt{2}\,a$ and $x + y = a$ (b) $x^2 + y^2 = 3a^2$ and $x^2 - y^2 = a^2$

Solution: – (a) $x - y = \sqrt{2}\,a$ and $x + y = a$ $\therefore 1 - \frac{dy}{dx} = 0$ and $1 + \frac{dy}{dx} = 0$ or $\frac{dy}{dx} = 1 = m_1$ and $\frac{dy}{dx} = -1 = m_2$

If θ be the angle between the curves, then $\tan\theta = \frac{m_1 - m_2}{1 + m_1m_2} = \frac{1+1}{1-1} = \frac{2}{0} = \infty$ $\therefore \cot\theta = 0$ or $\theta = \frac{\pi}{2}$ Ans.

(b) $x^2 + y^2 = 3a^2$ and $x^2 - y^2 = a^2$ $\therefore 2x + 2y\frac{dy}{dx} = 0$ and $2x - 2y\frac{dy}{dx} = 0$ or $\frac{dy}{dx} = -\frac{x}{y} = m_1$ and $\frac{dy}{dx} = \frac{x}{y} = m_2$

If θ be the angle between the curve then $\tan\theta = \frac{m_1 - m_2}{1 + m_1m_2} = \frac{-2xy}{y^2 - x^2}$ for points of intersection

or $x^2 = 2a^2, y^2 = a^2$ or $x = \sqrt{2}\,a$, $y = a$ $\therefore \tan\theta = \frac{-2\sqrt{2}\,a.a}{a^2 - 2a^2} = \frac{-2\sqrt{2}\,a^2}{-a^2} = 2\sqrt{2}$ $\therefore \tan\theta = 2\sqrt{2}$ or $\theta = \tan^{-1}(2\sqrt{2})$ Ans.

(20) Find the angle of intersection of the curve: – (a) $x^2y = 1$ and $xy = 1$ (b) $y^2 = x^2 + 4$ and $y^2 = 2x + 3$

(c) $x^2 + 2x = y^2$ and $x^2 + 1 = y^2$

Solution: – (a) $x^2y = 1 \ldots\ldots\ldots$ (i) and $xy = 1 \ldots\ldots\ldots\ldots$ (ii)

Differentiating, $x^2\frac{dy}{dx} + y.2x = 0$ and $x\frac{dy}{dx} + y.1 = 0$ or $\frac{dy}{dx} = -\frac{2y}{x} = m_1$ and $\frac{dy}{dx} = -\frac{y}{x} = m_2$

Solve the equatin (i) and (ii), we get $x = 1, y = 1$ then $m_1 = -\frac{2y}{x} = -2$, $m_2 = -\frac{y}{x} = -1$

If θ be the angle between the curves then $\tan\theta = \frac{m_1 - m_2}{1 + m_1m_2} = \frac{-2+1}{1+2} = -\frac{1}{3}$ or $\tan\theta = -\frac{1}{3}$ $\therefore \theta = \tan^{-1}\left(-\frac{1}{3}\right)$ Ans.

(b) $y^2 = x^2 + 4 \ldots\ldots\ldots\ldots\ldots\ldots$ (i) and $y^2 = 2x + 3 \ldots\ldots\ldots\ldots\ldots$ (ii)

Differentiating, $2y\frac{dy}{dx} = 2x$ and $2y\frac{dy}{dx} = 2$ or $\frac{dy}{dx} = \frac{x}{y} = m_1$ (say) and $\frac{dy}{dx} = \frac{1}{y} = m_2$ (say)

Solve the equation (i) and (ii), we get $y^2 = x^2 + 4$ and $y^2 = 2x + 3$

or $x^2 + 4 = 2x + 3$ or $x^2 - 2x + 1 = 0$ or $(x + 1)^2 = 0$

if $x = 1$ and $y^2 = x^2 + 4$ or $y^2 = 5$ $\therefore y = \pm\sqrt{5}$ then $m_1 = \frac{x}{y} = \frac{1}{\sqrt{5}}$, $m_2 = \frac{1}{y} = \frac{1}{\sqrt{5}}$

If θ be the angle between the curves then $\tan\theta = \frac{m_1 - m_2}{1 + m_1 m_2} = \frac{\frac{1}{\sqrt{5}} - \frac{1}{\sqrt{5}}}{1 + \frac{1}{\sqrt{5}}.\frac{1}{\sqrt{5}}} = 0$ or $\tan\theta = 0$ $\therefore \theta = 0$ Ans.

(c) $x^2 + 2x = y^2$ ……………….(i) and $x^2 + 1 = y^2$ …………………(ii)

Differentiating, $2x + 2 = 2y\frac{dy}{dx}$ and $2x = 2y\frac{dy}{dx}$ or $\frac{dy}{dx} = \frac{x+1}{y} = m_1$ and $\frac{dy}{dx} = \frac{x}{y} = m_2$

Solve the equation (i) and (ii), we get $x^2 + 2x = y^2$ and $x^2 + 1 = y^2$ $\therefore x^2 + 2x = x^2 + 1$

or $2x = 1$ $\therefore x = \frac{1}{2}$ At $x = \frac{1}{2}$ then $x^2 + 1 = y^2$ or $y^2 = \frac{1}{4} + 1 = \frac{5}{4}$ $\therefore y = \frac{\sqrt{5}}{2}$

If θ be the angle between the curves then $\tan\theta = \frac{m_1 - m_2}{1 + m_1 m_2} = \frac{\frac{x+1}{y} - \frac{x}{y}}{1 + \frac{x+1}{y}.\frac{x}{y}} = \frac{\frac{x+1-x}{y}}{\frac{y^2 + x^2 + x}{y^2}}$

$$\therefore \tan\theta = \frac{y}{y^2 + x^2 + x} = \frac{\frac{\sqrt{5}}{2}}{\frac{1}{4} + \frac{5}{4} + \frac{1}{2}} = \frac{\frac{\sqrt{5}}{2}}{\frac{1+5+2}{4}} = \frac{\sqrt{5}}{4} \quad \therefore \theta = \tan^{-1}\left(\frac{\sqrt{5}}{4}\right) \text{ Ans.}$$

(22) (a) The curves $x^2 + y^2 = a^2$ and $\frac{x}{a} + \frac{y}{b} = 1$ are intersect orthogonally then prove that $bx = -ay$.

Solution: – Given, $x^2 + y^2 = a^2$ and $\frac{x}{a} + \frac{y}{b} = 1$ Intersect orthogonally, $\left(\frac{dy}{dx}\right)_I \times \left(\frac{dy}{dx}\right)_{II} = -1$ or $m_1.m_2 = -1$

$x^2 + y^2 = a^2$ and $\frac{x}{a} + \frac{y}{b} = 1$ Differentiating, $2x + 2y\frac{dy}{dx} = 0$ and $\frac{1}{a} + \frac{1}{b}\frac{dy}{dx} = 0$ $\therefore$ $\frac{dy}{dx} = -\frac{x}{y} = m_1$ and $\frac{dy}{dx} = -\frac{b}{a} = m_2$

or $\left(\frac{dy}{dx}\right)_I \times \left(\frac{dy}{dx}\right)_{II} = -1$ or $-\frac{x}{y} \times -\frac{b}{a} = -1$ or $\frac{bx}{ay} = -1$ $\therefore$ $bx = -ay$ Proved.

(b) The curves $x^2y = a(x + 1)$ and $x^2 = by + 1$ are intersect orthogonally then prove that $x(a - b) = -2a$ and also prove that $\frac{a}{b} = \frac{x}{x+2}$.

Solution: – Given, $x^2y = a(x + 1)$ ……………..(i) and $x^2 = by + 1$ ………………..(ii)

Differentiating, $2x\frac{dy}{dx} + 2xy = a$ and $2x = b\frac{dy}{dx}$ or $\frac{dy}{dx} = \frac{a - 2xy}{2x} = m_1$ and $\frac{dy}{dx} = \frac{2x}{b} = m_2$

We know that, $m_1.m_2 = -1$ or $\frac{a - 2xy}{2x} \times \frac{2x}{b} = -1$ or $a + b = 2xy$ ……………..(A)

Equation (i), $x^2y = a(x + 1)$ or $xy = \frac{a(x+1)}{x}$ then put value xy in equation (A) we get

$\therefore$ $a + b = 2xy$ or $a + b = 2\frac{a(x+1)}{x}$ or $ax + bx = 2ax + 2a$ or $2ax - ax - bx = -2a$

or $ax - bx = -2a$ $\therefore$ $x(a - b) = -2a$ Proved.

Again, $ax + bx = 2ax + 2a$ or $ax - 2ax - 2a = -bx$ or $-(ax + 2a) = -bx$ or $a(x + 2) = bx$ $\therefore$ $\frac{a}{b} = \frac{x}{x+2}$ Proved.

(c) Two curves $\left(\frac{x}{b}\right)^3 + \left(\frac{y}{a}\right)^3 = 1$ and $x^3y^3 = ab$ are touch at the any point then prove that $\frac{y}{x} = \frac{b}{a}$.

Solution: – Given, $\left(\frac{x}{b}\right)^3 + \left(\frac{y}{a}\right)^3 = 1$ and $x^3y^3 = ab$

Differentiating, $3\left(\frac{x}{b}\right)^2.\frac{1}{b} + 3\left(\frac{y}{a}\right)^2.\frac{1}{a}.\frac{dy}{dx} = 0$ and $x^3.3y^2\frac{dy}{dx} + y^3.3x^2 = 0$

$$\therefore \frac{dy}{dx} = -\frac{3\left(\frac{x}{b}\right)^2 \cdot \frac{1}{b}}{3\left(\frac{y}{a}\right)^2 \cdot \frac{1}{a}} = -\frac{a^3}{b^3} \cdot \frac{x^2}{y^2} \text{ and } \frac{dy}{dx} = -\frac{y^3 . 3x^2}{x^3 . 3y^2} = -\frac{y}{x}$$

Two curve are touch then we know that $\left(\frac{dy}{dx}\right)_I = \left(\frac{dy}{dx}\right)_{II}$ or $-\frac{a^3}{b^3} \cdot \frac{x^2}{y^2} = -\frac{y}{x}$ $\therefore \frac{y^3}{x^3} = \frac{a^3}{b^3}$ $\therefore \frac{y}{x} = \frac{a}{b}$ Proved.

Exercise – A6

(1) (a)Find the tangent and normal to the curve $y = 3x + 2e^x$ at the point $x = 0$.

(b) Find the equation of the tangent and normal to the curve $y^2 = \sqrt{x}y + \frac{2}{\sqrt{x}}$ at the point $x = 1$.

(c) Find the equation of the tangent and normal to the curve $y(x + 1)(x - 2) + x - 7 = 0$ at the point where it cuts the axis of x.

(2) (a) Find the equation of the normal to the curve $y = (1 + x) \tan^{-1}(\tan^2 x)$ at $x = 0$.

(b) The equation of the normal to the curve $x^3 - y^3 = xy$ at the point (2,2).

(c) The equation of the tangent to the curve $e^{xy} = x + y$ at the point (1,1).

(3) (a) on the ellipse $16x^2 + 25y^2 = 1$ the points at which the tangents are perpendicular to the line $5x = 4y$.

(b) If the tangent to the curve $y = f(x)$ at the point (2,3) makes an angle $\frac{\pi}{3}$ with the positive x – axis then $f'(2)$ is equal to ……….

(c) The straight line $2x - 3y = 5a$ will be a tangent to the ellipse $\frac{x^2}{4} + \frac{y^2}{5} = 1$ then find the value of a.

(4) (a) Find the equation of the tangent and normal to the curve $x^2y^2 = x^2 + y^2$ at the point $x = \sin\theta$, $y = \cos\theta$ and angle $\theta = \frac{\pi}{4}$.

(b) Find the equation of the tangent and normal to the curve $x^2 + y^2 = a^2$ at the point $x = a\tan^2\theta$, $y = a\sec^2\theta$ and angle $\theta = \frac{\pi}{4}$.

(c) The equation of the tangent and normal to the curve $y = e^{x+y} + \log x$ at the point (1,0).

(5) (a) Tangents to $y = (x^2 - 1)(x + 2)$ at the points where the curve cuts the y – axis.

(b) Find the equation of the tangent to the curve $y = x^3 - 2$ of the point $(1, -1)$.

(c) Find the equation of the tangent and normal to the curve $y^2 = 2xy - 3$ at the point $y = 1$.

(6) (a) The point on the curve $y^2 + 3x = 12y$ where the tangent is vertical.

(b) The points on the curve $y^3 + 3x^2 = 12y$ where the tangent is horizantal.

(c) The points on the curves $2y^3 - 5x^2 = 24y$ where the tangent is vertical.

Answer

(1) (a) Tangent $5x - y + 2 = 0$, Normal $x + 5y = 10$

(b) Point are $(1, -1), (1,2)$, Tangent $2x - 5y = 7$ and $2x + 7y = 16$, Normal $5x + 2y = 3$ and $7x - 2y = 3$

(c) Tangent $x + 40y = 7$ and Normal $40x - y = 280$

(2) (a) Point $(x = 0, y = 0)$, $\boxed{x + y = 0}$ (b) $7x + 5y = 24$ (c) $x + y = 2$

(3) (a) Points $\left(\frac{1}{4\sqrt{2}}, \frac{1}{5\sqrt{2}}\right), \left(-\frac{1}{4\sqrt{2}}, -\frac{1}{5\sqrt{2}}\right)$ (b) $\sqrt{3}$ (c) $a = \pm\frac{2\sqrt{14}}{5}$

(4) (a) $x\sin^3\theta + y\cos^3\theta = \sin^4\theta + \cos^4\theta$ and $4x\cos^3\theta - 4y\sin^3\theta = \sin 4\theta$

(b) $x\sin^2\theta\cos^2\theta - y\cos^2\theta = a(\sin^4\theta - 1)$ and $x\cos^2\theta + y\sin^2\theta\cos^2\theta = 2a\sin^2\theta$ At $\theta = \frac{\pi}{4}$, $x - 2y = -3a$ and $2x + y = 4a$

(c) $(e+1)x - (1-e)y = e+1$ and $(e-1)x - (e+1)y = e-1$

(5) (a) $x + y = -2$ (b) $3x - y = 2$ (6) (a) Point $(12,6)$ (b) Points are $(0,0)$ & $(0, \pm 2\sqrt{3})$ (c) Points are $\left(\pm\frac{4\sqrt{2}}{\sqrt{5}}, -2\right)$ or $\left(\pm 4\sqrt{\frac{2}{5}}, -2\right)$

Differentiation

(1) Basic formulae: – (a) Definition: – Differential coefficient (or derivative) of a function.

If $y = f(x)$, then we define $\frac{dy}{dx} = \lim_{\Delta x \to 0} \frac{f(x+\Delta x) - f(x)}{\Delta x}$ we may define $\frac{dy}{dx}$ at $x = a$ as

$$\left(\frac{dy}{dx}\right)_{x=a} = \lim_{x \to a} \frac{f(x) - f(a)}{x - a} \quad \text{Also} \quad \left(\frac{dy}{dx}\right)_{x=a} = \lim_{\Delta x \to 0} \frac{f(a+\Delta x) - f(a)}{\Delta x}$$

In evaluating the limits the following should be noted

(i) $\lim_{\theta \to 0} \frac{\sin\theta}{\theta} = 1$ (ii) Limit of sum = sum of limits (iii) Limit of product = product of limits (iv) $\lim_{h \to 0} (1+h)^{\frac{1}{h}} = e$ and $\lim_{h \to 0}\left(1 + \frac{h}{2}\right)^{\frac{1}{h}} = e^{\frac{1}{2}}$

(2) Fundamental Theorems: – Let u, v, w, ……… be functions of x whose derivative exist.

(i) $\frac{d(k)}{dx} = 0$ where k is a constant. (ii) $\frac{d(ku)}{dx} = k\frac{du}{dx}$ (iii) $\frac{d(u \pm v)}{dx} = \frac{du}{dx} \pm \frac{dv}{dx}$ sum or difference (iv) $\frac{d(uv)}{dx} = u\frac{dv}{dx} + v\frac{du}{dx}$ Product

(v) $\frac{d\left(\frac{u}{v}\right)}{dx} = \frac{v\frac{du}{dx} - u\frac{dv}{dx}}{v^2}$ Quotient (vi) If $y = f(t)$ and $t = \phi(x)$ then $\frac{dy}{dx} = \frac{dy}{dt} \times \frac{dt}{dx}$ (function of a function)

(vii) If $u = f(y)$, then $\frac{du}{dx} = \frac{du}{dy}.\frac{dy}{dx}$ ……………..(A) $f'(y) = \frac{du}{dy}$

put value of $\frac{du}{dy}$ in equation (A), we get $\therefore \frac{du}{dx} = \frac{du}{dy}.\frac{dy}{dx}$ $\therefore \frac{du}{dx} = f'(y).\frac{dy}{dx}$

(viii) $\frac{dy}{dx}.\frac{dx}{dy} = 1$ or $\frac{dy}{dx} = \frac{1}{\left(\frac{dx}{dy}\right)}$

(ix) Differentiation of one function with respect to another function. Let $y = f(x)$, $z = \phi(x)$ and we have to differentiable

f(x) w. r. t $\phi(x)$ so that we have to find the value of $\frac{dy}{dz} = \frac{dy}{dx} + \frac{dz}{dx}$ or $\frac{dy}{dz} = \frac{dy}{dx}.\frac{dx}{dz}$

(x) Logarithmic differentiation: –If $y = [f_1(x)]^{f_2(x)}$ or $y = f_1(x)f_2(x)f_3(x)$ …….. or $y = \frac{f_1(x)f_2(x)\ldots\ldots}{\phi_1(x)\phi_2(x)\ldots\ldots}$

Then it will be convenient to take log of both sides before performing differentiation.

(3) (i) Parametric equation: – If $x = f(t)$, $y = \phi(t)$ then $\frac{dy}{dx} = \frac{dy}{dt} + \frac{dx}{dt}$

(ii) Implicit function: – $f(x, y) = c$, Differentiate each term with respect to x, we get $\frac{d[\phi(y)]}{dx} = \frac{d[\phi(y)]}{dx}.\frac{dy}{dx}$

Example: – $x^4 + y^4 - 5axy = 0$, Differentiating w. r. t x, we get

or $4x^3 + 4y^3\frac{dy}{dx} - 5a\left(y.1 + x\frac{dy}{dx}\right) = 0$ or $4x^3 + 4y^3\frac{dy}{dx} - 5ay - 5ax\frac{dy}{dx} = 0$ or $\frac{dy}{dx}(4y^3 - 5ax) = 5ay - 4x^3$ $\therefore \frac{dy}{dx} = \frac{5ay - 4x^3}{4y^3 - 5ax}$ Ans.

Partial Differentiation: – If $f(x, y) = c$ then we can find $\frac{dy}{dx}$ by the help of partial differentiation as under $\frac{dy}{dx} = -\frac{f_x}{f_y}$

where f_x is differential coefficient of $f(x, y)$ w. r. t. x treating y as constant. f_y is differentiation of $f(x, y)$ w. r. t. y treating x as constant.

Example: – If $f(x, y) = x^4 + y^4 - 5axy = 0$ then $f_x = 4x^3 - 5ay$, $f_y = 4y^3 - 5ax$

$$\therefore \frac{dy}{dx} = -\frac{f_x}{f_y} = -\frac{4x^3 - 5ay}{4y^3 - 5ax} \quad \therefore \frac{dy}{dx} = \frac{5ay - 4x^3}{4y^3 - 5ax} \quad \text{Ans.}$$

Some Standard Results

(i) $y = x^n,\ \frac{dy}{dx} = nx^{n-1}$ or $y = u^n,\ \frac{dy}{dx} = nu^{n-1}\frac{du}{dx}$

Note: $-\ y = \sqrt{x}$ then $\frac{dy}{dx} = \frac{1}{2\sqrt{x}}$ (Important result), $y = \frac{1}{x^n} = x^{-n} = -nx^{-n-1} = -nx^{-(n+1)} = -\frac{n}{x^{n+1}}$

(ii) $y = e^x$ then $\frac{dy}{dx} = e^x$ and $y = e^u$ then $\frac{dy}{dx} = e^u\frac{du}{dx}$ (iii) $y = \log x$ then $\frac{dy}{dx} = \frac{1}{x}$ and $y = \log u$ then $\frac{dy}{dx} = \frac{1}{u}\frac{du}{dx}$

(iv) $y = a^x$ then $\frac{dy}{dx} = a^x \log a$ and $y = a^u$ then $\frac{dy}{dx} = a^u \log a\frac{du}{dx}$ (v) $y = \sin x$ then $\frac{dy}{dx} = \cos x$ and $y = \sin u$ then $\frac{dy}{dx} = \cos u\frac{du}{dx}$

(vi) $y = \cos x$ then $\frac{dy}{dx} = -\sin x$ and $y = \cos u$ then $\frac{dy}{dx} = -\sin u\frac{du}{dx}$

(vii) $y = \tan x$ then $\frac{dy}{dx} = \sec^2 x$ and $y = \tan u$ then $\frac{dy}{dx} = \sec^2 u\frac{du}{dx}$

(viii) $y = \cot x$ then $\frac{dy}{dx} = -\operatorname{cosec}^2 x$ and $y = \cot u$ then $\frac{dy}{dx} = -\operatorname{cosec}^2 x\frac{du}{dx}$

(ix) $y = \sec x$ then $\frac{dy}{dx} = \sec x \tan x$ (x) $y = \operatorname{cosec} x$ then $\frac{dy}{dx} = -\operatorname{cosec} x \cot x$ (xi) $y = \sin^{-1} x$ then $\frac{dy}{dx} = \frac{1}{\sqrt{1-x^2}}$

(xii) $y = \cos^{-1} x$ then $\frac{dy}{dx} = -\frac{1}{\sqrt{1-x^2}}$ (xiii) $y = \tan^{-1} x$ then $\frac{dy}{dx} = \frac{1}{1+x^2}$ (xiv) $y = \cot^{-1} x$ then $\frac{dy}{dx} = -\frac{1}{1+x^2}$

(xv) $y = \sec^{-1} x$ then $\frac{dy}{dx} = \frac{1}{|x|\sqrt{x^2-1}},\ |x| > 0$ (xvi) $y = \operatorname{cosec}^{-1} x$ then $\frac{dy}{dx} = -\frac{1}{|x|\sqrt{x^2-1}},\ |x| > 0$

Trigonometrical Expansions and Formulae be Remembered

(i) $\sin x = x - \frac{x^3}{3!} + \frac{x^5}{5!} - \cdots \ldots\ldots\ldots\ldots\ldots\ldots\ldots\ldots\ldots\ldots\ldots\ldots$

$\cos x = 1 - \frac{x^2}{2!} + \frac{x^4}{4!} - \frac{x^6}{6!} + \cdots \ldots\ldots\ldots\ldots\ldots\ldots\ldots\ldots\ldots\ldots\ldots\ldots$

$\tan x = x - \frac{x^3}{3} + \frac{2}{15}x^5 + \cdots \ldots\ldots\ldots\ldots\ldots\ldots\ldots\ldots\ldots\ldots\ldots\ldots$

(ii) (a) $\tan^{-1} x - \tan^{-1} y = \tan^{-1}\frac{x-y}{1+xy}$ (b) $\tan^{-1} x + \tan^{-1} y = \tan^{-1}\frac{x+y}{1-xy}$ (c) $2\tan^{-1} x = \tan^{-1}\frac{2x}{1-x^2}$

(d) $\sin^{-1} x \pm \sin^{-1} y = \sin^{-1}\left[x\sqrt{1-y^2} \pm y\sqrt{1-x^2}\right]$ (e) $\cos^{-1} x \pm \cos^{-1} y = \cos^{-1}\left[xy \mp \sqrt{1-x^2}\sqrt{1-y^2}\right]$

(iii) $\sin^{-1} x + \cos^{-1} x = \tan^{-1} x + \cot^{-1} x = \sec^{-1} x + \operatorname{cosec}^{-1} x = \frac{\pi}{2}$

(iv) $\sin^{-1} x = \operatorname{cosec}^{-1}\left(\frac{1}{x}\right),\ \tan^{-1} x = \cot^{-1}\left(\frac{1}{x}\right)$ (v) $\sin^{-1}(\sin\theta) = \theta,\ \tan^{-1}(\tan\theta) = \theta,\ \cos^{-1}(\cos\theta) = \theta$

Example: $-\ \sin^{-1}(\cos\theta) = \sin^{-1}\sin\left(\frac{\pi}{2} - \theta\right) = \frac{\pi}{2} - \theta,\ \cos^{-1}(\sin\theta) = \cos^{-1}\cos\left(\frac{\pi}{2} - \theta\right) = \frac{\pi}{2} - \theta$

Important Results of Trigonometry

(vi) $\frac{1-\cos x}{1+\cos x}=\frac{2\sin^2\frac{x}{2}}{2\cos^2\frac{x}{2}}=\tan^2\frac{x}{2}$, $\frac{1+\cos x}{1-\cos x}=\cot^2\frac{x}{2}$, $\frac{1-\cos x}{\sin x}=\frac{2\sin^2\frac{x}{2}}{2\sin\frac{x}{2}\cos\frac{x}{2}}=\frac{\sin\frac{x}{2}}{\cos\frac{x}{2}}=\tan\left(\frac{x}{2}\right)$

or $\frac{1+\cos x}{\sin x}=\frac{2\cos^2\frac{x}{2}}{2\sin\frac{x}{2}\cos\frac{x}{2}}=\frac{\cos\frac{x}{2}}{\sin\frac{x}{2}}=\cot\left(\frac{x}{2}\right)$

or $\sqrt{1\pm\sin x}=\left[\cos^2\frac{x}{2}+\sin^2\frac{x}{2}\pm 2\sin\frac{x}{2}\cos\frac{x}{2}\right]^{\frac{1}{2}}=\left[\{\cos\frac{x}{2}\pm\sin\frac{x}{2}\}^2\right]^{\frac{1}{2}}=\cos\left(\frac{x}{2}\right)\pm\sin\left(\frac{x}{2}\right)$

$\tan A\pm\tan B=\frac{\sin A\cos B\pm\cos A\sin B}{\cos A\cos B}=\frac{\sin(A\pm B)}{\cos A\cos B}$ or $\tan\left(\frac{\pi}{4}+\theta\right)=\frac{1+\tan\theta}{1-\tan\theta}$, $\tan\left(\frac{\pi}{4}-\theta\right)=\frac{1-\tan\theta}{1+\tan\theta}$

(vii) $\cos x=2\cos^2\left(\frac{x}{2}\right)-1=1-2\sin^2\left(\frac{x}{2}\right)=\cos^2\left(\frac{x}{2}\right)-\sin^2\left(\frac{x}{2}\right)$.

(viii) $\sin x=2\sin\frac{x}{2}\cos\frac{x}{2}=\frac{2\tan\left(\frac{x}{2}\right)}{1+\tan^2\left(\frac{x}{2}\right)}$ (ix) $\cos x=\frac{1-\tan^2\left(\frac{x}{2}\right)}{1+\tan^2\left(\frac{x}{2}\right)}=1-2\sin^2\left(\frac{x}{2}\right)=2\cos^2\left(\frac{x}{2}\right)-1$

(x) $\tan x=\frac{2\tan\left(\frac{x}{2}\right)}{1-\tan^2\left(\frac{x}{2}\right)}$ (xi) $\sin 3x=3\sin x-4\sin^3 x$ (xii) $\cos 3x=4\cos^3 x-3\cos x$

Substitution for differentiation of algebric function

$\sqrt{a^2-x^2}$, put $x=a\sin\theta$ or $\sqrt{a^2+x^2}$, put $x=a\tan\theta$

$\sqrt{x^2-a^2}$, put $x=a\sec\theta$ or $\sqrt{\left(\frac{a+x}{a-x}\right)}$, put $x=a\cos 2\theta$

(xiii) $\cos 2x=2\cos^2 x-1=1-2\sin^2 x=\cos^2 x-\sin^2 x$ (xiv) $\sin 2x=2\sin x\cos x$

Differentiate from first Principles: –

(1) $\sin x$, Let $y=\sin x=f(x)$, $y+\delta y=\sin(x+\delta x)$ or $\delta y=\sin(x+\delta x)-y$ $\therefore\ \frac{\delta y}{\delta x}=\frac{\sin(x+\delta x)-\sin x}{\delta x}$

$$\frac{dy}{dx}=\lim_{\delta x\to 0}\frac{\sin(x+\delta x)-\sin x}{\delta x}=\lim_{\delta x\to 0}\frac{2\cos\left(x+\frac{\delta x}{2}\right).\sin\left(\frac{\delta x}{2}\right)}{\delta x}=\lim_{\delta x\to 0}\cos\left(x+\frac{\delta x}{2}\right).\left(\frac{\sin\left(\frac{\delta x}{2}\right)}{\frac{\delta x}{2}}\right)=\cos x.1$$

or $\frac{dy}{dx}=\cos x$ Ans. $\left[\because\ \sin C-\sin D=2\cos\left(\frac{C+D}{2}\right).\sin\left(\frac{C-D}{2}\right)\right]$ $\left[\because\ \lim_{\theta\to 0}\frac{\sin\theta}{\theta}=1\right]$

(2) $\cos x$, Ans: – $\frac{dy}{dx}=\frac{d(\cos x)}{dx}=-\sin x$ (Do yourself)

(3) $\tan x$, Let $y=\tan x$, $y+\delta y=\tan(x+\delta x)$ $\therefore\ \delta y=\tan(x+\delta x)-y=\tan(x+\delta x)-\tan x$

$$\therefore\ \frac{\delta y}{\delta x}=\frac{\tan(x+\delta x)-\tan x}{\delta x}$$

or $\frac{dy}{dx}=\lim_{\delta x\to 0}\frac{\tan(x+\delta x)-\tan x}{\delta x}$ (change to sin and cos) $=\lim_{\delta x\to 0}\frac{\frac{\sin(x+\delta x)}{\cos(x+\delta x)}-\frac{\sin x}{\cos x}}{\delta x}=\lim_{\delta x\to 0}\frac{\cos x.\sin(x+\delta x)-\sin x.\cos(x+\delta x)}{\delta x.\cos(x+\delta x).\cos x}$

$$=\lim_{\delta x\to 0}\frac{\sin(x+\delta x-x)}{\delta x.\cos(x+\delta x).\cos x}$$

or $\frac{dy}{dx}=\lim_{\delta x\to 0}\left(\frac{\sin\delta x}{\delta x}\right).\frac{1}{\cos(x+\delta x).\cos x}=1.\frac{1}{\cos^2 x}=\sec^2 x$ Ans. $[\because\ \sin x.\cos y-\sin y.\cos x=\sin(x-y)]$

(4) $\cot x$, Ans: – $\frac{dy}{dx}=\frac{d(\cot x)}{dx}=-\mathrm{cosec}^2 x$ (Do yourself)

(5) $\sec x$, Let $y = \sec x$, $y + \delta y = \sec(x + \delta x)$ or $\delta y = \sec(x + \delta x) - y = \sec(x + \delta x) - \sec x$

$$\therefore \frac{\delta y}{\delta x} = \frac{\sec(x + \delta x) - \sec x}{\delta x} \quad \therefore \frac{dy}{dx} = \lim_{\delta x \to 0} \frac{\sec(x + \delta x) - \sec x}{\delta x} \quad \text{(change to cos)}$$

$$\text{or } \frac{dy}{dx} = \lim_{\delta x \to 0} \frac{\left(\frac{1}{\cos(x + \delta x)} - \frac{1}{\cos x}\right)}{\delta x} = \lim_{\delta x \to 0} \frac{\cos x - \cos(x + \delta x)}{\delta x.\cos x.\cos(x + \delta x)} = \lim_{\delta x \to 0} \frac{2\sin\left(\frac{x + x + \delta x}{2}\right).\sin\left(\frac{x + \delta x - x}{2}\right)}{\delta x.\cos x.\cos(x + \delta x)} = \lim_{\delta x \to 0} \frac{2\sin\left(x + \frac{\delta x}{2}\right).\sin\frac{\delta x}{2}}{\delta x.\cos x.\cos(x + \delta x)}$$

$$= \lim_{\delta x \to 0} \frac{\sin\left(x + \frac{\delta x}{2}\right)}{\cos x.\cos(x + \delta x)} \left(\frac{\sin\frac{\delta x}{2}}{\frac{\delta x}{2}}\right) = \frac{\sin x.1}{\cos x.\cos x} = \tan x.\sec x$$

$$\therefore \frac{dy}{dx} = \tan x.\sec x \quad \text{Ans.} \qquad \left[\because \cos C - \cos D = 2\sin\left(\frac{C + D}{2}\right).\sin\left(\frac{D - C}{2}\right)\right]$$

(6) $\operatorname{cosec} x$, Ans:– $\frac{dy}{dx} = \frac{d(\operatorname{cosec} x)}{dx} = -\cot x.\operatorname{cosec} x$ Ans. (Do yourself)

(7) e^x, Let $y = e^x$, $y + \delta y = e^{x+\delta x}$ or $\delta y = e^{x+\delta x} - y$ or $\delta y = e^{x+\delta x} - e^x$

$$\therefore \frac{\delta y}{\delta x} = \frac{e^{x+\delta x} - e^x}{\delta x} \quad \therefore \frac{dy}{dx} = \lim_{\delta x \to 0} \frac{e^{x+\delta x} - e^x}{\delta x} = \lim_{\delta x \to 0} \frac{e^x.e^{\delta x} - e^x}{\delta x} = \lim_{\delta x \to 0} \frac{e^x(e^{\delta x} - 1)}{\delta x} = \lim_{\delta x \to 0} e^x\left(\frac{e^{\delta x} - 1}{\delta x}\right) = e^x.1$$

$$\therefore \frac{dy}{dx} = e^x \quad \text{Ans.} \quad \left[\because \lim_{x \to 0} \frac{e^x - 1}{x} = 1\right]$$

(8) $\log x$, Let $y = \log x$, $y + \delta y = \log(x + \delta x)$ or $\delta y = \log(x + \delta x) - y$ or $\delta y = \log(x + \delta x) - \log x$

$$\therefore \frac{\delta y}{\delta x} = \frac{\log(x + \delta x) - \log x}{\delta x}, \quad \frac{dy}{dx} = \lim_{\delta x \to 0} \frac{\log(x + \delta x) - \log x}{\delta x} = \lim_{\delta x \to 0} \frac{\log\left(\frac{x + \delta x}{x}\right)}{\delta x} = \lim_{\delta x \to 0} \frac{\log\left(1 + \frac{\delta x}{x}\right)}{\delta x}$$

$$\frac{dy}{dx} = \lim_{\delta x \to 0} \left(\frac{\frac{\delta x}{x} - \frac{1}{2}\left(\frac{\delta x}{x}\right)^2 + \cdots\ldots\ldots\ldots}{\delta x}\right) = \lim_{\delta x \to 0} \frac{\delta x\left[\frac{1}{x} - \frac{1}{2}.\frac{\delta x}{x^2} + \cdots\ldots\ldots\right]}{\delta x} = \lim_{\delta x \to 0} \left[\frac{1}{x} - \frac{1}{2}.\frac{\delta x}{x^2} + \cdots\ldots\ldots\right] = \frac{1}{x}$$

$$\therefore \frac{dy}{dx} = \frac{1}{x} \quad \text{Ans.} \quad \left[\because \log(1 + x) = x - \frac{1}{2}x^2 + \frac{1}{3}x^3 - \frac{1}{4}x^4 + \cdots\ldots\ldots\ldots\ldots\ldots\right]$$

(9) $\cos^{-1} x$, Let $y = \cos^{-1} x$ $\therefore \cos y = x$ or $x = \cos y$ $\therefore x + \delta x = \cos(y + \delta y)$

$$\therefore \delta x = \cos(y + \delta y) - \cos y \quad \text{or} \quad \frac{\delta x}{\delta y} = \frac{\cos(y + \delta y) - \cos y}{\delta y} \quad \text{or} \quad \frac{\delta y}{\delta x} = \frac{\delta y}{\cos(y + \delta y) - \cos y}$$

$$\frac{dy}{dx} = \lim_{\delta y \to 0} \frac{\delta y}{\cos(y + \delta y) - \cos y} = \lim_{\delta y \to 0} \frac{\delta y}{2\sin\left(\frac{y + \delta y + y}{2}\right).\sin\left(\frac{y - y - \delta y}{2}\right)} = \lim_{\delta y \to 0} \frac{\delta y}{2\sin\left(y + \frac{\delta y}{2}\right).\left[-\sin\left(\frac{\delta y}{2}\right)\right]}$$

$$= \lim_{\delta y \to 0} \frac{\delta y}{2\sin\left(y + \frac{\delta y}{2}\right).\left[-\left(\frac{\sin\frac{\delta y}{2}}{\frac{\delta y}{2}}\right).\frac{\delta y}{2}\right]} = \lim_{\delta y \to 0} \frac{1}{\sin\left(y + \frac{\delta y}{2}\right).(-1)} = -\frac{1}{\sin y}$$

$$\therefore \frac{dy}{dx} = -\frac{1}{\sqrt{1 - \cos^2 y}} \quad \text{put } \cos y = x \text{ then } \frac{dy}{dx} = -\frac{1}{\sqrt{1 - x^2}} \quad \text{Ans.} \quad [\because \sin^2 x + \cos^2 x = 1]$$

(10) $\sqrt{\cos x}$, Let $y = \sqrt{\cos x}$ $\therefore$ $y + \delta y = \sqrt{\cos(x + \delta x)}$ or $\delta y = \sqrt{\cos(x + \delta x)} - y$

$$\text{or } \delta y = \sqrt{\cos(x + \delta x)} - \sqrt{\cos x} \quad \therefore \frac{\delta y}{\delta x} = \frac{\sqrt{\cos(x + \delta x)} - \sqrt{\cos x}}{\delta x}$$

$$\frac{dy}{dx}=\lim_{\delta x\to 0}\frac{\sqrt{\cos(x+\delta x)}-\sqrt{\cos x}}{\delta x}=\lim_{\delta x\to 0}\frac{\left(\sqrt{\cos(x+\delta x)}-\sqrt{\cos x}\right).\left(\sqrt{\cos(x+\delta x)}+\sqrt{\cos x}\right)}{\delta x.\left(\sqrt{\cos(x+\delta x)}+\sqrt{\cos x}\right)}=\lim_{\delta x\to 0}\frac{\cos(x+\delta x)-\cos x}{\delta x.\left(\sqrt{\cos(x+\delta x)}+\sqrt{\cos x}\right)}$$

$$=\lim_{\delta x\to 0}\frac{2\sin\left(x+\frac{\delta x}{2}\right).\sin\left(-\frac{\delta x}{2}\right)}{\delta x.\left(\sqrt{\cos(x+\delta x)}+\sqrt{\cos x}\right)}$$

$$\frac{dy}{dx}=\lim_{\delta x\to 0}\frac{\sin\left(x+\frac{\delta x}{2}\right)}{\left(\sqrt{\cos(x+\delta x)}+\sqrt{\cos x}\right)}.\lim_{\delta x\to 0}-\left(\frac{\sin\frac{\delta x}{2}}{\frac{\delta x}{2}}\right)=\frac{\sin x.(-1)}{2\sqrt{\cos x}}=-\frac{\sin x}{2\sqrt{\cos x}}\quad \text{Ans.}$$

$$\left[\because \sin(-\theta)=-\sin\theta\ ,\ \ \sin C+\sin D=2\sin\left(\frac{C+D}{2}\right)\cos\left(\frac{C-D}{2}\right),\ \ \cos C+\cos D=2\cos\left(\frac{C+D}{2}\right)\cos\left(\frac{C-D}{2}\right)\right]$$

(11) $e^{\cos x}$, Let $y=e^{\cos x}$ $\therefore \log y=\log_e e^{\cos x}$ or $\log y=\cos x\log_e e$ or $\log y=\cos x$

or $\log(y+\delta y)=\cos(x+\delta x)$ or $\log(y+\delta y)-\log y=\cos(x+\delta x)-\cos x$

or $\dfrac{\log(y+\delta y)-\log y}{\delta x}=\dfrac{\cos(x+\delta x)-\cos x}{\delta x}$ or $\left(\dfrac{\log(y+\delta y)-\log y}{\delta x}\right)\dfrac{\delta y}{\delta x}=\dfrac{\cos(x+\delta x)-\cos x}{\delta x}$

or $\dfrac{\log\left(1+\frac{\delta y}{y}\right)}{\delta y}.\dfrac{\delta y}{\delta x}=\dfrac{2\sin\left(x+\frac{\delta x}{2}\right)\sin\left(-\frac{\delta x}{2}\right)}{\delta x}$ $\therefore \lim\limits_{\delta y\to 0}\dfrac{\log\left(1+\frac{\delta y}{y}\right)}{\delta y}.\dfrac{dy}{dx}=\lim\limits_{\delta x\to 0}\sin\left(x+\dfrac{\delta x}{2}\right).-\left(\dfrac{\sin\frac{\delta x}{2}}{\frac{\delta x}{2}}\right)$

or $\lim\limits_{\delta y\to 0}\dfrac{dy}{dx}\left[\dfrac{\frac{\delta y}{y}-\frac{1}{2}\left(\frac{\delta y}{y}\right)^2+\cdots\ldots\ldots\ldots}{\delta y}\right]=\sin x(-1)$ or $\dfrac{1}{y}\dfrac{dy}{dx}=-\sin x$ $\therefore$ $\dfrac{dy}{dx}=-y\sin x$

$$\therefore \frac{dy}{dx}=-e^{\cos x}\sin x\quad \text{Ans.}\quad [\text{Put } y=e^{\cos x}]$$

(12) $x\log x$, Let $y=x\log x$ $\therefore$ $y+\delta y=(x+\delta x)\log(x+\delta x)$ or $\delta y=(x+\delta x)\log(x+\delta x)-y$

$$\frac{dy}{dx}=\lim_{\delta x\to 0}\frac{(x+\delta x)\log(x+\delta x)-x\log x}{\delta x}=\lim_{\delta x\to 0}\frac{x\log(x+\delta x)+\delta x\log(x+\delta x)-x\log x}{\delta x}$$

$$\frac{dy}{dx}=\lim_{\delta x\to 0}\left\{\frac{x[\log(x+\delta x)-\log x]}{\delta x}+\frac{\delta x\log(x+\delta x)}{\delta x}\right\}=\lim_{\delta x\to 0}\frac{x\log\left(1+\frac{\delta x}{x}\right)}{\delta x}+\lim_{\delta x\to 0}\frac{\delta x\log(x+\delta x)}{\delta x}$$

$$\frac{dy}{dx}=x.\lim_{\delta x\to 0}\frac{\log\left(1+\frac{\delta x}{x}\right)}{\delta x}+\lim_{\delta x\to 0}\log(x+\delta x)=x.\lim_{\delta x\to 0}\frac{\frac{\delta x}{x}-\frac{1}{2}\left(\frac{\delta x}{x}\right)^2+\cdots\ldots\ldots\ldots}{\delta x}+\log x=x.\lim_{\delta x\to 0}\frac{\delta x\left[\frac{1}{x}-\frac{1}{2}.\frac{\delta x}{x^2}+\cdots\ldots\ldots\right]}{\delta x}+\log x$$

$$=x.\frac{1}{x}+\log x=1+\log x\quad \text{Ans.}$$

(13) x^2e^x , Let $y=x^2e^x$ $\therefore$ $y+\delta y=(x+\delta x)^2\,e^{x+\delta x}$ $\therefore$ $\delta y=(x+\delta x)^2\,e^{x+\delta x}-x^2e^x$

$$\frac{dy}{dx}=\lim_{\delta x\to 0}\frac{(x+\delta x)^2\,e^{x+\delta x}-x^2e^x}{\delta x}=\lim_{\delta x\to 0}\frac{(x^2+2x\delta x+\delta x^2)e^{x+\delta x}-x^2e^x}{\delta x}$$

$$\frac{dy}{dx}=\lim_{\delta x\to 0}\frac{x^2e^{x+\delta x}+2x\delta xe^{x+\delta x}+\delta x^2e^{x+\delta x}-x^2e^x}{\delta x}=\lim_{\delta x\to 0}\frac{x^2e^{x+\delta x}-x^2e^x}{\delta x}+\lim_{\delta x\to 0}\frac{2x\delta xe^{x+\delta x}}{\delta x}+\lim_{\delta x\to 0}\frac{\delta x^2e^{x+\delta x}}{\delta x}$$

$$\frac{dy}{dx}=\lim_{\delta x\to 0}\frac{x^2\left(e^{x+\delta x}-e^x\right)}{\delta x}+2xe^x+0=\lim_{\delta x\to 0}\frac{x^2e^x\left(e^{\delta x}-1\right)}{\delta x}+2xe^x=x^2e^x+2xe^x\quad\therefore\ \frac{dy}{dx}=xe^x(x+2)\ \text{Ans.}\quad\left[\because \lim_{x\to 0}\frac{e^x-1}{x}=1\right]$$

Solved Example

(1) Find the differential coefficient of the following: – (a) $y=\tan^{-1}x+x$ (b) $y=\tan^{-1}(x+1)-\tan^{-1}x$

(c) $\sin^{-1}y=\cos^{-1}x+x$ (d) $y=\dfrac{3x^2-6x+7}{x}$ (e) $y=\dfrac{x^2+2x+3}{x^2}$ (f) $y=\dfrac{(x+1)^2}{x}$ (g) $y=\dfrac{2x}{a}-\dfrac{2b}{x}$

(h) $y = 2\sqrt{x} + \frac{1}{\sqrt{x}}$ (i) $y = 3x^{\frac{4}{3}} - x^{\frac{1}{3}} + 2x^{-\frac{1}{3}} - 3x^{-\frac{2}{3}}$ (j) $y = \tan x + \cot x$ (k) $y = \sec x + \operatorname{cosec} x$ (l) $y = \tan x - \sec^2 x$

Solution: – (a) $y = \tan^{-1} x + x \quad \therefore \frac{dy}{dx} = \frac{1}{1+x^2} + 1 = \frac{1+1+x^2}{1+x^2} = \frac{2+x^2}{1+x^2}$ Ans.

(b) $y = \tan^{-1}(x+1) - \tan^{-1} x \quad \therefore \frac{dy}{dx} = \frac{1}{1+(x+1)^2} - \frac{1}{1+x^2} = \frac{1}{1+x^2+2x+1} - \frac{1}{1+x^2} = \frac{1}{x^2+2x+2} - \frac{1}{1+x^2}$

$$\therefore \frac{dy}{dx} = \frac{1+x^2-x^2-2x-2}{(x^2+2x+2)(1+x^2)} = \frac{-2x-1}{(x^2+2x+2)(1+x^2)} = -\frac{2x+1}{(x^2+2x+2)(1+x^2)} \text{ Ans.}$$

(c) $\sin^{-1} y = \cos^{-1} x + x$ or $\frac{1}{\sqrt{1-y^2}} \frac{dy}{dx} = \frac{1}{-\sqrt{1-x^2}} + 1 \quad \therefore \frac{dy}{dx} = \sqrt{1-y^2}\left(\frac{\sqrt{1-x^2}-1}{\sqrt{1-x^2}}\right) = \sqrt{1-\sin^2(\cos^{-1} x + x)}\left(\frac{\sqrt{1-x^2}-1}{\sqrt{1-x^2}}\right)$

$$= \cos(\cos^{-1} x + x)\left(\frac{\sqrt{1-x^2}-1}{\sqrt{1-x^2}}\right) \text{ Ans.}$$

(d) $y = \frac{3x^2-6x+7}{x}$ $\left[\text{formula,} \quad y = \frac{f(x)}{g(x)}, \quad \frac{dy}{dx} = \frac{g(x).f'(x) - f(x).g'(x)}{(g(x))^2}\right]$

Here $f(x) = 3x^2 - 6x + 7$ and $g(x) = x$ then $f'(x) = 6x - 6$ and $g'(x) = 1$

$$\therefore \frac{dy}{dx} = \frac{x.(6x-6) - (3x^2-6x+7).1}{x^2} = \frac{6x^2-6x-3x^2+6x-7}{x^2} = \frac{3x^2-7}{x^2} = 3 - \frac{7}{x^2} \text{ Ans.}$$

(e) $y = \frac{x^2+2x+3}{x^2}$, Let $f(x) = x^2+2x+3$ and $g(x) = x^2$ then $f'(x) = 2x+2$ and $g'(x) = 2x$

use formula, $\frac{dy}{dx} = \frac{g(x).f'(x) - f(x).g'(x)}{(g(x))^2} = \frac{x^2.(2x+2) - (x^2+2x+3).2x}{(x^2)^2}$

$$\frac{dy}{dx} = \frac{2x^3+2x^2-2x^3-4x^2-6x}{x^4} = \frac{-2x^2-6x}{x^4} = \frac{-2x(x+3)}{x^4} = \frac{-2(x+3)}{x^3} = -2\left[\frac{1}{x^2} + \frac{3}{x^3}\right] \text{ Ans.}$$

(f) $y = \frac{(x+1)^2}{x}$, Let $f(x) = (x+1)^2$ and $g(x) = x$ then $f'(x) = 2(x+1)$ and $g'(x) = 1$

use formula, $\frac{dy}{dx} = \frac{g(x).f'(x) - f(x).g'(x)}{(g(x))^2} = \frac{x.2(x+1) - (x+1)^2.1}{x^2} = \frac{2x^2+2x-(x^2+2x+1)}{x^2}$

$$\frac{dy}{dx} = \frac{2x^2+2x-x^2-2x-1}{x^2} = \frac{x^2-1}{x^2} = \frac{x^2}{x^2} - \frac{1}{x^2} = 1 - \frac{1}{x^2} \text{ Ans.}$$

(g) $y = \frac{2x}{a} - \frac{2b}{x} = \frac{2x^2-2ab}{ax}$, Let $f(x) = 2x^2 - 2ab$ and $g(x) = ax$ then $f'(x) = 4x$ and $g'(x) = a$

use formula, $\frac{dy}{dx} = \frac{g(x).f'(x) - f(x).g'(x)}{(g(x))^2} = \frac{ax.4x - (2x^2-2ab).a}{(ax)^2} = \frac{4ax^2-2ax^2+2a^2b}{a^2x^2} = \frac{2ax^2+2a^2b}{a^2x^2}$

$$\therefore \frac{dy}{dx} = \frac{2a(x^2+ab)}{a^2x^2} = \frac{2(x^2+ab)}{ax^2} = \frac{2}{a} + \frac{2b}{x^2} \text{ Ans.}$$

(h) $y = 2\sqrt{x} + \frac{1}{\sqrt{x}} = \frac{2x+1}{\sqrt{x}}$, Let $f(x) = 2x+1$ and $g(x) = \sqrt{x}$ then $f'(x) = 2$ and $g'(x) = \frac{1}{2\sqrt{x}}$

use formula, $\frac{dy}{dx} = \frac{g(x).f'(x) - f(x).g'(x)}{(g(x))^2} = \frac{\sqrt{x}.2 - (2x+1).\frac{1}{2\sqrt{x}}}{(\sqrt{x})^2} = \frac{\frac{4x-2x-1}{2\sqrt{x}}}{x} = \frac{2x-1}{2x\sqrt{x}}$ Ans.

(i) $y = 3x^{\frac{4}{3}} - x^{\frac{1}{3}} + 2x^{-\frac{1}{3}} - 3x^{-\frac{2}{3}}$ $\left[\text{formula,} \quad y = x^n, \quad \frac{dy}{dx} = nx^{n-1}\right]$

$$\frac{dy}{dx} = 3.\frac{4}{3}x^{\frac{4}{3}-1} - \frac{1}{3}x^{\frac{1}{3}-1} + 2.\frac{-1}{3}x^{-\frac{1}{3}-1} - 3.\frac{-2}{3}x^{-\frac{2}{3}-1} = 4x^{\frac{1}{3}} - \frac{1}{3}x^{-\frac{2}{3}} - \frac{2}{3}x^{-\frac{4}{3}} + 2x^{-\frac{5}{3}} \quad \text{Ans.}$$

(j) $y = \tan x + \cot x \quad \therefore \frac{dy}{dx} = \sec^2 x - \operatorname{cosec}^2 x = \frac{1}{\cos^2 x} - \frac{1}{\sin^2 x} = \frac{\sin^2 x - \cos^2 x}{\sin^2 x \cos^2 x} = \frac{-(\cos^2 x - \sin^2 x).4}{(2\sin x \cos x)^2}$

$$\frac{dy}{dx} = -\frac{4\cos 2x}{\sin^2 2x} = -4\cot 2x.\operatorname{cosec} 2x \quad \text{Ans.}$$

IInd Method:− $y = \tan x + \cot x = \frac{\sin x}{\cos x} + \frac{\cos x}{\sin x} = \frac{\sin^2 x + \cos^2 x}{\sin x \cos x} = \frac{1}{\sin x \cos x}$

$\left[\text{use formula } y = \frac{f(x)}{g(x)}, \frac{dy}{dx} = \frac{g(x).f'(x) - f(x).g'(x)}{(g(x))^2} \text{ and } y = f(x)g(x), \frac{dy}{dx} = f(x)g'(x) + g(x)f'(x)\right]$

Here Let $f(x) = 1$ and $g(x) = \sin x \cos x$ then $f'(x) = 0$ and $g'(x) = \sin x(-\sin x) + \cos x.\cos x = \cos 2x$

$$\frac{dy}{dx} = \frac{\sin x \cos x.0 - 1.\cos 2x}{(\sin x \cos x)^2} = -\frac{\cos 2x.4}{(2\sin x \cos x)^2} = -\frac{4\cos 2x}{\sin^2 2x} = -4\cot 2x.\operatorname{cosec} 2x \quad \text{Ans.}$$

(k) $y = \sec x + \operatorname{cosec} x = \frac{1}{\cos x} + \frac{1}{\sin x} = \frac{\sin x + \cos x}{\sin x \cos x}$

Let $f(x) = \sin x + \cos x$ and $g(x) = \sin x \cos x$

$f'(x) = \cos x - \sin x$ and $g'(x) = \sin x\frac{d(\cos x)}{dx} + \cos x\frac{d(\sin x)}{dx} = -\sin^2 x + \cos^2 x = \cos^2 x - \sin^2 x = \cos 2x$

$$\frac{dy}{dx} = \frac{g(x).f'(x) - f(x).g'(x)}{(g(x))^2} = \frac{\sin x \cos x(\cos x - \sin x) - (\sin x + \cos x).\cos 2x}{(\sin x \cos x)^2} = \frac{\sin x \cos x(\cos x - \sin x) - (\sin x + \cos x)(\cos^2 x - \sin^2 x)}{\sin^2 x \cos^2 x}$$

$$\frac{dy}{dx} = \frac{\sin x \cos^2 x - \sin^2 x \cos x - \sin x \cos^2 x + \sin^3 x - \cos^3 x + \sin^2 x \cos x}{\sin^2 x \cos^2 x} = \frac{\sin^3 x - \cos^3 x}{\sin^2 x \cos^2 x} = \frac{\sin^3 x}{\sin^2 x \cos^2 x} - \frac{\cos^3 x}{\sin^2 x \cos^2 x} = \frac{\sin x}{\cos^2 x} - \frac{\cos x}{\sin^2 x}$$
$$= \tan x \sec x - \cot x \operatorname{cosec} x \quad \text{Ans.}$$

or Direct Differentiate, $\frac{dy}{dx} = \frac{d(\sec x)}{dx} + \frac{d(\operatorname{cosec} x)}{dx} = \tan x \sec x + (-\cot x \operatorname{cosec} x) = \tan x \sec x - \cot x \operatorname{cosec} x$ Ans.

(l) $y = \tan x - \sec^2 x \quad \therefore \frac{dy}{dx} = \sec^2 x - 2\sec x \sec x \tan x = \sec^2 x(1 - 2\tan x) = (1 + \tan^2 x)(1 - 2\tan x)$

$$\frac{dy}{dx} = 1 + \tan^2 x - 2\tan x - 2\tan^3 x = (1 + \tan^2 x)\left(1 - 2\frac{\sin x}{\cos x}\right) = \sec^2 x\left(\frac{\cos x - 2\sin x}{\cos x}\right) = \frac{1}{\cos^2 x}\left(\frac{\cos x - 2\sin x}{\cos x}\right)$$

$$\therefore \frac{dy}{dx} = \frac{1}{\cos^3 x}(\cos x - 2\sin x) \quad \text{Ans.}$$

IInd Method:− $y = \tan x - \sec^2 x = \frac{\sin x}{\cos x} - \frac{1}{\cos^2 x} = \frac{\sin x \cos x - 1}{\cos^2 x}$

$$\frac{dy}{dx} = \frac{\cos^2 x.\frac{d(\sin x \cos x - 1)}{dx} - (\sin x \cos x - 1).\frac{d(\cos^2 x)}{dx}}{(\cos^2 x)^2} = \frac{\cos^2 x(\cos^2 x - \sin^2 x) - (\sin x \cos x - 1).2\cos x(-\sin x)}{\cos^4 x}$$

$$\frac{dy}{dx} = \frac{\cos^4 x - \sin^2 x \cos^2 x + 2\sin^2 x \cos^2 x - 2\sin x \cos x}{\cos^4 x} = 1 + \tan^2 x - 2\tan x \sec^2 x = \sec^2 x - 2\tan x \sec^2 x$$

$$\frac{dy}{dx} = \sec^2 x(1 - 2\tan x) = \frac{1}{\cos^3 x}(\cos x - 2\sin x) \quad \text{Ans.}$$

(2) Find the differential coefficient of the following:− (a) $y = \tan^{-1} x$ (b) $y = \sin^{-1} x$ (c) $y = \cos^{-1} x$

(d) $y = \cot^{-1} x$ (e) $y = \sec^{-1} x$ (f) $y = \operatorname{cosec}^{-1} x$ (g) $y = x\sin^{-1} x$ (h) $y = \frac{\cos^{-1} x}{x^2}$ (i) $y = \frac{\tan^{-1} x}{1 + x^2}$

(j) $y = \sqrt{1 - x^2}\,\sin^{-1} x$ (k) $y = x\sqrt{1 + x}$ (l) $y = x\sqrt{1 + x} + (x + 1)\sqrt{x - 3}$

Solution: − (a) $y = \tan^{-1}x$ or $\tan y = x$ or $\sec^2 y\frac{dy}{dx} = 1$ ∴ $\frac{dy}{dx} = \frac{1}{\sec^2 y} = \frac{1}{1+\tan^2 y} = \frac{1}{1+x^2}$ Ans.

(b) $y = \sin^{-1}x$ or $\sin y = x$ ∴ $\cos y\frac{dy}{dx} = 1$ or $\frac{dy}{dx} = \frac{1}{\cos y} = \frac{1}{\sqrt{1-\sin^2 y}} = \frac{1}{\sqrt{1-x^2}}$ Ans.

(c) $y = \cos^{-1}x$ ∴ $\frac{dy}{dx} = \frac{1}{-\sqrt{1-x^2}}$ Ans. (Do yourself), same as above question.

(d) $y = \cot^{-1}x$ or $\cot y = x$ or $-\operatorname{cosec}^2 y\frac{dy}{dx} = 1$ or $\frac{dy}{dx} = \frac{1}{-\operatorname{cosec}^2 y} = -\frac{1}{\sqrt{1+\cot^2 y}} = -\frac{1}{\sqrt{1+x^2}}$ Ans.

(e) $y = \sec^{-1}x$ or $\sec y = x$ or $\tan y\sec y\frac{dy}{dx} = 1$ or $\frac{dy}{dx} = \frac{1}{\tan y\sec y} = \frac{1}{\sqrt{\sec^2 y - 1}\,.\sec y} = \frac{1}{x\sqrt{x^2-1}}$ Ans.

(f) $y = \operatorname{cosec}^{-1}x$ or $\operatorname{cosec} y = x$ or $-\cot y\operatorname{cosec} y\frac{dy}{dx} = 1$ or $\frac{dy}{dx} = -\frac{1}{\cot y\operatorname{cosec} y} = -\frac{1}{\operatorname{cosec} y\sqrt{\operatorname{cosec}^2 y - 1}}$ ∴ $\frac{dy}{dx} = -\frac{1}{x\sqrt{x^2-1}}$ Ans.

(g) $y = x\sin^{-1}x$, Let $f(x) = x$ and $g(x) = \sin^{-1}x$ then $f'(x) = 1$ and $g'(x) = \frac{1}{\sqrt{1-x^2}}$

[use formula, $y = f(x).g(x)$ then $\frac{dy}{dx} = f(x).g'(x) + g(x).f'(x)$] ∴ $\frac{dy}{dx} = x.\frac{1}{\sqrt{1-x^2}} + \sin^{-1}x.1 = \frac{x + \sin^{-1}x.\sqrt{1-x^2}}{\sqrt{1-x^2}}$ Ans.

(h) $y = \frac{\cos^{-1}x}{x^2}$, Let $f(x) = \cos^{-1}x$ and $g(x) = x^2$ then $f'(x) = -\frac{1}{\sqrt{1-x^2}}$ and $g'(x) = 2x$

[use formula $y = \frac{f(x)}{g(x)}$, $\frac{dy}{dx} = \frac{g(x).f'(x) - f(x).g'(x)}{(g(x))^2}$]

$$\therefore \frac{dy}{dx} = \frac{x^2.\left(-\frac{1}{\sqrt{1-x^2}}\right) - \cos^{-1}x.2x}{(x^2)^2} = \frac{-(x^2 + 2x\cos^{-1}x\sqrt{1-x^2})}{x^4} = -\frac{x + 2\cos^{-1}x\sqrt{1-x^2}}{x^3} \quad \text{Ans.}$$

(i) $y = \frac{\tan^{-1}x}{1+x^2}$, Let $f(x) = \tan^{-1}x$ and $g(x) = 1+x^2$ then $f'(x) = \frac{1}{1+x^2}$ and $g'(x) = 2x$

use above formula, $$\frac{dy}{dx} = \frac{(1+x^2)\frac{1}{1+x^2} - \tan^{-1}x.2x}{(1+x^2)^2} = \frac{1 - 2x\tan^{-1}x}{(1+x^2)^2} \quad \text{Ans.}$$

(j) $y = \sqrt{1-x^2}\sin^{-1}x$, Let $f(x) = \sqrt{1-x^2}$ and $g(x) = \sin^{-1}x$ then $f'(x) = \frac{1}{2\sqrt{1-x^2}}(-2x)$ and $g'(x) = \frac{1}{\sqrt{1-x^2}}$

[use formula $y = f(x).g(x)$, $\frac{dy}{dx} = f(x)g'(x) + g(x)f'(x)$] ∴ $\frac{dy}{dx} = \sqrt{1-x^2}.\frac{1}{\sqrt{1-x^2}} + \sin^{-1}x.\left(-\frac{x}{\sqrt{1-x^2}}\right) = \frac{\sqrt{1-x^2} - x\sin^{-1}x}{\sqrt{1-x^2}}$ Ans.

(k) $y = x\sqrt{1+x}$, $\frac{dy}{dx} = \frac{3x+2}{2\sqrt{1+x}}$ Ans. (Same as above question)

(l) $y = x\sqrt{1+x} + (x+1)\sqrt{x-3}$, Let $u = x\sqrt{1+x}$ and $v = (x+1)\sqrt{x-3}$ then $y = u + v$ or $\frac{dy}{dx} = \frac{du}{dx} + \frac{dv}{dx}$

Now $u = x\sqrt{1+x}$, $\frac{du}{dx} = x.\frac{1}{2\sqrt{1+x}} + \sqrt{1+x}.1 = \frac{3x+2}{2\sqrt{1+x}}$

and $v = (x+1)\sqrt{x-3}$, $\frac{dv}{dx} = (x+1).\frac{1}{2\sqrt{x-3}} + \sqrt{x-3}.1 = \frac{3x-5}{2\sqrt{x-3}}$

$$\therefore \frac{dy}{dx} = \frac{du}{dx} + \frac{dv}{dx} = \frac{3x+2}{2\sqrt{1+x}} + \frac{3x-5}{2\sqrt{x-3}} = \frac{(3x+2)\sqrt{x-3} + (3x-5)\sqrt{1+x}}{2\sqrt{x+1}\sqrt{x-3}} \quad \text{Ans.}$$

(3) Find the differential coefficient of the following: − (a) $y = ax^3 + bx^2 + cx + d$ (b) $y = x^4 - \frac{4}{3}x^3 + 2x^2 - 4x + 5$

(c) $y = 3x^{-\frac{2}{3}} + 2x^{\frac{1}{2}} - x$ (d) $y = \sqrt{\dfrac{1}{x+1}}$ (e) $y = \sqrt{x - 2x^2}$ (f) $y = x^{-3} + 2\log x - 5e^x$ (g) $y = \tan x + \cot x - e^x$

(h) $y = x^5 + 3\log x - \operatorname{cosec} x$ (i) $y^2 = x^2 + 2x + 1$ (j) $\log y = \sin x + \cos x$ (k) $\sqrt{y} = x^2 + 2x$ (l) $\sin y = e^x$

(m) $\sqrt{\sin y} = (1+x)^{\frac{1}{2}}$ (n) $\cos^2 y = x + 1$ (o) $\sin x + y^2 = \tan x + 1$

Solution: – (a) $y = ax^3 + bx^2 + cx + d$, $\dfrac{dy}{dx} = 3ax^2 + 2bx + c$ Ans. $\left[\text{formula, } y = x^n \text{ then } \dfrac{dy}{dx} = nx^{n-1}\right]$

(b) $y = x^4 - \dfrac{4}{3}x^3 + 2x^2 - 4x + 5$, $\dfrac{dy}{dx} = 4x^3 - 4x^2 + 4x - 4 = 4(x^3 - x^2 + x - 1)$ Ans.

(c) $y = 3x^{-\frac{2}{3}} + 2x^{\frac{1}{2}} - x$, $\dfrac{dy}{dx} = -\dfrac{2}{3}3x^{-\frac{5}{3}} + 2\dfrac{1}{2}x^{-\frac{1}{2}} - 1 = -\dfrac{2}{x^{\frac{5}{3}}} + \dfrac{1}{x^{\frac{1}{2}}} - 1$ Ans.

(d) $y = \sqrt{\dfrac{1}{x+1}} = \dfrac{1}{\sqrt{x+1}}$, $\dfrac{dy}{dx} = \dfrac{\sqrt{x+1}.0 - 1.\dfrac{1}{2\sqrt{x+1}}}{\left(\sqrt{x+1}\right)^2} = \dfrac{-1}{2(x+1)\sqrt{x+1}} = -\dfrac{1}{2(x+1)^{\frac{3}{2}}}$ Ans.

$$\left[\text{use formula } y = \frac{f(x)}{g(x)},\quad \frac{dy}{dx} = \frac{g(x).f'(x) - f(x).g'(x)}{\left(g(x)\right)^2}\right]$$

(e) $y = \sqrt{x - 2x^2}$ squaring both side $\therefore$ $y^2 = x - 2x^2$ or $2y\dfrac{dy}{dx} = 1 - 4x$ $\therefore$ $\dfrac{dy}{dx} = \dfrac{1-4x}{2y} = \dfrac{1-4x}{2\sqrt{x-2x^2}}$ Ans.

(f) $y = x^{-3} + 2\log x - 5e^x$, $\dfrac{dy}{dx} = -3x^{-4} + \dfrac{2}{x} - 5e^x = -\dfrac{3}{x^4} + \dfrac{2}{x} - 5e^x$ Ans.

(g) $y = \tan x + \cot x - e^x$, $\dfrac{dy}{dx} = \sec^2 x - \operatorname{cosec}^2 x - e^x = \dfrac{1}{\cos^2 x} - \dfrac{1}{\sin^2 x} - e^x = \dfrac{\sin^2 x - \cos^2 x}{\sin^2 x \cos^2 x} - e^x$

or $\dfrac{dy}{dx} = \dfrac{-4\cos 2x}{(2\sin x\cos x)^2} - e^x = \dfrac{-4\cos 2x}{\sin^2 2x} - e^x = -4\cot 2x \operatorname{cosec} 2x - e^x$ Ans.

(h) $y = x^5 + 3\log x - \operatorname{cosec} x$, $\dfrac{dy}{dx} = 5x^4 + \dfrac{3}{x} - (-\cot x \operatorname{cosec} x) = 5x^4 + \dfrac{3}{x} + \cot x \operatorname{cosec} x$ Ans.

(i) $y^2 = x^2 + 2x + 1$, $2y\dfrac{dy}{dx} = 2x + 2$ $\therefore$ $\dfrac{dy}{dx} = \dfrac{2(x+1)}{2y} = \dfrac{x+1}{\sqrt{x^2+2x+1}} = \dfrac{x+1}{\sqrt{(x+1)^2}} = \dfrac{x+1}{x+1} = 1$ Ans.

IInd Method: – $y^2 = (x+1)^2$ $\therefore$ $y = \sqrt{(x+1)^2} = x + 1$ $\therefore$ $\dfrac{dy}{dx} = 1 + 0 = 1$ Ans.

(j) $\log y = \sin x + \cos x$ $\therefore y = e^{\sin x + \cos x}$ or $\dfrac{1}{y}\dfrac{dy}{dx} = \cos x - \sin x$ $\therefore$ $\dfrac{dy}{dx} = y(\cos x - \sin x)$

$$\frac{dy}{dx} = e^{\sin x + \cos x}(\cos x - \sin x) \text{ Ans.}$$

(k) $\sqrt{y} = x^2 + 2x$ squaring both side, we get $y = (x^2 + 2x)^2$ or $y = x^4 + 4x^3 + 4x^2$

or $\dfrac{dy}{dx} = 4x^3 + 12x^2 + 8x = 4x(x^2 + 3x + 2) = 4x(x^2 + 2x + x + 2) = 4x[x(x+2) + 1(x+2)] = 4x(x+1)(x+2)$ Ans.

(l) $\sin y = e^x$, $\cos y\dfrac{dy}{dx} = e^x$ $\therefore$ $\dfrac{dy}{dx} = \dfrac{e^x}{\cos y} = \dfrac{e^x}{\sqrt{1 - \sin^2 y}} = \dfrac{e^x}{\sqrt{1 - e^{2x}}}$ or $\dfrac{e^x\sqrt{1 + (e^x)^2}}{\sqrt{1 - (e^{2x})^2}}$ Ans.

(m) $\sqrt{\sin y} = (1+x)^{\frac{1}{2}}$ squaring both side we have $\sin y = 1 + x$ $\therefore$ $\cos y\dfrac{dy}{dx} = 1$

$$\text{or } \frac{dy}{dx} = \frac{1}{\cos y} = \frac{1}{\sqrt{1 - \sin^2 y}} \quad \therefore \frac{dy}{dx} = \frac{1}{\sqrt{1 - (1+x)^2}} = \frac{1}{\sqrt{1 - 1 - 2x - x^2}} = \frac{1}{\sqrt{-x(x+2)}} \text{ Ans.}$$

(n) $\cos^2 y = x + 1$, $2\cos y(-\sin y)\frac{dy}{dx} = 1$ or $-2\sin y\cos y\frac{dy}{dx} = 1$ $\therefore \frac{dy}{dx} = -\frac{1}{2\sin y\cos y}$

or $\cos^2 y = x + 1$ or $\cos y = \sqrt{x+1}$, $\sin y = \sqrt{1-\cos^2 y} = \sqrt{1-x-1} = \sqrt{-x}$ or $\frac{dy}{dx} = -\frac{1}{2\sin y\cos y} = -\frac{1}{2\sqrt{-x}\sqrt{x+1}}$ Ans.

(o) $\sin x + y^2 = \tan x + 1$ or $y^2 = \tan x - \sin x + 1$ $\therefore 2y\frac{dy}{dx} = \sec^2 x - \cos x$ $\therefore \frac{dy}{dx} = \frac{\sec^2 x - \cos x}{2y}$

$$y^2 = \tan x - \sin x + 1 \text{ or } y = \sqrt{\tan x - \sin x + 1} \quad \therefore \frac{dy}{dx} = \frac{\sec^2 x - \cos x}{2\sqrt{\tan x - \sin x + 1}} \quad \text{Ans.}$$

(4) (a) If $y = x^4 - 3x^3 + 2x^2 + 3x - 5$ then find $\frac{dy}{dx}$ at $x = 3$.

(b) If $y = \sqrt{\sin x + \cos x}$, find $\frac{dy}{dx}$ at $x = \frac{\pi}{4}$. (c) If $y = |x|^2 - 3|x| + 4$ then find $\frac{dy}{dx}$ at $x = 2$ and $x = -1$.

(d) If $y = |x|^3 - 2|x|^2 + 3|x| + 5$ then find $\frac{dy}{dx}$ at $x = 2$. (e) If $y = \log|x+1|$ then find $\frac{dy}{dx}$ at $x = 1$.

(f) If $y = x^3 + \sin x - \frac{3}{2}\log x$ then find $\frac{dy}{dx}$, when $x = \frac{\pi}{3}$. (g) If $y = |x-1| + |x-3|$ then find $\frac{dy}{dx}$.

(h) If $y = \log e^x + e^x$ then find $\frac{dy}{dx}$, when $x = \log_e 4$. (i) If $x + y = e^{x+y} + e^x + e^y$ then find $\frac{dy}{dx}$.

Solution: – (a) $y = x^4 - 3x^3 + 2x^2 + 3x - 5$, $\frac{dy}{dx} = 4x^3 - 9x^2 + 4x + 3$ at $x = 3$

$$\left(\frac{dy}{dx}\right)_{x=3} = 4(3)^3 - 9(3)^2 + 4(3) + 3 = 108 - 81 + 12 + 3 = 123 - 81 = 42 \quad \text{Ans.}$$

(b) $y = \sqrt{\sin x + \cos x}$, $\frac{dy}{dx} = \frac{1}{2\sqrt{\sin x + \cos x}}.(\cos x - \sin x) = \frac{(\cos x - \sin x)}{2\sqrt{\sin x + \cos x}}$ at $x = \frac{\pi}{4}$

$$\left(\frac{dy}{dx}\right)_{x=\frac{\pi}{4}} = \frac{(\cos x - \sin x)}{2\sqrt{\sin x + \cos x}} = \frac{\left(\cos\frac{\pi}{4} - \sin\frac{\pi}{4}\right)}{2\sqrt{\sin\frac{\pi}{4} + \cos\frac{\pi}{4}}} = 0 \quad \text{Ans.}$$

(c) $y = |x|^2 - 3|x| + 4$ $\therefore x = 0$ then $x > 0$, *x is positive and* $x < 0$, *x is negative*

$$\text{or } y = \begin{cases} x^2 - 3x + 4, & x > 0 \\ x^2 + 3x + 4, & x < 0 \end{cases} \quad \therefore \frac{dy}{dx} = \begin{cases} 2x - 3, & x > 0 \\ 2x + 3, & x < 0 \end{cases}$$

At $x = 2$, $\left(\frac{dy}{dx}\right)_{x=2} = \begin{cases} 2\times 2 - 3, & x > 0 \\ 2\times 2 + 3, & x < 0 \end{cases} = \begin{cases} 4 - 3, & x > 0 \\ 4 + 3, & x < 0 \end{cases} = \begin{cases} 1, & x > 0 \\ 7, & x < 0 \end{cases}$ Ans.

At $x = -1$, $\left(\frac{dy}{dx}\right)_{x=-1} = \begin{cases} 2(-1) - 3, & x > 0 \\ 2(-1) + 3, & x < 0 \end{cases} = \begin{cases} -2 - 3, & x > 0 \\ -2 + 3, & x < 0 \end{cases} = \begin{cases} -5, & x > 0 \\ 1, & x < 0 \end{cases}$ Ans.

(d) $y = |x|^3 - 2|x|^2 + 3|x| + 5$ at $x = 2$ is a positive integer. or $\frac{dy}{dx} = 3x^2 - 4x + 3$ at $x = 2$ $\therefore \left(\frac{dy}{dx}\right)_{x=2} = 12 - 8 + 3 = 7$ Ans.

(e) $y = \log|x+1|$ $\therefore x + 1 = 0$ or $x = -1$ $\therefore y = \begin{cases} \log(x+1), & x > -1 \\ \log -(x+1), & x < -1 \end{cases}$ *it is not possible.*

$$\therefore \frac{dy}{dx} = \frac{1}{x+1} \text{ at } x = 1 \quad \therefore \left(\frac{dy}{dx}\right)_{x=1} = \frac{1}{1+1} = \frac{1}{2} \quad \text{Ans.}$$

(f) $y = x^3 + \sin x - \frac{3}{2}\log x$, $\frac{dy}{dx} = 3x^2 + \cos x - \frac{3}{2}.\frac{1}{x} = 3x^2 + \cos x - \frac{3}{2x}$

At $x = \frac{\pi}{3}$ then $\frac{dy}{dx} = 3\left(\frac{\pi}{3}\right)^2 + \cos\frac{\pi}{3} - \frac{3}{2\frac{\pi}{3}} = \frac{3\pi^2}{9} + \frac{1}{2} - \frac{9}{2\pi} = \frac{\pi^2}{3} + \frac{1}{2} - \frac{9}{2\pi} = \frac{2\pi^3 + 3\pi - 27}{6\pi}$ Ans.

(g) $y = |x-1| + |x-3|$, Put $x-1=0$ and $x-3=0$ then $x = 1,3$

$-\infty$	+ ve	1	− ve	3	+ ve	∞

or $y = \begin{cases} (x-1)+(x-3) \text{ if } x<1 \text{ and } x \geq 3 \\ -(x-1)-(x-3) \text{ if } 1 \leq x < 3 \end{cases} = \begin{cases} 2x-4 \text{ if } x<1 \text{ and } x\geq 3 \\ -2x+4 \text{ if } 1\leq x<3 \end{cases}$

$\therefore \frac{dy}{dx} = \begin{cases} 2 \text{ if } x<1,\ x\geq 3 \\ -2 \text{ if } 1\leq x<3 \end{cases}$ or $\frac{dy}{dx} = 2$ and -2 $\therefore \frac{dy}{dx} = \{2,-2\}$ Ans.

(h) $y = \log e^x + e^x$, $\frac{dy}{dx} = \frac{1}{e^x} + e^x$ At $x = \log_e 4$ then $\frac{dy}{dx} = \frac{1}{e^{\log_e 4}} + e^{\log_e 4} = \frac{1}{4} + 4 = \frac{1+16}{4} = \frac{17}{4}$ Ans.

(i) $x + y = e^{x+y} + e^x + e^y$, $1 + \frac{dy}{dx} = e^{x+y}\left(1 + \frac{dy}{dx}\right) + e^x + e^y\frac{dy}{dx}$ or $\frac{dy}{dx}(1 - e^y - e^{x+y}) = e^{x+y} + e^x - 1$

$$\frac{dy}{dx} = \frac{e^{x+y} + e^x - 1}{1 - e^y - e^{x+y}} = \frac{-(1 - e^y - e^{x+y})}{1 - e^y - e^{x+y}} = -1 \text{ Ans.}$$

IInd Method:− $\log(x+y) = \log e^{x+y} + \log e^x + \log e^y = x + y + x + y = 2x + 2y$

or $\frac{1}{x+y}\left(1 + \frac{dy}{dx}\right) = 2 + 2\frac{dy}{dx}$ or $\frac{1}{x+y}\frac{dy}{dx} - 2\frac{dy}{dx} = 2 - \frac{1}{x+y}$ or $\frac{dy}{dx}\left(\frac{1}{x+y} - 2\right) = \frac{2x+2y-1}{x+y}$

$$\therefore \frac{dy}{dx} = \frac{\frac{2x+2y-1}{x+y}}{\frac{1-2x-2y}{x+y}} = \frac{2x+2y-1}{1-2x-2y} = \frac{-(1-2x-2y)}{1-2x-2y} = -1 \text{ Ans.}$$

(5) Find the differential coefficient of the following: −

(a) If $y = (x^2 - 2x + 3)(x^2 + 2)$ find $\frac{dy}{dx}$. (b) $y = \frac{1+x^3}{1-x^3}$ (c) $y = \frac{1}{x\tan x}$ (d) $y = \frac{\tan x - \cot x}{\tan x + \cot x}$ (e) $y = \frac{1+\sin x}{1-\sin x}$ (f) $y = \sqrt{\frac{1-x}{1+x}}$

(g) $y = \frac{x\sin x}{a+x}$ (h) $y = \frac{x\cot x}{\tan x + \cot x}$ (i) $y = x^3\sin x$ (j) $y = (x^2-1)\sec x$ (k) $y = \frac{1+\tan x}{1-\tan x}$ (l) $y = \frac{x^2 - \sec x}{1+\tan x}$ (m) $y = \frac{x^2\cos x}{1-\tan x}$

Solution:− (a) $y = (x^2 - 2x + 3)(x^2 + 2)$, Let $f(x) = x^2 - 2x + 3$ and $g(x) = x^2 + 2$

$f'(x) = 2x - 2$ and $g'(x) = 2x$ $\left[\text{use formula, } y = f(x)g.(x) \text{ then } \frac{dy}{dx} = f(x)g'(x) + g(x)f'(x)\right]$

$\frac{dy}{dx} = (x^2 - 2x + 3).2x + (x^2 + 2).(2x - 2) = 2x^3 - 4x^2 + 6x + 2x^3 + 4x - 2x^2 - 4 = 4x^3 - 6x^2 + 10x - 4$ Ans.

(b) $y = \frac{1+x^3}{1-x^3}$, Let $f(x) = 1 + x^3$ and $g(x) = 1 - x^3$ then $f'(x) = 3x^2$ and $g'(x) = -3x^2$

$$\left[\text{use formula } y = \frac{f(x)}{g(x)},\quad \frac{dy}{dx} = \frac{g(x).f'(x) - f(x).g'(x)}{(g(x))^2}\right]$$

$$\frac{dy}{dx} = \frac{(1-x^3).3x^2 - (1+x^3).(-3x^2)}{(1-x^3)^2} = \frac{3x^2 - 3x^5 + 3x^2 + 3x^3}{(1-x^3)^2} = \frac{6x^2}{(1-x^3)^2} \text{ Ans.}$$

(c) $y = \frac{1}{x\tan x}$, Let $f(x) = 1$ and $g(x) = x\tan x$ then $f'(x) = 0$ and $g'(x) = x.\sec^2 x + \tan x.1$

use above formula $\frac{dy}{dx} = \frac{x\tan x.0 - 1(x\sec^2 x + \tan x)}{x^2\tan^2 x} = \frac{-(x\sec^2 x + \tan x)}{x^2\tan^2 x} = \frac{-x(1+\tan^2 x) - \tan x}{x^2\tan^2 x} = \frac{-x - x\tan^2 x - \tan x}{x^2\tan^2 x}$

$$\frac{dy}{dx} = \frac{-x}{x^2\tan^2 x} - \frac{x\tan^2 x}{x^2\tan^2 x} - \frac{\tan x}{x^2\tan^2 x} = -\frac{1}{x\tan^2 x} - \frac{1}{x} - \frac{1}{x^2\tan x} = -\frac{\cos^2 x}{x\sin^2 x} - \frac{\cos x}{x^2\sin x} - \frac{1}{x} \text{ Ans.}$$

IInd Method:− $y = \frac{1}{x\tan x} = \frac{\cos x}{x\sin x}$, Let $f(x) = \cos x$ and $g(x) = x\sin x$ then $f'(x) = -\sin x$ and $g'(x) = x\cos x + \sin x$

use above formula, $\frac{dy}{dx}=\frac{x\sin x(-\sin x)-\cos x(x\cos x+\sin x)}{x^2\sin^2 x}=\frac{-x\sin^2 x-x\cos^2 x-\sin x\cos x}{x^2\sin^2 x}=\frac{-x(\sin^2 x+\cos^2 x)-\sin x\cos x}{x^2\sin^2 x}$

$=\frac{-x-\sin x\cos x}{x^2\sin^2 x}=-\frac{x}{x^2\sin^2 x}-\frac{\sin x\cos x}{x^2\sin^2 x}=-\frac{1}{x\sin^2 x}-\frac{\cos x}{x^2\sin x}=-\frac{\operatorname{cosec}^2 x}{x}-\frac{\cot x}{x^2}$ Ans.

(d) $y=\frac{\tan x-\cot x}{\tan x+\cot x}$, Let $f(x)=\tan x-\cot x$ and $g(x)=\tan x+\cot x$ then $f'(x)=\sec^2 x+\operatorname{cosec}^2 x$ and $g'(x)=\sec^2 x-\operatorname{cosec}^2 x$

use above formula, $\frac{dy}{dx}=\frac{(\tan x+\cot x).(\sec^2 x+\operatorname{cosec}^2 x)-(\tan x-\cot x).(\sec^2 x-\operatorname{cosec}^2 x)}{(\tan x+\cot x)^2}$

$\frac{dy}{dx}=\frac{(\tan x+\cot x).(\sec^2 x+\operatorname{cosec}^2 x)}{(\tan x+\cot x)^2}-\frac{(\tan x-\cot x).(\sec^2 x-\operatorname{cosec}^2 x)}{(\tan x+\cot x)^2}$

$=\frac{\sec^2 x+\operatorname{cosec}^2 x}{\tan x+\cot x}-\frac{(\tan x-\cot x).(\sec^2 x-\operatorname{cosec}^2 x)}{(\tan x+\cot x)^2}$ (change sin and cos)

$\frac{dy}{dx}=2\sin 2x$ Ans.

IInd Method:– $y=\frac{\tan x-\cot x}{\tan x+\cot x}$ (change sin and cos) or $y=-\cos 2x$ or $\frac{dy}{dx}=-(-\sin 2x).2=2\sin 2x$ Ans.

(e) $y=\frac{1+\sin x}{1-\sin x}=\frac{f(x)}{g(x)}$ (say) where $f(x)=1+\sin x$, $f'(x)=\cos x$ and $g(x)=1-\sin x$, $g'(x)=-\cos x$

or $\frac{dy}{dx}=\frac{g(x).f'(x)-f(x).g'(x)}{(g(x))^2}=\frac{(1-\sin x).\cos x-(1+\sin x).(-\cos x)}{(1-\sin x)^2}=\frac{\cos x-\sin x\cos x+\cos x+\sin x\cos x}{(1-\sin x)^2}=\frac{2\cos x}{(1-\sin x)^2}$ Ans.

(f) $y=\sqrt{\frac{1-x}{1+x}}=\frac{\sqrt{1-x}}{\sqrt{1+x}}=\frac{f(x)}{g(x)}$ (say) Here $f(x)=\sqrt{1-x}$, $f'(x)=\frac{1}{2\sqrt{1-x}}(-1)$ and $g(x)=\sqrt{1+x}$, $g'(x)=\frac{1}{2\sqrt{1+x}}$

$\frac{dy}{dx}=\frac{g(x).f'(x)-f(x).g'(x)}{(g(x))^2}=\frac{\sqrt{1+x}.\left(-\frac{1}{2\sqrt{1-x}}\right)-\sqrt{1-x}.\left(\frac{1}{2\sqrt{1+x}}\right)}{\left(\sqrt{1+x}\right)^2}=\frac{-\frac{\sqrt{1+x}}{2\sqrt{1-x}}-\frac{\sqrt{1-x}}{2\sqrt{1+x}}}{1+x}=\frac{-1-x-1+x}{2(1+x)\sqrt{(1-x)(1+x)}}$

$=-\frac{1}{(1+x)\sqrt{1-x^2}}$ Ans.

(g) $y=\frac{x\sin x}{a+x}=\frac{f(x)}{g(x)}$ (say) Here $f(x)=x\sin x$, $f'(x)=x\frac{d(\sin x)}{dx}+\sin x\frac{d(x)}{dx}=x\cos x+\sin x$ and $g(x)=a+x$, $g'(x)=1$

$\therefore \frac{dy}{dx}=\frac{g(x).f'(x)-f(x).g'(x)}{(g(x))^2}=\frac{(a+x)(x\cos x+\sin x)-x\sin x.1}{(a+x)^2}=\frac{ax\cos x+a\sin x+x^2\cos x+x\sin x-x\sin x}{(a+x)^2}$

$=\frac{ax\cos x+a\sin x+x^2\cos x}{(a+x)^2}$ Ans.

(h) $y=\frac{x\cot x}{\tan x+\cot x}=\frac{x\frac{\cos x}{\sin x}}{\frac{\sin x}{\cos x}+\frac{\cos x}{\sin x}}=\frac{\frac{x\cos x}{\sin x}}{\frac{\sin^2 x+\cos^2 x}{\sin x\cos x}}=x\cos^2 x$, $\frac{dy}{dx}=x\frac{d(\cos^2 x)}{dx}+\cos^2 x\frac{d(x)}{dx}$

$\frac{dy}{dx}=x.2\cos x(-\sin x)+\cos^2 x.1=-2x\sin x\cos x+\cos^2 x=-x\sin 2x+\cos^2 x=-x\sin 2x+\frac{1+\cos 2x}{2}=\frac{-2x\sin 2x+\cos 2x+1}{2}$ Ans.

(i) $y=x^3\sin x$, $\frac{dy}{dx}=x^3$(differential coefficient(d.c) of $\sin x$) + $\sin x$ (differential coefficient(d.c) of x^3)

$\therefore \frac{dy}{dx}=x^3\cos x+\sin x.3x^2=x^2(x\cos x+3\sin x)$ Ans.

(j) $y=(x^2-1)\sec x=f(x).g(x)$, $\frac{dy}{dx}=f(x)g'(x)+g(x)f'(x)$ where $f(x)=x^2-1$, $f'(x)=2x$ and $g(x)=\sec x$, $g'(x)=\tan x\sec x$

$\therefore \frac{dy}{dx}=(x^2-1).\tan x\sec x+\sec x.2x=\sec x(x^2\tan x-\tan x+2x)$ Ans.

(k) $y = \dfrac{1+\tan x}{1-\tan x} = \dfrac{f(x)}{g(x)}$ (say) here $f(x) = 1 + \tan x$, $f'(x) = \sec^2 x$ and $g(x) = 1 - \tan x$, $g'(x) = -\sec^2 x$

or $\dfrac{dy}{dx} = \dfrac{g(x).f'(x) - f(x).g'(x)}{(g(x))^2} = \dfrac{(1-\tan x).\sec^2 x - (1+\tan x).(-\sec^2 x)}{(1-\tan x)^2}$

$$\therefore \frac{dy}{dx} = \frac{\sec^2 x - \tan x \sec^2 x + \sec^2 x + \tan x \sec^2 x}{(1-\tan x)^2} = \frac{2\sec^2 x}{(1-\tan x)^2} \quad \text{Ans.}$$

(l) $y = \dfrac{x^2 - \sec x}{1+\tan x}$ (use above formula and solve this question) or $\dfrac{dy}{dx} = \dfrac{(2x - \tan x \sec x)}{1+\tan x} - \dfrac{(x^2 - \sec x)\sec^2 x}{(1+\tan x)^2}$ Ans.

(m) $y = \dfrac{x^2 \cos x}{1 - \tan x}$ (same as above question), $\dfrac{dy}{dx} = \dfrac{(2x\cos x - x^2 \sin x)}{(1-\tan x)} + \dfrac{x^2 \sec x}{(1-\tan x)^2}$ Ans.

(6) Find the differential coefficient of the following: – (a) $y = x^2$ (b) $y = 3x^3$

(c) $y = x + x^3 + 2x^5$ (d) $y + 1 = x^3 + 2x^2$ (e) $y = \dfrac{1}{x^3}$ (f) $y = \sqrt{x+1}$ (g) $y = \dfrac{1}{\sqrt{2x+1}}$

(h) $\sqrt{y+2} = x + 1$ (i) $y = \sin x + e^x$ (j) $\tan y = 2\sin x + x$ (k) $y = e^x \cos x + x \log x$ (l) $y = 2\sqrt{x} - \dfrac{1}{\sqrt{x}}$

Solution: – (a) $y = x^2$, $\dfrac{dy}{dx} = 2x$ Ans. $\left[\text{formula, } y = x^n, \dfrac{dy}{dx} = nx^{n-1} \text{ or } \dfrac{d(x^n)}{dx} = nx^{n-1}\right]$

(b) $y = 3x^3$, $\dfrac{dy}{dx} = 9x^2$ Ans. (c) $y = x + x^3 + 2x^5$, $\dfrac{dy}{dx} = 1 + 3x^2 + 10x^4$ Ans.

(d) $y + 1 = x^3 + 2x^2$, $\dfrac{dy}{dx} = 3x^2 + 4x$ Ans. (e) $y = \dfrac{1}{x^3} = x^{-3}$, $\dfrac{dy}{dx} = -3x^{-4} = -\dfrac{3}{x^4}$ Ans.

(f) $y = \sqrt{x+1}$ squaring both of sides $y^2 = x + 1$, $2y\dfrac{dy}{dx} = 1$ $\therefore \dfrac{dy}{dx} = \dfrac{1}{2y} = \dfrac{1}{2\sqrt{x+1}}$ Ans.

(g) $y = \dfrac{1}{\sqrt{2x+1}} = \dfrac{1}{(2x+1)^{\frac{1}{2}}} = (2x+1)^{-\frac{1}{2}}$, $\dfrac{dy}{dx} = -\dfrac{1}{2}.2\,(2x+1)^{-\frac{1}{2}-1} = -(2x+1)^{-\frac{3}{2}}$ Ans.

$$\left[\text{Use formula } y = (ax)^n, \frac{dy}{dx} = a^n.nx^{n-1} \text{ and } y = (ax+b)^n, \quad \frac{dy}{dx} = na(ax+b)^{n-1}\right]$$

(h) $\sqrt{y+2} = x + 1$ squaring both of sides $y + 2 = (x+1)^2$ or $y + 2 = x^2 + 2x + 1$ $\therefore \dfrac{dy}{dx} = 2x + 2 = 2(x+1)$ Ans.

(i) $y = \sin x + e^x$, $\dfrac{dy}{dx} = \cos x + e^x$ Ans.

(j) $\tan y = 2\sin x + x$, $\sec^2 y\dfrac{dy}{dx} = 2\cos x + 1$ $\therefore \dfrac{dy}{dx} = \dfrac{1 + 2\cos x}{\sec^2 y} = \dfrac{1 + 2\cos x}{\sec^2\{\tan(2\sin x + x)\}}$ Ans.

(k) $y = e^x \cos x + x \log x$, $\dfrac{dy}{dx} = e^x\dfrac{d(\cos x)}{dx} + \cos x\dfrac{d(e^x)}{dx} + x\dfrac{d(\log x)}{dx} + \log x\dfrac{d(x)}{dx}$

$$\frac{dy}{dx} = -e^x \sin x + e^x \cos x + x\frac{1}{x} + \log x.1 \quad \therefore \frac{dy}{dx} = e^x(\cos x - \sin x) + (1 + \log x) \quad \text{Ans.}$$

(l) $y = 2\sqrt{x} - \dfrac{1}{\sqrt{x}} = \dfrac{2x-1}{\sqrt{x}}$ or $\dfrac{dy}{dx} = \dfrac{g(x).f'(x) - f(x).g'(x)}{(g(x))^2} = \dfrac{\sqrt{x}\dfrac{d(2x-1)}{dx} - (2x-1)\dfrac{d(\sqrt{x})}{dx}}{(\sqrt{x})^2} = \dfrac{\sqrt{x}.2 - (2x-1).\dfrac{1}{2\sqrt{x}}}{x}$

$$\frac{dy}{dx} = \frac{\dfrac{4x - 2x + 1}{2\sqrt{x}}}{x} \quad \therefore \frac{dy}{dx} = \frac{2x+1}{2x\sqrt{x}} = \frac{2x+1}{2(x)^{\frac{3}{2}}} \quad \text{Ans.}$$

Exercise – A7

(1) Find the differential coefficient of the following: –

(a) $y = \sin(ax^2 + bx + c)$ (b) $y = \cos(2x - 3)$ (c) $y = \sqrt{x^2 - 2x + 3}$ (d) $y = \sin\sqrt{2x^2 - 3}$

(e) $y = \sqrt{\sin(2 + 3x^2)}$ (f) $y = \cos\sqrt{\sqrt{x} + 3}$ (g) $y = \sqrt{\cos(x - 2)}$ (h) $y = \sin\sqrt{\tan x}$

(i) $y = \tan\sqrt{\sin\sqrt{x} + \cos\sqrt{x}}$ (j) $y = \cos^2\sqrt{\sin x + 3}$ (k) $y = \sqrt{\sin\sqrt{\cos\sqrt{x} - 1}}$ (l) $y = \log\sqrt{\sin x + \cos x}$

(m) $y = \cot(x - 2)$ (n) $y = \sqrt{\cot\sqrt{2x + 1}}$ (o) $y = \sec\sqrt{\operatorname{cosec}(x^2 + 1)}$ (p) $y = \operatorname{cosec}(\sec\theta + \tan\theta)$

(q) $y = \sqrt{\sec(\tan\theta + 1)}$ (r) $y = \sqrt{\operatorname{cosec}(\sin\theta + \sec\theta)}$

(2) (a) $y = \sin^{-1}(\sec x + \tan x)$ (b) $y = \cos^{-1}(\cot\theta - \operatorname{cosec}\theta)$ (c) $y = \tan^{-1}(\sin\theta - \cos\theta)$

(d) $y = \cot^{-1}(\sec\theta + \operatorname{cosec}\theta)$ (e) $y = \sec^{-1}(\tan\theta - \theta)$ (f) $y = \operatorname{cosec}^{-1}(\sin\theta + \tan\theta)$ (g) $y = \sin^{-1}\sqrt{x^2 + 1}$

(h) $y = \sin^{-1}\sqrt{x^2 - 1} + \cos^{-1}\sqrt{x - 1}$ (i) $y = \tan^{-1}(x^2 - 5)$ (j) $y = \sin\sqrt{\cos\sqrt{x} - 1}$ (k) $y = \cos\sqrt{\sin\sqrt{x}}$

(l) $y = \tan\sqrt{\cot\sqrt{x^2 + 1}}$ (m) $y = \cot\sqrt{\cos\sqrt{1 + \sin\theta}}$ (n) $y = \sec\sqrt{\operatorname{cosec}\sqrt{x} + 1}$ (o) $y = \operatorname{cosec}\sqrt{\sin\sqrt{1 + \cos\theta}}$

(p) $y = \dfrac{\tan x}{x} - \sqrt{1 + x^2}$ (q) $y = \sin^2(x^2 - 1)$ (r) $y = \sin^2\left(\sqrt{x^2 - x + 1}\right)$ (s) $y = \cos^2(1 - x^2)$

(t) $y = \cos^2\sqrt{x^3 - 1}$ (u) $y = \tan^2(ax^3 + bx^2 + cx + d)$ (v) $y = \tan^2\sqrt{1 - ax^2}$ (w) $y = \sqrt{\tan(2x + 1)}$

(3) (a) $y = x\sin x^3$ (b) $y = x^2\cos(x^3 + 1)$ (c) $y = x^3\tan x^2$ (d) $y = (x - 1)\sec x^3$ (e) $y = \sqrt{x}\cot x^2$

(f) $y = (x^2 + 1)\operatorname{cosec}\sqrt{x + 1}$ (g) $y = \dfrac{\sin(x^2 + 1)}{x^3}$ (h) $y = \dfrac{\tan\sqrt{x + 1}}{x}$ (i) $y = \sin\left[\dfrac{1 + x}{1 - x}\right]$

(j) $y = \sin\sqrt{x^2 + ax + 1} \times \cos\sqrt{ax^2 + 1}$ (k) $y = \tan(ax^2 + bx + 1) + \cot\sqrt{ax^2 + bx + c}$ (l) $y = x^2\log\sqrt{x - 3}$

(m) $y = \sin^3\sqrt{5 - 2x + x^2}$ (n) $y = (x - 1)^2.(3x^2 - 1)$ (o) $y = \cos(\log x^2)$ (p) $y = \sin(e^x + 1)$

(q) $y = \tan[\log(x - 5)]$ (r) $y = \log[\log(3 - x)]$ (s) $y = e^{\sqrt{x}}.\log x$ (t) $y = e^{2x}.x^3$ (u) $y = \sqrt{1 + \tan\theta}$ (v) $y = \sqrt{\sin\sqrt{x}}$

(4) (a) $y = \dfrac{\tan\left(\frac{1 - x}{1 + x}\right)}{x} + \dfrac{\cot\left(\frac{1 + x}{1 - x}\right)}{x}$ (b) $y = \cos\sqrt{\sin\sqrt{ax + b}}$ (c) $y = \dfrac{\sqrt{\tan x}}{\sqrt{x^2 - 3}}$ (d) $y = \dfrac{x^2 + \sin x^3}{x^3 + \cos x^2}$ (e) $y = x^{\frac{3}{2}}\cos(ax^2 + bx + c)$

(5) Find the differentiate the following: – (a) $y = \tan^{-1}\left(\dfrac{\sqrt{x} - 1}{1 + \sqrt{x}}\right)$ (b) $y = \tan^{-1}\left(\dfrac{\sin x + \cos x}{1 - \sin x\cos x}\right)$ (c) $y = \tan^{-1}\left(\dfrac{6x}{1 - x^2}\right)$

(d) $y = \tan^{-1}\left(\dfrac{x + 3}{1 + x^2}\right) + \tan^{-1}\left(\dfrac{2x + 1}{2 - 3x}\right)$ (e) $y = \cot^{-1}\left(\dfrac{\sqrt{1 + x} - \sqrt{1 - x}}{\sqrt{1 - x} + \sqrt{1 + x}}\right)$ (f) $y = \sin^{-1}\dfrac{1}{\sqrt{1 + x^2}} + \tan^{-1}\dfrac{x}{\sqrt{1 - x^2}}$

(6) (a) $y = \sin^{-1}\sqrt{x} + \sin^{-1}\sqrt{1 - x}$ (b) $y = \sin^{-1}(x^2 - 1) - \sin^{-1}(2x^2)$ (c) $y = \cos^{-1}\dfrac{1}{1 + x} + \cos^{-1}(1 - x)$

(d) $y = \cos^{-1}\dfrac{x}{\sqrt{1 + x}} - \cos^{-1}\dfrac{1}{\sqrt{1 - x}}$ (e) $y = \sin^{-1}[\sqrt{x}.\sqrt{2 - x} + \sqrt{x - 1}.\sqrt{1 - x}]$ (f) $y = \cos^{-1}\left[\sqrt{x}.\dfrac{1}{\sqrt{1 + x^2}} - \sqrt{1 - x}.\sqrt{\dfrac{x^2}{1 + x^2}}\right]$

(7) (a) Differentiate $\tan^{-1}\frac{\sqrt{1-x^2}}{x}$ with respect to $\tan^{-1}x$ (b) Differentiate $\sin^{-1}\sqrt{\frac{1+x}{1-x}}$ with respect to $\sqrt{x}$

(c) Differentiate $x^{\cos^{-1}\sqrt{x}}$ with respect to $\cos^{-1}\sqrt{x}$ (d) Differentiate $\cot^{-1}\frac{x}{\sqrt{1+x^2}}$ with respect to $\text{cosec}^{-1}\frac{1}{x^2+1}$

(8) (a) Find the differential coefficient of $\log_{(1-x)}\cos^{-1}(1-x)$ with respect to $2^{2(1-x)}$ and also find its value at $x=\frac{1}{2}$.

(b) If $\sqrt{1+4x^2}-\sqrt{1-9y^2}=a(2x-3y)$ then find $\frac{dy}{dx}$. (c) If $y=\tan^{-1}\frac{1}{1+x}+\cot^{-1}(1+x)$ find $\frac{d^2y}{dx^2}$.

(d) If $x=a(1-\cos t)$, $y=a(1+\sin t)$ find $\frac{dy}{dx}$ and independent of t. (e) If $x=a\left[\sin t-\log\left(\tan\frac{t}{2}\right)\right]$, $y=a\cos t$ find $\frac{dy}{dx}$ at $t=\frac{\pi}{4}$.

(9) (a) If $x=t^2+\frac{1}{t^2}$ and $y=t^2-\frac{1}{t^2}$ find $\frac{d^2y}{dx^2}$ and $\frac{dy}{dx}$ is independent of t. (b) If $x=\frac{1-t^2}{t^3}$, $y=\frac{3}{2t^3}-\frac{2}{t^2}$ find $\frac{dy}{dx}$ at $t=2$.

(c) If $x=2\sin t+\sin 2t$, $y=2\cos t+\cos 2t$ find the value of $\frac{dy}{dx}$ at $x=\frac{\pi}{2}$.

(d) If $x=\sin^{-1}\frac{1}{\sqrt{1+t^2}}$, $y=\cos^{-1}\frac{t}{\sqrt{1+t^2}}$ show that $\frac{dy}{dx}$ is independent of t.

(10) (a) If $x=\frac{\sin^2 t}{\sqrt{\cos 2t}}$, $y=\frac{\cos^2 t}{\sqrt{\sin 2t}}$ find $\frac{dy}{dx}$ at $t=\frac{\pi}{4}$ (b) If $x=\tan\theta.\sqrt{\cos 2\theta}$, $y=\cot\theta.\sqrt{\sin 2\theta}$ find $\frac{dy}{dx}$ at $\theta=\frac{\pi}{6}$

(c) If $x^3-2xy+y^3=a^3$ find $\frac{d^2y}{dx^2}$. (d) If $x=\cos^{-1}\left(2t\sqrt{1-t^2}\right)$ and $y=\frac{\pi}{2}+\sin^{-1}t$ then find the value of $\frac{d^2y}{dx^2}$ at $t=\frac{\pi}{6}$.

(11) (a) $y=(\cos^{-1}x)^x$ (b) $y=x^{x^2}$ (c) $y=(\sin x)^{\log x}$ (d) $y=(\cos x)^{\log x}$ (e) $y=(\tan x)^{\log x}$ (f) $y=(\log x)^{\log(\tan x)}$

(g) $y=(\sin x)^{\cos x}+(\tan x)^{\cot x}$ (h) $y=(\cot x)^{\cos x}$ (i) $y=(\tan x)^{\sin x}$ (j) $y=e^{x\cos x^2}+(\cot x)^x$

(12) (a) $y=e^{x+x^2+x^3+\cdots\cdots\infty}$ (b) $y=e^{(\sin x+\cos x)}$ (c) $y=\sqrt{\sin x\sqrt{\sin x\sqrt{\sin \ldots\ldots\ldots\ldots\infty}}}$

(d) $y=\sqrt{\sin x+\sqrt{\sin x+\sqrt{\sin x+\sqrt{\ldots\ldots\ldots\ldots\infty}}}}$ (e) $y=e^{\sin x+\sin^2 x+\sin^3 x+\cdots\cdots\infty}$ (f) $y=(\sin x)^{(\sin x)^{\sin x\cdots\cdots\infty}}$

(g) $y=(\log x)^{(\log x)^{\log x\cdots\cdots\infty}}$ (h) $y=e^{(\sqrt{x})^{e^{(\sqrt{x})^{\cdots\cdots\infty}}}}$ (i) $y=\log(\log(\log(\cos x)))$ (j) $y=x^x+x^{\sin x}$ (k) $y=a^x+a^{\sin x}$

(l) $e^y=e^{x-y}$ (m) $x^my^n=(x-y)^{m-n}$ (n) $x^y+y^x=(x-y)^{x-y}$

(13) (a) If $y=x^2\log\left(\frac{a+x}{a-x}\right)$ then $\frac{dy}{dx}=2x\left(\frac{ax}{a^2-x^2}+\log\left(\frac{a+x}{a-x}\right)\right)$ (b) If $y=\left(x-\sqrt{x^2-a^2}\right)^n$ then $\frac{dy}{dx}$.

(c) $y=\frac{1-\sqrt{x-1}}{1+\sqrt{x-1}}+\frac{\sqrt{x-1}}{\sqrt{x}}$, find $\frac{dy}{dx}$. (d) $\cos y=a\cos(x+y)$, find $\frac{dy}{dx}$. (e) $y=1+\frac{Ax^2}{x-A}+\frac{Bx}{x-B}+\frac{C}{x-C}$, find $\frac{dy}{dx}$.

(f) $y=\tan^{-1}\left(\frac{\sqrt{a^2+b^2}\cos x}{b-a\cos x}\right)$, find $\frac{dy}{dx}$. (g) $y\sqrt{x^2+1}=\log\left(\sqrt{x^2+1}-x\right)$, find $\frac{dy}{dx}$.

(h) If $y=2\tan^{-1}\left(\frac{x}{\sqrt{1-x^2}}\right)+\log\left(\frac{1+2\sqrt{x}+x}{1-2\sqrt{x}+x}\right)$ then find $\frac{dy}{dx}$.

(14) (a) $y=x+\cfrac{1}{x+\cfrac{1}{x+\cfrac{1}{x+\cdots\ldots\infty}}}$ (b) $y=\sin x+\cfrac{1}{\sin x+\cfrac{1}{\sin x+\cdots\ldots\ldots\infty}}$

(c) If $x^2+y^2+2xy+x+y+5=0$ then find the value of $\frac{d^2y}{dx^2}$. (d) $y=(\log_{\sin x}\cos x)(\log_{\tan x}\cot x)+\cos^{-1}\frac{x}{1+x}$ at $x=\frac{\pi}{4}$.

(e) $f(x)=\log_x\cos x^2+\cos x^2$, find $\frac{dy}{dx}$ with respect to $\sqrt{x+1}$ (f) If $y=(\cos^{-1}x+\sin^{-1}x)x$ then prove that $\frac{d^2y}{dx^2}=0$ and $\frac{dy}{dx}=\frac{\pi}{2}$.

(g) If $Ax^2+By^2=1$ then prove that $By^2y''-Axy'+Ay=0$ or $(1-Ax^2)y''-Axy'+Ay=0$

(h) If $y=(\cos^{-1}x)^3$ then prove that $\frac{d^2y}{dx^2}+\frac{x}{1-x^2}\frac{dy}{dx}-\frac{6}{1-x^2}y^{\frac{1}{3}}=0$. (i) If $y=e^{a\cos^{-1}x}$ then prove that $\left(\sqrt{1-x^2}\right)y_1+ay=0$.

(j) If $y=\tan^{-1}\sqrt{x-1}$ then prove that $y_2+\frac{1}{x}y_1=0$ or $xy_2=-y_1$.

(15) (a) $y=\cos x.\cos 3x.\cos 5x$ (b) $y=8\sin x.\sin 2x.\sin 4x$ (c) $y=\cos 5x.\sin 3x$ (d) $y=\sin 2x+\cos 3x$

(e) $y=2\tan 2\theta+3\cot 2\theta$ (f) $y=\sec 2\theta.\operatorname{cosec} 3\theta$ (g) $y=\cos 3\theta+\sec 2\theta$ (h) $y=\sin 2\theta+\operatorname{cosec} 3\theta$ (i) $y=\log_2\log_2\log_2 x^2$

(j) $y=4\log_e\log_e\log_e\log_e x$ (k) $f(x)=\log_{x^3}\log x^2$ then find $f'(x)$ at $x=e$.

(16) (a) $y=\sin^m x.\cos^n x$ (b) $y=\sin^n x.\cos mx$ (c) $y=f(f(\log x))$ where $f(x)=\log x$.

(d) $y=f(f(\sin x))$, where $f(x)=\sin x$ then prove that $\frac{dy}{dx}=\cos x.\cos(\sin x).\cos(\sin(\sin x))$.

(e) If $x\sqrt{1-y}-y\sqrt{1-x}=0$ then $\frac{dy}{dx}=-\frac{1}{(x-1)^2}$.

(f) If $\cos^{-1}\left(\frac{x^2+y^2}{x^2-y^2}\right)=\log a$ then show that $\frac{dy}{dx}=\frac{y}{x}$ or $xy'-y=0$ or $x\frac{dy}{dx}-y=0$.

(g) If $\sqrt{x+y}+\sqrt{2x-y}=c$ then show that $\frac{d^2y}{dx^2}$.

(17) (a) Prove that the value of $6+\log_{\frac{3}{2}}\left[\frac{1}{3\sqrt{2}}\sqrt{4-\frac{1}{3\sqrt{2}}\sqrt{4-\frac{1}{3\sqrt{2}}\sqrt{4-\frac{1}{3\sqrt{2}}\cdots\cdots}}}\right]$ is 4.

Answer

Solution:− (1) (a) $y=\sin(ax^2+bx+c)$, Let $u=ax^2+bx+c$, $\frac{du}{dx}=2ax+b$

Now, $y=\sin u$, $\frac{dy}{du}=\cos u$ $\therefore \frac{dy}{dx}=\frac{dy}{du}\times\frac{du}{dx}=\cos u.(2ax+b)=(2ax+b)\cos(ax^2+bx+c)$ Ans.

IInd Method:− $y=\sin(ax^2+bx+c)$

or $\frac{dy}{dx}=\cos(ax^2+bx+c).\frac{d(ax^2+bx+c)}{dx}=\cos(ax^2+bx+c)(2ax+b)=(2ax+b)\cos(ax^2+bx+c)$ Ans.

(b) Do yourself. $y=\cos(2x-3)$, Let $u=2x-3$ $\therefore \frac{dy}{dx}=-2\sin(2x-3)$ Ans.

(c) $y=\sqrt{x^2-2x+3}$, Let $u=x^2-2x+3$ $\therefore \frac{du}{dx}=2x-2$ or $\left[\text{use formula, } y=\sqrt{f(x)} \therefore \frac{dy}{dx}=\frac{1}{2\sqrt{f(x)}}.f'(x)\right]$

or $y=\sqrt{u}$ $\therefore \frac{dy}{du}=\frac{1}{2\sqrt{u}}$ $\therefore \frac{dy}{dx}=\frac{dy}{du}\times\frac{du}{dx}=\frac{1}{2\sqrt{u}}(2x-2)=\frac{2(x-1)}{2\sqrt{x^2-2x+3}}=\frac{(x-1)}{\sqrt{x^2-2x+3}}$ Ans.

(d) $y=\sin\sqrt{2x^2-3}$, Let $u=2x^2-3$ $\therefore \frac{du}{dx}=4x$ or $y=\sin\sqrt{u}$, Let $v=\sqrt{u}$ $\therefore \frac{dv}{du}=\frac{1}{2\sqrt{u}}$ or $y=\sin v$ $\therefore \frac{dy}{dv}=\cos v$

$$\therefore \frac{dy}{dx} = \frac{dy}{dv} \times \frac{dv}{du} \times \frac{du}{dx} = \cos v . \frac{1}{2\sqrt{u}} . 4x = \frac{2x \cos\sqrt{2x^2-3}}{\sqrt{2x^2-3}} \quad \text{Ans.}$$

Direct: $- \; y = \sin\sqrt{2x^2-3}\,,\; \frac{dy}{dx} = \cos\sqrt{2x^2-3}\,.\frac{d(\sqrt{2x^2-3})}{dx} = \cos\sqrt{2x^2-3}\,.\frac{1}{2\sqrt{2x^2-3}}.4x = \frac{2x\cos\sqrt{2x^2-3}}{\sqrt{2x^2-3}}$ Ans.

(e) $y = \sqrt{\sin(2+3x^2)}$, Let $u = 2+3x^2 \quad \therefore \frac{du}{dx} = 6x$ or $y = \sqrt{\sin u}$, Let $v = \sin u \quad \therefore \frac{dv}{du} = \cos u$

$y = \sqrt{v}\,,\; \frac{dy}{dv} = \frac{1}{2\sqrt{v}} \quad \therefore \frac{dy}{dx} = \frac{dy}{dv} \times \frac{dv}{du} \times \frac{du}{dx} = \frac{1}{2\sqrt{v}}.\cos u.6x = \frac{3x\cos(2+3x^2)}{\sqrt{\sin u}} = \frac{3x\cos(2+3x^2)}{\sqrt{\sin(2+3x^2)}}$ Ans.

(f) $y = \cos\sqrt{\sqrt{x}+3}$, Let $u = \sqrt{x} \quad \therefore \frac{du}{dx} = \frac{1}{2\sqrt{x}}$ or $y = \cos\sqrt{u+3}$, Let $v = u+3 \quad \therefore \frac{dv}{du} = 1$

$y = \cos\sqrt{v}$, Let $w = \sqrt{v} \quad \therefore \frac{dw}{dv} = \frac{1}{2\sqrt{v}} = \frac{1}{2\sqrt{u+3}} = \frac{1}{2\sqrt{\sqrt{x}+3}}$

$y = \cos w,\; \frac{dy}{dw} = -\sin w = -\sin\sqrt{v} = -\sin\sqrt{u+3} = -\sin\sqrt{\sqrt{x}+3}$

$$\therefore \frac{dy}{dx} = \frac{dy}{dw} \times \frac{dw}{dv} \times \frac{dv}{du} \times \frac{du}{dx} = -\sin\sqrt{\sqrt{x}+3}\,.\frac{1}{2\sqrt{\sqrt{x}+3}}.1.\frac{1}{2\sqrt{x}} = -\frac{1}{4}.\frac{\sin\sqrt{\sqrt{x}+3}}{\sqrt{\sqrt{x}+3}.\sqrt{x}} \quad \text{Ans.}$$

IInd Method: $- \; y = \cos\sqrt{\sqrt{x}+3}\,,\; \frac{dy}{dx} = -\sin\sqrt{\sqrt{x}+3}\,.\frac{d\left(\sqrt{\sqrt{x}+3}\right)}{dx}.\frac{d(\sqrt{x})}{dx} = -\sin\sqrt{\sqrt{x}+3}\,.\frac{1}{2\sqrt{\sqrt{x}+3}}.\frac{1}{2\sqrt{x}} = -\frac{1}{4}.\frac{\sin\sqrt{\sqrt{x}+3}}{\sqrt{\sqrt{x}+3}.\sqrt{x}}$ Ans.

(g) solve same as above question. $y = \sqrt{\cos(x-2)}\,,\; \frac{dy}{dx} = -\frac{1}{2}.\frac{\sin(x-2)}{\sqrt{\cos(x-2)}}$ Ans.

(h) $y = \sin\sqrt{\tan x}$, Let $u = \tan x \quad \therefore \frac{du}{dx} = \sec^2 x$

$y = \sin\sqrt{u}$, Let $v = \sqrt{u} \quad \therefore \frac{dv}{du} = \frac{1}{2\sqrt{u}} = \frac{1}{2\sqrt{\tan x}}$ or $y = \sin v \quad \therefore \frac{dy}{dv} = \cos v = \cos\sqrt{u} = \cos\sqrt{\tan x}$

$\therefore \frac{dy}{dx} = \frac{dy}{dv} \times \frac{dv}{du} \times \frac{du}{dx} = \cos\sqrt{\tan x}\,.\frac{1}{2\sqrt{\tan x}}.\sec^2 x = \frac{\sec^2 x.\cos\sqrt{\tan x}}{2\sqrt{\tan x}} = \frac{\cos\sqrt{\tan x}}{2\cos^2 x.\sqrt{\tan x}}$ Ans.

(i) $y = \tan\sqrt{\sin\sqrt{x}+\cos\sqrt{x}}$, Let $u = \sqrt{x} \quad \therefore \frac{du}{dx} = \frac{1}{2\sqrt{x}}$

Now, $y = \tan\sqrt{\sin u + \cos u}$, Let $v = \sin u + \cos u \quad \therefore \frac{dv}{du} = \cos u - \sin u = \cos\sqrt{x} - \sin\sqrt{x}$ put $u = \sqrt{x}$

Now, $y = \tan\sqrt{v}$, Let $w = \sqrt{v} \quad \therefore \frac{dw}{dv} = \frac{1}{2\sqrt{v}} = \frac{1}{2\sqrt{\sin u + \cos u}} = \frac{1}{2\sqrt{\sin\sqrt{x}+\cos\sqrt{x}}}$

put $v = \sin u + \cos u,\, u = \sqrt{x}$

Now, $y = \tan w \quad \therefore \frac{dy}{dw} = \sec^2 w = \sec^2\sqrt{v} = \sec^2\sqrt{\sin u + \cos u} = \sec^2\sqrt{\sin\sqrt{x}+\cos\sqrt{x}}$

(put $w = \sqrt{v}$, $v = \sin u + \cos u$ and $u = \sqrt{x}$)

Then, $\frac{dy}{dx} = \frac{dy}{dw} \times \frac{dw}{dv} \times \frac{dv}{du} \times \frac{du}{dx} = \sec^2\sqrt{\sin\sqrt{x}+\cos\sqrt{x}}\,.\frac{1}{2\sqrt{\sin\sqrt{x}+\cos\sqrt{x}}}.\left(\cos\sqrt{x}-\sin\sqrt{x}\right).\frac{1}{2\sqrt{x}}$

$\therefore \frac{dy}{dx} = \frac{1}{4}.\frac{\left(\cos\sqrt{x}-\sin\sqrt{x}\right)}{\cos^2\sqrt{\sin\sqrt{x}+\cos\sqrt{x}}.\sqrt{\sin\sqrt{x}+\cos\sqrt{x}}.\sqrt{x}}$ Ans.

(j) $y = \cos^2\sqrt{\sin x + 3}$, Let $u = \sin x + 3$ $\therefore \frac{du}{dx} = \cos x$

Now, $y = \cos^2\sqrt{u}$, Let $v = \sqrt{u}$ $\therefore \frac{dv}{du} = \frac{1}{2\sqrt{u}} = \frac{1}{2\sqrt{\sin x + 3}}$ (put $u = \sin x + 3$)

Now, $y = \cos^2 v = (\cos v)^2$, Let $w = \cos v$ $\therefore \frac{dw}{dv} = -\sin v = -\sin\sqrt{u}$

$$\frac{dw}{dv} = -\sin\sqrt{\sin x + 3} \quad (\text{put } v = \sqrt{u} \text{ and } u = \sin x + 3)$$

Now, $y = w^2$ $\therefore \frac{dy}{dw} = 2w = 2\cos v = 2\cos\sqrt{u} = 2\cos\sqrt{\sin x + 3}$ (put $w = \cos v, v = \sqrt{u}$ and $u = \sin x + 3$)

Then, $\frac{dy}{dx} = \frac{dy}{dw} \times \frac{dw}{dv} \times \frac{dv}{du} \times \frac{du}{dx} = 2\cos\sqrt{\sin x + 3}\,.\left(-\sin\sqrt{\sin x + 3}\,\right).\frac{1}{2\sqrt{\sin x + 3}}.\cos x$

$$\therefore \frac{dy}{dx} = \frac{-2\sin\sqrt{\sin x + 3}\,.\cos\sqrt{\sin x + 3}\,.\cos x}{2\sqrt{\sin x + 3}} = -\frac{\sin\{2(\sqrt{\sin x + 3}\,)\}.\cos x}{2\sqrt{\sin x + 3}} \quad \text{Ans.} \quad [\sin 2x = 2\sin x\cos x]$$

(k) Do yourself. $y = \sqrt{\sin\sqrt{\cos\sqrt{x-1}}}$, $\frac{dy}{dx} = -\frac{1}{8}.\frac{\sin\sqrt{x-1}\,.\cos\sqrt{\cos\sqrt{x-1}}}{\sqrt{\sin\sqrt{\cos\sqrt{x-1}}}\,.\sqrt{\cos\sqrt{x-1}}\,.\sqrt{x-1}}$ Ans.

(l) Do yourself. $y = \log\sqrt{\sin x + \cos x}$, $\frac{dy}{dx} = \frac{1}{2}[\sec 2x - \tan 2x]$ Ans.

(m) $y = \cot(x-2)$, Let $u = x - 2$ $\therefore \frac{du}{dx} = 1$

Now, $y = \cot u = \frac{\cos u}{\sin u} = \frac{f(x)}{g(x)}$ (say) $\therefore \frac{dy}{du} = \frac{g(x).f'(x) - f(x).g'(x)}{\left(g(x)\right)^2} = \frac{\sin u.\frac{d(\cos u)}{du} - \cos u.\frac{d(\sin u)}{du}}{(\sin u)^2}$

$$\therefore \frac{dy}{du} = \frac{\sin u.(-\sin u) - \cos u.\cos u}{(\sin u)^2} = \frac{-\sin^2 u - \cos^2 u}{\sin^2 u} = \frac{-(\sin^2 u + \cos^2 u)}{\sin^2 u} = -\frac{1}{\sin^2 u} = -\operatorname{cosec}^2 u$$

$$\therefore \frac{dy}{dx} = \frac{dy}{du} \times \frac{du}{dx} = -\operatorname{cosec}^2 u.1 = -\operatorname{cosec}^2(x-2) \quad \text{Ans.} \quad (\text{put } u = x - 2)$$

(n) $y = \sqrt{\cot\sqrt{2x+1}}$, Let $u = \sqrt{2x+1}$ $\therefore \frac{du}{dx} = \frac{1}{2\sqrt{2x+1}}.2 = \frac{1}{\sqrt{2x+1}}$

Now, $y = \sqrt{\cot u}$, $\frac{dy}{du} = \frac{1}{2\sqrt{\cot u}}.\frac{d(\cot u)}{du} = \frac{1}{2\sqrt{\cot u}}.(-\operatorname{cosec}^2 u) = -\frac{\operatorname{cosec}^2\sqrt{2x+1}}{2\sqrt{\cot\sqrt{2x+1}}}$ (put $u = \sqrt{2x+1}$)

$$\therefore \frac{dy}{dx} = \frac{dy}{du} \times \frac{du}{dx} = -\frac{\operatorname{cosec}^2\sqrt{2x+1}}{2\sqrt{\cot\sqrt{2x+1}}}.\frac{1}{\sqrt{2x+1}} = -\frac{\operatorname{cosec}^2\sqrt{2x+1}}{2\sqrt{2x+1}.\sqrt{\cot\sqrt{2x+1}}} \quad \text{Ans.}$$

(o) $y = \sec\sqrt{\operatorname{cosec}(x^2+1)}$, Let $u = x^2 + 1$ $\therefore \frac{du}{dx} = 2x$

$y = \sec\sqrt{\operatorname{cosec} u}$, Let $v = \sqrt{\operatorname{cosec} u}$ $\therefore \frac{dv}{du} = \frac{1}{2\sqrt{\operatorname{cosec} u}}.\frac{d(\operatorname{cosec} u)}{du} = \frac{1}{2\sqrt{\operatorname{cosec} u}}.(-\cot u.\operatorname{cosec} u)$

or $\frac{dv}{du} = -\frac{\cot u.\operatorname{cosec} u}{2\sqrt{\operatorname{cosec} u}} = -\frac{\cot(x^2+1).\operatorname{cosec}(x^2+1)}{2\sqrt{\operatorname{cosec}(x^2+1)}}$ (put $u = x^2 + 1$)

$y = \sec v$ $\therefore \frac{dy}{dv} = \tan v.\sec v = \tan\sqrt{\operatorname{cosec} u}.\sec\sqrt{\operatorname{cosec} u} = \tan\sqrt{\operatorname{cosec}(x^2+1)}.\sec\sqrt{\operatorname{cosec}(x^2+1)}$ (put $v = \sqrt{\operatorname{cosec} u}$ and $u = x^2 + 1$)

$$\therefore \frac{dy}{dx} = \frac{dy}{dv} \times \frac{dv}{du} \times \frac{du}{dx} = \tan\sqrt{\operatorname{cosec}(x^2+1)}.\sec\sqrt{\operatorname{cosec}(x^2+1)}.-\frac{\cot(x^2+1).\operatorname{cosec}(x^2+1)}{2\sqrt{\operatorname{cosec}(x^2+1)}}.2x$$

$$\frac{dy}{dx} = -\tan\sqrt{\operatorname{cosec}(x^2+1)}\,.\sec\sqrt{\operatorname{cosec}(x^2+1)}\,.\frac{\cot(x^2+1)\,.\sqrt{\operatorname{cosec}(x^2+1)}\,.\sqrt{\operatorname{cosec}(x^2+1)}}{\sqrt{\operatorname{cosec}(x^2+1)}}\,.x$$

$$= -\tan\sqrt{\operatorname{cosec}(x^2+1)}\,.\sec\sqrt{\operatorname{cosec}(x^2+1)}\,.\cot(x^2+1)\,.\sqrt{\operatorname{cosec}(x^2+1)}\,.x$$

put $\theta = x^2+1$, $\frac{dy}{dx} = -\tan\sqrt{\operatorname{cosec}\theta}\,.\sec\sqrt{\operatorname{cosec}\theta}\,.\cot\theta\,.\sqrt{\operatorname{cosec}\theta}\,.\sqrt{\theta-1}$ Ans. where $\theta = x^2+1, x = \sqrt{\theta-1}$

(p) Do yourself. Hint – $y = \operatorname{cosec}(\sec\theta+\tan\theta)$, $\frac{dy}{d\theta} = -\sec(\tan\theta+\sec\theta)\,.\cot(\tan\theta+\sec\theta)\,.\operatorname{cosec}(\tan\theta+\sec\theta)$ Ans.

(q) $y = \sqrt{\sec(1+\tan\theta)}$, Let $u = 1+\tan\theta$ $\therefore \frac{du}{d\theta} = \sec^2\theta$

Now, $y = \sqrt{\sec u}$, Let $v = \sec u$ $\therefore \frac{dv}{du} = \tan u\,.\sec u = \tan(1+\tan\theta)\,.\sec(1+\tan\theta)$ [put $u = 1+\tan\theta$]

Now, $y = \sqrt{v}$, $\frac{dy}{dv} = \frac{1}{2\sqrt{v}} = \frac{1}{2\sqrt{\sec u}} = \frac{1}{2\sqrt{\sec(1+\tan\theta)}}$ [put $v = \sec u$ and $u = 1+\tan\theta$]

Then, $\frac{dy}{dx} = \frac{dy}{dv}\times\frac{dv}{du}\times\frac{du}{d\theta} = \frac{1}{2\sqrt{\sec(1+\tan\theta)}}\,.\tan(1+\tan\theta)\,.\sec(1+\tan\theta)\,.\sec^2\theta = \frac{1}{2}\,.\sec^2\theta\,.\sqrt{\sec(1+\tan\theta)}\,.\tan(1+\tan\theta)$ Ans.

(r) Do yourself. $y = \sqrt{\operatorname{cosec}(\sin\theta+\sec\theta)}$, $\frac{dy}{d\theta} = -\frac{\cot(\sin\theta+\sec\theta)\,.\operatorname{cosec}(\sin\theta+\sec\theta).(\cos\theta+\tan\theta\,.\sec\theta)}{2\sqrt{\operatorname{cosec}(\sin\theta+\sec\theta)}}$ Ans.

(2) (a) $y = \sin^{-1}(\sec x+\tan x)$, Let $u = \sec x+\tan x$ $\therefore \frac{du}{dx} = \sec x\,.\tan x+\sec^2 x = \sec x\,(\tan x+\sec x)$

Now, $y = \sin^{-1}u$ $\therefore \frac{dy}{du} = \frac{1}{\sqrt{1-u^2}} = \frac{1}{\sqrt{1-(\sec x+\tan x)^2}}$ [put $u = \sec x+\tan x$]

Then, $\frac{dy}{dx} = \frac{dy}{du}\times\frac{du}{dx} = \frac{1}{\sqrt{1-(\sec x+\tan x)^2}}\,.\sec x\,(\tan x+\sec x)$ Ans.

(b) Do yourself. $y = \cos^{-1}(\cot\theta-\operatorname{cosec}\theta)$, $\frac{dy}{dx} = \frac{(1-\cos\theta)}{\sqrt{2}\sin\theta\,.\sqrt{\cos\theta\,(1-\cos\theta)}}$ Ans.

(c) Do yourself. Ans: – $\frac{\sin\theta+\cos\theta}{2(1-\sin\theta\cos\theta)}$ (d) Ans: – $\frac{(\cos\theta-\sin\theta)(1+\sin\theta\cos\theta)}{1+\sin\theta\cos\theta\,(2+\sin\theta\cos\theta)}$

(e) Ans: – $\frac{(\sec^2\theta-1)}{(\tan\theta-\theta)\sqrt{(\tan\theta-\theta)^2-1}}$ or $\frac{\tan^2\theta}{(\tan\theta-\theta)\sqrt{\tan^2\theta-2\tan\theta-1+\theta^2}}$

(f) Ans: – $\frac{1}{(\sin\theta+\tan\theta)\sqrt{(\sin\theta+\tan\theta)^2-1}}\times(\cos\theta+\sec^2\theta)$

(g) $y = \sin^{-1}\sqrt{x^2+1}$, Let $u = \sqrt{x^2+1}$ $\therefore \frac{du}{dx} = \frac{1}{2\sqrt{x^2+1}}\,.\frac{d(x^2)}{dx} = \frac{1}{2\sqrt{x^2+1}}\,.2x = \frac{x}{\sqrt{x^2+1}}$

Now, $y = \sin^{-1}u$ $\therefore \frac{dy}{du} = \frac{1}{\sqrt{1-u^2}} = \frac{1}{\sqrt{1-\left(\sqrt{x^2+1}\right)^2}} = \frac{1}{\sqrt{1-x^2-1}} = \frac{1}{\sqrt{-x^2}} = \frac{1}{|x|}$ $\left[\text{put } u = \sqrt{x^2+1}\right]$

Then, $\frac{dy}{dx} = \frac{dy}{du}\times\frac{du}{dx} = \frac{1}{|x|}\,.\frac{x}{\sqrt{x^2+1}} = \frac{x}{|x|.\sqrt{x^2+1}}$ Ans.

(h) $y = \sin^{-1}\sqrt{x^2-1}+\cos^{-1}\sqrt{x-1}$

Let $u = \sqrt{x^2-1}$, $\frac{du}{dx} = \frac{1}{2\sqrt{x^2-1}}\,.2x = \frac{x}{\sqrt{x^2-1}}$ and $v = \sqrt{x-1}$, $\frac{dv}{dx} = \frac{1}{\sqrt{x-1}}$

Now, $y = \sin^{-1}u+\cos^{-1}v$ $\therefore \frac{dy}{dx} = \frac{1}{\sqrt{1-u^2}}\,.\frac{du}{dx} - \frac{1}{\sqrt{1-v^2}}\,.\frac{dv}{dx}$

$$\therefore \frac{dy}{dx} = \frac{1}{\sqrt{1-\left(\sqrt{x^2-1}\right)^2}}.\frac{x}{\sqrt{x^2-1}} - \frac{1}{\sqrt{1-\left(\sqrt{x-1}\right)^2}}.\frac{1}{\sqrt{x-1}} = \frac{x}{\sqrt{x^2-1}.\sqrt{1-x^2+1}} - \frac{1}{\sqrt{x-1}}.\frac{1}{\sqrt{1-x+1}}$$

$$= \frac{x}{\sqrt{x^2-1}.\sqrt{2-x^2}} - \frac{1}{\sqrt{x-1}.\sqrt{2-x}} \quad \text{Ans.}$$

(i) $y = \tan^{-1}\sqrt{x^2-5}$, Let $u = \sqrt{x^2-5}$ $\therefore \frac{du}{dx} = \frac{1}{2\sqrt{x^2-5}}.\frac{d(x^2-5)}{dx} = \frac{1}{2\sqrt{x^2-5}}.2x = \frac{x}{\sqrt{x^2-5}}$

Now, $y = \tan^{-1}u$ $\therefore \frac{dy}{du} = \frac{1}{1+u^2} = \frac{1}{1+\left(\sqrt{x^2-5}\right)^2} = \frac{1}{1+x^2-5} = \frac{1}{x^2-4}$ $\left[\text{Put } u = \sqrt{x^2-5}\right]$

Then, $\frac{dy}{dx} = \frac{dy}{du} \times \frac{du}{dx} = \frac{1}{x^2-4}.\frac{x}{\sqrt{x^2-5}} = \frac{x}{(x-2)(x+2)\sqrt{x^2-5}}$ Ans.

(j) see question no. (1) (k), Ans: $- \ -\frac{\sin\sqrt{x-1}.\cos\sqrt{\cos\sqrt{x-1}}}{4\sqrt{x-1}.\sqrt{\cos\sqrt{x-1}}}$ (k) Ans: $- \ -\frac{1}{4}.\frac{\sin\sqrt{\sin\sqrt{x}}.\cos\sqrt{x}}{\sqrt{x}.\sqrt{\sin\sqrt{x}}}$

(l) $y = \tan\sqrt{\cot\sqrt{x^2+1}}$, $\frac{dy}{dx} = -\frac{1}{2}.\frac{x.\sec^2\sqrt{\cot\sqrt{x^2+1}}.\mathrm{cosec}^2\sqrt{x^2+1}}{\sqrt{x^2+1}.\sqrt{\cot\sqrt{x^2+1}}}$ Ans.

(m) $y = \cot\sqrt{\cos\sqrt{1+\sin\theta}}$, $\frac{dy}{d\theta} = \frac{\cos\theta.\sin\sqrt{1+\sin\theta}.\mathrm{cosec}^2\sqrt{\cos\sqrt{1+\sin\theta}}}{4\sqrt{1+\sin\theta}.\sqrt{\cos\sqrt{1+\sin\theta}}}$ Ans.

(n) $y = \sec\sqrt{\mathrm{cosec}\sqrt{x+1}}$, $\frac{dy}{dx} = \frac{\tan\sqrt{\mathrm{cosec}\sqrt{x+1}}.\cot\sqrt{x+1}.\sec\sqrt{\mathrm{cosec}\sqrt{x+1}}.\mathrm{cosec}\sqrt{x+1}}{4\sqrt{x+1}.\sqrt{\mathrm{cosec}\sqrt{x+1}}}$ Ans.

(o) $y = \mathrm{cosec}\sqrt{\sin\sqrt{1+\cos\theta}}$, $\frac{dy}{d\theta} = \frac{(2x-x^2).\cos\sqrt{x}.\cot\sqrt{\sin\sqrt{x}}.\mathrm{cosec}\sqrt{\sin\sqrt{x}}}{4\sqrt{x}.\sqrt{\sin\sqrt{x}}}$ (where $1+\cos\theta = x$) Ans.

(p) $y = \frac{\tan x}{x} - \sqrt{1+x^2}$, $\frac{dy}{dx} = \frac{x\frac{d(\tan x)}{dx} - \tan x\frac{d(x)}{dx}}{x^2} - \frac{1}{2\sqrt{1+x^2}}.\frac{d(1+x^2)}{dx} = \frac{x\sec^2 x - \tan x}{x^2} - \frac{1}{2\sqrt{1+x^2}}.2x$

$$\therefore \frac{dy}{dx} = \frac{x.\sqrt{1+x^2}[(1+\tan^2 x) - \tan x] - x^3}{x^2\sqrt{1+x^2}} = \frac{x.\sqrt{1+x^2}(\tan^2 x - \tan x + 1) - x^3}{x^2\sqrt{1+x^2}} \quad \text{Ans.}$$

(q) $y = \sin^2(x^2-1)$, Let $u = x^2-1$ $\therefore \frac{du}{dx} = 2x$

Now, $y = \sin^2 u = (\sin u)^2$, Let $v = \sin u$ $\therefore \frac{dv}{du} = \cos u = \cos(x^2-1)$ [Put $u = x^2-1$]

Now, $y = v^2$ $\therefore \frac{dy}{dv} = 2v = 2\sin u = 2\sin(x^2-1)$ [Put $v = \sin u$ and $u = x^2-1$]

Then, $\frac{dy}{dx} = \frac{dy}{dv} \times \frac{dv}{du} \times \frac{du}{dx} = 2\sin(x^2-1).\cos(x^2-1).2x = 2x\sin 2(x^2-1)$ Ans.

(r) $y = \sin^2\left(\sqrt{x^2-x+1}\right)$, Let $u = \sqrt{x^2-x+1}$ $\therefore \frac{du}{dx} = \frac{1}{2\sqrt{x^2-x+1}}.\frac{d(x^2-x+1)}{dx} = \frac{2x-1}{2\sqrt{x^2-x+1}}$

Now, $y = \sin^2 u = (\sin u)^2$, Let $v = \sin u$ $\therefore \frac{dv}{du} = \cos u = \cos\left(\sqrt{x^2-x+1}\right)$ $\left[\text{Put } u = \sqrt{x^2-x+1}\right]$

Now, $y = v^2$ $\therefore \frac{dy}{dv} = 2v = 2\sin u = 2\sin\left(\sqrt{x^2-x+1}\right)$ $\left[\text{Put } v = \sin u \text{ and } u = \sqrt{x^2-x+1}\right]$

Then, $\frac{dy}{dx} = \frac{dy}{dv} \times \frac{dv}{du} \times \frac{du}{dx} = 2\sin\left(\sqrt{x^2-x+1}\right).\cos\left(\sqrt{x^2-x+1}\right).\frac{2x-1}{2\sqrt{x^2-x+1}} = \frac{(2x-1).\sin\left(2\sqrt{x^2-x+1}\right)}{2\sqrt{x^2-x+1}}$ Ans.

(s) $y = \cos^2(1 - x^2)$, Let $u = 1 - x^2$ $\therefore \frac{du}{dx} = -2x$

or $y = (\cos u)^2$, Let $v = \cos u$ $\therefore \frac{dv}{du} = -\sin u = -\sin(1 - x^2)$ [Put $u = 1 - x^2$]

$\therefore y = v^2$, $\frac{dy}{dv} = 2v = 2\cos u = 2\cos(1 - x^2)$ [Put $v = \cos u$ and $u = 1 - x^2$]

or $\frac{dy}{dx} = \frac{dy}{dv} \times \frac{dv}{du} \times \frac{du}{dx} = 2\cos(1 - x^2).[-\sin(1 - x^2)].(-2x) = 2x.\sin[2(1 - x^2)]$ Ans.

(t) $y = \cos^2\sqrt{x^3 - 1}$, $\frac{dy}{dx} = \frac{-3x^2.\sin[2\sqrt{x^3 - 1}]}{2\sqrt{x^3 - 1}}$ Ans. [See Question No. $-(2)(q)$.]

(u) $y = \tan^2(ax^3 + bx^2 + cx + d)$, Let $u = ax^3 + bx^2 + cx + d$ $\therefore \frac{du}{dx} = 3ax^2 + 2bx + c$

or $y = (\tan u)^2$, Let $v = \tan u$ $\therefore \frac{dv}{du} = \sec^2 u = \sec^2(ax^3 + bx^2 + cx + d)$ [Put $u = ax^3 + bx^2 + cx + d$]

or $y = v^2$, $\frac{dy}{dv} = 2v = 2\tan u = 2\tan(ax^3 + bx^2 + cx + d)$ [Put $v = \tan u$ and $u = ax^3 + bx^2 + cx + d$]

or $\frac{dy}{dx} = \frac{dy}{dv} \times \frac{dv}{du} \times \frac{du}{dx} = 2\tan(ax^3 + bx^2 + cx + d).\sec^2(ax^3 + bx^2 + cx + d).(3ax^2 + 2bx + c)$

$$\therefore \frac{dy}{dx} = (3ax^2 + 2bx + c).[2\tan(ax^3 + bx^2 + cx + d) + 2\tan^3(ax^3 + bx^2 + cx + d)] \quad \text{Ans.}$$

(v) $y = \tan^2\sqrt{1 - ax^2}$, Let $u = \sqrt{1 - ax^2}$ $\therefore \frac{du}{dx} = \frac{1}{2\sqrt{1 - ax^2}}.2ax = \frac{ax}{\sqrt{1 - ax^2}}$

$y = \tan^2 u = (\tan u)^2$, Let $v = \tan u$ $\therefore \frac{dv}{du} = \sec^2 u = \sec^2\sqrt{1 - ax^2}$ $\left[\text{Put } u = \sqrt{1 - ax^2}\right]$

$y = v^2$ $\therefore \frac{dy}{dv} = 2v = 2\tan u = 2\tan\sqrt{1 - ax^2}$ $\left[\text{Put } v = \tan u \text{ and } u = \sqrt{1 - ax^2}\right]$

or $\frac{dy}{dx} = \frac{dy}{dv} \times \frac{dv}{du} \times \frac{du}{dx} = 2\tan\sqrt{1 - ax^2}.\sec^2\sqrt{1 - ax^2}.\frac{ax}{\sqrt{1 - ax^2}} = \frac{-2ax(\tan\sqrt{1 - ax^2} + \tan^3\sqrt{1 - ax^2})}{\sqrt{1 - ax^2}}$ Ans.

(w) $y = \sqrt{\tan(2x + 1)}$, Let $u = 2x + 1$ $\therefore \frac{du}{dx} = 2$

$y = \sqrt{\tan u}$, $\frac{dy}{du} = \frac{1}{2\sqrt{\tan u}}.\frac{d(\tan u)}{du} = \frac{1}{2\sqrt{\tan u}}.\sec^2 u = \frac{\sec^2(2x + 1)}{2\sqrt{\tan(2x + 1)}}$ [Put $u = 2x + 1$]

or $\frac{dy}{dx} = \frac{dy}{du} \times \frac{du}{dx} = \frac{\sec^2(2x + 1)}{2\sqrt{\tan(2x + 1)}}.2 = \frac{\sec^2(2x + 1)}{\sqrt{\tan(2x + 1)}} = \frac{1 + \tan^2(2x + 1)}{\sqrt{\tan(2x + 1)}} = \frac{1}{\sqrt{\tan(2x + 1)}} + \frac{\tan^2(2x + 1)}{\sqrt{\tan(2x + 1)}}$

$= [\tan(2x + 1)]^{-\frac{1}{2}} + [\tan(2x + 1)]^{\frac{3}{2}}$ Ans.

(3) (a) $y = x\sin x^3$, Let $u = x^3$ or $x = u^{\frac{1}{3}}$ $\therefore \frac{du}{dx} = 3x^2$

$y = u^{\frac{1}{3}}.\sin u$, $\frac{dy}{du} = u^{\frac{1}{3}}.\frac{d(\sin u)}{du} + \sin u.\frac{d\left(u^{\frac{1}{3}}\right)}{du} = u^{\frac{1}{3}}.\cos u + \sin u.\frac{1}{3}u^{\frac{1}{3}-1} = u^{\frac{1}{3}}.\cos u + \sin u.\frac{u^{-\frac{2}{3}}}{3} = u^{\frac{1}{3}}.\cos u + \frac{\sin u}{3u^{\frac{2}{3}}} = \frac{3u\cos u + \sin u}{3u^{\frac{2}{3}}}$

$= \frac{3x^3\cos x^3 + \sin x^3}{3(x^3)^{\frac{2}{3}}} = \frac{3x^3\cos x^3 + \sin x^3}{3x^2}$ [Put $u = x^3$]

$\therefore \frac{dy}{dx} = \frac{dy}{du} \times \frac{du}{dx} = \frac{3x^3\cos x^3 + \sin x^3}{3x^2}.3x^2 = 3x^3\cos x^3 + \sin x^3$ Ans.

(b) $y = x^2\cos(x^3 + 1) = f(x).g(x)$,

or $\frac{dy}{dx} = f(x).g'(x) + g(x).f'(x) = x^2.\frac{d[\cos(x^3+1)]}{dx} + \cos(x^3+1).\frac{d(x^2)}{dx} = x^2.(-\sin(x^3+1)).3x^2 + \cos(x^3+1).2x$

$= 2x.\cos(x^3+1) - 3x^2.\sin(x^3+1)$ Ans.

(c) $y = x^3\tan x^2$, $\frac{dy}{dx} = x^3.\sec^2 x^2.2x + \tan x^2.3x^2 = 2x^4\sec^2 x^2 + 3x^2\tan x^2 = 2x^4(1+\tan^2 x^2) + 3x^2\tan x^2$

$= 2x^4 + 2x^4\tan^2 x^2 + 3x^2\tan x^2 = 2x^4 + x^2\tan x^2\,(3 + 2x^2\tan x^2)$ Ans.

(d) $y = (x-1)\sec x^3$, Let $f(x) = x-1$, $f'(x) = 1$ and $g(x) = \sec x^3$, $g'(x) = \tan x^3.\sec x^3.3x^2$

$\therefore \frac{dy}{dx} = f(x).g'(x) + g(x).f'(x) = (x-1).\tan x^3.\sec x^3.3x^2 + \sec x^3.1 = 3x^2(x-1).\frac{\sin x^3}{\cos x^3}.\frac{1}{\cos x^3} + \frac{1}{\cos x^3}$

$\therefore \frac{dy}{dx} = \frac{3x^2(x-1)\sin x^3 + \cos x^3}{\cos^2 x^3} = \frac{3x^2\sin x^3\,(x-1) + \cos x^3}{\cos^2 x^3}$ Ans.

(e) $y = \sqrt{x}\,\cot x^2$, Let $f(x) = \sqrt{x}$, $f'(x) = \frac{1}{2\sqrt{x}}$ and $g(x) = \cot x^2 = \frac{\cos x^2}{\sin x^2}$ Put $z = x^2$, $\frac{dz}{dx} = 2x$

$g(x) = \frac{\cos x^2}{\sin x^2} = \frac{\cos z}{\sin z} = \frac{u}{v}$ (say) $u = \cos z$, $\frac{du}{dz} = -\sin z$ and $v = \sin z$, $\frac{dv}{dz} = \cos z$

$g'(x) = \frac{v.\frac{du}{dz}.\frac{dz}{dx} - u.\frac{dv}{dz}.\frac{dz}{dx}}{v^2} = \frac{\sin z.(-\sin z)\frac{dz}{dx} - \cos z.\cos z.\frac{dz}{dx}}{(\sin z)^2} = \frac{-(\sin^2 z + \cos^2 z).\frac{dz}{dx}}{\sin^2 z} = -\frac{1.\frac{dz}{dx}}{\sin^2 z} = -\frac{dz}{dx}.\operatorname{cosec}^2 z = -2x.\operatorname{cosec}^2 x^2$

$\therefore \frac{dy}{dx} = f(x).g'(x) + g(x).f'(x) = -\sqrt{x}.2x.\operatorname{cosec}^2 x^2 + \cot x^2.\frac{1}{2\sqrt{x}} = \frac{-2x\sqrt{x}}{\sin^2 x^2} + \frac{\cos x^2}{2\sqrt{x}.\sin x^2} = \frac{-4x^2 + \sin x^2.\cos x^2}{2\sqrt{x}.\sin^2 x^2} = \frac{-4x^2 + \frac{2\sin x^2.\cos x^2}{2}}{2\sqrt{x}.\sin^2 x^2}$

$= \frac{-8x^2 + \sin 2x^2}{4\sqrt{x}.\sin^2 x^2} = \frac{\sin 2x^2 - 8x^2}{4\sqrt{x}.\sin^2 x^2}$ Ans.

(f) $y = (x^2+1)\operatorname{cosec}\sqrt{x+1}$, $\frac{dy}{dx} = \frac{4x\sqrt{x+1}\sin\sqrt{x+1} - (x^2+1)\cos\sqrt{x+1}}{2\sqrt{x+1}.\sin^2\sqrt{x+1}}$ Ans.

(g) $y = \frac{\sin(x^2+1)}{x^3}$, $\frac{dy}{dx} = \frac{2\cos(x^2+1)}{x^2} - \frac{3\sin(x^2+1)}{x^4}$ Ans. (h) $y = \frac{\tan\sqrt{x+1}}{x}$, $\frac{dy}{dx} = \frac{\sec^2\sqrt{x+1}}{2x\sqrt{x+1}} - \frac{\tan\sqrt{x+1}}{x^2}$ Ans.

(i) $y = \sin\left(\frac{1+x}{1-x}\right)$, Let $u = \frac{1+x}{1-x}$ $\therefore \frac{du}{dx} = \frac{(1-x).1 - (1+x).(-1)}{(1-x)^2} = \frac{1-x+1+x}{(1-x)^2} = \frac{2}{(1-x)^2}$

or $y = \sin u$, $\frac{dy}{du} = \cos u = \cos\left(\frac{1+x}{1-x}\right)$ $\left[\text{Put } u = \frac{1+x}{1-x}\right]$ $\therefore \frac{dy}{dx} = \frac{dy}{du}\times\frac{du}{dx} = \cos\left(\frac{1+x}{1-x}\right).\frac{2}{(1-x)^2} = \frac{2\cos\left(\frac{1+x}{1-x}\right)}{(1-x)^2}$ Ans.

(j) $y = \sin\sqrt{x^2+ax+1}.\cos\sqrt{ax^2+1} = u.v$ (say) Let $u = \sin\sqrt{x^2+ax+1}$ and $v = \cos\sqrt{ax^2+1}$

$\frac{du}{dx} = \cos\sqrt{x^2+ax+1}.\frac{1}{2\sqrt{x^2+ax+1}}.(2x+a)$ and $\frac{dv}{dx} = -\sin\sqrt{ax^2+1}.\frac{1}{2\sqrt{ax^2+1}}.2ax$

$\therefore \frac{dy}{dx} = u.\frac{dv}{dx} + v.\frac{du}{dx}$ or $\frac{dy}{dx} = -\sin\sqrt{x^2+ax+1}.\frac{ax.\sin\sqrt{ax^2+1}}{\sqrt{ax^2+1}} + \cos\sqrt{ax^2+1}.\frac{(2x+a).\cos\sqrt{x^2+ax+1}}{2\sqrt{x^2+ax+1}}$ Ans.

(k) $y = \tan(ax^2+bx+1) + \cot\sqrt{ax^2+bx+c}$

$\frac{dy}{dx} = \sec^2(ax^2+bx+1).(2ax+b) + \left(-\operatorname{cosec}^2\sqrt{ax^2+bx+c}.\frac{1}{2\sqrt{ax^2+bx+c}}.(2ax+b)\right)$

$\frac{dy}{dx} = \frac{(2ax+b)}{2\sqrt{ax^2+bx+c}}\left[2\sqrt{ax^2+bx+c}.\sec^2(ax^2+bx+1) - \operatorname{cosec}^2\sqrt{ax^2+bx+c}\right]$ Ans.

(l) $y = x^2\log\sqrt{x-3}$, Let $u = x^2$, $\frac{du}{dx} = 2x$ and $v = \log\sqrt{x-3}$, $\frac{dv}{dx} = \frac{1}{\sqrt{x-3}}.\frac{1}{2\sqrt{x-3}}.1 = \frac{1}{2(x-3)}$

$$\therefore \frac{dy}{dx} = u.\frac{dv}{dx} + v.\frac{du}{dx} = x^2.\frac{1}{2(x-3)} + \log\sqrt{x-3}\,.\,2x = x\left[\frac{x}{2(x-3)} + \log(x-3)\right] \quad \text{Ans.}$$

(m) $y = \sin^3\sqrt{5-2x+x^2}$, Let $u = \sqrt{5-2x+x^2}$ $\therefore \frac{du}{dx} = \frac{1}{2\sqrt{5-2x+x^2}}.(2x-2)$

$\therefore y = \sin^3 u = (\sin u)^3$, Let $v = \sin u$ $\therefore \frac{dv}{du} = \cos u = \cos\sqrt{5-2x+x^2}$ $\left[\text{Put } u = \sqrt{5-2x+x^2}\right]$

$\therefore y = v^3$, $\frac{dy}{dv} = 3v^2 = 3(\sin u)^2 = 3\left(\sin\sqrt{5-2x+x^2}\right)^2$ $\left[\text{Put } v = \sin u \text{ and } u = \sqrt{5-2x+x^2}\right]$

$$\therefore \frac{dy}{dx} = \frac{dy}{dv}\times\frac{dv}{du}\times\frac{du}{dx} = 3\left(\sin\sqrt{5-2x+x^2}\right)^2.\cos\sqrt{5-2x+x^2}\,.\frac{1}{2\sqrt{5-2x+x^2}}.(2x-2)$$
$$= \frac{3(x-1).\sin^2\sqrt{5-2x+x^2}\,.\cos\sqrt{5-2x+x^2}}{\sqrt{5-2x+x^2}} \quad \text{Ans.}$$

(n) $y = (x-1)^2.(3x^2-1)$, Let $u = (x-1)^2$ $\therefore \frac{du}{dx} = 2(x-1)$ and $v = (3x^2-1)$ $\therefore \frac{dv}{dx} = 6x$

$$\therefore \frac{dy}{dx} = u.v = u.\frac{dv}{dx} + v.\frac{du}{dx} = (x-1)^2.6x + (3x^2-1).2(x-1) = 2(x-1)[3x(x-1) + 3x^2 - 1] = 2(x-1)[3x^2 - 3x + 3x^2 - 1]$$
$$= 2(x-1)(6x^2-3x-1) \quad \text{Ans.}$$

(o) $y = \cos(\log x^2)$, Let $u = x^2$ $\therefore \frac{du}{dx} = 2x$ $\therefore y = \cos(\log u)$ Now, Let $v = \log u$ $\therefore \frac{dv}{du} = \frac{1}{u} = \frac{1}{x^2}$ [Put $u = x^2$]

$\therefore y = \cos v$, $\frac{dy}{dv} = -\sin v = -\sin(\log u) = -\sin(\log x^2)$ [Put $v = \log u$ and $u = x^2$]

$$\therefore \frac{dy}{dx} = \frac{dy}{dv}\times\frac{dv}{du}\times\frac{du}{dx} = -\sin(\log x^2)\,.\frac{1}{x^2}.2x = -\frac{2\sin(\log x^2)}{x} \quad \text{Ans.}$$

(p) $y = \sin(e^x+1)$, $\frac{dy}{dx} = \cos(e^x+1)\,.\frac{d(e^x+1)}{dx} = e^x.\cos(e^x+1)$ Ans.

(q) $y = \tan[\log(x-5)]$, $\frac{dy}{dx} = \sec^2[\log(x-5)]\,.\frac{d(\log(x-5))}{dx} = \sec^2[\log(x-5)]\,.\frac{1}{(x-5)}.\frac{d(x-5)}{dx} = \frac{\sec^2[\log(x-5)]}{(x-5)}$
$$= \frac{1+\tan^2[\log(x-5)]}{(x-5)} \quad \text{Ans.}$$

(r) $y = \log[\log(3-x)]$, Let $u = 3-x$ $\therefore \frac{du}{dx} = -1$

$\Rightarrow y = \log[\log u]$, Let $v = \log u$ $\therefore \frac{dv}{du} = \frac{1}{u} = \frac{1}{(3-x)}$ [Put $u = 3-x$]

$\Rightarrow y = \log v$ $\therefore \frac{dy}{dv} = \frac{1}{v} = \frac{1}{\log u} = \frac{1}{\log(3-x)}$ [Put $v = \log u$ and $u = 3-x$]

$$\therefore \frac{dy}{dx} = \frac{dy}{dv}\times\frac{dv}{du}\times\frac{du}{dx} = \frac{1}{\log(3-x)}.\frac{1}{(3-x)}.(-1) = -\frac{1}{(3-x).\log(3-x)} = \frac{1}{(x-3).\log(3-x)} \quad \text{Ans.}$$

(s) $y = e^{\sqrt{x}}.\log x$, Let $u = e^{\sqrt{x}}$ $\therefore \frac{du}{dx} = e^{\sqrt{x}}.\frac{1}{2\sqrt{x}}$ and $v = \log x$ $\therefore \frac{dv}{dx} = \frac{1}{x}$

$\Rightarrow y = u.v$, $\frac{dy}{dx} = u.\frac{dv}{dx} + v.\frac{du}{dx} = e^{\sqrt{x}}.\frac{1}{x} + \log x\,.e^{\sqrt{x}}.\frac{1}{2\sqrt{x}} = \frac{2\sqrt{x}.e^{\sqrt{x}} + x.e^{\sqrt{x}}.\log x}{2x.\sqrt{x}}$

$$\therefore \frac{dy}{dx} = \frac{\sqrt{x}.e^{\sqrt{x}}(2+\sqrt{x}.\log x)}{2x.\sqrt{x}} = \frac{e^{\sqrt{x}}(2+\sqrt{x}.\log x)}{2x} \quad \text{Ans.}$$

(t) $y = e^{2x}.x^3$, $\frac{dy}{dx} = e^{2x}.\frac{d(x^3)}{dx} + x^3.\frac{d(e^{2x})}{dx} = 3x^2.e^{2x} + x^3.e^{2x}.2 = x^2.e^{2x}(3+2x)$ Ans.

(u) $y = \sqrt{1+\tan\theta}$, $\frac{dy}{d\theta} = \frac{1}{2\sqrt{1+\tan\theta}} . \frac{d(1+\tan\theta)}{d\theta} = \frac{1}{2\sqrt{1+\tan\theta}} . \sec^2\theta = \frac{1+\tan^2\theta}{2\sqrt{1+\tan\theta}}$ Ans.

(v) $y = \sqrt{\sin\sqrt{x}}$, Let $u = \sqrt{x}$ $\therefore \frac{du}{dx} = \frac{1}{2\sqrt{x}}$

$\Rightarrow y = \sqrt{\sin u}$, $\frac{dy}{du} = \frac{1}{2\sqrt{\sin u}} . \frac{d(\sin u)}{du} = \frac{1}{2\sqrt{\sin u}} . \cos u = \frac{\cos\sqrt{x}}{2\sqrt{\sin\sqrt{x}}}$ [Put $u = \sqrt{x}$]

$\therefore \frac{dy}{dx} = \frac{dy}{du} \times \frac{du}{dx} = \frac{\cos\sqrt{x}}{2\sqrt{\sin\sqrt{x}}} . \frac{1}{2\sqrt{x}} = \frac{\cos\sqrt{x}}{4\sqrt{x}.\sqrt{\sin\sqrt{x}}} = \frac{\cos\sqrt{x}}{\sqrt{16x.\sin\sqrt{x}}}$ Ans.

(4) (a) $y = \frac{\tan\left(\frac{1-x}{1+x}\right)}{x} + \frac{\cot\left(\frac{1+x}{1-x}\right)}{x}$, Let $u = \frac{\tan\left(\frac{1-x}{1+x}\right)}{x}$ and $v = \frac{\cot\left(\frac{1+x}{1-x}\right)}{x}$

$\Rightarrow \; u = \frac{\tan\left(\frac{1-x}{1+x}\right)}{x}$, $\frac{du}{dx} = \frac{x.\sec^2\left(\frac{1-x}{1+x}\right).\frac{d}{dx}\left(\frac{1-x}{1+x}\right) - \tan\left(\frac{1-x}{1+x}\right).\frac{d(x)}{dx}}{x^2}$

or $\frac{d}{dx}\left(\frac{1-x}{1+x}\right) = \frac{(1+x).(-1)-(1-x).1}{(1+x)^2} = \frac{-1-x-1+x}{(1+x)^2} = -\frac{2}{(1+x)^2}$

$\therefore \frac{du}{dx} = \frac{x.\sec^2\left(\frac{1-x}{1+x}\right).\left(-\frac{2}{(1+x)^2}\right) - \tan\left(\frac{1-x}{1+x}\right).1}{x^2} = -\frac{2x.\sec^2\left(\frac{1-x}{1+x}\right) + (1+x)^2.\tan\left(\frac{1-x}{1+x}\right)}{x^2.(1+x)^2}$

$\Rightarrow \; v = \frac{\cot\left(\frac{1+x}{1-x}\right)}{x}$, $\frac{d}{dx}\left(\frac{1+x}{1-x}\right) = \frac{(1-x).1-(1+x).(-1)}{(1-x)^2} = \frac{1-x+1+x}{(1-x)^2} = \frac{2}{(1-x)^2}$

or $\frac{dv}{dx} = \frac{-x.\operatorname{cosec}^2\left(\frac{1+x}{1-x}\right).\frac{2}{(1-x)^2} - \cot\left(\frac{1+x}{1-x}\right).1}{x^2} = -\frac{2x.\operatorname{cosec}^2\left(\frac{1+x}{1-x}\right) + (1-x)^2.\cot\left(\frac{1+x}{1-x}\right)}{x^2.(1-x)^2}$

$\therefore \frac{dy}{dx} = u + v = \frac{du}{dx} + \frac{dv}{dx}$

$$\frac{dy}{dx} = -\frac{2x.\sec^2\left(\frac{1-x}{1+x}\right) + (1+x)^2.\tan\left(\frac{1-x}{1+x}\right)}{x^2.(1+x)^2} - \frac{2x.\operatorname{cosec}^2\left(\frac{1+x}{1-x}\right) + (1-x)^2.\cot\left(\frac{1+x}{1-x}\right)}{x^2.(1-x)^2}$$

$$= \frac{-2x\left[\sec^2\left(\frac{1-x}{1+x}\right) + (1+x)^2.\tan\left(\frac{1-x}{1+x}\right) + \operatorname{cosec}^2\left(\frac{1+x}{1-x}\right) + (1-x)^2.\cot\left(\frac{1+x}{1-x}\right)\right]}{x^2.(1+x)^2}$$ Ans.

(b) $y = \cos\sqrt{\sin\sqrt{ax+b}}$, Let $u = \sqrt{ax+b}$ $\therefore \frac{du}{dx} = \frac{1}{2\sqrt{ax+b}} . \frac{d}{dx}(ax+b) = \frac{a}{2\sqrt{ax+b}}$

$\Rightarrow y = \cos\sqrt{\sin u}$, Let $v = \sqrt{\sin u}$ $\therefore \frac{dv}{du} = \frac{1}{2\sqrt{\sin u}} . \frac{d}{du}(\sin u) = \frac{\cos u}{2\sqrt{\sin u}} = \frac{\cos(ax+b)}{2\sqrt{\sin\sqrt{ax+b}}}$ [Put $u = \sqrt{ax+b}$]

$\Rightarrow \; y = \cos v$, $\frac{dy}{dv} = -\sin v = -\sin\sqrt{\sin u} = -\sin\sqrt{\sin\sqrt{ax+b}}$ [Put $v = \sqrt{\sin u}$ and $u = \sqrt{ax+b}$]

$\therefore \frac{dy}{dx} = \frac{dy}{dv} \times \frac{dv}{du} \times \frac{du}{dx} = -\sin\sqrt{\sin\sqrt{ax+b}} . \frac{\cos(ax+b)}{2\sqrt{\sin\sqrt{ax+b}}} . \frac{a}{2\sqrt{ax+b}} = \frac{-a\sin\sqrt{\sin\sqrt{ax+b}}.\cos(ax+b)}{4.\sqrt{\sin\sqrt{ax+b}}.\sqrt{ax+b}}$ Ans.

(c) $y = \frac{\sqrt{\tan x}}{\sqrt{x^2-3}} = \frac{u}{v}$ (say) $u = \sqrt{\tan x}$, $\frac{du}{dx} = \frac{1}{2\sqrt{\tan x}} . \sec^2 x$ and $v = \sqrt{x^2-3}$, $\frac{dv}{dx} = \frac{1}{2\sqrt{x^2-3}} . 2x$

or $\frac{dy}{dx} = \frac{v.\frac{du}{dx} - u.\frac{dv}{dx}}{v^2} = \frac{\sqrt{x^2-3}.\frac{1}{2\sqrt{\tan x}}.\sec^2 x - \sqrt{\tan x}.\frac{1}{2\sqrt{x^2-3}}.2x}{\left(\sqrt{x^2-3}\right)^2} = \frac{\frac{(x^2-3).\sec^2 x - 2x.\tan x}{2\sqrt{\tan x}.\sqrt{x^2-3}}}{(x^2-3)} = \frac{(x^2-3).\sec^2 x - 2x.\tan x}{2.(x^2-3).\sqrt{x^2-3}.\sqrt{\tan x}}$ Ans.

(d) $y = \dfrac{x^2 + \sin x^3}{x^3 + \cos x^2} = \dfrac{u}{v}$ (say) $u = x^2 + \sin x^3$ and $v = x^3 + \cos x^2$ or $\dfrac{du}{dx} = 2x + \cos x^3 . 3x^2$, $\dfrac{dv}{dx} = 3x^2 - \sin x^2 . 2x$

$$\text{or } \frac{dy}{dx} = \frac{v.\frac{du}{dx} - u.\frac{dv}{dx}}{v^2} = \frac{(x^3 + \cos x^2)(2x + 3x^2.\cos x^3) - (x^2 + \sin x^3)(3x^2 - 2x.\sin x^2)}{(x^3 + \cos x^2)^2}$$

$$= \frac{x[(x^3 + \cos x^2)(2 + 3x\cos x^3) - (x^2 + \sin x^3)(3x - 2\sin x^2)]}{(x^3 + \cos x^2)^2} \quad \text{Ans.}$$

(e) $y = x^{\frac{3}{2}}.\cos(ax^2 + bx + c) = u.v$ (say), $u = x^{\frac{3}{2}}$ $\therefore \dfrac{du}{dx} = \dfrac{3}{2}x^{\frac{3}{2}-1} = \dfrac{3}{2}\sqrt{x}$ $[y = x^n, \; y' = nx^{n-1}]$

and $v = \cos(ax^2 + bx + c)$ $\therefore \dfrac{dv}{dx} = -\sin(ax^2 + bx + c).\dfrac{d}{dx}(ax^2 + bx + c) = -\sin(ax^2 + bx + c).(2ax + b)$

$$\therefore \frac{dy}{dx} = u.\frac{dv}{dx} + v.\frac{du}{dx} = x^{\frac{3}{2}}.[-\sin(ax^2 + bx + c).(2ax + b)] + \cos(ax^2 + bx + c).\frac{3}{2}\sqrt{x}$$

$$= \frac{3}{2}\sqrt{x}.\cos(ax^2 + bx + c) - x^{\frac{3}{2}}.(2ax + b).\sin(ax^2 + bx + c) \quad \text{Ans.}$$

(5) (a) $y = \tan^{-1}\left(\dfrac{\sqrt{x} - 1}{1 + \sqrt{x}}\right) = \tan^{-1}\sqrt{x} - \tan^{-1}1$ $\left[\text{formula}, \; \tan^{-1}\left(\dfrac{A - B}{1 + AB}\right) = \tan^{-1}A - \tan^{-1}B\right]$

$$\therefore \frac{dy}{dx} = \frac{1}{1 + (\sqrt{x})^2}.\frac{d}{dx}(\sqrt{x}) - 0 = \frac{1}{1 + x}.\frac{1}{2\sqrt{x}} = \frac{1}{2\sqrt{x}\,(1 + x)} \quad \text{Ans.} \quad \left[\text{formula}, \; y = \tan^{-1}x, \; \frac{dy}{dx} = \frac{1}{1 + x^2}.\frac{d}{dx}(x)\right]$$

IInd Method: $-$ $y = \tan^{-1}\left(\dfrac{\sqrt{x} - 1}{1 + \sqrt{x}}\right)$, Let $u = \dfrac{\sqrt{x} - 1}{1 + \sqrt{x}} = \dfrac{f(x)}{g(x)}$ (say) $f'(x) = \dfrac{1}{2\sqrt{x}}$ and $g'(x) = \dfrac{1}{2\sqrt{x}}$

$$\text{or } \frac{du}{dx} = \frac{g(x).f'(x) - f(x).g'(x)}{(g(x))^2} = \frac{(1 + \sqrt{x}).\frac{1}{2\sqrt{x}} - (\sqrt{x} - 1).\frac{1}{2\sqrt{x}}}{(1 + \sqrt{x})^2} = \frac{\frac{1 + \sqrt{x} - \sqrt{x} + 1}{2\sqrt{x}}}{(1 + \sqrt{x})^2} = \frac{2}{2\sqrt{x}.(1 + \sqrt{x})^2} = \frac{1}{\sqrt{x}.(1 + \sqrt{x})^2}$$

$$\Rightarrow y = \tan^{-1}u, \; \frac{dy}{du} = \frac{1}{1 + u^2} = \frac{1}{1 + \left(\frac{\sqrt{x} - 1}{1 + \sqrt{x}}\right)^2} = \frac{(1 + \sqrt{x})^2}{(1 + \sqrt{x})^2 + (\sqrt{x} - 1)^2} = \frac{(1 + \sqrt{x})^2}{1 + 2\sqrt{x} + x + x - 2\sqrt{x} + 1} = \frac{(1 + \sqrt{x})^2}{2x + 2} = \frac{(1 + \sqrt{x})^2}{2(x + 1)}$$

$$\therefore \frac{dy}{dx} = \frac{dy}{du} \times \frac{du}{dx} = \frac{(1 + \sqrt{x})^2}{2(x + 1)}.\frac{1}{\sqrt{x}.(1 + \sqrt{x})^2} = \frac{1}{2\sqrt{x}\,(x + 1)} \quad \text{Ans.}$$

(b) $y = \tan^{-1}\left(\dfrac{\sin x + \cos x}{1 - \sin x \cos x}\right) = \tan^{-1}(\sin x) + \tan^{-1}(\cos x)$ $\left[\text{formula}, \; \tan^{-1}\left(\dfrac{A + B}{1 - AB}\right) = \tan^{-1}A + \tan^{-1}B\right]$

$$\text{or } \frac{dy}{dx} = \frac{1}{1 + (\sin x)^2}.\frac{d}{dx}(\sin x) + \frac{1}{1 + (\cos x)^2}.\frac{d}{dx}(\cos x) \quad \left[\text{formula}, \; y = \tan^{-1}u, \; \frac{dy}{dx} = \frac{1}{1 + u^2}.\frac{du}{dx}\right]$$

$$\text{or } \frac{dy}{dx} = \frac{1}{1 + \sin^2 x}.\cos x + \frac{1}{1 + \cos^2 x}.(-\sin x) = \frac{\cos x\,(1 + \cos^2 x) - \sin x\,(1 + \sin^2 x)}{(1 + \sin^2 x)(1 + \cos^2 x)} = \frac{\cos x + \cos^3 x - \sin x - \sin^3 x}{1 + \cos^2 x + \sin^2 x + \sin^2 x\cos^2 x}$$

$$= \frac{(\cos x - \sin x) + (\cos^3 x - \sin^3 x)}{(1 + 1 + \sin^2 x\cos^2 x)}$$

$$\text{or } \frac{dy}{dx} = \frac{(\cos x - \sin x) + [(\cos x - \sin x)^3 + 3\sin x\cos x\,(\cos x - \sin x)]}{(2 + \sin^2 x\cos^2 x)} = \frac{(\cos x - \sin x)[1 + (\cos x - \sin x)^2 + 3\sin x\cos x]}{(2 + \sin^2 x\cos^2 x)}$$

$$\text{or } \frac{dy}{dx} = \frac{(\cos x - \sin x)[1 + \cos^2 x + \sin^2 x - 2\sin x\cos x + 3\sin x\cos x]}{(2 + \sin^2 x\cos^2 x)}$$

$$= \frac{(\cos x - \sin x)[2 + \sin x\cos x]}{(2 + \sin^2 x\cos^2 x)} \quad \text{Ans.} \quad [(a - b)^3 = a^3 - b^3 - 3ab(a - b)]$$

IInd Method: $-$ $y = \tan^{-1}\left(\dfrac{\sin x + \cos x}{1 - \sin x\cos x}\right)$ Let $u = \dfrac{\sin x + \cos x}{1 - \sin x\cos x}$ $\therefore \dfrac{du}{dx} = \dfrac{(\cos x - \sin x)[2 + \sin x\cos x]}{(1 - \sin x\cos x)^2}$

$\Rightarrow y=\tan^{-1}u \quad \therefore \dfrac{dy}{du}=\dfrac{1}{1+u^2}=\dfrac{1}{1+\left(\dfrac{\sin x+\cos x}{1-\sin x\cos x}\right)^2}=\dfrac{(1-\sin x\cos x)^2}{(1-\sin x\cos x)^2+(\sin x+\cos x)^2}$

$\therefore \dfrac{dy}{dx}=\dfrac{dy}{du}\times\dfrac{du}{dx}=\dfrac{(1-\sin x\cos x)^2}{(1-\sin x\cos x)^2+(\sin x+\cos x)^2}\cdot\dfrac{(\cos x-\sin x)[2+\sin x\cos x]}{(1-\sin x\cos x)^2}=\dfrac{(\cos x-\sin x)[2+\sin x\cos x]}{(1-\sin x\cos x)^2+(\sin x+\cos x)^2}$ (solve)

$=\dfrac{(\cos x-\sin x)[2+\sin x\cos x]}{(2+\sin^2 x\cos^2 x)}$ Ans.

(c) $y=\tan^{-1}\left(\dfrac{6x}{1-x^2}\right)$, Let $u=\dfrac{6x}{1-x^2} \quad \therefore \dfrac{du}{dx}=\dfrac{(1-x^2).\frac{d}{dx}(6x)-6x.\frac{d}{dx}(1-x^2)}{(1-x^2)^2}$

$\therefore \dfrac{dy}{dx}=\dfrac{6(1-x^2)-6x.(-2x)}{(1-x^2)^2}=\dfrac{6-6x^2+12x^2}{(1-x^2)^2}=\dfrac{6(1-x^2)}{(1-x^2)^2}=\dfrac{6}{1-x^2}$

Now, $y=\tan^{-1}u \quad \therefore \dfrac{dy}{du}=\dfrac{1}{1+u^2}=\dfrac{1}{1+\left(\dfrac{6x}{1-x^2}\right)^2}=\dfrac{(1-x^2)^2}{(1-x^2)^2+36x^2}=\dfrac{(1-x^2)^2}{1+x^4-2x^2+36x^2}=\dfrac{(1-x^2)^2}{x^4+34x^2+1}$

$\therefore \dfrac{dy}{dx}=\dfrac{dy}{du}\times\dfrac{du}{dx}=\dfrac{(1-x^2)^2}{(x^4+34x^2+1)}\cdot\dfrac{6}{1-x^2}=\dfrac{6(1-x^2)}{(x^4+34x^2+1)}$ Ans.

(d) $y=\tan^{-1}\left(\dfrac{x+3}{1+x^2}\right)+\tan^{-1}\left(\dfrac{2x+1}{2-3x}\right)=\tan^{-1}\left(\dfrac{\frac{x+3}{1+x^2}+\frac{2x+1}{2-3x}}{1-\frac{x+3}{1+x^2}\cdot\frac{2x+1}{2-3x}}\right)=\tan^{-1}\left(\dfrac{2x^2-2x^3+5x-7}{3x^3+10x+1}\right)$

$\left[\text{use formula, } \tan^{-1}\left(\dfrac{A+B}{1-AB}\right)=\tan^{-1}A+\tan^{-1}B \text{ and } y=\tan^{-1}u\,,\ \dfrac{dy}{dx}=\dfrac{1}{1+u^2}\cdot\dfrac{du}{dx}\right]$

$\therefore \dfrac{dy}{dx}=\dfrac{(3x^3+10x+1)(5+4x-6x^2)-(2x^2-2x^3+5x-7)(9x^2+10)}{(3x^3+10x+1)^2+(2x^2-2x^3+5x-7)^2}$ Ans.

(e) $y=\cot^{-1}\left(\dfrac{\sqrt{1+x}-\sqrt{1-x}}{\sqrt{1-x}+\sqrt{1+x}}\right)=\cot^{-1}\left(\dfrac{\sqrt{1+x}-\sqrt{1-x}}{\sqrt{1-x}+\sqrt{1+x}}\times\dfrac{\sqrt{1+x}+\sqrt{1-x}}{\sqrt{1+x}+\sqrt{1-x}}\right)$

$y=\cot^{-1}\left(\dfrac{1+x-1+x}{1+x+1-x-2\sqrt{1-x^2}}\right)=\cot^{-1}\left(\dfrac{2x}{2-2\sqrt{1-x^2}}\right)=\cot^{-1}\left(\dfrac{x}{1-\sqrt{1-x^2}}\right)$

Put $x=\sin\theta \quad \therefore \dfrac{dx}{d\theta}=\cos\theta$ then $y=\cot^{-1}\left(\dfrac{\sin\theta}{1-\sqrt{1-\sin^2\theta}}\right)=\cot^{-1}\left(\dfrac{\sin\theta}{1-\cos\theta}\right)$

$y=\cot^{-1}\left(\dfrac{2\sin\frac{\theta}{2}\cos\frac{\theta}{2}}{1-\left(\cos^2\frac{\theta}{2}-\sin^2\frac{\theta}{2}\right)}\right)=\cot^{-1}\left(\dfrac{2\sin\frac{\theta}{2}\cos\frac{\theta}{2}}{1-\cos^2\frac{\theta}{2}+\sin^2\frac{\theta}{2}}\right)=\cot^{-1}\left(\dfrac{2\sin\frac{\theta}{2}\cos\frac{\theta}{2}}{\sin^2\frac{\theta}{2}+\sin^2\frac{\theta}{2}}\right)=\cot^{-1}\left(\dfrac{2\sin\frac{\theta}{2}\cos\frac{\theta}{2}}{2\sin^2\frac{\theta}{2}}\right)$

$y=\cot^{-1}\left(\dfrac{\cos\frac{\theta}{2}}{\sin\frac{\theta}{2}}\right)=\cot^{-1}\left(\cot\frac{\theta}{2}\right)=\dfrac{\theta}{2} \quad \therefore \dfrac{dy}{d\theta}=\dfrac{1}{2}$

$\therefore \dfrac{dy}{dx}=\dfrac{\frac{dy}{d\theta}}{\frac{dx}{d\theta}}=\dfrac{dy}{d\theta}\div\dfrac{dx}{d\theta}=\dfrac{\frac{1}{2}}{\cos\theta}=\dfrac{1}{2\cos\theta}=\dfrac{1}{2\sqrt{1-\sin^2\theta}}=\dfrac{1}{2\sqrt{1-x^2}}$ Ans. $\left[\text{Put } x=\sin\theta,\ \cos^2\theta=1-\sin^2\theta \text{ or } \cos\theta=\sqrt{1-\sin^2\theta}\right]$

(f) $y=\sin^{-1}\dfrac{1}{\sqrt{1+x^2}}+\tan^{-1}\dfrac{x}{\sqrt{1-x^2}}=u+v$ (say)

$u=\sin^{-1}\dfrac{1}{\sqrt{1+x^2}}$, Put $x=\tan\theta$ then $\dfrac{dx}{d\theta}=\sec^2\theta$

$u=\sin^{-1}\left(\dfrac{1}{\sqrt{1+(\tan\theta)^2}}\right)=\sin^{-1}\left(\dfrac{1}{\sec\theta}\right)=\sin^{-1}(\cos\theta)=\sin^{-1}\left(\sin\left(\dfrac{\pi}{2}-\theta\right)\right)=\dfrac{\pi}{2}-\theta$

$$\therefore \frac{du}{d\theta} = -1, \quad \frac{du}{dx} = \frac{du}{d\theta} \div \frac{dx}{d\theta} = -\frac{1}{\sec^2\theta} = -\frac{1}{1+\tan^2\theta} = -\frac{1}{1+x^2} \quad [\text{Put } \tan\theta = x, \ \sec^2\theta = 1+\tan^2\theta]$$

$$v = \tan^{-1}\frac{x}{\sqrt{1-x^2}}, \quad \text{Put } x = \sin\theta \text{ then } \frac{dx}{d\theta} = \cos\theta$$

$$v = \tan^{-1}\frac{x}{\sqrt{1-x^2}} = \tan^{-1}\left(\frac{\sin\theta}{\sqrt{1-(\sin\theta)^2}}\right) = \tan^{-1}\left(\frac{\sin\theta}{\sqrt{1-\sin^2\theta}}\right) = \tan^{-1}\left(\frac{\sin\theta}{\cos\theta}\right) = \tan^{-1}(\tan\theta) = \theta$$

$$\therefore \frac{dv}{d\theta} = 1, \quad \frac{dv}{dx} = \frac{dv}{d\theta} \div \frac{dx}{d\theta} = \frac{1}{\cos\theta} = \frac{1}{\sqrt{1-\sin^2\theta}} = \frac{1}{\sqrt{1-x^2}} \quad \left[\text{Put } \sin\theta = x, \ \cos\theta = \sqrt{1-\sin^2\theta}\right]$$

$$y = u+v, \quad \frac{dy}{dx} = \frac{du}{dx} + \frac{dv}{dx} = -\frac{1}{1+x^2} + \frac{1}{\sqrt{1-x^2}} = \frac{1}{\sqrt{1-x^2}} - \frac{1}{1+x^2} \quad \text{Ans.}$$

(6) (a) $y = \sin^{-1}\sqrt{x} + \sin^{-1}\sqrt{1-x}$ $\quad \left\{\text{formula}, \ \sin^{-1}A \pm \sin^{-1}B = \sin^{-1}\left[A\sqrt{1-B^2} \pm B\sqrt{1-A^2}\right]\right\}$

$$y = \sin^{-1}\left[\sqrt{x}.\sqrt{1-\left(\sqrt{1-x}\right)^2} + \sqrt{1-x}.\sqrt{1-\left(\sqrt{x}\right)^2}\right] = \sin^{-1}\left[\sqrt{x}.\sqrt{1-1+x} + \sqrt{1-x}.\sqrt{1-x}\right] = \sin^{-1}[x+1-x] = \sin^{-1}1 = \frac{\pi}{2}$$

$$y = \frac{\pi}{2}, \quad \frac{dy}{dx} = 0 \quad \text{Ans.}$$

(b) $y = \sin^{-1}(x^2-1) - \sin^{-1}(2x^2)$ $\quad \left\{\text{formula}, \quad y = \sin^{-1}u \quad \therefore \frac{dy}{dx} = \frac{1}{\sqrt{1-u^2}}.\frac{du}{dx}\right\}$

$$\frac{dy}{dx} = \frac{1}{\sqrt{1-(x^2-1)^2}}.2x - \frac{1}{\sqrt{1-(2x^2)^2}}.4x = \frac{2x}{\sqrt{1-(x^4-2x^2+1)}} - \frac{4x}{\sqrt{1-4x^4}} = \frac{2x}{\sqrt{1-x^4+2x^2-1}} - \frac{4x}{\sqrt{1-4x^4}} = \frac{2x}{\sqrt{2x^2-x^4}} - \frac{4x}{\sqrt{1-4x^4}}$$
$$= \frac{2}{\sqrt{2-x^2}} - \frac{4x}{\sqrt{1-4x^4}} \quad \text{Ans.}$$

(c) $y = \cos^{-1}\frac{1}{1+x} + \cos^{-1}(1-x)$, $\quad \frac{dy}{dx} = \frac{1}{-\sqrt{1-\left(\frac{1}{1+x}\right)^2}}.\frac{d}{dx}\left(\frac{1}{1+x}\right) + \frac{1}{-\sqrt{1-(1-x)^2}}.\frac{d}{dx}(1-x)$

$$\therefore \frac{dy}{dx} = \frac{1+x}{-\sqrt{(1+x)^2-1}}.-\frac{1}{(1+x)^2} + \frac{1}{-\sqrt{1-(1-2x+x^2}}.(-1) = \frac{1}{(1+x).\sqrt{1+2x+x^2-1}} + \frac{1}{\sqrt{1-1+2x-x^2}}$$
$$= \frac{1}{(1+x).\sqrt{2x+x^2}} + \frac{1}{\sqrt{2x-x^2}} \quad \text{Ans.}$$

(d) $y = \cos^{-1}\frac{x}{\sqrt{1+x}} - \cos^{-1}\frac{1}{\sqrt{1-x}}$, $\quad \left\{\text{formula}, \quad y = \cos^{-1}u \quad \therefore \frac{dy}{dx} = \frac{1}{-\sqrt{1-u^2}}.\frac{du}{dx}\right\}$

$$\frac{d}{dx}\left(\frac{x}{\sqrt{1+x}}\right) = \frac{\sqrt{1+x}.\frac{d(x)}{dx} - x.\frac{d\left(\sqrt{1+x}\right)}{dx}}{\left(\sqrt{1+x}\right)^2} = \frac{\sqrt{1+x} - x.\frac{1}{2\sqrt{1+x}}.\frac{d(1+x)}{dx}}{\left(\sqrt{1+x}\right)^2} = \frac{2+2x-x}{2\sqrt{1+x}.(1+x)} = \frac{2+x}{2(1+x)\sqrt{1+x}}$$

$$\frac{d}{dx}\left(\frac{1}{\sqrt{1-x}}\right) = \frac{\sqrt{1-x}.\frac{d(1)}{dx} - 1.\frac{d\left(\sqrt{1-x}\right)}{dx}}{\left(\sqrt{1-x}\right)^2} = \frac{0 - 1.\frac{1}{2\sqrt{1-x}}.\frac{d(1-x)}{dx}}{(1-x)} = \frac{1}{2(1-x)\sqrt{1-x}}$$

$$\frac{dy}{dx} = \frac{1}{2}\left[\frac{1}{\sqrt{1-x}.\sqrt{1-2x}} - \frac{2+x}{\sqrt{1+x}}\right] \quad \text{Ans.} \quad (\text{see above question and solve it.})$$

(e) $y = \sin^{-1}\left[\sqrt{x}.\sqrt{2-x} + \sqrt{x-1}.\sqrt{1-x}\right]$ $\left\{\text{formula}, \ \sin^{-1}A + \sin^{-1}B = \sin^{-1}\left[A\sqrt{1-B^2} + B\sqrt{1-A^2}\right]\right\}$

$$y = \sin^{-1}\left[\sqrt{x}.\sqrt{1-\left(\sqrt{x-1}\right)^2} + \sqrt{x-1}.\sqrt{1-\left(\sqrt{x}\right)^2}\right] \quad \text{here } A = \sqrt{x} \text{ and } B = \sqrt{x-1}$$

$$y = \sin^{-1}\sqrt{x} + \sin^{-1}\sqrt{x-1}, \quad \frac{dy}{dx} = \frac{1}{\sqrt{1-\left(\sqrt{x}\right)^2}}.\frac{d\left(\sqrt{x}\right)}{dx} + \frac{1}{\sqrt{1-\left(\sqrt{x-1}\right)^2}}.\frac{d\left(\sqrt{x-1}\right)}{dx}$$

$$\therefore \frac{dy}{dx}=\frac{1}{\sqrt{1-x}}.\frac{1}{2\sqrt{x}}+\frac{1}{\sqrt{2-x}}.\frac{1}{2\sqrt{x-1}}=\frac{1}{2.\sqrt{x}.\sqrt{1-x}}+\frac{1}{2.\sqrt{x-1}.\sqrt{2-x}} \quad \text{Ans.}$$

(f) $y=\cos^{-1}\left[\sqrt{x}.\frac{1}{\sqrt{1+x^2}}-\sqrt{1-x}.\sqrt{\frac{x^2}{1+x^2}}\right]$ $\left[\text{formula, } \cos^{-1}\left(AB-\sqrt{1-A^2}.\sqrt{1-B^2}\right)=\cos^{-1}A+\cos^{-1}B\right]$

$$y=\cos^{-1}\left(\sqrt{x}.\frac{1}{\sqrt{1+x^2}}-\sqrt{1-(\sqrt{x})^2}.\sqrt{1-\left(\frac{1}{\sqrt{1+x^2}}\right)^2}\right)=\cos^{-1}\sqrt{x}+\cos^{-1}\left(\frac{1}{\sqrt{1+x^2}}\right)$$

Put $x=\tan\theta$, $\frac{dx}{d\theta}=\sec^2\theta$ then $y=\cos^{-1}\sqrt{\tan\theta}+\cos^{-1}\left(\frac{1}{\sqrt{1+\tan^2\theta}}\right)=\cos^{-1}\sqrt{\tan\theta}+\cos^{-1}\left(\frac{1}{\sqrt{\sec^2\theta}}\right)$

$$y=\cos^{-1}\sqrt{\tan\theta}+\cos^{-1}\left(\frac{1}{\sec\theta}\right)=\cos^{-1}\sqrt{\tan\theta}+\cos^{-1}(\cos\theta)=\cos^{-1}\sqrt{\tan\theta}+\theta$$

$$\frac{dy}{d\theta}=\frac{1}{-\sqrt{1-\left(\sqrt{\tan\theta}\right)^2}}.\frac{d\left(\sqrt{\tan\theta}\right)}{d\theta}+1=-\frac{1}{\sqrt{1-\tan\theta}}.\frac{1}{2\sqrt{\tan\theta}}.\frac{d(\tan\theta)}{d\theta}+1=-\frac{1}{\sqrt{1-\tan\theta}}.\frac{1}{2\sqrt{\tan\theta}}.\sec^2\theta+1$$

$$\frac{dy}{d\theta}=-\frac{1+\tan^2\theta}{2\sqrt{\tan\theta}.\sqrt{1-\tan\theta}}+1=-\frac{1+x^2}{2\sqrt{x}.\sqrt{1-x}}+1 \quad \text{Put } \tan\theta=x, \frac{dx}{d\theta}=\sec^2\theta=1+\tan^2\theta=1+x^2$$

$$\frac{dy}{dx}=\frac{dy}{d\theta}\div\frac{dx}{d\theta}=\frac{-\frac{1+x^2}{2\sqrt{x}.\sqrt{1-x}}+1}{1+x^2}=\frac{2\sqrt{x}.\sqrt{1-x}-(1+x^2)}{2\sqrt{x}.\sqrt{1-x}.(1+x^2)}=\frac{1}{1+x^2}-\frac{1}{2\sqrt{x}.\sqrt{1-x}} \quad \text{Ans.}$$

(7) (a) Differentiate $\tan^{-1}\frac{\sqrt{1-x^2}}{x}$ with respect to $\tan^{-1}x$

Let $y=\tan^{-1}\frac{\sqrt{1-x^2}}{x}$ and $z=\tan^{-1}x$ then find $\frac{dy}{dz}=\frac{dy}{dx}\div\frac{dz}{dx}$ Put $x=\cos\theta$ and $\frac{dz}{dx}=\frac{1}{1+x^2}$

$$y=\tan^{-1}\frac{\sqrt{1-\cos^2\theta}}{\cos\theta}=\tan^{-1}\left(\frac{\sin\theta}{\cos\theta}\right)=\tan^{-1}(\tan\theta)=\theta=\cos^{-1}x, \quad \frac{dy}{dx}=-\frac{1}{\sqrt{1-x^2}}$$

$$\frac{dy}{dz}=\frac{dy}{dx}\div\frac{dz}{dx}=\frac{-\frac{1}{\sqrt{1-x^2}}}{\frac{1}{1+x^2}}=-\frac{1+x^2}{\sqrt{1-x^2}} \quad \text{Ans.}$$

(b) Differentiate $\sin^{-1}\sqrt{\frac{1+x}{1-x}}$ with respect to $\sqrt{x}$.

Let $y=\sin^{-1}\sqrt{\frac{1+x}{1-x}\times\frac{1+x}{1+x}}=\sin^{-1}\left(\frac{1+x}{\sqrt{1-x^2}}\right)$ and $z=\sqrt{x}$ then find $\frac{dy}{dz}$. $\therefore \frac{dy}{dz}=\frac{dy}{dx}\div\frac{dz}{dx}$

$$\text{Put } x=\cos\theta, \ \frac{dx}{d\theta}=-\sin\theta=-\sqrt{1-\cos^2\theta}=-\sqrt{1-x^2}$$

$$y=\sin^{-1}\left(\frac{1+\cos\theta}{\sqrt{1-\cos^2\theta}}\right)=\sin^{-1}\left(\frac{1+\cos\theta}{\sin\theta}\right)=\sin^{-1}\left(\frac{1+\cos^2\frac{\theta}{2}-\sin^2\frac{\theta}{2}}{2\sin\frac{\theta}{2}\cos\frac{\theta}{2}}\right)=\sin^{-1}\left(\frac{2\cos^2\frac{\theta}{2}}{2\sin\frac{\theta}{2}\cos\frac{\theta}{2}}\right)$$

$$y=\sin^{-1}\left(\frac{\cos\frac{\theta}{2}}{\sin\frac{\theta}{2}}\right)=\sin^{-1}\left(\cot\frac{\theta}{2}\right) \text{ or } \sin y=\cot\frac{\theta}{2}, \ \cos y\frac{dy}{d\theta}=-\frac{1}{2}\operatorname{cosec}^2\left(\frac{\theta}{2}\right) \ \therefore \frac{dy}{d\theta}=-\frac{\operatorname{cosec}^2\left(\frac{\theta}{2}\right)}{2\cos y}$$

To be solve, $\frac{dy}{d\theta}=\frac{-(1+\cos\theta)}{\sin\theta.\sqrt{-2\cos^2\theta-2\cos\theta}}$ then $\frac{dy}{dx}=\frac{dy}{d\theta}\times\frac{d\theta}{dx}=\frac{-(1+\cos\theta)}{\sin\theta.\sqrt{-2\cos^2\theta-2\cos\theta}}.-\frac{1}{\sin\theta}$

$$\frac{dy}{dx}=\frac{(1+\cos\theta)}{(1-\cos^2\theta)\sqrt{-2\cos^2\theta-2\cos\theta}}=\frac{(1+x)}{(1-x^2).\sqrt{-2x^2-2x}}=\frac{(1+x)}{(1-x^2).\sqrt{-2x(x+1)}}=\frac{(1+x)}{\sqrt{2}.\sqrt{x}(1-x^2)\sqrt{-x-1}}$$

$$z=\sqrt{x}\,,\ \frac{dz}{dx}=\frac{1}{2\sqrt{x}}\quad\therefore\ \frac{dy}{dz}=\frac{dy}{dx}\div\frac{dz}{dx}=\frac{(1+x)}{\sqrt{2}.\sqrt{x}(1-x^2)\sqrt{-x-1}}.2\sqrt{x}=\frac{\sqrt{2}.(1+x)}{(1+x)(1-x).\sqrt{-x-1}}$$

$$\frac{dy}{dz}=\frac{\sqrt{2}}{(1-x).\sqrt{-x-1}}\quad \text{Ans.}$$

(c) Differentiate $x^{\cos^{-1}\sqrt{x}}$ with respect to $\cos^{-1}\sqrt{x}$.

Let $y=x^{\cos^{-1}\sqrt{x}}$ and $z=\cos^{-1}\sqrt{x}$ or $\cos z=\sqrt{x}$ or $x=\cos^2 z$

$y=x^z=(\cos^2 z)^z\quad\therefore\ \log y=\log(\cos^2 z)^z=z\log(\cos z)^2=2z\log(\cos z)=u.v$ (say)

then $\frac{1}{y}.\frac{dy}{dz}=2z.\frac{d}{dz}(\log(\cos z))+\log(\cos z).\frac{d}{dz}(2z)=2z.\frac{1}{\cos z}.(-\sin z)+2\log(\cos z)=2\log(\cos z)-2z.\tan z$

$\frac{dy}{dz}=y[2\log(\cos z)-2z.\tan z]=2(\cos^2 z)^z.[\log(\cos z)-z.\tan z]$ Ans.

(d) Differentiate $\cot^{-1}\frac{x}{\sqrt{1+x^2}}$ with respect to $\operatorname{cosec}^{-1}\frac{1}{x^2+1}$.

Let $y=\cot^{-1}\frac{x}{\sqrt{1+x^2}}$ Put $x=\tan\theta$, $\frac{dx}{d\theta}=\sec^2\theta=1+\tan^2\theta=1+x^2$

$$y=\cot^{-1}\left(\frac{\tan\theta}{\sqrt{1+\tan^2\theta}}\right)=\cot^{-1}\left(\frac{\tan\theta}{\sec\theta}\right)=\cot^{-1}\left(\frac{\frac{\sin\theta}{\cos\theta}}{\frac{1}{\cos\theta}}\right)=\cot^{-1}(\sin\theta),\ \frac{dy}{d\theta}=-\frac{1}{1+\sin^2\theta}.\cos\theta=-\frac{\cos\theta}{1+\sin^2\theta}$$

$\therefore\ \frac{dy}{dx}=\frac{dy}{d\theta}\div\frac{dx}{d\theta}=-\frac{\cos\theta}{1+\sin^2\theta}.\frac{1}{1+\tan^2\theta}$ if $x=\tan\theta$ then $\cos\theta=\frac{1}{\sqrt{1+x^2}}$ and $\sin\theta=\frac{x}{\sqrt{1+x^2}}$

Put the value of $\cos\theta,\sin\theta$ and $\tan\theta$ in $\frac{dy}{dx}=-\frac{\cos\theta}{1+\sin^2\theta}.\frac{1}{1+\tan^2\theta}=-\frac{1}{(1+2x^2).\sqrt{1+x^2}}$

Now, Let $z=\operatorname{cosec}^{-1}\frac{1}{x^2+1}\quad\therefore\ \frac{dz}{dx}=-\frac{1}{\left|\frac{1}{x^2+1}\right|.\sqrt{\left(\frac{1}{x^2+1}\right)^2-1}}.\frac{d}{dx}\left(\frac{1}{x^2+1}\right)=-\frac{(x^2+1).|x^2+1|}{\sqrt{-x^4-2x^2}}.-\frac{2x}{(x^2+1)^2}$

$$\frac{dz}{dx}=\frac{2(x^2+1).|x^2+1|}{\sqrt{-x^2-2}}.\frac{1}{(x^2+1)^2}=\frac{2}{\sqrt{-x^2-2}}\ \text{ or }\ \frac{dy}{dz}=\frac{dy}{dx}\div\frac{dz}{dx}=-\frac{1}{(1+2x^2).\sqrt{1+x^2}}.\frac{\sqrt{-x^2-2}}{2}=-\frac{\sqrt{-x^2-2}}{2(1+2x^2).\sqrt{1+x^2}}\quad\text{Ans.}$$

(8) (a) Let $y=\log_{(1-x)}\cos^{-1}(1-x)$ and $z=2^{2(1-x)}$ Put $1-x=t$, $\frac{dx}{dt}=-1$

or $y=\log_t\cos^{-1}t$ and $z=2^{2t}=4^t\quad\therefore\ \frac{dy}{dz}=\frac{dy}{dt}\times\frac{dt}{dz}$ or $\frac{dy}{dt}\div\frac{dz}{dt}$

$$y=\log_t\cos^{-1}t=\frac{\log(\cos^{-1}t)}{\log t},\ \frac{dy}{dt}=\frac{\log t.\frac{d}{dt}[\log(\cos^{-1}t)]-\log(\cos^{-1}t).\frac{d}{dt}(\log t)}{(\log t)^2}$$

$$\frac{dy}{dt}=\frac{\log t.\frac{1}{\cos^{-1}t}.\frac{1}{-\sqrt{1-t^2}}-\log(\cos^{-1}t).\frac{1}{t}}{(\log t)^2}=\frac{-t\log t-\sqrt{1-t^2}.\cos^{-1}t.\log(\cos^{-1}t)}{t.\cos^{-1}t.\sqrt{1-t^2}.(\log t)^2}=\frac{-1}{\cos^{-1}t.\sqrt{1-t^2}.\log t}-\frac{\log(\cos^{-1}t)}{t.(\log t)^2}$$

or $z=4^t$, $\frac{dz}{dt}=4^t\log 4$ or $\frac{dt}{dz}=\frac{1}{4^t\log 4}$

$$\frac{dy}{dz}=\frac{dy}{dt}\times\frac{dt}{dz}=\left[\frac{-1}{\cos^{-1}t.\sqrt{1-t^2}.\log t}-\frac{\log(\cos^{-1}t)}{t.(\log t)^2}\right]\frac{1}{4^t\log 4}=\frac{1}{4^t\log 4}\left[\frac{-1}{\cos^{-1}t.\sqrt{1-t^2}.\log t}-\frac{\log(\cos^{-1}t)}{t.(\log t)^2}\right]\quad\text{Ans.}$$

At $x = \frac{1}{2}$ then $t = 1 - x = 1 - \frac{1}{2} = \frac{1}{2}$ $\quad \therefore \log t = \log\left(\frac{1}{2}\right) = -\log 2\,,\ \cos^{-1} t = \cos^{-1}\left(\frac{1}{2}\right) = \frac{\pi}{3}$

$$\left(\frac{dy}{dz}\right)_{t=\frac{1}{2}} = \frac{1}{4^{\frac{1}{2}}.\log 4}\left\{\frac{-1}{\cos^{-1}\left(\frac{1}{2}\right).\sqrt{1-\left(\frac{1}{2}\right)^2}\,.\log\left(\frac{1}{2}\right)} - \frac{\log\left(\cos^{-1}\left(\frac{1}{2}\right)\right)}{\frac{1}{2}.\left[\log\left(\frac{1}{2}\right)\right]^2}\right\} = \frac{1}{2\log 4}\left\{\frac{-1}{\frac{\pi}{3}.\frac{\sqrt{3}}{2}.(-\log 2)} - \frac{\log\left(\frac{\pi}{3}\right)}{\frac{1}{2}.(\log 2)^2}\right\}$$

$$= \frac{1}{\log 16}\left\{\frac{6}{\sqrt{3}.\pi.\log 2} - \frac{2.\log\left(\frac{\pi}{3}\right)}{(\log 2)^2}\right\} \quad \text{Ans.}$$

(b) If $\sqrt{1+4x^2} - \sqrt{1-9y^2} = a(2x-3y)$ then find $\frac{dy}{dx}$.

or $\sqrt{1+4x^2} - \sqrt{1-9y^2} = a(2x-3y)$ $\quad \therefore \frac{1}{2\sqrt{1+4x^2}}.8x - \frac{1}{2\sqrt{1-9y^2}}.18y.\frac{dy}{dx} = 2a - 3a.\frac{dy}{dx}$

$$\Rightarrow \frac{dy}{dx}\left(3a - \frac{9y}{\sqrt{1-9y^2}}\right) = \left(2a - \frac{4x}{\sqrt{1+4x^2}}\right) \text{ or } \frac{dy}{dx} = \frac{\left(2a - \frac{4x}{\sqrt{1+4x^2}}\right)}{\left(3a - \frac{9y}{\sqrt{1-9y^2}}\right)} = \frac{2a.\sqrt{1+4x^2} - 4x}{\sqrt{1+4x^2}} \times \frac{\sqrt{1-9y^2}}{3a.\sqrt{1-9y^2} - 9y}$$

$$\therefore \frac{dy}{dx} = \frac{2\sqrt{1-9y^2}.\left(a\sqrt{1+4x^2} - 2x\right)}{3\sqrt{1+4x^2}.\left(a\sqrt{1-9y^2} - 3y\right)} \quad \text{Ans.}$$

(c) $y = \tan^{-1}\frac{1}{1+x} + \cot^{-1}(1+x) = \tan^{-1}\frac{1}{1+x} + \tan^{-1}\frac{1}{1+x} = 2\tan^{-1}\frac{1}{1+x}$ $\quad \left[\text{Put } \tan^{-1} x = \cot^{-1}\left(\frac{1}{x}\right)\right]$

$$\text{or } \frac{dy}{dx} = 2.\frac{1}{1+\left(\frac{1}{1+x}\right)^2}.\frac{d}{dx}\left(\frac{1}{1+x}\right) = \frac{2(1+x)^2}{(1+x)^2+1}.\left(-\frac{1}{(1+x)^2}\right) = \frac{-2}{(1+x)^2+1} = \frac{-2}{1+2x+x^2+1} = \frac{-2}{x^2+2x+2}$$

Aganin, Differentiate with respect to x, we get

$$\frac{dy}{dx} = \frac{-2}{x^2+2x+2},\ \frac{d^2y}{dx^2} = \frac{(x^2+2x+2).\frac{d(-2)}{dx} - (-2).\frac{d}{dx}(x^2+2x+2)}{(x^2+2x+2)^2} = \frac{0+2(2x+2)}{(x^2+2x+2)^2} = \frac{4(x+1)}{(x^2+2x+2)^2} \quad \text{Ans.}$$

(d) $x = a(1-\cos t),\ y = a(1+\sin t)$

$\frac{dx}{dt} = -a(-\sin t) = a\sin t$ and $\frac{dy}{dt} = a\cos t$ $\quad \therefore \frac{dy}{dx} = \frac{dy}{dt} \div \frac{dx}{dt} = \frac{a\cos t}{a\sin t} = \frac{\cos t}{\sin t} = \cot t$ Ans.

Independent of t, $x = a(1-\cos t)$ or $\frac{x}{a} = 1 - \cos t$ or $\cos t = 1 - \frac{x}{a}$ $\quad \therefore \cos t = \frac{a-x}{a}$

and $y = a(1+\sin t)$ or $\frac{y}{a} = 1 + \sin t$ or $\sin t = \frac{y}{a} - 1$ $\quad \therefore \sin t = \frac{y-a}{a}$

Put value of $\sin t, \cos t$ in $\frac{dy}{dx}$ then $\frac{dy}{dx} = \cot t = \frac{\cos t}{\sin t} = \frac{\frac{a-x}{a}}{\frac{y-a}{a}} = \frac{a-x}{y-a}$ (independent of t) Ans.

(e) $x = a\left[\sin t - \log\left(\tan\frac{t}{2}\right)\right],\ y = a\cos t$

$$\frac{dx}{dt} = a\left[\cos t - \frac{1}{\tan(^t/_2)}.\sec^2(^t/_2).\frac{1}{2}\right] = a\left[\cos t - \frac{\frac{1}{\cos^2(^t/_2)}}{2\frac{\sin(^t/_2)}{\cos(^t/_2)}}\right] = a\left[\cos t - \frac{1}{2\sin(^t/_2).\cos(^t/_2)}\right] = a\left[\cos t - \frac{1}{\sin t}\right] = a[\cos t - \operatorname{cosec} t]$$

and $y = a\cos t$ $\quad \therefore \frac{dy}{dt} = -a\sin t$

$$\frac{dy}{dx} = \frac{dy}{dt} \div \frac{dx}{dt} = \frac{dy}{dt} \times \frac{dt}{dx} = -a\sin t.\frac{1}{a(\cos t - \operatorname{cosec} t)} = -\frac{\sin t}{(\cos t - \operatorname{cosec} t)} \text{ or } \frac{\sin t}{\operatorname{cosec} t - \cos t} \quad \text{Ans.}$$

At $t = \frac{\pi}{4}$ then $\frac{dy}{dx} = \frac{\sin t}{\operatorname{cosec} t - \cos t} = \frac{\sin\frac{\pi}{4}}{\operatorname{cosec}\frac{\pi}{4} - \cos\frac{\pi}{4}} = \frac{\frac{1}{\sqrt{2}}}{\sqrt{2} - \frac{1}{\sqrt{2}}} = \frac{\frac{1}{\sqrt{2}}}{\frac{2-1}{\sqrt{2}}} = 1$ Ans.

(9) (a) $x = t^2 + \frac{1}{t^2}$ ……… (i) and $y = t^2 - \frac{1}{t^2}$ ………… (ii)

Adding equation (i) & (ii), we have $\Rightarrow x + y = t^2 + \frac{1}{t^2} + t^2 - \frac{1}{t^2} = 2t^2 \quad \therefore t^2 = \frac{x+y}{2}$ or $t = \sqrt{\frac{x+y}{2}}$

Now, $x = t^2 + \frac{1}{t^2} \quad \therefore \frac{dx}{dt} = 2t - \frac{2}{t^3} = \frac{2t^4 - 2}{t^3}$ and $y = t^2 + \frac{1}{t^2} \quad \therefore \frac{dy}{dt} = 2t + \frac{2}{t^3} = \frac{2t^4 + 2}{t^3}$

Then, $\frac{dy}{dx} = \frac{dy}{dt} \div \frac{dx}{dt} = \frac{dy}{dt} \times \frac{dt}{dx} = \frac{2t^4+2}{t^3} \times \frac{t^3}{2t^4-2} = \frac{2(t^4+1)}{2(t^4-1)} = \frac{t^4+1}{t^4-1} \quad \therefore \frac{d^2y}{dx^2} = 0$ Ans.

$\therefore \frac{dy}{dx}$ is independent of t, $\frac{dy}{dx} = \frac{t^4+1}{t^4-1}$ Put $t^2 = \frac{x+y}{2}$ or $t = \sqrt{\frac{x+y}{2}}$

Then, $\frac{dy}{dx} = \frac{(t^2)^2+1}{(t^2)^2-1} = \frac{\left(\frac{x+y}{2}\right)^2+1}{\left(\frac{x+y}{2}\right)^2-1} = \frac{\frac{(x+y)^2+4}{4}}{\frac{(x+y)^2-4}{4}} = \frac{(x+y)^2+4}{(x+y)^2-4}$ (independent of t) Ans.

(b) Do yourself. $x = \frac{1-t^2}{t^3}$ and $y = \frac{3}{2t^3} - \frac{2}{t^2} \quad \therefore \frac{dx}{dt} = \frac{t^2-3}{t^4}, \ \frac{dy}{dt} = \frac{8t-9}{2t^4}$

$\therefore \frac{dy}{dx} = \frac{dy}{dt} \div \frac{dx}{dt} = \frac{dy}{dt} \times \frac{dt}{dx} = \frac{8t-9}{2t^4} \times \frac{t^4}{t^2-3} = \frac{8t-9}{2(t^2-3)}$ at $t = 2 \quad \therefore \frac{dy}{dx} = \frac{16-9}{2(4-3)} = \frac{7}{2}$ Ans.

(c) $x = 2\sin t + \sin 2t$, $y = 2\cos t + \cos 2t$

$\therefore \frac{dx}{dt} = 2\cos t + 2\cos 2t = 2(\cos t + \cos 2t)$ and $\frac{dy}{dt} = -2\sin t - 2\sin 2t = -2(\sin t + \sin 2t)$

$\therefore \frac{dy}{dx} = \frac{dy}{dt} \div \frac{dx}{dt} = \frac{dy}{dt} \times \frac{dt}{dx} = \frac{-2(\sin t + \sin 2t)}{2(\cos t + \cos 2t)} = \frac{-(\sin t + \sin 2t)}{(\cos t + \cos 2t)}$

At $t = \frac{\pi}{2}$ then $\frac{dy}{dx} = \frac{-\left(\sin\left(\frac{\pi}{2}\right) + \sin\left(2.\frac{\pi}{2}\right)\right)}{\left(\cos\left(\frac{\pi}{2}\right) + \cos\left(2.\frac{\pi}{2}\right)\right)} \quad \therefore \left(\frac{dy}{dx}\right)_{t=\frac{\pi}{2}} = \frac{-(1+\sin\pi)}{0+\cos\pi} = \frac{-(1+0)}{0+(-1)} = \frac{-1}{-1} = 1$ Ans.

(d) $x = \sin^{-1}\frac{1}{\sqrt{1+t^2}}$, $y = \cos^{-1}\frac{t}{\sqrt{1+t^2}}$ Put $t = \cot\theta \quad \therefore \frac{dt}{d\theta} = -\operatorname{cosec}^2\theta$

$x = \sin^{-1}\frac{1}{\sqrt{1+(\cot\theta)^2}} = \sin^{-1}(\sin\theta) = \theta$ and $y = \cos^{-1}\frac{\cot\theta}{\sqrt{1+(\cot\theta)^2}} = \cos^{-1}(\cos\theta) = \theta$

$\frac{dx}{d\theta} = 1$ and $\frac{dy}{d\theta} = 1 \quad \therefore \frac{dy}{dx} = \frac{dy}{d\theta} \div \frac{dx}{d\theta} = \frac{dy}{d\theta} \times \frac{d\theta}{dx} = \frac{1}{1} = 1$ (independent of t) Proved.

(10) (a) $x = \frac{\sin^2 t}{\sqrt{\cos 2t}}$ and $y = \frac{\cos^2 t}{\sqrt{\sin 2t}}$ $\left[\text{formula,} \quad y = \frac{u}{v} \quad \therefore \frac{dy}{dx} = \frac{v.\frac{du}{dx} - u.\frac{dv}{dx}}{v^2}\right]$

$$\frac{dx}{dt} = \frac{\sqrt{\cos 2t}.\frac{d}{dt}(\sin^2 t) - \sin^2 t.\frac{d}{dt}\left(\sqrt{\cos 2t}\right)}{\left(\sqrt{\cos 2t}\right)^2} = \frac{\sqrt{\cos 2t}.2\sin t.\cos t - \sin^2 t.\frac{1}{2\sqrt{\cos 2t}}.(-\sin 2t).2}{\cos 2t} = \frac{\frac{\sin 2t.\cos 2t + \sin 2t.\sin^2 t}{\sqrt{\cos 2t}}}{\cos 2t}$$
$$= \frac{\sin 2t.(\cos 2t + \sin^2 t)}{\cos 2t.\sqrt{\cos 2t}}$$

$$\frac{dy}{dt}=\frac{\sqrt{\sin 2t}.\frac{d}{dt}(\cos^2 t)-\cos^2 t.\frac{d}{dt}\left(\sqrt{\sin 2t}\right)}{\left(\sqrt{\sin 2t}\right)^2}=\frac{\sqrt{\sin 2t}.2.\cos t.(-\sin t)-\cos^2 t.\frac{1}{2.\sqrt{\sin 2t}}.\cos 2t.2}{\sin 2t}$$

$$\frac{dy}{dt}=\frac{\frac{-\sin 2t.\sin 2t-\cos 2t.\cos^2 t}{\sqrt{\sin 2t}}}{\sin 2t}=\frac{-(\sin^2 2t+\cos 2t.\cos^2 t)}{\sin 2t.\sqrt{\sin 2t}}$$

$$\frac{dy}{dx}=\frac{dy}{dt}\div\frac{dx}{dt}=\frac{dy}{dt}\times\frac{dt}{dx}=\frac{-(\sin^2 2t+\cos 2t.\cos^2 t)}{\sin 2t.\sqrt{\sin 2t}}\times\frac{\cos 2t.\sqrt{\cos 2t}}{\sin 2t.(\cos 2t+\sin^2 t)}$$

At $t=\frac{\pi}{4}$ then $\frac{dy}{dx}=\frac{-\left(\sin^2\left(2.\frac{\pi}{4}\right)+\cos\left(2.\frac{\pi}{4}\right).\cos^2\left(\frac{\pi}{4}\right)\right)}{\sin\left(2.\frac{\pi}{4}\right).\sqrt{\sin\left(2.\frac{\pi}{4}\right)}}\times\frac{\cos\left(2.\frac{\pi}{4}\right).\sqrt{\cos\left(2.\frac{\pi}{4}\right)}}{\sin\left(2.\frac{\pi}{4}\right).\left(\cos\left(2.\frac{\pi}{4}\right)+\sin^2\left(\frac{\pi}{4}\right)\right)}=\frac{-(1+0).0}{1\left(0+\frac{1}{2}\right)}$

$$\frac{dy}{dx}=\frac{2.0}{1}=0 \quad \text{Ans.}$$

(b) Do yourself. $x=\tan\theta.\sqrt{\cos 2\theta}$ and $y=\cot\theta.\sqrt{\sin 2\theta}$ $\left[\text{formula, } y=u.v \quad \therefore \frac{dy}{dx}=u.\frac{dv}{dx}+v.\frac{du}{dx}\right]$

see above question and solve it, $\frac{dx}{d\theta}=\frac{\cos 2\theta.\sec^2\theta-\sin 2\theta.\tan\theta}{\sqrt{\cos 2\theta}}$ and $\frac{dy}{d\theta}=\frac{\cos 2\theta.\cot\theta-\sin 2\theta.\mathrm{cosec}^2\theta}{\sqrt{\sin 2\theta}}$

$$\frac{dy}{dx}=\frac{dy}{d\theta}\div\frac{dx}{d\theta}=\frac{dy}{d\theta}\times\frac{d\theta}{dx}=\frac{\cos 2\theta.\cot\theta-\sin 2\theta.\mathrm{cosec}^2\theta}{\sqrt{\sin 2\theta}}\times\frac{\sqrt{\cos 2\theta}}{\cos 2\theta.\sec^2\theta-\sin 2\theta.\tan\theta}$$

At $\theta=\frac{\pi}{6}$ then $\frac{dy}{dx}=\frac{\sqrt{\cos\left(\frac{\pi}{3}\right)}}{\sqrt{\sin\left(\frac{\pi}{3}\right)}}\left[\frac{\cos\left(\frac{\pi}{3}\right).\cot\left(\frac{\pi}{6}\right)-\sin\left(\frac{\pi}{3}\right).\mathrm{cosec}^2\left(\frac{\pi}{6}\right)}{\cos\left(\frac{\pi}{3}\right).\sec^2\left(\frac{\pi}{6}\right)-\sin\left(\frac{\pi}{3}\right).\tan\left(\frac{\pi}{6}\right)}\right]=-\frac{9.\sqrt{3}}{3^{\frac{1}{4}}}=-\left(3^2.3^{\frac{1}{2}}.3^{-\frac{1}{4}}\right)=-(3)^{\frac{9}{4}}$ Ans.

(c) $x^3-2xy+y^3=a^3 \quad \Rightarrow \quad 3x^2-2\left(x.\frac{dy}{dx}+y.1\right)+3y^2.\frac{dy}{dx}=0$

$\Rightarrow 3y^2.\frac{dy}{dx}-2x.\frac{dy}{dx}=2y-3x^2$ or $\frac{dy}{dx}(3y^2-2x)=(2y-3x^2) \quad \therefore \frac{dy}{dx}=\frac{2y-3x^2}{3y^2-2x}$ Ans.

$$\frac{d^2y}{dx^2}=\frac{(3y^2-2x).\frac{d}{dx}(2y-3x^2)-(2y-3x^2).\frac{d}{dx}(3y^2-2x)}{(3y^2-2x)^2}=\frac{(3y^2-2x).\left(2.\frac{dy}{dx}-6x\right)-(2y-3x^2).\left(6y.\frac{dy}{dx}-2\right)}{(3y^2-2x)^2}$$

$\therefore$ Put $\frac{dy}{dx}=\frac{2y-3x^2}{3y^2-2x}$ in $\frac{d^2y}{dx^2}=\frac{(3y^2-2x).\left(2.\frac{2y-3x^2}{3y^2-2x}-6x\right)-(2y-3x^2).\left(6y.\frac{2y-3x^2}{3y^2-2x}-2\right)}{(3y^2-2x)^2}$

$$\therefore \frac{d^2y}{dx^2}=\frac{(3y^2-2x)(6x^2-18xy^2+4y)-(2y-3x^2)(6y^2-18x^2y+4x)}{(3y^2-2x)^3} \quad \text{Ans.}$$

(d) $x=\cos^{-1}\left(2t\sqrt{1-t^2}\right)$ and $y=\frac{\pi}{2}+\sin^{-1}t$ Put $t=\sin\theta$

Now, $x=\cos^{-1}\left(2\sin\theta.\sqrt{1-\sin^2\theta}\right)$ and $y=\frac{\pi}{2}+\sin^{-1}(\sin\theta)=\frac{\pi}{2}+\theta$ ………………….(i)

$x=\cos^{-1}(2\sin\theta.\cos\theta)=\cos^{-1}(\sin 2\theta)=\cos^{-1}\left(\cos\left(\frac{\pi}{2}-2\theta\right)\right)=\frac{\pi}{2}-2\theta$ ………………...(ii)

solve the equation (i) & (ii), we have

$x=\frac{\pi}{2}-2\theta$ and $y=\frac{\pi}{2}+\theta$ or $\theta=y-\frac{\pi}{2}$ $\therefore x=\frac{\pi}{2}-2\left(y-\frac{\pi}{2}\right)=\frac{\pi}{2}-2y+\pi$ or $x+2y=\frac{\pi}{2}+\pi$

or $x+2y=\frac{3\pi}{2}$ $\therefore 1+2\frac{dy}{dx}=0$ or $2\frac{dy}{dx}=-1$ $\therefore \frac{dy}{dx}=-\frac{1}{2}$ Ans. $\therefore \frac{d^2y}{dx^2}=0$ Ans.

(11) (a) $y = (\cos^{-1} x)^x$ $\therefore$ $\log y = x \log(\cos^{-1} x)$ $\therefore$ $\frac{1}{y}.\frac{dy}{dx} = x.\frac{1}{\cos^{-1} x}.\frac{1}{-\sqrt{1-x^2}} + \log(\cos^{-1} x).1$

$\therefore \frac{dy}{dx} = y\left[\log(\cos^{-1} x) - \frac{x}{\cos^{-1} x.\sqrt{1-x^2}}\right] = (\cos^{-1} x)^x.\log(\cos^{-1} x) - (\cos^{-1} x)^x.\frac{x}{\cos^{-1} x.\sqrt{1-x^2}}$ Ans.

(b) $y = x^{x^2}$ $\Rightarrow$ $\log y = \log x^{x^2} = x^2 \log x$ $\therefore$ $\frac{1}{y}.\frac{dy}{dx} = x^2.\frac{1}{x} + \log x.2x = x + 2x.\log x = x(1 + 2\log x)$

$\therefore \frac{dy}{dx} = y.x(1 + 2\log x) = x.x^{x^2}(1 + 2\log x) = x^{x^2+1}(1 + 2\log x)$ or $x^{x^2+1}(1 + \log x^2)$ Ans.

(c) $y = (\sin x)^{\log x}$ $\Rightarrow$ $\log y = \log x.\log(\sin x)$ $\therefore$ $\frac{1}{y}.\frac{dy}{dx} = \log x.\frac{1}{\sin x}.\cos x + \log(\sin x).\frac{1}{x}$

$\therefore \frac{dy}{dx} = y\left[\log x.\cot x + \frac{\log(\sin x)}{x}\right] = (\sin x)^{\log x}\left[\log x.\cot x + \frac{\log(\sin x)}{x}\right]$ Ans.

(d) Do yourself (see above question), $y = (\cos x)^{\log x}$ $\therefore$ $\frac{dy}{dx} = (\cos x)^{\log x}\left[\frac{\log(\cos x)}{x} - \log x.\tan x\right]$ Ans.

(e) Do yourself (see above question), $y = (\tan x)^{\log x}$ $\therefore$ $\frac{dy}{dx} = (\tan x)^{\log x}\left[\frac{2x.\log x + \sin 2x.\log(\tan x)}{x.\sin 2x}\right]$ Ans.

(f) $y = (\log x)^{\log(\tan x)}$ $\Rightarrow$ $\log y = \log(\tan x).\log(\log x)$ $\therefore$ $\frac{1}{y}.\frac{dy}{dx} = \log(\tan x).\frac{1}{\log x}.\frac{1}{x} + \log(\log x).\frac{1}{\tan x}.\sec^2 x$

or $\frac{dy}{dx} = y\left[\log(\tan x).\frac{1}{x.\log x} + \log(\log x).\frac{1}{\sin x.\cos x}\right] = (\log x)^{\log(\tan x)}\left[\frac{\log(\tan x)}{x.\log x} + \frac{2\log(\log x)}{2\sin x.\cos x}\right]$

or $\frac{dy}{dx} = (\log x)^{\log(\tan x)}\left[\frac{\log(\tan x)}{x.\log x} + \frac{2\log(\log x)}{\sin 2x}\right] = (\log x)^{\log(\tan x)}\left[\frac{\sin 2x.\log(\tan x) + 2x.\log x.\log(\log x)}{x.\sin 2x.\log x}\right]$ Ans.

(g) $y = (\sin x)^{\cos x} + (\tan x)^{\cot x} = u + v$ (say) where $u = (\sin x)^{\cos x}$ and $v = (\tan x)^{\cot x}$

$\Rightarrow$ $u = (\sin x)^{\cos x}$ $\therefore$ $\log u = \cos x.\log(\sin x)$ $\therefore$ $\frac{1}{u}.\frac{du}{dx} = \cos x.\frac{1}{\sin x}.\cos x + \log(\sin x).(-\sin x)$

$\therefore \frac{du}{dx} = u\left[\frac{\cos^2 x}{\sin x} - \sin x.\log(\sin x)\right] = (\sin x)^{\cos x}\left[\frac{\cos^2 x - \sin^2 x.\log(\sin x)}{\sin x}\right]$

$\Rightarrow$ $v = (\tan x)^{\cot x}$ $\therefore$ $\log v = \cot x.\log(\tan x)$ $\therefore$ $\frac{1}{v}.\frac{dv}{dx} = \cot x.\frac{1}{\tan x}.\sec^2 x + \log(\tan x).(-\operatorname{cosec}^2 x)$

$\therefore \frac{dv}{dx} = v\left[\frac{\cos^2 x}{\sin^2 x}.\frac{1}{\cos^2 x} - \frac{1}{\sin^2 x}.\log(\tan x)\right] = (\tan x)^{\cot x}\left[\frac{1 - \log(\tan x)}{\sin^2 x}\right]$

$\therefore \frac{dy}{dx} = \frac{du}{dx} + \frac{dv}{dx} = (\sin x)^{\cos x}\left[\frac{\cos^2 x - \sin^2 x.\log(\sin x)}{\sin x}\right] + (\tan x)^{\cot x}\left[\frac{1 - \log(\tan x)}{\sin^2 x}\right]$ Ans.

(h) $y = (\cot x)^{\cos x}$ $\therefore$ $\log y = \cos x.\log(\cot x)$ $\therefore$ $\frac{1}{y}.\frac{dy}{dx} = \cos x.\frac{1}{\cot x}.(-\operatorname{cosec}^2 x) + \log(\cot x).(-\sin x)$

$\therefore \frac{dy}{dx} = -y\left[\frac{1}{\sin x} + \sin x.\log(\cot x)\right] = -(\cot x)^{\cos x}\left[\frac{1 + \sin^2 x.\log(\cot x)}{\sin x}\right]$ Ans. or $\frac{dy}{dx} = -(\cot x)^{\cos x}[\operatorname{cosec} x + \sin x.\log(\cot x)]$ Ans.

(i) $y = (\tan x)^{\sin x}$ $\therefore$ $\log y = \sin x.\log(\tan x)$ $\therefore$ $\frac{1}{y}.\frac{dy}{dx} = \sin x.\frac{1}{\tan x}.\sec^2 x + \log(\tan x).\cos x$

$\therefore \frac{dy}{dx} = y\left[\frac{1}{\cos x} + \cos x.\log(\tan x)\right] = (\tan x)^{\sin x}\left[\frac{1 + \cos^2 x.\log(\tan x)}{\cos x}\right]$ or $(\tan x)^{\sin x}[\sec x + \cos x.\log(\tan x)]$ Ans.

(j) $y = e^{x\cos x^2} + (\cot x)^x = u + v$ (say) where $u = e^{x\cos x^2}$ and $v = (\cot x)^x$

$\Rightarrow\ u = e^{x\cos x^2} \quad \therefore\ \dfrac{du}{dx} = e^{x\cos x^2}[\cos x^2 . 1 - x.\sin x^2 . 2x] = e^{x\cos x^2}[\cos x^2 - 2x^2.\sin x^2]$

$\Rightarrow\ v = (\cot x)^x \quad \therefore\ \log v = x\log(\cot x) \quad \therefore\ \dfrac{1}{v}.\dfrac{dv}{dx} = x.\dfrac{1}{\cot x}.(-\operatorname{cosec}^2 x) + \log(\cot x).1$

$\therefore\ \dfrac{dv}{dx} = v\left[\log(\cot x) - \dfrac{x}{\sin x\cos x}\right] = (\cot x)^x\left[\dfrac{\sin 2x.\log(\cot x) - 2x}{\sin 2x}\right]$

$\therefore\ \dfrac{dy}{dx} = \dfrac{du}{dx} + \dfrac{dv}{dx} = e^{x\cos x^2}[\cos x^2 - 2x^2.\sin x^2] + (\cot x)^x\left[\dfrac{\sin 2x.\log(\cot x) - 2x}{\sin 2x}\right]$ Ans.

(12) (a) $y = e^{x+x^2+x^3+\cdots\cdots\infty} = e^{\left(\frac{x}{1-x}\right)}$ $\left[\begin{array}{c}\text{formula, } x + x^2 + x^3 + \cdots\ldots\ldots\ldots\infty \text{ are in G.P, common ratio } (r) = \dfrac{x^2}{x} = \dfrac{x^3}{x^2} = \cdots\ldots = x \\ \text{and first term } (a) = x \quad \text{then } s_n = \dfrac{x}{1-x}\end{array}\right]$

$\therefore y = e^{\left(\frac{x}{1-x}\right)}$ or $\log y = \log e^{\left(\frac{x}{1-x}\right)} = \dfrac{x}{1-x} \quad \therefore\ \dfrac{1}{y}.\dfrac{dy}{dx} = \dfrac{(1-x) - x(-1)}{(1-x)^2} = \dfrac{1-x+x}{(1-x)^2} = \dfrac{1}{(1-x)^2}$

$\therefore\ \dfrac{dy}{dx} = y\left[\dfrac{1}{(1-x)^2}\right] = e^{\left(\frac{x}{1-x}\right)}\left[\dfrac{1}{(1-x)^2}\right] = \dfrac{e^{\left(\frac{x}{1-x}\right)}}{(1-x)^2}$ Ans.

(b) $y = e^{(\sin x + \cos x)} \quad \therefore\ \log y = \log e^{(\sin x+\cos x)} = \sin x + \cos x \quad \therefore\ \dfrac{1}{y}.\dfrac{dy}{dx} = \cos x - \sin x$

$\therefore\ \dfrac{dy}{dx} = y(\cos x - \sin x) = e^{(\sin x+\cos x)}(\cos x - \sin x)$ Ans.

(c) $y = \sqrt{\sin x\sqrt{\sin x\sqrt{\sin\ldots\ldots\ldots\ldots\infty}}}$, Let $y = \sqrt{\sin x\sqrt{\sin\ldots\ldots\ldots\ldots\infty}}$ then $y = \sqrt{\sin xy}$

squaring both of side, we have

$\Rightarrow\ y^2 = \sin xy \quad \therefore\ 2y.\dfrac{dy}{dx} = \cos xy.\left(x.\dfrac{dy}{dx} + y.1\right) \quad \therefore\ \dfrac{dy}{dx}(2y - x\cos xy) = y\cos xy \quad \therefore\ \dfrac{dy}{dx} = \dfrac{y.\cos xy}{(2y - x\cos xy)}$ Ans.

(d) $y = \sqrt{\sin x + \sqrt{\sin x + \sqrt{\sin x + \sqrt{\ldots\ldots\ldots\infty}}}}$, Let $y = \sqrt{\sin x + \sqrt{\sin x + \sqrt{\ldots\ldots\ldots\infty}}}$

then $y = \sqrt{\sin x + y}$ squaring both of side, we have $\therefore\ y^2 = \sin x + y$

$\therefore\ 2y.\dfrac{dy}{dx} = \cos x + \dfrac{dy}{dx} \quad \therefore\ \dfrac{dy}{dx}(2y-1) = \cos x \quad \therefore\ \dfrac{dy}{dx} = \dfrac{\cos x}{2y-1}$ Ans.

(e) $y = e^{\sin x+\sin^2 x+\sin^3 x+\cdots\cdots\infty}$, Let $s_n = \sin x + \sin^2 x + \sin^3 x + \cdots\ldots\ldots\ldots\infty$ are in G.P, $a = \sin x$, $r = \sin x$

then $s_n = \dfrac{a}{1-r} = \dfrac{\sin x}{1-\sin x} \quad \therefore\ y = e^{\left(\frac{\sin x}{1-\sin x}\right)} \quad \therefore\ \log y = \dfrac{\sin x}{1-\sin x}.\log e = \dfrac{\sin x}{1-\sin x}$

$\therefore\ \dfrac{1}{y}.\dfrac{dy}{dx} = \dfrac{(1-\sin x).\cos x - \sin x.(-\cos x)}{(1-\sin x)^2} = \dfrac{\cos x - \sin x\cos x + \sin x\cos x}{(1-\sin x)^2} = \dfrac{\cos x}{(1-\sin x)^2}$

$\therefore\ \dfrac{dy}{dx} = y\left[\dfrac{\cos x}{(1-\sin x)^2}\right] = e^{\left(\frac{\sin x}{1-\sin x}\right)}\left[\dfrac{\cos x}{(1-\sin x)^2}\right]$ Ans.

(f) $y = (\sin x)^{(\sin x)^{\sin x\cdots\cdots\infty}}$, Let $y = (\sin x)^{\sin x\cdots\cdots\infty}$ then $y = (\sin x)^y \quad \therefore\ \log y = y.\log(\sin x)$

$\therefore\ \dfrac{1}{y}.\dfrac{dy}{dx} = y.\dfrac{1}{\sin x}.\cos x + \log(\sin x).\dfrac{dy}{dx} \quad \therefore\ \dfrac{dy}{dx}\left(\dfrac{1}{y} - \log(\sin x)\right) = y\cot x \quad \therefore\ \dfrac{dy}{dx} = \dfrac{y\cot x}{\dfrac{1}{y} - \log(\sin x)}$

$$\therefore \frac{dy}{dx} = \frac{y^2 \cot x}{1 - y\log(\sin x)} = \frac{\left[(\sin x)^{(\sin x)^{\sin x \cdots\cdots\infty}}\right]^2 . \cot x}{1 - \left[(\sin x)^{(\sin x)^{\sin x \cdots\cdots\infty}}\right] . \log(\sin x)} \quad \text{Ans.}$$

(g) $y = (\log x)^{(\log x)^{\log x \cdots\cdots\infty}}$, Let $y = (\log x)^{\log x \cdots\cdots\infty}$ then $y = (\log x)^y$ $\therefore \log y = y\log(\log x)$

$\therefore \frac{1}{y}.\frac{dy}{dx} = y.\frac{1}{\log x}.\frac{1}{x} + \log(\log x).\frac{dy}{dx}$ $\therefore \frac{dy}{dx}\left(\frac{1}{y} - \log(\log x)\right) = \frac{y}{x\log x}$ $\therefore \frac{dy}{dx} = \frac{\frac{y}{x\log x}}{\frac{1}{y} - \log(\log x)}$

$$\therefore \frac{dy}{dx} = \frac{y^2}{x\log x . [1 - y\log(\log x)]} \quad \text{Ans.}$$

(h) $y = e^{(\sqrt{x})^{e^{(\sqrt{x})^{\cdots\cdots\infty}}}}$, Let $y = e^{(\sqrt{x})^{\cdots\cdots\infty}}$ then $y = e^{(\sqrt{x})^y}$ $\therefore \log y = (\sqrt{x})^y . \log_e e = (\sqrt{x})^y$

$\therefore \log(\log y) = y\log(\sqrt{x})$ $\therefore \frac{1}{\log y}.\frac{1}{y}.\frac{dy}{dx} = y.\frac{1}{\sqrt{x}}.\frac{1}{2\sqrt{x}} + \log(\sqrt{x}).\frac{dy}{dx}$

$\therefore \frac{dy}{dx}\left(\frac{1}{y\log y} - \log(\sqrt{x})\right) = \frac{y}{2x}$ $\therefore \frac{dy}{dx} = \frac{\frac{y}{2x}}{\frac{1 - y\log y . \log(\sqrt{x})}{y\log y}} = \frac{y^2\log y}{2x[1 - y\log y . \log(\sqrt{x})]}$ Ans.

(i) $y = \log(\log(\log(\cos x)))$, Let $u = \log(\cos x)$ $\therefore \frac{du}{dx} = \frac{1}{\cos x}.(-\sin x) = -\tan x$

$\Rightarrow y = \log(\log u)$, Let $v = \log u$ $\therefore \frac{dv}{du} = \frac{1}{u} = \frac{1}{\log(\cos x)}$ [Put $u = \log(\cos x)$]

$\Rightarrow y = \log v$ $\therefore \frac{dy}{dv} = \frac{1}{v} = \frac{1}{\log u} = \frac{1}{\log(\log(\cos x))}$ [Put $v = \log u$ and $u = \log(\cos x)$]

$\therefore \frac{dy}{dx} = \frac{dy}{dv} \times \frac{dv}{du} \times \frac{du}{dx} = \frac{1}{\log(\log(\cos x))} \times \frac{1}{\log(\cos x)} \times (-\tan x) = -\frac{\tan x}{\log(\cos x) . \log(\log(\cos x))}$ Ans.

(j) Do yourself (see question no. 11. (g)) .

$y = x^x + x^{\sin x} = u + v$ (say) where $u = x^x$ and $v = x^{\sin x}$

$\therefore \frac{du}{dx} = x^x(1 + \log x)$ and $\frac{dv}{dx} = x^{\sin x}\left(\log x . \cos x + \frac{\sin x}{x}\right)$ $\therefore \frac{dy}{dx} = \frac{du}{dx} + \frac{dv}{dx} = x^x(1 + \log x) + x^{\sin x}\left(\log x . \cos x + \frac{\sin x}{x}\right)$ Ans.

(k) Do yourself (see above question). $y = a^x + a^{\sin x} = u + v$ (say) where $u = a^x$ and $v = a^{\sin x}$

$\therefore \frac{du}{dx} = a^x \log a$ and $\frac{dv}{dx} = a^{\sin x} . \log a . \cos x$ $\therefore \frac{dy}{dx} = \frac{du}{dx} + \frac{dv}{dx} = a^x \log a + a^{\sin x} . \log a . \cos x$ $\therefore \frac{dy}{dx} = \log a\left(a^x + a^{\sin x} . \cos x\right)$ Ans.

(l) $e^y = e^{x-y}$ or $e^y = \frac{e^x}{e^y}$ or $e^{2y} = e^x$ $\therefore e^{2y}.2.\frac{dy}{dx} = e^x$ $\therefore \frac{dy}{dx} = \frac{e^x}{2e^{2y}} = \frac{e^x}{2e^x}$ [Put $e^{2y} = e^x$] $\therefore \frac{dy}{dx} = \frac{1}{2}$ Ans.

(m) $x^m y^n = (x - y)^{m-n}$ $\therefore x^m.\frac{d(y^n)}{dx} + y^n.\frac{d(x^m)}{dx} = (m - n).(x - y)^{m-n-1}\left(1 - \frac{dy}{dx}\right)$

$\therefore x^m . ny^{n-1}.\frac{dy}{dx} + y^n . mx^{m-1} = (m - n)(x - y)^{m-n-1} - (m - n).(x - y)^{m-n-1}.\frac{dy}{dx}$

$\therefore \frac{dy}{dx}(x^m . ny^{n-1} + (m - n).(x - y)^{m-n-1}) = (m - n)(x - y)^{m-n-1} - y^n . mx^{m-1}$

$$\therefore \frac{dy}{dx} = \frac{(m - n)(x - y)^{m-n-1} - y^n . mx^{m-1}}{x^m . ny^{n-1} + (m - n).(x - y)^{m-n-1}} \text{ or } \frac{y(my - nx)}{x(nx + my - 2ny)} \quad \text{Ans.}$$

IInd Method: – $x^m y^n = (x - y)^{m-n}$ $\therefore \log(x^m y^n) = \log(x - y)^{m-n}$

$\therefore \log x^m + \log y^n = (m - n)\log(x - y)$ Differentiate with respect to x, we have

$\therefore\ \frac{m}{x}+\frac{n}{y}.\frac{dy}{dx}=\frac{m-n}{x-y}.\left(1-\frac{dy}{dx}\right)\quad \therefore\ \frac{dy}{dx}\left(\frac{n}{y}+\frac{m-n}{x-y}\right)=\frac{m-n}{x-y}-\frac{m}{x}\quad \therefore\ \frac{dy}{dx}=\frac{\frac{m-n}{x-y}-\frac{m}{x}}{\frac{n}{y}+\frac{m-n}{x-y}}$

$$\therefore\ \frac{dy}{dx}=\frac{\frac{mx-nx-mx+my}{(x-y)x}}{\frac{nx-ny+my-ny}{y(x-y)}}=\frac{my-nx}{nx+my-2ny}\times\frac{y}{x}=\frac{y}{x}\left[\frac{my-nx}{nx+my-2ny}\right]\quad \text{Ans.}$$

(n) $x^y+y^x=(x-y)^{x-y}\quad \therefore\ yx^{y-1}+xy^{x-1}.\frac{dy}{dx}=(x-y)(x-y)^{x-y-1}\left(1-\frac{dy}{dx}\right)$

$\therefore\ xy^{x-1}.\frac{dy}{dx}+(x-y)(x-y)^{x-y-1}.\frac{dy}{dx}=(x-y)(x-y)^{x-y-1}-yx^{y-1}$

$\therefore\ \frac{dy}{dx}(xy^{x-1}+(x-y)(x-y)^{x-y-1})=(x-y)(x-y)^{x-y-1}-yx^{y-1}\quad \therefore\ \frac{dy}{dx}=\frac{(x-y)(x-y)^{x-y-1}-yx^{y-1}}{[xy^{x-1}+(x-y)(x-y)^{x-y-1}]}\quad$ Ans.

(13) (a) $y=x^2\log\left(\frac{a+x}{a-x}\right)=u.v$ (say), where $u=x^2$ and $v=\log\left(\frac{a+x}{a-x}\right)$

or $\frac{du}{dx}=2x$ and $\frac{dv}{dx}=\frac{1}{\frac{a+x}{a-x}}.\frac{(a-x).1-(a+x)(-1)}{(a-x)^2}=\frac{(a-x)2a}{(a+x)(a-x)^2}=\frac{2a}{a^2-x^2}$

$$\therefore\ \frac{dy}{dx}=u.\frac{dv}{dx}+v.\frac{du}{dx}=x^2.\frac{2a}{a^2-x^2}+\log\left(\frac{a+x}{a-x}\right).2x=2x\left[\frac{ax}{a^2-x^2}+\log\left(\frac{a+x}{a-x}\right)\right]\quad \text{Proved.}$$

(b) $y=\left(x-\sqrt{x^2-a^2}\right)^n$, Put $x=a\sec\theta\quad \therefore\ \frac{dx}{d\theta}=a\tan\theta\sec\theta$

$\Rightarrow\ y=\left(a\sec\theta-\sqrt{(a\sec\theta)^2-a^2}\right)^n=\left(a\sec\theta-\sqrt{a^2(\sec^2\theta-1)}\right)^n=\left(a\sec\theta-\sqrt{a^2.\tan^2\theta}\right)^n=(a\sec\theta-a\tan\theta)^n=[a(\sec\theta-\tan\theta)]^n$
$=a^n.(\sec\theta-\tan\theta)^n$

$\therefore\ \frac{dy}{d\theta}=n.[a(\sec\theta-\tan\theta)]^{n-1}.a(\tan\theta\sec\theta-\sec^2\theta)=na.a^{n-1}(\sec\theta-\tan\theta)^{n-1}.\sec\theta\,(\tan\theta-\sec\theta)$

$\therefore\ \frac{dy}{dx}=\frac{dy}{d\theta}\times\frac{d\theta}{dx}=na^n.\sec\theta.(\sec\theta-\tan\theta)^{n-1}.(\tan\theta-\sec\theta)\times\frac{1}{a\tan\theta\sec\theta}=\frac{na^{n-1}.(\sec\theta-\tan\theta)^{n-1}.(\tan\theta-\sec\theta)}{\tan\theta}$
$=\frac{-na^{n-1}.(\sec\theta-\tan\theta)^{n-1}.(\sec\theta-\tan\theta)}{\tan\theta}$

$\therefore\ \frac{dy}{dx}=-\frac{na^{n-1}.(\sec\theta-\tan\theta)^n.(\sec\theta-\tan\theta)}{\tan\theta.(\sec\theta-\tan\theta)}=-\frac{na^{n-1}.(\sec\theta-\tan\theta)^n}{\tan\theta}\quad$ Ans.

(c) $y=\frac{1-\sqrt{x-1}}{1+\sqrt{x-1}}+\frac{\sqrt{x-1}}{\sqrt{x}}$, Put $x=\sec^2\theta\quad \therefore\ \frac{dx}{d\theta}=2\sec\theta.\tan\theta.\sec\theta=2\sec^2\theta.\tan\theta$

$\therefore\ y=\frac{1-\sqrt{\sec^2\theta-1}}{1+\sqrt{\sec^2\theta-1}}+\frac{\sqrt{\sec^2\theta-1}}{\sqrt{\sec^2\theta}}=\frac{1-\tan\theta}{1+\tan\theta}+\frac{\tan\theta}{\sec\theta}=\frac{1-\frac{\sin\theta}{\cos\theta}}{1+\frac{\sin\theta}{\cos\theta}}+\frac{\frac{\sin\theta}{\cos\theta}}{\frac{1}{\cos\theta}}=\frac{\cos\theta-\sin\theta}{\cos\theta+\sin\theta}+\sin\theta$

$\therefore\ \frac{dy}{d\theta}=\frac{(\cos\theta+\sin\theta)(-\sin\theta-\cos\theta)-(\cos\theta-\sin\theta)(-\sin\theta+\cos\theta)}{(\cos\theta+\sin\theta)^2}+\cos\theta=\frac{-(\cos\theta+\sin\theta)^2-(\cos\theta-\sin\theta)^2}{(\cos\theta+\sin\theta)^2}+\cos\theta$

$\therefore\ \frac{dy}{d\theta}=\frac{-(\cos^2\theta+\sin^2\theta+2\sin\theta\cos\theta)-(\cos^2\theta+\sin^2\theta+2\sin\theta\cos\theta)}{(\cos\theta+\sin\theta)^2}+\cos\theta$
$=\frac{-\cos^2\theta-\sin^2\theta-2\sin\theta\cos\theta-\cos^2\theta-\sin^2\theta+2\sin\theta\cos\theta}{(\cos\theta+\sin\theta)^2}+\cos\theta$

$\therefore\ \frac{dy}{d\theta}=\frac{-2(\sin^2\theta+\cos^2\theta)}{(\cos\theta+\sin\theta)^2}+\cos\theta=\frac{-2}{(\cos\theta+\sin\theta)^2}+\cos\theta=\frac{-2}{(\sin^2\theta+\cos^2\theta+2\cos\theta\sin\theta)}+\cos\theta=\frac{-2}{1+\sin2\theta}+\cos\theta$
$=\frac{-2+\cos\theta+\sin2\theta\cos\theta}{1+\sin2\theta}$

$$\therefore\ \frac{dy}{dx}=\frac{dy}{d\theta}\times\frac{d\theta}{dx}=\frac{-2+\cos\theta+\sin 2\theta\cos\theta}{1+\sin 2\theta}\times\frac{1}{2\sec^2\theta.\tan\theta}=\frac{\cos\theta+\sin 2\theta\cos\theta-2}{1+\sin 2\theta}\times\frac{1}{2\frac{1}{\cos^2\theta}\cdot\frac{\sin\theta}{\cos\theta}}=\frac{(\cos\theta+\sin 2\theta\cos\theta-2)}{1+\sin 2\theta}\times\frac{\cos^3\theta}{2\sin\theta}$$

$$=\frac{\cos^4\theta\,(1+\sin 2\theta)-2\cos^3\theta}{2\sin\theta\,(1+\sin 2\theta)}=\frac{\cos^3\theta\,[\cos\theta+\sin 2\theta.\cos\theta-2]}{2\sin\theta\,(1+\sin 2\theta)}=\frac{\cos^3\theta}{\sin\theta}\left[\frac{\cos\theta}{2}-\frac{1}{1+\sin 2\theta}\right]\quad \text{Ans.}$$

(d) $\cos y = a\cos(x+y)$ $\therefore \cos y = a(\cos x\cos y - \sin x\sin y)$ Divide both of sides by $\cos y$, we have

$$\therefore\ 1=a\cos x-a\sin x.\tan y\quad \therefore\ -a\sin x-a\left[\sin x.\sec^2 y.\frac{dy}{dx}+\tan y.\cos x\right]=0$$

$$\text{or } a\sin x.\sec^2 y.\frac{dy}{dx}=-a\tan y.\cos x-a\sin x\quad \therefore\ \frac{dy}{dx}=\frac{-a(\tan y.\cos x+\sin x)}{a\sin x.\sec^2 y}=\frac{-\left(\frac{\sin y}{\cos y}.\cos x+\sin x\right)}{\sin x.\frac{1}{\cos^2 y}}$$

$$\therefore\ \frac{dy}{dx}=\frac{-(\sin y\cos x+\sin x\cos y)}{\sin x}\times\cos y=\frac{-\cos y\,(\sin x\cos y+\cos x\sin y)}{\sin x}=-\frac{\cos y.\sin(x+y)}{\sin x}\quad \text{Ans.}$$

$$\text{or } \frac{dy}{dx}=-\cos^2 y-\cot x.\sin y\cos y\quad \text{or}\quad \frac{dy}{dx}=-\frac{a\cos(x+y).\sin(x+y)}{\sin x}=-\frac{a\sin 2(x+y)}{2\sin x}\quad \text{Ans.}$$

(e) $y=1+\frac{Ax^2}{x-A}+\frac{Bx}{x-B}+\frac{C}{x-C}$, $\frac{dy}{dx}=0+\frac{(x-A).2Ax-Ax^2}{(x-A)^2}+\frac{(x-B).B-Bx}{(x-B)^2}+\frac{(x-C).0-C}{(x-C)^2}$

$$\therefore\ \frac{dy}{dx}=\frac{2Ax^2-2xA^2-Ax^2}{(x-A)^2}+\frac{Bx-B^2-Bx}{(x-B)^2}-\frac{C}{(x-C)^2}=\frac{Ax(x-2A)}{(x-A)^2}-\frac{B^2}{(x-B)^2}-\frac{C}{(x-C)^2}\quad \text{Ans.}$$

(f) Do yourself. $y=\tan^{-1}\left(\frac{\sqrt{a^2+b^2}\cos x}{b-a\cos x}\right)$ $\therefore \tan y=\frac{\sqrt{a^2+b^2}\cos x}{b-a\cos x}$ $\therefore \frac{dy}{dx}=\frac{-b.\sqrt{a^2+b^2}.\sin x}{(b-a\cos x)^2+(a^2+b^2).\cos^2 x}$ Ans.

(g) $y\sqrt{x^2+1}=\log\left(\sqrt{x^2+1}-x\right)$ or $y=\frac{\log(\sqrt{x^2+1}-x)}{\sqrt{x^2+1}}$ Put $x=\tan\theta$ $\therefore \frac{dx}{d\theta}=\sec^2\theta=1+\tan^2\theta=1+x^2$

$$\Rightarrow\ y=\frac{\log\left(\sqrt{(\tan\theta)^2+1}-\tan\theta\right)}{\sqrt{(\tan\theta)^2+1}}=\frac{\log(\sqrt{\tan^2\theta+1}-\tan\theta)}{\sqrt{\tan^2\theta+1}}=\frac{\log(\sec\theta-\tan\theta)}{\sec\theta}$$

$$\therefore\ \frac{dy}{d\theta}=\frac{\sec\theta.\frac{1}{(\sec\theta-\tan\theta)}.(\sec\theta\tan\theta-\sec^2\theta)-\log(\sec\theta-\tan\theta).\sec\theta\tan\theta}{\sec^2\theta}=\frac{\frac{\sec^2\theta\,(\tan\theta-\sec\theta)}{(\sec\theta-\tan\theta)}-\log(\sec\theta-\tan\theta).\sec\theta\tan\theta}{\sec^2\theta}$$

$$=\frac{-\sec^2\theta-\log(\sec\theta-\tan\theta).\sec\theta\tan\theta}{\sec^2\theta}=\frac{-\sec\theta\,[\sec\theta+\log(\sec\theta-\tan\theta).\tan\theta]}{\sec^2\theta}$$

$$\therefore\ \frac{dy}{d\theta}=-\frac{[\sec\theta+\log(\sec\theta-\tan\theta).\tan\theta]}{\sec\theta}\quad \text{Put } x=\tan\theta,\ 1+\tan^2\theta=\sec^2\theta\quad \therefore\ \sec\theta=\sqrt{1+x^2}$$

$$\therefore\ \frac{dy}{dx}=\frac{dy}{d\theta}\times\frac{d\theta}{dx}=-\frac{[\sec\theta+\log(\sec\theta-\tan\theta).\tan\theta]}{\sec\theta}\times\frac{1}{\sec^2\theta}=-\frac{[\sec\theta+\log(\sec\theta-\tan\theta).\tan\theta]}{\sec\theta.\sec^2\theta}$$

$$\therefore\ \frac{dy}{dx}=-\frac{[\sqrt{1+x^2}+x.\log(\sqrt{1+x^2}-x)]}{(1+x^2).\sqrt{1+x^2}}\quad \text{Ans.}$$

(h) $y=2\tan^{-1}\left(\frac{x}{\sqrt{1-x^2}}\right)+\log\left(\frac{1+2\sqrt{x}+x}{1-2\sqrt{x}+x}\right)=u+v$ (say)

where $u=2\tan^{-1}\left(\frac{x}{\sqrt{1-x^2}}\right)$, Put $x=\sin\theta$ $\therefore \frac{dx}{d\theta}=\cos\theta$ and $v=\log\left(\frac{1+2\sqrt{x}+x}{1-2\sqrt{x}+x}\right)$

$$u=2\tan^{-1}\left(\frac{\sin\theta}{\sqrt{1-(\sin\theta)^2}}\right)=2\tan^{-1}\left(\frac{\sin\theta}{\sqrt{1-\sin^2\theta}}\right)=2\tan^{-1}\left(\frac{\sin\theta}{\sqrt{\cos^2\theta}}\right)=2\tan^{-1}\left(\frac{\sin\theta}{\cos\theta}\right)=2\tan^{-1}(\tan\theta)=2\theta$$

$$\therefore\ \frac{du}{d\theta}=2\quad \therefore\ \frac{du}{dx}=\frac{du}{d\theta}\times\frac{d\theta}{dx}=2.\frac{1}{\cos\theta}=\frac{2}{\sqrt{1-\sin^2\theta}}=\frac{2}{\sqrt{1-x^2}}\quad \left[\text{Put } x=\sin\theta,\ \cos\theta=\sqrt{1-\sin^2\theta}\right]$$

Now, $v=\log\left(\frac{1+2\sqrt{x}+x}{1-2\sqrt{x}+x}\right)=\log\left[\frac{(1+\sqrt{x})}{(1-\sqrt{x})}\right]^2=2\log\left(\frac{1+\sqrt{x}}{1-\sqrt{x}}\right)$

$$\therefore\ \frac{dv}{dx}=2.\frac{1}{\left(\frac{1+\sqrt{x}}{1-\sqrt{x}}\right)}.\frac{(1-\sqrt{x}).\frac{1}{2\sqrt{x}}-(1+\sqrt{x}).\left(-\frac{1}{2\sqrt{x}}\right)}{(1-\sqrt{x})^2}=\frac{\frac{2[1-\sqrt{x}+1+\sqrt{x}]}{2\sqrt{x}}}{(1+\sqrt{x})(1-\sqrt{x})}=\frac{2}{\sqrt{x}.(1-x)}$$

$$\therefore\ \frac{dy}{dx}=\frac{du}{dx}+\frac{dv}{dx}=\frac{2}{\sqrt{1-x^2}}+\frac{2}{\sqrt{x}.(1-x)}=2\left(\frac{1}{\sqrt{1-x^2}}+\frac{1}{\sqrt{x}.(1-x)}\right)\quad \text{Ans.}$$

(14) (a) $y=x+\cfrac{1}{x+\cfrac{1}{x+\cfrac{1}{x+\cdots\ldots\infty}}}$, Let $y=x+\cfrac{1}{x+\cfrac{1}{x+\cdots\ldots\infty}}$ then $y=x+\frac{1}{y}$ or $y^2-xy=1$

$$\therefore\ 2y.\frac{dy}{dx}-\left[x.\frac{dy}{dx}+y\right]=0 \text{ or } 2y.\frac{dy}{dx}-x.\frac{dy}{dx}=y \text{ or } \frac{dy}{dx}(2y-x)=y \text{ or } \frac{dy}{dx}=\frac{y}{2y-x}\quad \text{Ans.}$$

$$\therefore\ \frac{dy}{dx}=\frac{y}{2y-x}\times\frac{y}{y}=\frac{y^2}{2y^2-xy}=\frac{y^2}{2y^2-y^2+1}=\frac{y^2}{y^2+1}\quad \text{Ans.}\quad [\text{Put } y^2-xy=1,\ xy=y^2-1]$$

(b) $y=\sin x+\cfrac{1}{\sin x+\cfrac{1}{\sin x+\cdots\ldots\ldots\infty}}$, Let $y=\sin x+\cfrac{1}{\sin x+\cdots\ldots\ldots\infty}$ then $y=\sin x+\frac{1}{y}$ or $y^2-y\sin x=1$

$$\therefore\ 2y\frac{dy}{dx}-\left[y\cos x+\sin x\frac{dy}{dx}\right]=0 \text{ or } 2y\frac{dy}{dx}-\sin x\frac{dy}{dx}=y\cos x \text{ or } \frac{dy}{dx}(2y-\sin x)=y\cos x$$

$$\therefore\ \frac{dy}{dx}=\frac{y\cos x}{2y-\sin x}\times\frac{y}{y}=\frac{y^2\cos x}{2y^2-y\sin x}=\frac{y^2\cos x}{2y^2-(y^2-1)}=\frac{y^2\cos x}{2y^2-y^2+1}=\frac{y^2\cos x}{y^2+1}\quad \text{Ans.}$$

(c) $x^2+y^2+2xy+x+y+5=0$ or $2x+2y\frac{dy}{dx}+2\left(x\frac{dy}{dx}+y\right)+1+\frac{dy}{dx}=0$

or $\frac{dy}{dx}(2y+2x+1)=-(2y+2x+1)$ $\quad\therefore\ \frac{dy}{dx}=-\frac{(2y+2x+1)}{(2y+2x+1)}=-1$ $\quad\therefore\ \frac{d^2y}{dx^2}=0$ Ans.

(d) $y=(\log_{\sin x}\cos x)(\log_{\tan x}\cot x)+\cos^{-1}\frac{x}{1+x}=\frac{\log(\cos x)}{\log(\sin x)}.\frac{\log(\cot x)}{\log(\tan x)}+\cos^{-1}\frac{x}{1+x}$

$\Rightarrow\ y=\frac{\log(\cos x).\log(\cot x)}{\log(\sin x).\log(\tan x)}+\cos^{-1}\frac{x}{1+x}=u+v$ (say) where $u=\frac{\log(\cos x).\log(\cot x)}{\log(\sin x).\log(\tan x)}$ and $v=\cos^{-1}\frac{x}{1+x}$

$$\therefore\ \frac{d}{dx}[\log(\cos x).\log(\cot x)]=\log(\cos x).\frac{1}{\cot x}.(-\operatorname{cosec}^2 x)+\log(\cot x).\frac{1}{\cos x}.(-\sin x)=-\left[\frac{\log(\cos x)}{\sin x\cos x}+\tan x\log(\cot x)\right]$$

$$\therefore\ \frac{d}{dx}[\log(\sin x).\log(\tan x)]=\log(\sin x).\frac{1}{\tan x}.\sec^2 x+\log(\tan x).\frac{1}{\sin x}.\cos x=\left\{\frac{\log(\cos x)}{\sin x\cos x}+\cot x\log(\tan x)\right\}$$

$$\frac{du}{dx}=\frac{-[\log(\sin x).\log(\tan x)].\left[\frac{\log(\cos x)}{\sin x\cos x}+\tan x\log(\cot x)\right]-[\log(\cos x).\log(\cot x)]\left[\frac{\log(\cos x)}{\sin x\cos x}+\cot x\log(\tan x)\right]}{[\log(\sin x).\log(\tan x)]^2}$$

and $v=\cos^{-1}\frac{x}{1+x}$ $\quad\therefore\ \frac{dv}{dx}=\frac{1}{-\sqrt{1-\left(\frac{x}{1+x}\right)^2}}.\frac{(1+x).1-x.1}{(1+x)^2}=-\frac{(1+x)}{\sqrt{2x+1}.(1+x)^2}=-\frac{1}{(1+x).\sqrt{2x+1}}$

$$\therefore\ \frac{dy}{dx}=\frac{du}{dx}+\frac{dv}{dx}$$

$$\frac{dy}{dx}=\frac{-[\log(\sin x).\log(\tan x)].\left[\frac{\log(\cos x)}{\sin x\cos x}+\tan x\log(\cot x)\right]-[\log(\cos x).\log(\cot x)]\left[\frac{\log(\cos x)}{\sin x\cos x}+\cot x\log(\tan x)\right]}{[\log(\sin x).\log(\tan x)]^2}-\frac{1}{(1+x).\sqrt{2x+1}}$$

At $x = \frac{\pi}{4}$, $\quad \frac{dy}{dx} = 0 - \frac{1}{\left(1+\frac{\pi}{4}\right).\sqrt{2.\frac{\pi}{4}+1}} = -\frac{1}{\left(\frac{4+\pi}{4}\right).\sqrt{\frac{\pi+2}{2}}} = -\frac{4\sqrt{2}}{(4+\pi).\sqrt{\pi+2}}$ Ans.

(e) Differentiate $f(x) = \log_x \cos x^2 + \cos x^2$ with respect to $\sqrt{x+1}$

Let $y = f(x) = \log_x \cos x^2 + \cos x^2 = \frac{\log(\cos x^2)}{\log x} + \cos x^2$ and $z = g(x) = \sqrt{x+1}$

$$\frac{dy}{dx} = \frac{\log x.\frac{1}{\cos x^2}.(-\sin x^2).2x - \log(\cos x^2).\frac{1}{x}}{(\log x)^2} + (-\sin x^2).2x = \frac{-[2x^2.\tan x^2.\log x + \log(\cos x^2)]}{x(\log x)^2} - 2x\sin x^2$$
$$= \frac{-[2x^2.\tan x^2.\log x + \log(\cos x^2) + 2x^2\sin x^2.(\log x)^2]}{x(\log x)^2}$$

now $\frac{dz}{dx} = \frac{1}{2\sqrt{x+1}}$ $\quad \therefore \frac{dy}{dz} = \frac{dy}{dx}\times\frac{dx}{dz} = \frac{-[2x^2.\tan x^2.\log x + \log(\cos x^2) + 2x^2\sin x^2.(\log x)^2]}{x(\log x)^2} \times 2\sqrt{x+1}$ Ans.

(f) $y = (\cos^{-1}x + \sin^{-1}x)x = \frac{\pi}{2}x$ $\left[\text{formula, } \sin^{-1}x + \cos^{-1}x = \frac{\pi}{2}\right]$ $\therefore \frac{dy}{dx} = \frac{\pi}{2}$ Proved. $\quad \therefore \frac{d^2y}{dx^2} = 0$ Proved.

(g) $Ax^2 + By^2 = 1$ or $Ax^2 + By^2 - 1 = 0$ $\quad \therefore 2Ax + 2By\frac{dy}{dx} = 0$ or $2By\frac{dy}{dx} = -2Ax$

$\therefore \frac{dy}{dx} = -\frac{2Ax}{2By} = -\frac{Ax}{By}$ Again differentiate with respect to x, we have

$\therefore \frac{d^2y}{dx^2} = -\frac{By.A - Ax.B\frac{dy}{dx}}{(By)^2}$ or $\frac{d^2y}{dx^2} = \frac{AB\left(x\frac{dy}{dx} - y\right)}{B^2y^2} = \frac{Ax.\frac{dy}{dx} - Ay}{By^2}$ or $By^2\frac{d^2y}{dx^2} = Ax.\frac{dy}{dx} - Ay$

$\therefore By^2\frac{d^2y}{dx^2} - Ax.\frac{dy}{dx} + Ay = 0$ or $By^2y'' - Axy' + Ay = 0$ Proved.

or $(1 - Ax^2)y'' - Axy' + Ay = 0$ Proved. [Put $Ax^2 + By^2 = 1$ or $By^2 = 1 - Ax^2$]

(h), (i) and (j) Do yourself.

(15) (a) $y = \cos x.\cos 3x.\cos 5x = \frac{2\cos x.\cos 3x.\cos 5x}{2}$ or $2y = 2\cos x.\cos 3x.\cos 5x$

[use formula, $2\cos A\cos B = \cos(A+B) + \cos(A-B)$ and $\cos(-\theta) = \cos\theta$]

or $2y = (2\cos x.\cos 3x).\cos 5x = [\cos(x+3x) + \cos(x-3x)].\cos 5x = \cos 4x\cos 5x + \cos 2x\cos 5x$

$$\therefore 2\frac{dy}{dx} = \cos 4x.(-\sin 5x).5 + \cos 5x.(-\sin 4x).4 + \cos 2x.(-\sin 5x).5 + \cos 5x.(-\sin 2x).2$$
$$= -5\cos 4x\sin 5x - 4\cos 5x\sin 4x - 5\cos 2x\sin 5x - 2\cos 5x\sin 2x$$

$$\therefore \frac{dy}{dx} = -\frac{1}{2}[5\cos 4x\sin 5x + 5\cos 2x\sin 5x + 4\cos 5x\sin 4x + 2\cos 5x\sin 2x]$$
$$= -\frac{1}{2}[5\sin 5x(\cos 4x + \cos 2x) + 2\cos 5x(2\sin 4x + \sin 2x)] \quad \text{Ans.}$$

or $\frac{dy}{dx} = -\frac{5}{2}\sin 5x.2\cos\left(\frac{4x+2x}{2}\right).\cos\left(\frac{4x-2x}{2}\right) - 2\cos 5x\sin 4x - \cos 5x\sin 2x$

$= -5\sin 5x.\cos 3x.\cos x - 2\cos 5x.\sin 4x - \cos 5x.\sin 2x$ Ans.

Useful formula, $2\sin A\sin B = \cos(A-B) - \cos(A+B)$, $2\sin A\cos B = \sin(A+B) + \sin(A-B)$

$2\cos A\cos B = \cos(A+B) + \cos(A-B)$, $2\cos A\sin B = \sin(A+B) - \sin(A-B)$

$\sin C + \sin D = 2\sin\frac{C+D}{2}.\cos\frac{C-D}{2}$, $\sin C + \sin D = 2\cos\frac{C+D}{2}.\sin\frac{C-D}{2}$

$$\cos C + \cos D = 2\cos\frac{C+D}{2}.\cos\frac{C-D}{2}\ ,\quad \cos C + \cos D = 2\sin\frac{C+D}{2}.\sin\frac{D-C}{2}$$

$$\sin(-\theta) = -\sin\theta \text{ and } \cos(-\theta) = \cos\theta$$

(b) $y = 8\sin x.\sin 2x.\sin 4x = 4(2\sin x.\sin 2x).\sin 4x = 4\sin 4x[\cos(x-2x) - \cos(x+2x)]$ (above formula)

or $y = 4\sin 4x[\cos x - \cos 3x] = 4\sin 4x\cos x - 4\sin 4x\cos 3x = 2(2\sin 4x\cos x) - 2(2\sin 4x\cos 3x)$

or $y = 2[\sin(4x+x) + \sin(4x-x)] - 2[\sin(4x+3x) + \sin(4x-3x)] = 2(\sin 5x + \sin 3x) - 2(\sin 7x + \sin x)$

or $y = 2\sin 5x + 2\sin 3x - 2\sin 7x - 2\sin x$

$\therefore\ \frac{dy}{dx} = 2\cos 5x.5 + 2\cos 3x.3 - 2\cos 7x.7 - 2\cos x = 10\cos 5x + 6\cos 3x - 14\cos 7x - 2\cos x$ Ans.

(c) $y = \cos 5x.\sin 3x = \frac{2\cos 5x.\sin 3x}{2} = \frac{\sin(5x+3x) - \sin(5x-3x)}{2} = \frac{\sin 8x - \sin 2x}{2} = \frac{1}{2}(\sin 8x - \sin 2x)$

$\therefore\ \frac{dy}{dx} = \frac{1}{2}(\cos 8x.8 - \cos 2x.2) = \frac{1}{2}(8\cos 8x - 2\cos 2x) = \frac{1}{2}.2(4\cos 8x - \cos 2x) = 4\cos 8x - \cos 2x$ Ans.

(d) $y = \sin 2x + \cos 3x,\ \frac{dy}{dx} = \cos 2x.2 + (-\sin 3x.3) = 2\cos 2x - 3\cos 3x$ Ans.

(e) $y = 2\tan 2\theta + 3\cot 2\theta \quad \therefore\ \frac{dy}{d\theta} = 2\sec^2 2\theta.2 + 3(-\mathrm{cosec}^2 2\theta).2 = 4\sec^2 2\theta - 6\,\mathrm{cosec}^2 2\theta$ Ans.

or $\frac{dy}{d\theta} = \frac{4}{\cos^2 2\theta} - \frac{6}{\sin^2 2\theta} = \frac{2(2\sin^2 2\theta - 3\cos^2 2\theta)}{\sin^2 2\theta.\cos^2 2\theta}$ Ans.

(f) Do yourself. $y = \sec 2\theta.\mathrm{cosec}\,3\theta \ \therefore\ \frac{dy}{d\theta} = \sec 2\theta.\mathrm{cosec}\,3\theta\,(2\tan 2\theta - 3\cot 3\theta) = y(2\tan 2\theta - 3\cot 3\theta)$ Ans.

(g) $y = \cos 3\theta + \sec 2\theta \quad \therefore\ \frac{dy}{d\theta} = -\sin 3\theta.3 + 2\tan 2\theta\sec 2\theta = 2\tan 2\theta\sec 2\theta - 3\sin 3\theta$ Ans.

(h) $y = \sin 2\theta + \mathrm{cosec}\,3\theta \quad \therefore\ \frac{dy}{d\theta} = \cos 2\theta.2 + 3(-\cot 3\theta\,\mathrm{cosec}\,3\theta) = 2\cos 2\theta - 3\cot 3\theta\,\mathrm{cosec}\,3\theta$ Ans.

(i) $y = \log_2\log_2\log_2 x^2$, Let $u = \log_2 x^2 \quad \therefore\ \frac{du}{dx} = \frac{1}{x^2}.2x = \frac{2}{x}$

Now, $y = \log_2\log_2 u$ Let $v = \log_2 u \quad \therefore\ \frac{dv}{du} = \frac{1}{u} = \frac{1}{\log_2 x^2}$ [Put $u = \log_2 x^2$]

Now, $y = \log_2 v \quad \therefore\ \frac{dy}{dv} = \frac{1}{v} = \frac{1}{\log_2 u} = \frac{1}{\log_2\log_2 x^2}$ [Put $v = \log_2 u$ and $u = \log_2 x^2$]

Then, $\frac{dy}{dx} = \frac{dy}{dv}\times\frac{dv}{du}\times\frac{du}{dx} = \frac{1}{\log_2\log_2 x^2}\times\frac{1}{\log_2 x^2}\times\frac{2}{x} = \frac{2}{x.\log_2 x^2.\log_2\log_2 x^2}$ Ans.

(j) $y = 4\log_e\log_e\log_e\log_e x$, Let $u = \log_e x \quad \therefore\ \frac{du}{dx} = \frac{1}{x}$

Now, $y = 4\log_e\log_e\log_e u$ Let $v = \log_e u \quad \therefore\ \frac{dv}{du} = \frac{1}{u} = \frac{1}{\log_e x}$ [Put $u = \log_e x$]

Now, $y = 4\log_e\log_e v$ Let $w = \log_e v \quad \therefore\ \frac{dw}{dv} = \frac{1}{v} = \frac{1}{\log_e u} = \frac{1}{\log_e\log_e x}$ [Put $v = \log_e u$ and $u = \log_e x$]

Now, $y = 4\log_e w \quad \therefore\ \frac{dy}{dw} = \frac{4}{w} = \frac{4}{\log_e v} = \frac{4}{\log_e\log_e\log_e x}$ [Put $w = \log_e v$, $v = \log_e u$ and $u = \log_e x$]

Then, $\frac{dy}{dx} = \frac{dy}{dw}\times\frac{dw}{dv}\times\frac{dv}{du}\times\frac{du}{dx} = \frac{4}{\log_e\log_e\log_e x}\times\frac{1}{\log_e\log_e x}\times\frac{1}{\log_e x}\times\frac{1}{x} = \frac{4}{x.\log_e x.\log_e\log_e x.\log_e\log_e\log_e x}$

$$\therefore \frac{dy}{dx} = \frac{4}{x.\log_e x.\log x.\log_e \log x} \quad \text{Ans.} \quad [e^{\log_e x} = x]$$

(k) $f(x) = \log_{x^3} \log x^2 = \frac{1}{3}\log_x \log x^2$, Let $f(u) = \log x^2$ $\quad \therefore f'(u) = \frac{1}{x^2}.2x = \frac{2}{x}$

Now, $f(x) = \frac{1}{3}\log_x f(u) = \frac{\log f(u)}{3\log x}$ $\quad \therefore f'(x) = \frac{1}{3}\left[\frac{\log x.\frac{1}{f(u)}.f'(u) - \log f(u).\frac{1}{x}}{(\log x)^2}\right]$ Put value of $f(u)$ and $f'(u)$

$$\therefore f'(x) = \frac{1}{3}\left[\frac{\log x.\frac{1}{\log x^2}.\frac{2}{x} - \log\log x^2.\frac{1}{x}}{(\log x)^2}\right] = \frac{1}{3}\left[\frac{\log x.\frac{1}{2\log x}.\frac{2}{x} - \log\log x^2.\frac{1}{x}}{(\log x)^2}\right] = \frac{1}{3}\left[\frac{1-\log\log x^2}{x(\log x)^2}\right]$$

At $x = e$ then $f'(e) = \frac{1}{3}\left[\frac{1-\log\log e^2}{e(\log e)^2}\right] = \left(\frac{1-\log 2}{3e}\right)$ Ans.

(16) (a) $y = \sin^m x.\cos^n x$ $\quad \therefore \frac{dy}{dx} = \sin^m x.n(\cos x)^{n-1}(-\sin x) + \cos^n x.m(\sin x)^{m-1}.\cos x$

$$\therefore \frac{dy}{dx} = m(\sin x)^{m-1}(\cos x)^{n+1} - n(\sin x)^{m+1}(\cos x)^{n-1} \quad \text{Ans.}$$

(b) $y = \sin^n x.\cos mx$ $\quad \therefore \frac{dy}{dx} = \sin^n x.(-\sin mx).m + \cos mx.n(\sin x)^{n-1}.\cos x$

$\therefore \frac{dy}{dx} = n.\cos x.\cos mx.(\sin x)^{n-1} - m.\sin mx.\sin^n x$ Ans.

(c) $y = f(f(\log x))$ where $f(x) = \log x$, $f(\log x) = \log\log x$, $f(f(\log x)) = \log[\log(\log x)]$

Now, $y = \log[\log(\log x)]$ Let $u = \log x$ $\quad \therefore \frac{du}{dx} = \frac{1}{x}$

Now, $y = \log(\log u)$ Let $v = \log u$ $\quad \therefore \frac{dv}{du} = \frac{1}{u} = \frac{1}{\log x}$ [Put $u = \log x$]

Now, $y = \log v$ $\quad \therefore \frac{dy}{dv} = \frac{1}{v} = \frac{1}{\log u} = \frac{1}{\log\log x}$ [Put $v = \log u$ and $u = \log x$]

Then, $\frac{dy}{dx} = \frac{dy}{dv} \times \frac{dv}{du} \times \frac{du}{dx} = \frac{1}{\log\log x} \times \frac{1}{\log x} \times \frac{1}{x} = \frac{1}{x.\log x.\log(\log x)}$ Ans.

(d) $y = f(f(\sin x))$, where $f(x) = \sin x$, $f(\sin x) = \sin(\sin x)$, $f(f(\sin x)) = \sin[\sin(\sin x)]$

Now, $y = f(f(\sin x)) = \sin[\sin(\sin x)]$ Let $u = \sin x$ $\quad \therefore \frac{du}{dx} = \cos x$

Now, $y = \sin(\sin u)$ Let $v = \sin u$ $\quad \therefore \frac{dv}{du} = \cos u = \cos(\sin x)$ [Put $u = \sin x$]

Now, $y = \sin v$ $\quad \therefore \frac{dy}{dv} = \cos v = \cos(\sin u) = \cos[\sin(\sin x)]$ [Put $v = \sin u$ and $u = \sin x$]

Then, $\frac{dy}{dx} = \frac{dy}{dv} \times \frac{dv}{du} \times \frac{du}{dx} = \cos[\sin(\sin x)] \times \cos(\sin x) \times \cos x = \cos x.\cos(\sin x).\cos[\sin(\sin x)]$ Proved.

(16) Do yourself. Question No.– (e), (f) and (g).

Integration

Indefinite Integral: – Some Basic Integrals in Standard Form: –

(i) $\int x^n\,dx = \frac{x^{n+1}}{n+1}$, $\frac{d}{dx}\left(\frac{x^{n+1}}{n+1}\right) = x^n$ (ii) $\int \sin x\,dx = -\cos x$, $\frac{d}{dx}(\cos x) = -\sin x$

(iii) $\int \cos x\,dx = \sin x$, $\frac{d}{dx}(\sin x) = \cos x$ (iv) $\int \sec^2 x\,dx = \tan x$, $\frac{d}{dx}(\tan x) = \sec^2 x$

(v) $\int \operatorname{cosec}^2 x\,dx = -\cot x$, $\frac{d}{dx}(\cot x) = -\operatorname{cosec}^2 x$

(vi) $\int \sec x \tan x\,dx = \sec x$, $\frac{d}{dx}(\sec x) = \sec x \tan x$

(vii) $\int \operatorname{cosec} x \cot x\,dx = -\operatorname{cosec} x$, $\frac{d}{dx}(\operatorname{cosec} x) = \operatorname{cosec} x \cot x$

(viii) $\int \frac{1}{x}\,dx = \log x$, $\frac{d}{dx}(\log x) = \frac{1}{x}$ (ix) $\int e^x\,dx = e^x$, $\frac{d}{dx}(e^x) = e^x$

(x) $\int \frac{dx}{\sqrt{1-x^2}} = \sin^{-1} x$, $\frac{d}{dx}(\sin^{-1} x) = \frac{1}{\sqrt{1-x^2}}$

(xi) $\int \frac{dx}{-\sqrt{1-x^2}} = \cos^{-1} x$, $\frac{d}{dx}(\cos^{-1} x) = \frac{1}{-\sqrt{1-x^2}}$

(xii) $\int \frac{dx}{1+x^2} = \tan^{-1} x$, $\frac{d}{dx}(\tan^{-1} x) = \frac{1}{1+x^2}$

(xiii) $\int \frac{dx}{-(1+x^2)} = \cot^{-1} x$, $\frac{d}{dx}(\cot^{-1} x) = -\frac{1}{1+x^2}$

(xiv) $\int \frac{dx}{x\sqrt{x^2-1}} = \sec^{-1} x$, $\frac{d}{dx}(\sec^{-1} x) = \frac{1}{x\sqrt{x^2-1}}$

(xv) $\int \frac{dx}{-x\sqrt{x^2-1}} = \operatorname{cosec}^{-1} x$, $\frac{d}{dx}(\operatorname{cosec}^{-1} x) = -\frac{1}{x\sqrt{x^2-1}}$

Some Standard Formula: –

(i) $\int \frac{dx}{\sqrt{x^2+a^2}} = \log\left|x+\sqrt{x^2+a^2}\right|$ or $\sinh^{-1}\left(\frac{x}{a}\right)$

(ii) $\int \frac{dx}{\sqrt{x^2-a^2}} = \log\left|x+\sqrt{x^2-a^2}\right|$ or $\cosh^{-1}\left(\frac{x}{a}\right)$ (iii) $\int \frac{dx}{\sqrt{a^2-x^2}} = \sin^{-1}\frac{x}{a}$

(iv) $\int \frac{dx}{x^2-a^2} = \frac{1}{2a}\log\left|\frac{x-a}{x+a}\right|$, when $x > a$ (v) $\int \frac{dx}{a^2-x^2} = \frac{1}{2a}\log\left|\frac{a+x}{a-x}\right|$, when $x < a$

(vi) $\int \frac{dx}{x^2+a^2} = \frac{1}{a}\tan^{-1}\frac{x}{a}$

(vii) $\int \sqrt{x^2+a^2}\,dx = \frac{x}{2}\sqrt{x^2+a^2} + \frac{a^2}{2}\log\left|x+\sqrt{x^2+a^2}\right|$ or $\frac{x}{2}\sqrt{x^2+a^2} + \frac{a^2}{2}\sinh^{-1}\left(\frac{x}{a}\right)$

(viii) $\int \sqrt{x^2-a^2}\,dx = \frac{x}{2}\sqrt{x^2-a^2} - \frac{a^2}{2}\log\left|x+\sqrt{x^2-a^2}\right|$ or $\frac{x}{2}\sqrt{x^2-a^2} - \frac{a^2}{2}\cosh^{-1}\left(\frac{x}{a}\right)$

(ix) $\int \sqrt{a^2-x^2}\,dx = \frac{x}{2}\sqrt{a^2-x^2} + \frac{a^2}{2}\sin^{-1}\frac{x}{a}$

Example: – (1) (a) $I=\int\frac{dx}{2x^2+3}=\int\frac{dx}{(\sqrt{2}\,x)^2+(\sqrt{3})^2}$ $\left[\text{use formula}, \int\frac{dx}{x^2+a^2}=\frac{1}{a}\tan^{-1}\frac{x}{a}\right]$

$\therefore\ I=\int\frac{dx}{(\sqrt{2}\,x)^2+(\sqrt{3})^2}=\frac{1}{\sqrt{3}}\tan^{-1}\frac{\sqrt{2}\,x}{\sqrt{3}}\cdot\frac{1}{\sqrt{2}}=\frac{1}{\sqrt{6}}\tan^{-1}\left(\frac{\sqrt{2}\,x}{\sqrt{3}}\right)+c$ Ans.

(b) $I=\int\frac{dx}{(3)^2+(5-x)^2}$ (solve same as above question) $\therefore\ I=\frac{1}{3}\tan^{-1}\left(\frac{5-x}{3}\right)\cdot\frac{1}{-1}$ $\therefore\ I=-\frac{1}{3}\tan^{-1}\left(\frac{5-x}{3}\right)+c$ Ans.

(c) $I=\int\frac{dx}{\sqrt{(3)^2+(4-x)^2}}$ $\left[\text{formula}, \int\frac{dx}{\sqrt{x^2+a^2}}=\log\left|x+\sqrt{x^2+a^2}\right|\right]$

$\therefore\ I=\int\frac{dx}{\sqrt{(3)^2+(4-x)^2}}=\log\left|(4-x)+\sqrt{(3)^2+(4-x)^2}\right|\cdot\frac{1}{-1}=-\log\left|(4-x)+\sqrt{(3)^2+(4-x)^2}\right|$ Ans.

(d) $I=\int\frac{dx}{\sqrt{(x+3)^2-4}}=\int\frac{dx}{\sqrt{(x+3)^2-(2)^2}}$ $\left[\text{formula}, \int\frac{dx}{\sqrt{x^2-a^2}}=\log\left|x+\sqrt{x^2-a^2}\right|\right]$

$\therefore\ I=\int\frac{dx}{\sqrt{(x+3)^2-(2)^2}}=\log\left|(x+3)+\sqrt{(x+3)^2-(2)^2}\right|\cdot\frac{1}{1}=\log\left|(x+3)+\sqrt{(x+3)^2-4}\right|$ Ans.

(e) $I=\int\sqrt{x^2-16}\,dx=\int\sqrt{x^2-4^2}\,dx$ $\left[\text{formula}, \int\sqrt{x^2-a^2}\,dx=\frac{x}{2}\sqrt{x^2-a^2}-\frac{a^2}{2}\log\left|x+\sqrt{x^2-a^2}\right|\right]$

$\therefore\ I=\int\sqrt{x^2-4^2}\,dx=\frac{x}{2}\sqrt{x^2-4^2}-\frac{4^2}{2}\log\left|x+\sqrt{x^2-4^2}\right|+c=\frac{x}{2}\sqrt{x^2-16}-\frac{16}{2}\log\left|x+\sqrt{x^2-16}\right|+c$

$=\frac{x}{2}\sqrt{x^2-16}-8\log\left|x+\sqrt{x^2-16}\right|+c$ Ans.

(f) $I=\int\frac{dx}{16-3x^2}=\int\frac{dx}{(4)^2-(\sqrt{3}\,x)^2}$ $\left[\text{formula}, \int\frac{dx}{a^2-x^2}=\frac{1}{2a}\log\left|\frac{a+x}{a-x}\right|\right]$

$\therefore\ I=\int\frac{dx}{(4)^2-(\sqrt{3}\,x)^2}=\frac{1}{2.4}\log\left|\frac{4+\sqrt{3}x}{4-\sqrt{3}x}\right|\cdot\frac{1}{\sqrt{3}}+c=\frac{1}{8\sqrt{3}}\log\left|\frac{4+\sqrt{3}x}{4-\sqrt{3}x}\right|+c$ Ans.

(g) $I=\int\sqrt{9-x^2}\,dx=\int\sqrt{(3)^2-x^2}\,dx$ $\left[\text{formula}, \int\sqrt{a^2-x^2}\,dx=\frac{x}{2}\sqrt{a^2-x^2}+\frac{a^2}{2}\sin^{-1}\frac{x}{a}\right]$

or $I=\int\sqrt{(3)^2-x^2}\,dx=\frac{x}{2}\sqrt{9-x^2}+\frac{9}{2}\sin^{-1}\frac{x}{3}$ Ans.

$\mathbf{I=\int\frac{f'(x)}{f(x)}dx=\log|f(x)|}$

Proof: – Let $f(x)=z$ $\therefore\ f'(x)\,dx=dz$ then $\int\frac{f'(x)}{f(x)}dx=\int\frac{dz}{z}=\log|z|=\log|f(x)|$ Proved.

Example: – (a) $I=\int\frac{(4x+9)}{2x^2+9x+14}dx$, Let $f(x)=2x^2+9x+14$, $f'(x)=4x+9$

or $I=\int\frac{f'(x)}{f(x)}dx=\log|f(x)|+c=\log|2x^2+9x+14|+c$ Ans.

(b) $I=\int\cot x\,dx=\int\frac{\cos x}{\sin x}dx$, Let $f(x)=\sin x$ $\therefore\ f'(x)=\cos x$ $\therefore\ I=\int\frac{\cos x}{\sin x}dx=\int\frac{f'(x)}{f(x)}dx=\log|f(x)|+c=\log|\sin x|+c$ Ans.

(c) $I=\int\frac{x-2}{x^2-4x+8}dx$, Let $f(x)=x^2-4x+8$ $\therefore\ f'(x)=2x-4=2(x-2)$

$$I=\int\frac{2x-4}{2(x^2-4x+8)}dx=\frac{1}{2}\int\frac{2x-4}{x^2-4x+8}dx=\frac{1}{2}\int\frac{f'(x)}{f(x)}dx=\frac{1}{2}\log|f(x)|+c=\frac{1}{2}\log|x^2-4x+8|+c \quad \text{Ans.}$$

(d) $$I=\int\frac{(\cos x-\sin x)}{\sin\left(x+\frac{\pi}{4}\right)}dx=\int\frac{\cos x-\sin x}{\sin x.\cos\frac{\pi}{4}+\cos x.\sin\frac{\pi}{4}}dx=\int\frac{\cos x-\sin x}{\frac{1}{\sqrt{2}}(\sin x+\cos x)}dx=\sqrt{2}\int\frac{\cos x-\sin x}{\sin x+\cos x}dx$$

Let $f(x)=\sin x+\cos x$, $f'(x)=\cos x-\sin x$ then $\therefore\ I=\sqrt{2}\int\frac{f'(x)}{f(x)}dx=\sqrt{2}\log|f(x)|+c$ $\therefore\ I=\sqrt{2}\log|\sin x+\cos x|+c$ Ans.

$\mathbf{I=\int\{f(x)\}^n.f'(x)\,dx=\frac{\{f(x)\}^{n+1}}{n+1}}$, $\mathbf{n\neq-1}$

Proof: – Let $f(x)=z$ $\therefore\ f'(x)\,dx=dz$ then $I=\int\{f(x)\}^n.f'(x)\,dx=\int z^n.dz=\frac{z^{n+1}}{n+1}$ $\therefore\ I=\frac{z^{n+1}}{n+1}=\frac{\{f(x)\}^{n+1}}{n+1}$ Proved

Example: – (a) $I=\int\tan x.\sec^2 x\,dx$ Let $f(x)=\tan x,\ f'(x)=\sec^2 x$

or $I=\int\{f(x)\}^n.f'(x)\,dx=\frac{\{f(x)\}^{n+1}}{n+1}$ or $I=\int\tan x.\sec^2 x\,dx=\int(\tan x)^1.\sec^2 x\,dx=\frac{(\tan x)^{1+1}}{1+1}+c=\frac{(\tan x)^2}{2}+c=\frac{\tan^2 x}{2}+c$ Ans.

IInd Method: – $I=\int\tan x.\sec^2 x\,dx=\int\frac{\sin x}{\cos x}.\frac{1}{\cos^2 x}dx=\int\frac{\sin x}{\cos^3 x}dx$, Let $f(x)=\cos x,\ f'(x)=-\sin x$

$$\therefore\ I=\int\cos^{-3}x.\sin x\ dx=-\int[f(x)]^{-3}.(-f'(x))\,dx=-\frac{\{f(x)\}^{n+1}}{n+1}=-\frac{\{\cos x\}^{-3+1}}{-3+1}=\frac{\cos^{-2}x}{2}=\frac{1}{2\cos^2 x}+c=\frac{1}{2}\sec^2 x+c$$
$$=\frac{1}{2}(1+\tan^2 x)+c=\frac{\tan^2 x}{2}+\frac{1}{2}+c \quad \text{Ans.}$$

IIIrd Method: – $I=\int\tan x.\sec^2 x\,dx=\int\frac{\sin x}{\cos x}.\frac{1}{\cos^2 x}\,dx=\int\frac{\sin x}{\cos^3 x}\,dx$ (solve above Method) $\therefore\ I=\frac{1}{2}\sec^2 x=\frac{\tan^2 x}{2}+\frac{1}{2}$ Ans.

IVth Method: – $I=\int\tan x.\sec^2 x\,dx$ Put $z=\tan x$ $\therefore\ dz=\sec^2 x\ dx$ $\therefore\ I=\int z.dz=\frac{z^2}{2}+c=\frac{\tan^2 x}{2}+c$ Ans.

(b) $I=\int\frac{x^2\,dx}{(2x^3-5)^4}=\int(2x^3-5)^{-4}.x^2\ dx$ Let $f(x)=2x^3-5$, $f'(x)=6x^2$

$\therefore\ I=\frac{1}{6}\int(2x^3-5)^{-4}.6x^2\ dx$ $\left[\text{formula, } I=\int\{f(x)\}^n.f'(x)\,dx=\frac{\{f(x)\}^{n+1}}{n+1}\right]$

$\therefore\ I=\frac{1}{6}.\frac{\{f(x)\}^{-4+1}}{-4+1}=\frac{1}{6}.\frac{(2x^3-5)^{-3}}{-3}=-\frac{1}{18}(2x^3-5)^{-3}=-\frac{1}{18(2x^3-5)^3}$ Ans.

(c) $I=\int\sin^4 x.\cos x\ dx$, Let $f(x)=\sin x$ $\therefore\ f'(x)=\cos x$ and $n=4$ $\therefore\ I=\frac{[f(x)]^{4+1}}{4+1}=\frac{(\sin x)^5}{5}+c=\frac{\sin^5 x}{5}+c$ Ans.

IInd Method: – $I=\int\sin^4 x.\cos x\ dx$, Put $\sin x=z$ $\therefore\ \cos x\,dx=dz$ $\therefore\ I=\int z^4.dz=\frac{z^{4+1}}{4+1}=\frac{z^5}{5}=\frac{(\sin x)^5}{5}+c=\frac{\sin^5 x}{5}+c$ Ans.

$\mathbf{I=\int\frac{f'(x)}{\sqrt{f(x)}}dx=2\sqrt{f(x)}}$

Proof: – Let $f(x)=z$ $\therefore\ f'(x)\,dx=dz$ then $I=\int\frac{dz}{\sqrt{z}}=\int z^{-\frac{1}{2}}.dz=\frac{z^{-\frac{1}{2}+1}}{-\frac{1}{2}+1}=\frac{z^{\frac{1}{2}}}{\frac{1}{2}}=2\sqrt{z}=2\sqrt{f(x)}$ Proved.

Example: – (a) $I=\int\frac{2ax+b}{\sqrt{ax^2+bx+c}}\,dx$ Let $f(x)=ax^2+bx+c$ $\therefore\ f'(x)=2ax+b$

or $I=\int\frac{f'(x)}{\sqrt{f(x)}}\,dx=2\sqrt{f(x)}=2\sqrt{ax^2+bx+c}+d$ Ans.

(b) $I = \int \frac{\cos x}{\sqrt{\sin x}} dx$ Let $f(x) = \sin x$ $\therefore$ $f'(x) = \cos x$ or $I = \int \frac{f(x)}{\sqrt{f(x)}} dx = 2\sqrt{f(x)}$ $\therefore$ $I = \int \frac{\cos x}{\sqrt{\sin x}} dx = 2\sqrt{\sin x} + c$ Ans.

(c) $I = \int \frac{\sec^2 x}{\sqrt{3 - \tan x}} dx$ Let $f(x) = 3 - \tan x$ $\therefore$ $f'(x) = -\sec^2 x$

or $I = \int \frac{f(x)}{\sqrt{f(x)}} dx = 2\sqrt{f(x)}$ $\therefore$ $I = \int \frac{\sec^2 x}{\sqrt{3 - \tan x}} dx = -\int \frac{\sec^2 x}{\sqrt{3 - \tan x}} dx = -2\sqrt{3 - \tan x} + c$ Ans.

$\mathbf{\frac{d}{dx}\int f(x)\, dx = f(x)}$ **or** $\mathbf{\int \frac{d}{dx}[f(x)]\, dx = f(x)}$

(i) $\frac{d}{dx}\int \frac{\log(2+x)}{3x} dx = \frac{\log(2+x)}{3x}$ Ans. (ii) $\int \frac{d}{dx}\left(\frac{x^2+3x+5}{2x-3}\right) dx = \frac{x^2+3x+5}{2x-3}$ Ans.

Form of the integrals: –

$\int \frac{dx}{ax^2+bx+c}$ or $\int \frac{dx}{\sqrt{ax^2+bx+c}}$ or $\int \sqrt{ax^2+bx+c}\, dx$ where $a \neq 0$

write $ax^2 + bx + c$ in the form $a\left\{\left(x+\frac{b}{2a}\right)^2 - \frac{D}{4a^2}\right\}$ and then use standard result of integration.

Example: – (a) $I = \int \frac{dx}{x^2+2x+3} = \int \frac{dx}{ax^2+bx+c}$ Here $a = 1, b = 2$ and $c = 3$ then $D = b^2 - 4ac = 4 - 12 = -8$

$$\text{or } I = \int \frac{dx}{a\left\{\left(x+\frac{b}{2a}\right)^2 - \frac{D}{4a^2}\right\}} = \int \frac{dx}{1\left\{\left(x+\frac{2}{2}\right)^2 - \frac{-8}{4.1}\right\}} = \int \frac{dx}{(x+1)^2+2} = \int \frac{dx}{(x+1)^2+\left(\sqrt{2}\right)^2}$$

$$\therefore\ I = \int \frac{dx}{x^2+a^2} = \frac{1}{a}\tan^{-1}\frac{x}{a} \quad \therefore\ I = \int \frac{dx}{(x+1)^2+\left(\sqrt{2}\right)^2} = \frac{1}{\sqrt{2}}\tan^{-1}\left(\frac{x+1}{\sqrt{2}}\right) + c \quad \text{Ans.}$$

(b) $I = \int \frac{dx}{\sqrt{5+4x-3x^2}} = \int \frac{dx}{\sqrt{ax^2+bx+c}}$, here $a = -3, b = 4$ and $c = 5$ then $D = 16 + 60 = 76$

$$\text{or } I = \int \frac{dx}{\sqrt{a\left\{\left(x+\frac{b}{2a}\right)^2 - \frac{D}{4a^2}\right\}}} = \int \frac{dx}{\sqrt{-3\left\{\left(x+\frac{4}{-6}\right)^2 - \frac{76}{36}\right\}}} = \int \frac{dx}{\sqrt{3\left\{\frac{19}{9} - \left(x-\frac{2}{3}\right)^2\right\}}} = \frac{1}{\sqrt{3}}\int \frac{dx}{\sqrt{\left(\frac{\sqrt{19}}{3}\right)^2 - \left(x-\frac{2}{3}\right)^2}}$$

$$\therefore\ I = \frac{1}{\sqrt{3}}\int \frac{dx}{\sqrt{a^2-x^2}} = \frac{1}{\sqrt{3}}\sin^{-1}\frac{x}{a} = \frac{1}{\sqrt{3}}\sin^{-1}\left(\frac{x-\frac{2}{3}}{\frac{\sqrt{19}}{3}}\right) = \frac{1}{\sqrt{3}}\sin^{-1}\left(\frac{3x-2}{\sqrt{19}}\right) \quad \text{Ans.}$$

(c) $I = \int \sqrt{-x^2+3x+8}\, dx = \int \sqrt{ax^2+bx+c}\, dx$, here $a = -1, b = 3$ and $c = 8$ then $D = 9 + 32 = 41$

$$\text{or } I = \int \sqrt{a\left\{\left(x+\frac{b}{2a}\right)^2 - \frac{D}{4a^2}\right\}}\, dx = \int \sqrt{-1\left\{\left(x+\frac{3}{-2}\right)^2 - \frac{41}{4}\right\}}\, dx = \int \sqrt{\frac{41}{4} - \left(x-\frac{3}{2}\right)^2}\, dx$$

$$I = \int \sqrt{\left(\frac{\sqrt{41}}{2}\right)^2 - \left(x-\frac{3}{2}\right)^2}\, dx \qquad \text{formula, } \int \sqrt{a^2-x^2}\, dx = \frac{x}{2}\sqrt{a^2-x^2} + \frac{a^2}{2}\sin^{-1}\frac{x}{a}$$

$$\text{or } I = \int \sqrt{\left(\frac{\sqrt{41}}{2}\right)^2 - \left(x-\frac{3}{2}\right)^2} \text{ or } dx = \frac{\left(x-\frac{3}{2}\right)}{2}\sqrt{\frac{41}{4} - \left(x-\frac{3}{2}\right)^2} + \frac{41}{8}\sin^{-1}\left(\frac{x-\frac{3}{2}}{\frac{\sqrt{41}}{2}}\right) + c$$

$$= \frac{(2x-3)}{4}\sqrt{\frac{41}{4} - \left(x-\frac{3}{2}\right)^2} + \frac{41}{8}\sin^{-1}\left(\frac{2x-3}{\sqrt{41}}\right) + c \quad \text{Ans.}$$

Form of the integrals: $-\quad \int \frac{dx}{ax^2+bx+c}$ or $\int \frac{dx}{\sqrt{ax^2+bx+c}}$ or $\int \sqrt{ax^2+bx+c}\ dx$

Express ax^2+bx+c as sum of difference of two square and apply formulae.

Example: $-$ (1) (a) $I=\int \frac{dx}{x^2+2x+1}$ (b) $I=\int \frac{dx}{\sqrt{x^2+2x+1}}$ (c) $I=\int \sqrt{x^2+2x+1}\ dx$

Solution: $-$ (a) $I=\int \frac{dx}{x^2+2x+1}=\int \frac{dx}{(x+1)^2}$ Let $x+1=t$ or $x=t-1$ $\therefore$ $dx=dt$

$$\therefore\ I=\int \frac{dt}{t^2}=\int t^{-2}\ dt=\frac{t^{-2+1}}{-2+1}+c=-t^{-1}+c=-\frac{1}{t}+c=-\frac{1}{x+1}+c \quad \text{Ans.}$$

$$\text{(b) } I=\int \frac{dx}{\sqrt{x^2+2x+1}}=\int \frac{dx}{\sqrt{(x+1)^2}}=\int \frac{dx}{x+1}=\log(x+1)+c \quad \text{Ans.}$$

$$\text{(c) } I=\int \sqrt{x^2+2x+1}\ dx=\int \sqrt{(x+1)^2}\ dx=\int (x+1)\ dx=\int x\ dx+\int dx=\frac{x^2}{2}+x+c \quad \text{Ans.}$$

Example: $-$ (2) (a) $I=\int \frac{dx}{4x^2+4x+5}$ (b) $I=\int \frac{dx}{\sqrt{4x^2+4x+5}}$ (c) $I=\int \sqrt{4x^2+4x+5}\ dx$

Solution: $-$ (a) $I=\int \frac{dx}{4x^2+4x+5}=\int \frac{dx}{ax^2+bx+c}$, here $a=4, b=4$ and $c=5$ then $D=16-80=-64$

$$\therefore\ I=\int \frac{dx}{a\left\{\left(x+\frac{b}{2a}\right)^2-\frac{D}{4a^2}\right\}}=\int \frac{dx}{4\left\{\left(x+\frac{4}{8}\right)^2-\frac{-64}{64}\right\}}=\int \frac{dx}{(2x+1)^2+4}=\int \frac{dx}{(2x+1)^2+(2)^2}$$

use formula, $\therefore\ I=\int \frac{dx}{x^2+a^2}=\frac{1}{a}\tan^{-1}\frac{x}{a}$ $\quad\therefore\ I=\int \frac{dx}{(2x+1)^2+(2)^2}$

Let $t=2x+1$, $dt=2\ dx$ $\quad\therefore\ I=\frac{1}{2}\int \frac{dt}{(t)^2+(2)^2}=\frac{1}{2}\cdot\frac{1}{2}\tan^{-1}\left(\frac{t}{2}\right)+c=\frac{1}{4}\tan^{-1}\left(\frac{2x+1}{2}\right)+c$ Ans.

(b) $I=\int \frac{dx}{\sqrt{4x^2+4x+5}}=\int \frac{dx}{\sqrt{ax^2+bx+c}}=\int \frac{dx}{\sqrt{(2x+1)^2+(2)^2}}$ formula, $\int \frac{dx}{\sqrt{x^2+a^2}}=\log\left|x+\sqrt{x^2+a^2}\right|$

$$\therefore\ I=\int \frac{dx}{\sqrt{(2x+1)^2+(2)^2}}=\log\left[(2x+1)+\sqrt{4x^2+4x+5}\right].\frac{1}{\frac{d}{dx}(2x+1)}+c=\frac{1}{2}\log\left[(2x+1)+\sqrt{4x^2+4x+5}\right]+c \quad \text{Ans.}$$

$$\text{(c) } I=\int \sqrt{4x^2+4x+5}\ dx=\int \sqrt{ax^2+bx+c}\ dx=\int \sqrt{(2x+1)^2+(2)^2}\ dx$$

Let $t=2x+1$ $\therefore$ $dt=2\ dx$ use formula, $\int \sqrt{x^2+a^2}\,dx=\frac{x}{2}\sqrt{x^2+a^2}+\frac{a^2}{2}\log\left|x+\sqrt{x^2+a^2}\right|$

$\therefore\ I=\frac{1}{2}\int \sqrt{(t)^2+(2)^2}\ dt=\frac{1}{2}\left[\frac{t}{2}\sqrt{t^2+4}+2\log\left|t+\sqrt{t^2+4}\right|\right]$ Put $t=2x+1$

$$\therefore\ I=\frac{(2x+1)}{4}\sqrt{(2x+1)^2+(2)^2}+\log\left|(2x+1)+\sqrt{(2x+1)^2+(2)^2}\right|+c \quad \text{Ans.}$$

\# **Form of the integrals**: $-\int \frac{Ax+B}{ax^2+bx+c}dx$, $\int \frac{Ax+B}{\sqrt{ax^2+bx+c}}dx$, $\int (Ax+B)\sqrt{ax^2+bx+c}\ dx$

Express $Ax+B=l[\text{d.c of }(ax^2+bx+c)]+m$ ……………………………..(i)

find the value of l, m by comparing the coefficient of x and constant term on both sides of (i)

$\int \frac{f'(x)}{f(x)}\,dx = \log[f(x)]$ by putting $f(x) = t$ or $\int \frac{f'(x)}{\sqrt{f(x)}}\,dx = 2\sqrt{f(x)}$ by putting $f(x) = t$

$$\int f'(x).\sqrt{f(x)}\,dx = \frac{2}{3}[f(x)]^{\frac{3}{2}} \text{ by putting } f(x) = t$$

Example: – (1) (a) $I = \int \frac{2x+3}{4x^2+4x+5}\,dx$ [use form of the integral $Ax + B = l(\text{d.c of } ax^2 + bx + c) + m$]

$$\therefore\ 2x + 3 = l(\text{d.c of } 4x^2 + 4x + 5) + m \quad \text{or} \quad 2x + 3 = l(8x + 4) + m$$

find the value of l, m by comparing the coefficient of x and constant term on both of sides.

$2x + 3 = l(8x + 4) + m$ ………….(i) $\therefore\ 8l = 2$ and $4l + m = 3$ $\therefore\ l = \frac{1}{4}$ and $4.\frac{1}{4} + m = 3$ or $m = 2$

Divide $4x^2 + 4x + 5$ both of sides in equation (i), we have $\frac{2x+3}{4x^2+4x+5} = \frac{l(8x+4)}{4x^2+4x+5} + \frac{m}{4x^2+4x+5}$

Integrating both of sides and put value of l and m, we have

$$\int \frac{2x+3}{4x^2+4x+5}\,dx = \frac{1}{4}\int \frac{8x+4}{4x^2+4x+5}\,dx + \int \frac{2}{4x^2+4x+5}\,dx$$

or $I = I_1 + I_2$ where $I_1 = \frac{1}{4}\int \frac{8x+4}{4x^2+4x+5}\,dx$ and $I_2 = \int \frac{2}{4x^2+4x+5}\,dx$

$I_1 = \frac{1}{4}\int \frac{8x+4}{4x^2+4x+5}\,dx$ Let $f(x) = 4x^2 + 4x + 5$ $\therefore\ f'(x) = 8x + 4$ $\left[\text{formula}, \int \frac{f'(x)}{f(x)}\,dx = \log[f(x)]\right]$

$$I_1 = \frac{1}{4}\int \frac{f'(x)}{f(x)}\,dx = \frac{1}{4}\int \frac{8x+4}{4x^2+4x+5}\,dx = \frac{1}{4}\log[4x^2+4x+5] + c$$

$I_2 = \int \frac{2}{4x^2+4x+5}\,dx = 2\int \frac{dx}{(2x+1)^2+(2)^2} = 2\int \frac{dx}{x^2+a^2}$ $\left[\text{formula}, \int \frac{dx}{x^2+a^2} = \frac{1}{a}\tan^{-1}\frac{x}{a}\right]$

$$I_2 = 2.\frac{1}{2}\tan^{-1}\left(\frac{2x+1}{2}\right).\frac{1}{\text{d.c of }(2x+1)} + c = \tan^{-1}\left(\frac{2x+1}{2}\right).\frac{1}{2} + c = \frac{1}{2}\tan^{-1}\left(\frac{2x+1}{2}\right) + c$$

or $I = I_1 + I_2 = \frac{1}{4}\log[4x^2+4x+5] + c + \frac{1}{2}\tan^{-1}\left(\frac{2x+1}{2}\right) + c = \frac{1}{4}\log[4x^2+4x+5] + \frac{1}{2}\tan^{-1}\left(\frac{2x+1}{2}\right) + 2c$ Ans.

(b) $I = \int \frac{2x+3}{\sqrt{4x^2+4x+5}}\,dx$ (see above question) $l = \frac{1}{4}$ and $m = 2$

$$\int \frac{2x+3}{\sqrt{4x^2+4x+5}}\,dx = l\int \frac{8x+4}{\sqrt{4x^2+4x+5}}\,dx + m\int \frac{dx}{\sqrt{4x^2+4x+5}}$$

$\left[\text{formula}, \int \frac{f'(x)}{\sqrt{f(x)}}\,dx = 2\sqrt{f(x)} \text{ and } \int \frac{dx}{\sqrt{x^2+a^2}} = \log\left|x + \sqrt{x^2+a^2}\right|\right]$

or $I = \int \frac{2x+3}{\sqrt{4x^2+4x+5}}\,dx = \frac{1}{4}\int \frac{8x+4}{\sqrt{4x^2+4x+5}}\,dx + 2\int \frac{dx}{\sqrt{(2x+1)^2+(2)^2}} = \frac{1}{4}.2\sqrt{4x^2+4x+5} + 2.\frac{1}{2}\log\left|(2x+1) + \sqrt{4x^2+4x+5}\right| + c$

$= \frac{1}{2}\sqrt{4x^2+4x+5} + \log\left|(2x+1) + \sqrt{4x^2+4x+5}\right| + c$ Ans.

(c) $I = \int (2x+3)\sqrt{4x^2+4x+5}\,dx$ (see above question) $l = \frac{1}{4}$ and $m = 2$

$$\int (2x+3)\sqrt{4x^2+4x+5}\,dx = l\int (8x+4)\sqrt{4x^2+4x+5}\,dx + m\int \frac{dx}{\sqrt{4x^2+4x+5}}$$

$$\therefore\ I = \int (2x+3)\sqrt{4x^2+4x+5}\,dx = \frac{1}{4}\int (8x+4)\sqrt{4x^2+4x+5}\,dx + 2\int \frac{dx}{\sqrt{4x^2+4x+5}}$$

$$\left[\text{formula } \int \sqrt{x^2+a^2}\,dx = \frac{x}{2}\sqrt{x^2+a^2} + \frac{a^2}{2}\log\left|x+\sqrt{x^2+a^2}\right| \text{ and } \int f'(x).\sqrt{f(x)}\,dx = \frac{2}{3}[f(x)]^{\frac{3}{2}}\right]$$

$$I = \frac{1}{4}.\frac{2}{3}[4x^2+4x+5]^{\frac{3}{2}} + 2\int \frac{dx}{\sqrt{(2x+1)^2+(2)^2}}$$

$$I = \frac{1}{6}[4x^2+4x+5]^{\frac{3}{2}} + \frac{1}{2}\left\{2.\frac{(2x+1)}{2}\sqrt{4x^2+4x+5} + 2.\frac{4}{2}\log\left[(2x+1)+\sqrt{4x^2+4x+5}\right]\right\} + c$$

$$\therefore\ I = \frac{1}{6}[4x^2+4x+5]^{\frac{3}{2}} + \frac{(2x+1)}{2}\sqrt{4x^2+4x+5} + 2\log\left[(2x+1)+\sqrt{4x^2+4x+5}\right] + c \quad \text{Ans.}$$

(2) (a) $I = \int \frac{1+\sin 4x}{\sin 4x + \cos 4x + 1}\,dx = \int \frac{\sin^2 2x + \cos^2 2x + 2\sin 2x \cos 2x}{\cos^2 2x - \sin^2 2x + 2\sin 2x\cos 2x + 1}\,dx$

$$I = \int \frac{(\sin 2x+\cos 2x)^2}{\cos^2 2x + 2\sin 2x\cos 2x + (1-\sin^2 2x)}\,dx = \int \frac{(\sin 2x+\cos 2x)^2}{2\cos^2 2x + 2\sin 2x\cos 2x}\,dx = \int \frac{(\sin 2x+\cos 2x)^2}{2\cos 2x(\cos 2x+\sin 2x)}\,dx = \int \frac{\sin 2x+\cos 2x}{2\cos 2x}\,dx$$
$$= \frac{1}{2}\int\left(\frac{\sin 2x}{\cos 2x}+1\right)dx$$

$I = \frac{1}{2}\int \frac{\sin 2x}{\cos 2x}\,dx + \frac{1}{2}\int dx = I_1 + I_2$ where $I_1 = \frac{1}{2}\int \frac{\sin 2x}{\cos 2x}\,dx$ and $I_2 = \frac{1}{2}\int dx$

$I_1 = \frac{1}{2}\int \frac{\sin 2x}{\cos 2x}\,dx$ Let $\cos 2x = t$ $\therefore -\sin 2x.2\,dx = dt$ $\therefore \sin 2x\,dx = -\frac{dt}{2}$

$$I_1 = \frac{1}{2}\int \frac{\sin 2x}{\cos 2x}\,dx = -\frac{1}{4}\int \frac{dt}{t} = -\frac{1}{4}\log t = -\frac{1}{4}\log(\cos 2x) + c$$

$$I_2 = \frac{1}{2}\int dx = \frac{1}{2}x + c$$

$\therefore\ I = I_1 + I_2 = -\frac{1}{4}\log(\cos 2x) + c + \frac{1}{2}x + c = -\frac{1}{4}\log(\cos 2x) + \frac{1}{2}x + k$ where $k = 2c$ Ans.

(b) $I = \int \frac{1+\sin(4\log x)}{x(\log x+1)}\,dx$ Let $\log x = t$ $\therefore \frac{1}{x}dx = dt$

or $I = \int \frac{1+\sin(4\log x)}{(\log x+1)}.\frac{1}{x}dx = \int \frac{1+\sin 4t}{(t+1)}\,dt = \int \frac{1}{1+t}\,dt + \int \frac{\sin 4t}{1+t}\,dt = I_1 + I_2$ (say)

where $I_1 = \int \frac{1}{1+t}\,dt = \log(1+t) + c = \log(1+\log x) + c$

$\therefore\ I_2 = \int \frac{\sin 4t}{1+t}\,dt$

 # Integrating by part formula, $\int \mathbf{u.v\,dx = v.\int u\,dx - \int\left[\frac{dv}{dx}\int u\,dx\right]dx}$

or $I_2 = \int \frac{\sin 4t}{1+t}\,dt = \int \frac{1}{1+t}.\sin 4t\,dt = \sin 4t.\int \frac{1}{1+t}\,dt - \int\left[\frac{d}{dx}(\sin 4t)\int \frac{1}{1+t}\,dt\right]dt = \sin 4t.\log(1+t) - \int \log(1+t).\cos 4t.4\,dt$
$= \sin 4t.\log(1+t) - 4\int \cos 4t\log(1+t)\,dt$

Let $I_3 = \int \cos 4t\log(1+t)\,dt$, Again use integration by part formula

or $I_3 = \int \cos 4t\log(1+t)\,dt = \cos 4t.\int \log(1+t)\,dt - \int\left[\frac{d}{dx}(\cos 4t)\int \log(1+t)\,dt\right]dt = \frac{\cos 4t}{1+t} + 4\int \sin 4t.\frac{1}{1+t}\,dt = \frac{\cos 4t}{1+t} + 4I_2$

or $I_2 = \sin 4t.\log(1+t) - 4\left[\frac{\cos 4t}{1+t} + 4I_2\right] = \sin 4t.\log(1+t) - 4.\frac{\cos 4t}{1+t} - 16I_2$

or $17I_2 = \sin 4t.\log(1+t) - \frac{4\cos 4t}{1+t} = \frac{(1+t).\sin 4t.\log(1+t) - 4\cos 4t}{1+t}$

or $I_2 = \dfrac{(1+t).\sin 4t.\log(1+t) - 4\cos 4t}{17(1+t)} + c$

Now, $I = I_1 + I_2 = \log(1+t) + c + \dfrac{(1+t).\sin 4t.\log(1+t) - 4\cos 4t}{17(1+t)} + c$

$= \dfrac{17(1+t).\log(1+t) + (1+t).\sin 4t.\log(1+t) - 4\cos 4t}{17(1+t)} + k$ where $k = 2c$

$\therefore\ I = \dfrac{(1+t).\log(1+t)\,[17+\sin 4t] - 4\cos 4t}{17(1+t)} + k$ Ans.

(3) (a) $I = \displaystyle\int \frac{1+\tan 2\theta}{\cot\theta - \tan\theta}\,d\theta = \int \frac{1+\frac{\sin 2\theta}{\cos 2\theta}}{\frac{\cos\theta}{\sin\theta} - \frac{\sin\theta}{\cos\theta}}\,d\theta = \int \frac{\frac{\cos 2\theta + \sin 2\theta}{\cos 2\theta}}{\frac{\cos^2\theta - \sin^2\theta}{\sin\theta\cos\theta}}\,d\theta = \int \frac{(\cos 2\theta + \sin 2\theta)\sin\theta\cos\theta}{\cos 2\theta.\cos 2\theta}\,d\theta$

or $I = \displaystyle\int \frac{(\cos 2\theta + \sin 2\theta)2\sin\theta\cos\theta}{2\cos^2 2\theta}\,d\theta = \int \frac{(\cos 2\theta + \sin 2\theta)\sin 2\theta}{2\cos^2 2\theta}\,d\theta = \int \frac{\sin 2\theta.\cos 2\theta + \sin^2 2\theta}{2\cos^2 2\theta}\,d\theta$

or $I = \displaystyle\int \frac{\sin 2\theta.\cos 2\theta}{2\cos^2 2\theta}\,d\theta + \int \frac{\sin^2 2\theta}{2\cos^2 2\theta}\,d\theta = \frac{1}{2}\int \tan 2\theta\ d\theta + \frac{1}{2}\int \tan^2 2\theta\ d\theta = \frac{1}{2}I_1 + \frac{1}{2}I_2$ (say)

where $I_1 = \displaystyle\int \tan 2\theta\ d\theta = \int \frac{\sin 2\theta}{\cos 2\theta}\,d\theta$ Let $\cos 2\theta = t$ $\therefore\ -2\sin 2\theta\,d\theta = dt$ $\therefore\ 2\sin 2\theta\,d\theta = -dt$

$\therefore\ I_1 = -\dfrac{1}{2}\displaystyle\int \frac{dt}{t} = -\frac{1}{2}\log t + c = -\frac{1}{2}\log(\cos 2\theta) + c$

where $I_2 = \displaystyle\int \tan^2 2\theta\ d\theta = \int (\sec^2 2\theta - 1)\,d\theta = \int \sec^2 2\theta\,d\theta - \int d\theta$

Let $2\theta = t$ $\therefore 2\,d\theta = dt$ or $d\theta = \dfrac{dt}{2}$ $\therefore\ I_2 = \dfrac{1}{2}\displaystyle\int \sec^2 t\,dt - \theta = \frac{1}{2}\tan t - \theta = \frac{1}{2}\tan 2\theta - \theta + c$

$\therefore\ I = \dfrac{1}{2}I_1 + \dfrac{1}{2}I_2 = \dfrac{-\log(\cos 2\theta)}{4} + \dfrac{c}{2} + \dfrac{1}{4}\tan 2\theta - \dfrac{\theta}{2} + \dfrac{c}{2} = -\dfrac{1}{4}\log(\cos 2\theta) + \dfrac{1}{4}\tan 2\theta - \dfrac{1}{2}\theta + c$ Ans.

(b) $I = \displaystyle\int \frac{\sec x + \tan x}{1+\operatorname{cosec} x}\,dx = \int \frac{\frac{1}{\cos x} + \frac{\sin x}{\cos x}}{1 + \frac{1}{\sin x}}\,dx = \int \frac{\frac{1+\sin x}{\cos x}}{\frac{\sin x + 1}{\sin x}}\,dx = \int \frac{1+\sin x}{\cos x} \times \frac{\sin x}{1+\sin x}\,dx = \int \frac{\sin x}{\cos x}\,dx$

Let $\cos x = t$ $\therefore\ -\sin x\,dx = dt$ then $I = \displaystyle\int \frac{\sin x}{\cos x}\,dx = -\int \frac{1}{t}dt = -\log t + c = -\log(\cos x) + c$ Ans.

(c) $I = \displaystyle\int \frac{e^{\sqrt{x}}(\sin e^{\sqrt{x}} + \cos e^{\sqrt{x}})}{\sqrt{x}(1+e^{\sqrt{x}})}\,dx$ Let $e^{\sqrt{x}} = t$ $\therefore\ e^{\sqrt{x}}.\dfrac{1}{2\sqrt{x}}dx = dt$ or $\dfrac{e^{\sqrt{x}}}{\sqrt{x}}dx = 2dt$

$\therefore\ I = \displaystyle\int \frac{(\sin e^{\sqrt{x}} + \cos e^{\sqrt{x}})}{(1+e^{\sqrt{x}})}.\frac{e^{\sqrt{x}}}{\sqrt{x}}\,dx = \int \frac{(\sin t + \cos t)}{(1+t)}.2\,dt = 2\int \frac{\sin t}{1+t}dt + 2\int \frac{\cos t}{1+t}dt = 2I_1 + 2I_2$ (say)

Solve I_1 and I_2 seprately,

where $I_1 = \displaystyle\int \frac{\sin t}{1+t}dt$ and $I_2 = \displaystyle\int \frac{\cos t}{1+t}dt$ (using integration by part formula)

or $I_1 = \displaystyle\int \frac{\sin t}{1+t}dt = \sin t.\int \frac{dt}{1+t} - \int \left[\frac{d}{dt}(\sin t)\int \frac{dt}{1+t}\right]dt = \sin t.\log(1+t) - \int \log(1+t).\cos t\ dt$

or $I_1 = \sin t.\log(1+t) - \left\{\cos t.\displaystyle\int \log(1+t)\,dt - \int \left[\frac{d}{dt}(\cos t)\int \log(1+t)\,dt\right]dt\right\} = \sin t.\log(1+t) - \left[\frac{\cos t}{1+t} + \int \sin t.\frac{1}{1+t}dt\right]$

$= \sin t.\log(1+t) - \dfrac{\cos t}{1+t} - I_1$

or $2I_1 = \sin t.\log(1+t) - \dfrac{\cos t}{1+t} = \dfrac{(1+t)\sin t.\log(1+t) - \cos t}{1+t} + c$ $\therefore\ I_1 = \dfrac{(1+t)\sin t.\log(1+t) - \cos t}{2(1+t)} + c$

where $I_2 = \int \frac{\cos t}{1+t} dt$ using integration by part formula

$$\text{or } I_2 = \int \frac{\cos t}{1+t} dt = \cos t . \int \frac{dt}{1+t} - \int \left[\frac{d}{dt}(\cos t) \int \frac{dt}{1+t}\right] dt = \sin t . \log(1+t) + \int \sin t . \log(1+t)\, dt$$

$$= \sin t . \log(1+t) + \left\{\sin t . \frac{1}{1+t} - \int \cos t . \frac{1}{1+t}\, dt\right\} = \sin t . \log(1+t) + \frac{\sin t}{1+t} - I_2$$

$$\text{or } 2I_2 = \sin t . \log(1+t) + \frac{\sin t}{(1+t)} + c = \frac{(1+t)\sin t . \log(1+t) + \sin t}{(1+t)} + c$$

$$\text{or } I = 2I_1 + 2I_2 = \frac{(1+t)\sin t . \log(1+t) - \cos t}{(1+t)} + c + \frac{(1+t)\sin t . \log(1+t) + \sin t}{(1+t)} + c$$

$$= \frac{(1+t)\sin t . \log(1+t) - \cos t + (1+t)\sin t . \log(1+t) + \sin t}{(1+t)} + k$$

$$\therefore I = \frac{(1+t)\log(1+t)\,[\sin t + \cos t] + (\sin t - \cos t)}{(1+t)} + k = \log(1+t)\,[\sin t + \cos t] + \frac{(\sin t - \cos t)}{(1+t)} + k \quad \text{Ans.}$$

Form of the integrals: – $\int \frac{dx}{a\cos^2 x + 2b\sin x\cos x + c\sin^2 x}$, $\int \frac{dx}{a\cos^2 x + b}$, $\int \frac{dx}{a + b\sin^2 x}$

In above type of question divide above and below by $\cos^2 x$.

Example: – (a) $I = \int \frac{dx}{3\cos^2 x + 2\sin x\cos x + \sin^2 x}$, Divide above and below by $\cos^2 x$

or $I = \int \frac{\sec^2 x}{3 + 2\tan x + \tan^2 x} dx$ Let $\tan x = t \ \therefore \sec^2 x\, dx = dt$

or $I = \int \frac{dt}{3 + 2t + t^2} = \int \frac{dt}{(t+1)^2 + (\sqrt{2})^2}$ $\left[\text{formula, } \int \frac{dx}{x^2 + a^2} = \frac{1}{a}\tan^{-1}\frac{x}{a}\right]$

$$\therefore I = \int \frac{dt}{(t+1)^2 + (\sqrt{2})^2} = \frac{1}{\sqrt{2}}\tan^{-1}\left(\frac{t+1}{\sqrt{2}}\right) = \frac{1}{\sqrt{2}}\tan^{-1}\left(\frac{\tan x + 1}{\sqrt{2}}\right) + c \quad \text{Ans.}$$

(b) $I = \int \frac{dx}{2\cos^2 x + 3}$ Divide above and below by $\cos^2 x$, we have

or $I = \int \frac{\sec^2 x}{2 + 3\sec^2 x} dx = \int \frac{\sec^2 x}{2 + 3(1 + \tan^2 x)} dx = \int \frac{\sec^2 x}{5 + 3\tan^2 x} dx$ Let $\tan x = t \ \therefore \sec^2 x\, dx = dt$

or $I = \int \frac{dt}{5 + 3t^2} = \int \frac{dt}{(\sqrt{5})^2 + (\sqrt{3}t)^2}$ $\left[\text{formula, } \int \frac{dx}{x^2 + a^2} = \frac{1}{a}\tan^{-1}\frac{x}{a}\right]$

$$\therefore I = \int \frac{dt}{(\sqrt{5})^2 + (\sqrt{3}t)^2} = \frac{1}{\sqrt{5}}\tan^{-1}\left(\frac{\sqrt{3}t}{\sqrt{5}}\right) . \frac{1}{\sqrt{3}} = \frac{1}{\sqrt{15}}\tan^{-1}\left(\frac{\sqrt{3}t}{\sqrt{5}}\right) + c = \frac{1}{\sqrt{15}}\tan^{-1}\left(\frac{\sqrt{3}\tan x}{\sqrt{5}}\right) + c \quad \text{Ans.}$$

(c) $I = \int \frac{dx}{3 - 5\sin^2 x}$ Divide above and below by $\cos^2 x$, we have

$$\text{or } I = \int \frac{\sec^2 x}{3\sec^2 x - 5\tan^2 x} dx = \int \frac{\sec^2 x}{3(1 + \tan^2 x) - 5\tan^2 x} dx = \int \frac{\sec^2 x}{3 + 3\tan^2 x - 5\tan^2 x} dx = \int \frac{\sec^2 x}{3 - 2\tan^2 x} dx$$

Let $\tan x = t \ \therefore \sec^2 x\, dx = dt \ \therefore I = \int \frac{\sec^2 x}{3 - 2\tan^2 x} dx = \int \frac{dt}{3 - 2t^2} = \int \frac{dt}{(\sqrt{3})^2 - (\sqrt{2}t)^2}$

Let $\sqrt{2}t = z \ \therefore \sqrt{2}\, dt = dz$ or $dt = \frac{dz}{\sqrt{2}}$ $\therefore I = \int \frac{dt}{(\sqrt{3})^2 - (\sqrt{2}t)^2} = \frac{1}{\sqrt{2}}\int \frac{dz}{(\sqrt{3})^2 - (z)^2}$

using formula, $\int \frac{dx}{a^2 - x^2} = \frac{1}{2a}\log\left|\frac{a+x}{a-x}\right|$, when $x < a$ *and put* $\tan x = t, z = \sqrt{2}t$

or $I = \frac{1}{\sqrt{2}}\int \frac{dz}{(\sqrt{3})^2 - (z)^2} = \frac{1}{\sqrt{2}}.\frac{1}{2.\sqrt{3}}\log\left|\frac{\sqrt{3}+z}{\sqrt{3}-z}\right| + c = \frac{1}{2\sqrt{6}}\log\left(\frac{\sqrt{3}+\sqrt{2}t}{\sqrt{3}-\sqrt{2}t}\right) + c = \frac{1}{2\sqrt{6}}\log\left(\frac{\sqrt{3}+\sqrt{2}\tan x}{\sqrt{3}-\sqrt{2}\tan x}\right) + c$ Ans.

Form of the integrals: – $\int \frac{dx}{a\cos x + b\sin x + c}$, $\int \frac{dx}{a + b\cos x}$ and $\int \frac{dx}{a + b\sin x}$

write $\cos x = \cos^2\frac{x}{2} - \sin^2\frac{x}{2}$, $\sin x = 2\sin\frac{x}{2}\cos\frac{x}{2}$ Then, Divide above and below by $\cos^2\frac{x}{2}$.

Form of the integrals: – $\int \frac{p\cos x + q\sin x + r}{a\cos x + b\sin x + c}dx$

Express, $p\cos x + q\sin x + r = l(a\cos x + b\sin x + c) + m(\text{d.c of } a\cos x + b\sin x + c) + n$

Divide both of sides by $(a\cos x + b\sin x + c)$ and integrating,

$$\int \frac{p\cos x + q\sin x + r}{a\cos x + b\sin x + c}dx = l\int \frac{a\cos x + b\sin x + c}{a\cos x + b\sin x + c}dx + m\int \frac{\text{d.c of }(a\cos x + b\sin x + c)}{a\cos x + b\sin x + c}dx + n\int \frac{dx}{a\cos x + b\sin x + c}$$

$$\int \frac{p\cos x + q\sin x + r}{a\cos x + b\sin x + c}dx = l\int dx + m\int \frac{\text{d.c of }(a\cos x + b\sin x + c)}{a\cos x + b\sin x + c}dx + n\int \frac{dx}{a\cos x + b\sin x + c}$$

find l, m and n by comparing the coefficient of $\sin x$, $\cos x$ and constant term.

Form of the integrals: – $\int \frac{p\cos x + q\sin x}{a\cos x + b\sin x}dx$

Express, $p\cos x + q\sin x = l(a\cos x + b\sin x) + m(\text{d.c of } a\cos x + b\sin x)$ and find l and m by comparing the coefficient of $\sin x \cos x$.

Divide both of side by $a\cos x + b\sin x$ (D^r) and integrating.

$$\int \frac{p\cos x + q\sin x}{a\cos x + b\sin x}dx = l\int \frac{a\cos x + b\sin x}{a\cos x + b\sin x}dx + m\int \frac{\text{d.c of } a\cos x + b\sin x}{a\cos x + b\sin x}dx = l\int dx + m\int \frac{\text{d.c of } a\cos x + b\sin x}{a\cos x + b\sin x}dx$$

Example: – (1) (a) $I = \int \frac{dx}{3\cos x + 5\sin x + 2} = \int \frac{dx}{3\left(\cos^2\frac{x}{2} - \sin^2\frac{x}{2}\right) + 5.2\sin\frac{x}{2}\cos\frac{x}{2} + 2} = \int \frac{dx}{3\cos^2\frac{x}{2} - 3\sin^2\frac{x}{2} + 10\sin\frac{x}{2}\cos\frac{x}{2} + 2}$

Divide above and below by $\cos^2\frac{x}{2}$, we have $\therefore I = \int \frac{\sec^2 x/2}{3 - 3\tan^2 x/2 + 10\tan x/2 + 2\sec^2 x/2}dx$

or $I = \int \frac{\sec^2 x/2\, dx}{3 - 3\tan^2 x/2 + 10\tan x/2 + 2(1 + \tan^2 x/2)} = \int \frac{\sec^2 x/2\, dx}{5 + 10\tan x/2 - \tan^2 x/2}$

Let $\tan x/2 = t$ $\therefore \sec^2 x/2.\frac{1}{2}dx = dt$ or $\sec^2 x/2\, dx = 2\,dt$

or $I = \int \frac{\sec^2 x/2\, dx}{5 + 10\tan x/2 - \tan^2 x/2} = \int \frac{2\,dt}{5 + 10t - t^2} = -2\int \frac{dt}{t^2 - 10t - 5} = -2\int \frac{dt}{(t-5)^2 - (\sqrt{30})^2}$

using formula, $\int \frac{dx}{x^2 - a^2} = \frac{1}{2a}\log\left|\frac{x-a}{x+a}\right|$, when $x > a$

$\therefore I = -2\int \frac{dt}{(t-5)^2 - (\sqrt{30})^2} = -2.\frac{1}{2.\sqrt{30}}\log\left|\frac{(t-5) - \sqrt{30}}{(t-5) + \sqrt{30}}\right| + c = -\frac{1}{\sqrt{30}}\log\left[\frac{(\tan x/2 - 5) - \sqrt{30}}{(\tan x/2 - 5) + \sqrt{30}}\right] + c$ Ans.

(b) $I = \int \frac{dx}{3 - 5\sin x} = \int \frac{dx}{3 - 5.2\sin\frac{x}{2}\cos\frac{x}{2}} = \int \frac{dx}{3 - 10\sin\frac{x}{2}\cos\frac{x}{2}}$ Divide above and below by $\cos^2 x/2$, we have

or $I = \int \frac{\sec^2 x/2\, dx}{3\sec^2 x/2 - 10\tan x/2} = \int \frac{\sec^2 x/2\, dx}{3(1 + \tan^2 x/2) - 10\tan x/2} = \int \frac{\sec^2 x/2\, dx}{3 + 3\tan^2 x/2 - 10\tan x/2}$

Let $\tan x/2 = t$ $\therefore \sec^2 x/2.\frac{1}{2}dx = dt$ or $\sec^2 x/2\, dx = 2\,dt$

$$\text{or } I=\int\frac{\sec^2{}^x/_2\,dx}{3+3\tan^2{}^x/_2-10\tan{}^x/_2}=\int\frac{2\,dt}{3+3t^2-10t}=2\int\frac{dt}{3t^2-10t+3}$$

$$\text{using }\left[ax^2+bx+c=0\ \therefore\ a\left\{\left(x+\frac{b}{2a}\right)^2-\frac{D}{4a^2}\right\}=0\text{ and }\int\frac{dx}{x^2-a^2}=\frac{1}{2a}\log\left|\frac{x-a}{x+a}\right|,\ \text{when } x>a\right]$$

$$\text{or } I=2\int\frac{dt}{3t^2-10t+3}=2\int\frac{dt}{\left(\sqrt{3}t-\frac{5}{\sqrt{3}}\right)^2-\left(\frac{4}{\sqrt{3}}\right)^2}=2.\frac{1}{2.\frac{4}{\sqrt{3}}}\log\left|\frac{\sqrt{3}t-\frac{5}{\sqrt{3}}-\frac{4}{\sqrt{3}}}{\sqrt{3}t-\frac{5}{\sqrt{3}}+\frac{4}{\sqrt{3}}}\right|.\frac{1}{\sqrt{3}}+c=\frac{1}{4}\log\left(\frac{3t-5-4}{3t-5+4}\right)+c$$

$$\therefore\ I=\frac{1}{4}\log\left(\frac{3t-9}{3t-1}\right)+c=\frac{1}{4}\log\left(\frac{3\tan{}^x/_2-9}{3\tan{}^x/_2-1}\right)+c\quad\text{Ans.}$$

(c) $I=\int\frac{dx}{3+\cos x}=\int\frac{dx}{3+\cos^2{}^x/_2-\sin^2{}^x/_2}$, Divide above and below by $\cos^2{}^x/_2$ we have

$$\text{or } I=\int\frac{\sec^2{}^x/_2\,dx}{3\sec^2{}^x/_2+1-\tan^2{}^x/_2}=\int\frac{\sec^2{}^x/_2\,dx}{3(1+\tan^2{}^x/_2)+1-\tan^2{}^x/_2}=\int\frac{\sec^2{}^x/_2\,dx}{3+3\tan^2{}^x/_2+1-\tan^2{}^x/_2}=\int\frac{\sec^2{}^x/_2\,dx}{2\tan^2{}^x/_2+4}$$

$$=\int\frac{\sec^2{}^x/_2\,dx}{2(\tan^2{}^x/_2+2)}=\frac{1}{2}\int\frac{\sec^2{}^x/_2\,dx}{(\tan^2{}^x/_2+2)}$$

Let $\tan{}^x/_2=t\quad\therefore\ \sec^2{}^x/_2.\frac{1}{2}dx=dt$ or $\sec^2{}^x/_2\,dx=2\,dt$

$$\text{or } I=2.\frac{1}{2}\int\frac{dt}{2+t^2}=\int\frac{dt}{(\sqrt{2})^2+t^2}=\frac{1}{\sqrt{2}}\tan^{-1}\left(\frac{t}{\sqrt{2}}\right)+c\quad\left[\text{using formula, }\int\frac{dx}{x^2+a^2}=\frac{1}{a}\tan^{-1}\frac{x}{a}\right]$$

$$\therefore\ I=\frac{1}{\sqrt{2}}\tan^{-1}\left(\frac{t}{\sqrt{2}}\right)+c=\frac{1}{\sqrt{2}}\tan^{-1}\left(\frac{\tan{}^x/_2}{\sqrt{2}}\right)+c\quad\text{Ans.}$$

$$\text{(d) } I=\int\frac{1+\cos x}{\sin x+\cos x}dx=\int\frac{1+\cos^2{}^x/_2-\sin^2{}^x/_2}{2\sin{}^x/_2\cos{}^x/_2+\cos^2{}^x/_2-\sin^2{}^x/_2}dx=\int\frac{2\cos^2{}^x/_2}{2\sin{}^x/_2\cos{}^x/_2+\cos^2{}^x/_2-\sin^2{}^x/_2}dx$$

Divide above and below by $\cos^2{}^x/_2$, we have

$$\text{or } I=\int\frac{2\cos^2{}^x/_2}{2\sin{}^x/_2\cos{}^x/_2+\cos^2{}^x/_2-\sin^2{}^x/_2}dx=\int\frac{2}{2\tan{}^x/_2+1-\tan^2{}^x/_2}dx=2\int\frac{dx}{1+2\tan{}^x/_2-\tan^2{}^x/_2}$$

Let $\tan{}^x/_2=t\quad\therefore\ \sec^2{}^x/_2.\frac{1}{2}dx=dt$ or $\sec^2{}^x/_2\,dx=2\,dt$

$$\text{or } I=4\int\frac{dt}{1+2t-t^2}=-4\int\frac{dt}{t^2-2t-1}=-4\int\frac{dt}{(t-1)^2-(\sqrt{2})^2}\quad\left[\int\frac{dx}{x^2-a^2}=\frac{1}{2a}\log\left|\frac{x-a}{x+a}\right|,\ \text{when } x>a\right]$$

$$\therefore\ I=-4.\frac{1}{2.\sqrt{2}}\log\left|\frac{(t-1)-\sqrt{2}}{(t-1)+\sqrt{2}}\right|+c=-\sqrt{2}\log\left|\frac{[(\tan{}^x/_2-1)-\sqrt{2}]}{[(\tan{}^x/_2-1)+\sqrt{2}]}\right|+c\quad\text{Ans.}$$

(2) (a) $I=\int\frac{3\cos x+5\sin x+7}{2\cos x+3\sin x+8}dx=\int\frac{p\cos x+q\sin x+r}{a\cos x+b\sin x+c}dx$ (integral of the form)

using $\int\frac{p\cos x+q\sin x+r}{a\cos x+b\sin x+c}dx$ Express, $p\cos x+q\sin x+r=l(a\cos x+b\sin x+c)+m(\text{d.c of } a\cos x+b\sin x+c)+n$

Divide both of sides by $(a\cos x+b\sin x+c)$ and integrating,

Solution: $-\quad I=\int\frac{3\cos x+5\sin x+7}{2\cos x+3\sin x+8}dx$

$$3\cos x+5\sin x+7=l(2\cos x+3\sin x+8)+m(\text{d.c of }2\cos x+3\sin x+8)+n$$

find l, m and n by comparing the coefficient of $\sin x$, $\cos x$ and constant.

$$3\cos x + 5\sin x + 7 = l(2\cos x + 3\sin x + 8) + m(3\cos x - 2\sin x) + n \ldots\ldots\ldots\ldots (A)$$

$\therefore\ 2l + 3m = 3 \ldots\ldots\ldots (i)$, $3l - 2m = 5 \ldots\ldots\ldots\ldots (ii)$ and $8l + n = 7 \ldots\ldots\ldots\ldots (iii)$

solve (i) & (ii), we get $2l + 3m = 3$ and $3l - 2m = 5$ or $2m = 3l - 5$ or $m = \frac{3l-5}{2}$

Put value of m in equation (i), we get $2l + 3\left(\frac{3l-5}{2}\right) = 3$ or $4l + 9l - 15 = 6$ or $13l = 21$ $\therefore\ l = \frac{21}{13}$

Put value of l in equation (i) and (iii) then find value of m and n, we get

or $2l + 3m = 3$ or $2.\frac{21}{13} + 3m = 3$ or $3m = 3 - \frac{42}{13} = -\frac{3}{13}$ $\therefore\ m = -\frac{1}{13}$

and $8l + n = 7$ or $n = 7 - 8.\frac{21}{13} = \frac{91-168}{13} = -\frac{77}{13}$

Put value of l, m and n in equation (A), we get

$$3\cos x + 5\sin x + 7 = l(2\cos x + 3\sin x + 8) + m(3\cos x - 2\sin x) + n = \frac{21}{13}(2\cos x + 3\sin x + 8) - \frac{1}{13}(3\cos x - 2\sin x) - \frac{77}{13}$$

Divide both of sides by $(2\cos x + 3\sin x + 8)$ and integrate

$$\int \frac{3\cos x + 5\sin x + 7}{2\cos x + 3\sin x + 8}dx = \frac{21}{13}\int \frac{2\cos x + 3\sin x + 8}{2\cos x + 3\sin x + 8}dx - \frac{1}{13}\int \frac{3\cos x - 2\sin x}{2\cos x + 3\sin x + 8}dx - \frac{77}{13}\int \frac{dx}{2\cos x + 3\sin x + 8}$$

$$\text{or } I = \int \frac{3\cos x + 5\sin x + 7}{2\cos x + 3\sin x + 8}dx = \frac{21}{13}\int dx - \frac{1}{13}\int \frac{3\cos x - 2\sin x}{2\cos x + 3\sin x + 8}dx - \frac{77}{13}\int \frac{dx}{2\cos x + 3\sin x + 8}$$

Let $I_1 = \int \frac{3\cos x - 2\sin x}{2\cos x + 3\sin x + 8}dx$ and $I_2 = \int \frac{dx}{2\cos x + 3\sin x + 8}$

$I_1 = \int \frac{3\cos x - 2\sin x}{2\cos x + 3\sin x + 8}dx$ Let $2\cos x + 3\sin x + 8 = t$ $\therefore (-2\sin x + 3\cos x)dx = dt$ $\therefore (3\cos x - 2\sin x)dx$ $= dt$ $\left[\text{formula, } \int \frac{f'(x)}{f(x)}dx = \log[f(x)]\right]$

$$\therefore\ I_1 = \int \frac{dt}{t} = \log t + c = \log(2\cos x + 3\sin x + 8) + c$$

$$\text{Now, } I_2 = \int \frac{dx}{2\cos x + 3\sin x + 8} = \int \frac{dx}{2(\cos^2 {}^x/_2 - \sin^2 {}^x/_2) + 3.2\sin {}^x/_2 \cos {}^x/_2 + 8}$$

Divide above and below by $\cos^2 {}^x/_2$

$$I_2 = \int \frac{\sec^2 {}^x/_2\ dx}{2 - 2\tan^2 {}^x/_2 + 6\tan {}^x/_2 + 8\sec^2 {}^x/_2} = \int \frac{\sec^2 {}^x/_2\ dx}{2 - 2\tan^2 {}^x/_2 + 6\tan {}^x/_2 + 8(1 + \tan^2 {}^x/_2)} = \int \frac{\sec^2 {}^x/_2\ dx}{6\tan^2 {}^x/_2 + 6\tan {}^x/_2 + 10}$$

Let $\tan {}^x/_2 = t$ $\therefore\ \sec^2 {}^x/_2 . \frac{1}{2}dx = dt$ or $\sec^2 {}^x/_2\ dx = 2\,dt$

$$I_2 = \int \frac{2\,dt}{6t^2 + 6t + 10} = \int \frac{2\,dt}{2(3t^2 + 3t + 5)} = \int \frac{dt}{3t^2 + 3t + 5} = \int \frac{dt}{\left(\sqrt{3}t + \frac{\sqrt{3}}{2}\right)^2 + \left(\frac{\sqrt{17}}{2}\right)^2} \quad \left[\text{use formula, } \int \frac{dx}{x^2 + a^2} = \frac{1}{a}\tan^{-1} {}^x/_a . \frac{1}{\text{d. c of x}}\right]$$

$$I_2 = \int \frac{dt}{\left(\sqrt{3}t + \frac{\sqrt{3}}{2}\right)^2 + \left(\frac{\sqrt{17}}{2}\right)^2} = \frac{1}{\frac{\sqrt{17}}{2}}\tan^{-1}\left(\sqrt{3}t \Big/ \frac{\sqrt{17}}{2}\right).\frac{1}{\sqrt{3}} = \frac{2}{\sqrt{51}}\tan^{-1}\left(\frac{2\sqrt{3}t}{\sqrt{17}}\right) + c = \frac{2}{\sqrt{51}}\tan^{-1}\left(\frac{2\sqrt{3}\tan {}^x/_2}{\sqrt{17}}\right) + c$$

$$\text{or } I = \frac{21}{13}\int dx - \frac{1}{13}I_1 - \frac{77}{13}I_2$$

$$\text{or } I = \frac{21}{13}x + c - \frac{1}{13}[\log(2\cos x + 3\sin x + 8) + c] - \frac{77}{13}\left[\frac{2}{\sqrt{51}}\tan^{-1}\left(\frac{2\sqrt{3}\tan {x}/{2}}{\sqrt{17}}\right) + c\right]$$

$$\therefore\ I = \frac{21}{13}x - \frac{1}{13}[\log(2\cos x + 3\sin x + 8)] - \frac{77}{13}\left[\frac{2}{\sqrt{51}}\tan^{-1}\left(\frac{2\sqrt{3}\tan {x}/{2}}{\sqrt{17}}\right)\right] + k \quad \text{Ans.}$$

$$\text{where } k = \left(c - \frac{c}{13} - \frac{77c}{13}\right)$$

(b) $I = \int \frac{7\cos x + 4\sin x}{2\cos x + 3\sin x}dx$, use form of integral $\int \frac{p\cos x + q\sin x}{a\cos x + b\sin x}dx$

Express, $p\cos x + q\sin x = l(a\cos x + b\sin x) + m(\text{d.c of } a\cos x + b\sin x)$

and find l and m by comparing the coefficient of $\sin x \cos x$.

Divide both of side by $a\cos x + b\sin x$ (D^r) and integrating.

Solution: – $I = \int \frac{7\cos x + 4\sin x}{2\cos x + 3\sin x}dx$

$$7\cos x + 4\sin x = l(2\cos x + 3\sin x) + m(\text{d.c of } 2\cos x + 3\sin x) = l(2\cos x + 3\sin x) + m(3\cos x - 2\sin x) \ldots\ldots\ldots\ldots\ldots\ldots (A)$$

find l and m by comparing the coefficient of $\sin x, \cos x$.

or $2l + 3m = 7 \ldots\ldots\ldots\ldots\ldots\ldots$ (i) and $3l - 2m = 4 \ldots\ldots\ldots\ldots\ldots\ldots\ldots$ (ii)

solve equation (i) and (ii) and find the value of l and m, we have $\therefore\ l = 2$ and $m = 1$

Now, Put value of l and m in equation (A), we have

$$7\cos x + 4\sin x = 2(2\cos x + 3\sin x) + 1(3\cos x - 2\sin x)$$

Divide both of sides by $(2\cos x + 3\sin x)$ and integrating

$$\int \frac{7\cos x + 4\sin x}{2\cos x + 3\sin x}dx = 2\int \frac{2\cos x + 3\sin x}{2\cos x + 3\sin x}dx + \int \frac{3\cos x - 2\sin x}{2\cos x + 3\sin x}dx = 2\int dx + \int \frac{3\cos x - 2\sin x}{2\cos x + 3\sin x}dx = 2\int dx + I_1$$

$I_1 = \int \frac{3\cos x - 2\sin x}{2\cos x + 3\sin x}dx$ Let $2\cos x + 3\sin x = t$ $\therefore (-2\sin x + 3\cos x)dx = dt$ $\therefore\ (3\cos x - 2\sin x)dx = dt$

$$I_1 = \int \frac{dt}{t} = \log t + c = \log(2\cos x + 3\sin x) + c$$

$$\therefore\ I = 2\int dx + I_1 = 2x + c + \log(2\cos x + 3\sin x) + c = 2x + \log(2\cos x + 3\sin x) + k \quad \text{Ans.} \quad \text{where } 2c = k$$

(c) $I = \int \frac{\tan x + \cot x}{\sec x - \operatorname{cosec} x}dx = \int \frac{\frac{\sin x}{\cos x} + \frac{\cos x}{\sin x}}{\frac{1}{\cos x} - \frac{1}{\sin x}}dx = \int \frac{\frac{\sin^2 x + \cos^2 x}{\sin x \cos x}}{\frac{\sin x - \cos x}{\sin x \cos x}}dx = \int \frac{1}{\sin x - \cos x}dx = \int \frac{1}{2\sin {x}/{2}\cos {x}/{2} - (\cos^2 {x}/{2} - \sin^2 {x}/{2})}dx$

$$= \int \frac{1}{2\sin {x}/{2}\cos {x}/{2} - \cos^2 {x}/{2} + \sin^2 {x}/{2}}dx$$

Divide above and below by $\cos^2 {x}/{2}$

$I = \int \frac{\sec^2 {x}/{2}\ dx}{2\tan {x}/{2} - 1 + \tan^2 {x}/{2}}$ Let $\tan {x}/{2} = t$ $\therefore\ \sec^2 {x}/{2} . \frac{1}{2}dx = dt$ or $\sec^2 {x}/{2}\ dx = 2\ dt$

$$I = \int \frac{2\ dt}{2t - 1 + t^2} = 2\int \frac{dt}{t^2 + 2t - 1} = 2\int \frac{dt}{(t+1)^2 - (\sqrt{2})^2}$$

$$\left[\text{use formula, } \int \frac{dx}{x^2 - a^2} = \frac{1}{2a}\log\left|\frac{x-a}{x+a}\right|\right]$$

$$I = 2.\frac{1}{2.\sqrt{2}}\log\left[\frac{(t+1)-\sqrt{2}}{(t+1)+\sqrt{2}}\right] + c = \frac{1}{\sqrt{2}}\log\left[\frac{(\tan{}^{x}/_{2}+1)-\sqrt{2}}{(\tan{}^{x}/_{2}+1)+\sqrt{2}}\right] + c \quad \text{Ans.}$$

(d) $I = \int \frac{\cos 2\theta}{\sin^4\theta - \cos^4\theta} d\theta = \int \frac{\cos^2\theta - \sin^2\theta}{\sin^4\theta - \cos^4\theta} d\theta$

Divide above and below by $\cos^4\theta$

$$I = \int \frac{\frac{\cos^2\theta - \sin^2\theta}{\cos^4\theta}}{\frac{\sin^4\theta - \cos^4\theta}{\cos^4\theta}} d\theta = \int \frac{\sec^2\theta - \tan^2\theta\sec^2\theta}{\tan^4\theta - 1} d\theta = \int \frac{\sec^2\theta\,(1-\tan^2\theta)}{\tan^4\theta - 1} d\theta$$

Let $\tan\theta = t \;\therefore\; \sec^2\theta\, d\theta = dt$

$$I = \int \frac{(1-t^2)}{t^4-1} dt = \int \frac{-(t^2-1)}{(t^2)^2-1} dt = -\int \frac{(t^2-1)}{(t^2-1)(t^2+1)} dt = -\int \frac{dt}{(t^2+1)}$$

$$\left[\text{formula} \quad \int \frac{dx}{x^2+a^2} = \frac{1}{a}\tan^{-1}{}^{x}/_{a}\right]$$

$$I = -\int \frac{dt}{1+t^2} = -\frac{1}{1}\tan^{-1}{}^{t}/_{1} + c = -\tan^{-1}t + c = -\tan^{-1}(\tan\theta) + c = -\theta + c \quad \text{Ans.}$$

\# **Form of the integrals**: – (a) $\int \sqrt{\sec^2 x \pm a^2}\, dx$ (b) $\int \sqrt{\operatorname{cosec}^2 x \pm a^2}\, dx$

(c) $\int \sqrt{\tan^2 x \pm a^2}\, dx$ (d) $\int \sqrt{\cot^2 x \pm a^2}\, dx$

Solution: – (a) Let $I = \int \sqrt{\sec^2 x \pm a^2}\, dx = \int \frac{\sqrt{\sec^2 x \pm a^2}}{\sqrt{\sec^2 x \pm a^2}} \times \sqrt{\sec^2 x \pm a^2}\, dx = \int \frac{\sec^2 x \pm a^2}{\sqrt{\sec^2 x \pm a^2}} dx$

$$I = \int \frac{\sec^2 x \pm a^2}{\sqrt{\sec^2 x \pm a^2}} dx = \int \frac{\sec^2 x\, dx}{\sqrt{\sec^2 x \pm a^2}} \pm \int \frac{a^2}{\sqrt{\sec^2 x \pm a^2}} dx = \int \frac{\sec^2 x\, dx}{\sqrt{(1+\tan^2 x) \pm a^2}} \pm a^2 \int \frac{dx}{\sqrt{\frac{1}{\cos^2 x} \pm a^2}}$$

$$= \int \frac{\sec^2 x\, dx}{\sqrt{(1\pm a^2) + \tan^2 x}} \pm a^2 \int \frac{\cos x\, dx}{\sqrt{1 \pm a^2\cos^2 x}} = \int \frac{\sec^2 x\, dx}{\sqrt{(1\pm a^2) + \tan^2 x}} \pm a^2 \int \frac{\cos x\, dx}{\sqrt{1 \pm a^2(1-\sin^2 x)}}$$

In first part putting $t = \tan x$ and in second part putting $z = \sin x$

$\therefore\; t = \tan x \quad \therefore\; dt = \sec^2 x\, dx \quad$ and $\quad z = \sin x \quad \therefore\; dz = \cos x\, dx$

$$I = \int \frac{dt}{\sqrt{(1\pm a^2) + t^2}} \pm a^2 \int \frac{dz}{\sqrt{1 \pm a^2(1-z^2)}} = \int \frac{dt}{\sqrt{\left(\sqrt{(1\pm a^2)}\right)^2 + t^2}} \pm a^2 \int \frac{dz}{\sqrt{\left(\sqrt{(1\pm a^2)}\right)^2 - (az)^2}}$$

Using formula, $\int \frac{dx}{\sqrt{a^2+x^2}} = \log\left(x + \sqrt{a^2+x^2}\right)$ or $\sinh^{-1}\left(\frac{x}{a}\right)$ and $\int \frac{dx}{\sqrt{a^2-x^2}} = \sin^{-1}\left(\frac{x}{a}\right)$

$$I = \int \frac{dt}{\sqrt{\left(\sqrt{(1\pm a^2)}\right)^2 + t^2}} \pm a^2 \int \frac{dz}{\sqrt{\left(\sqrt{(1\pm a^2)}\right)^2 - (az)^2}} = \log\left(t + \sqrt{\left(\sqrt{(1\pm a^2)}\right)^2 + t^2}\right) \pm \frac{a^2}{a}\sin^{-1}\left(\frac{az}{\sqrt{(1\pm a^2)}}\right)$$

$$I = \log\left(\tan x + \sqrt{\left(\sqrt{(1\pm a^2)}\right)^2 + (\tan x)^2}\right) \pm a\sin^{-1}\left(\frac{a\sin x}{\sqrt{(1\pm a^2)}}\right) + c \quad \text{Ans.}$$

(b), (c) and (d) Do yourself. (solve same as above question)

Integral of the form: – $\int \frac{dx}{P\sqrt{Q}}$ and $\int \frac{f(x)}{P\sqrt{Q}}dx$

Linear	Quadratic	Substitution
P, Q	– – – – – –	$Q = z^2$
Q	P	$Q = z^2$
P	Q	$P = {}^1/_z$
– – – – – –	P, Q	$z^2 = {}^Q/_P$

Example: – (a) $\int \frac{dx}{(x+2)\sqrt{x-3}}$ (b) $\int \frac{dx}{(x^2+2x+3)\sqrt{2-x}}$ (c) $\int \frac{dx}{(1+x)\sqrt{2x^2+2x+1}}$

(d) $\int \frac{dx}{x^2\sqrt{x^2+1}}$ (e) $\int \frac{dx}{(x^2+1)\sqrt{x^2-1}}$

Solution: – (a) Let $I = \int \frac{dx}{(x+2)\sqrt{x-3}}$ $\left[(x+2) \text{ and } (x-3) \text{ are linear then use formula } \int \frac{dx}{P\sqrt{Q}}, \text{ put } Q = z^2\right]$

Putting $Q = z^2$ $\therefore x - 3 = z^2$ or $x = z^2 + 3$ or $z = \sqrt{x-3}$ $\therefore dx = 2z\,dz$

$$I = \int \frac{dx}{(x+2)\sqrt{x-3}} = \int \frac{2z\,dz}{(z^2+3+2)\sqrt{z^2}} = \int \frac{2z\,dz}{(z^2+5)z} = 2\int \frac{dz}{z^2+5} \left[\text{using formula}, \int \frac{dx}{x^2+a^2} = \frac{1}{a}\tan^{-1}\left(\frac{x}{a}\right)\right]$$

$$I = 2\int \frac{dz}{z^2+(\sqrt{5})^2} = 2.\frac{1}{\sqrt{5}}\tan^{-1}\left(\frac{z}{\sqrt{5}}\right) + c = \frac{2}{\sqrt{5}}\tan^{-1}\left(\frac{\sqrt{x-3}}{\sqrt{5}}\right) + c \quad \text{Ans.} \quad \text{put } z = \sqrt{x-3}$$

(b) Let $I = \int \frac{dx}{(x^2+2x+3)\sqrt{2-x}} = \int \frac{dx}{P\sqrt{Q}}$ Here P is quadratic and Q is linear then putting $Q = z^2$

Putting $Q = z^2$ or $2 - x = z^2$ or $z = \sqrt{2-x}$ or $x = 2 - z^2$ $\therefore dx = -2z\,dz$

$$I = -\int \frac{2z\,dz}{[(2-z^2)^2 + 2(2-z^2) + 3]\sqrt{z^2}} = -\int \frac{2z\,dz}{z[4 - 4z^2 + z^4 + 4 - 2z^2 + 3]} = -2\int \frac{dz}{z^4 - 6z^2 + 11}$$

$$I = -2\int \frac{dz}{(z^2-3)^2+2} = -2\int \frac{dz}{(z^2-3)^2+(\sqrt{2})^2} \left[\text{using formula}, \int \frac{dx}{x^2+a^2} = \frac{1}{a}\tan^{-1}\left(\frac{x}{a}\right).\frac{1}{\text{d.c of x}}\right]$$

$$I = -2.\frac{1}{\sqrt{2}}\tan^{-1}\left[\frac{(z^2-3)}{\sqrt{2}}\right].\frac{1}{\text{d.c of }(z^2-3)} = -2.\frac{1}{\sqrt{2}}\tan^{-1}\left[\frac{(z^2-3)}{\sqrt{2}}\right].\frac{1}{2z} = -\frac{1}{\sqrt{2}}\tan^{-1}\left[\frac{2-x-3}{\sqrt{2}}\right].\frac{1}{\sqrt{2-x}} + c$$

$$I = -\frac{1}{\sqrt{2}.\sqrt{2-x}}\tan^{-1}\left[\frac{-(x+1)}{\sqrt{2}}\right] + c \quad \text{Ans.} \quad \left[\text{put } 2 - x = z^2 \ \therefore z = \sqrt{2-x}\right]$$

(c) Let $I = \int \frac{dx}{(1+x)\sqrt{2x^2+2x+1}} = \int \frac{dx}{P\sqrt{Q}}$ Here P is linear and Q is quadratic then putting $P = {}^1/_z$

Put $1 + x = \frac{1}{z}$ $\therefore dx = -\frac{1}{z^2}dz$ or $x = \frac{1}{z} - 1 = \frac{1-z}{z}$ or $z = \frac{1}{1+x}$

$$I = \int \frac{-\frac{1}{z^2}dz}{\frac{1}{z}.\sqrt{2\left(\frac{1-z}{z}\right)^2 + 2\left(\frac{1-z}{z}\right) + 1}} = -\int \frac{dz}{z.\sqrt{\frac{2(1-2z+z^2)}{z^2} + \frac{2-2z}{z} + 1}} = -\int \frac{dz}{z.\sqrt{\frac{2 - 4z + 2z^2 + 2z - 2z^2 + z^2}{z^2}}} = -\int \frac{dz}{\sqrt{z^2-2z+2}}$$

$$= -\int \frac{dz}{\sqrt{(z-1)^2+1}} \quad \left[\text{using formula}, \int \frac{dx}{x^2+a^2} = \frac{1}{a}\tan^{-1}\left(\frac{x}{a}\right)\right]$$

$$I = -\int \frac{dz}{\sqrt{(z-1)^2 + 1^2}} = -\frac{1}{1}\tan^{-1}\left(\frac{z-1}{1}\right) = -\tan^{-1}\left(\frac{1}{1+x} - 1\right) + c = -\tan^{-1}\left(\frac{1-1-x}{1+x}\right) + c = -\tan^{-1}\left(\frac{-x}{1+x}\right) + c \quad \text{Ans.}$$

(d) Let $I = \int \frac{dx}{x^2\sqrt{x^2+1}}$

Particular Cases: – $\int \frac{dx}{(ax^2+b)\sqrt{cx^2+d}}$

Putting $x = \frac{1}{t}$ and simplify and then putting expression under radical sign $= z^2$.

Solution: – $I = \int \frac{dx}{x^2\sqrt{x^2+1}}$ Put $x = \frac{1}{t}$ $\therefore dx = -\frac{1}{t^2}dt$ and $t = \frac{1}{x}$

$$I = \int \frac{-\frac{1}{t^2}dt}{\left(\frac{1}{t^2}\right)\sqrt{\left(\frac{1}{t}\right)^2 + 1}} = -\int \frac{dt}{\sqrt{\frac{1+t^2}{t^2}}} = -\int \frac{t\,dt}{\sqrt{1+t^2}} \quad \text{Again put } 1 + t^2 = z^2 \;\therefore\; 2t\,dt = 2z\,dz \;\therefore\; t\,dt = z\,dz$$

$$I = -\int \frac{z\,dz}{\sqrt{z^2}} = -\int \frac{z\,dz}{z} = -\int dz = -z + c = -\sqrt{1+t^2} + c = -\sqrt{1+\frac{1}{x^2}} + c = -\frac{\sqrt{1+x^2}}{x} + c \quad \text{Ans.}$$

(e) Let $I = \int \frac{dx}{(x^2+1)\sqrt{x^2-1}}$ Put $x = \frac{1}{t}$ $\therefore dx = -\frac{1}{t^2}dt$ and $t = \frac{1}{x}$

$$I = \int \frac{-\frac{1}{t^2}dt}{\left(\frac{1}{t^2}+1\right)\sqrt{\frac{1}{t^2}-1}} = -\int \frac{dt}{t^2.\left(\frac{1+t^2}{t^2}\right)\sqrt{\frac{1-t^2}{t^2}}} = -\int \frac{t\,dt}{(1+t^2)\sqrt{1-t^2}}$$

Again put $1 - t^2 = z^2$ $\therefore -2t\,dt = 2z\,dz$ $\therefore -t\,dt = z\,dz$ and $t^2 = 1 - z^2$

$$I = \int \frac{-t\,dt}{(1+t^2)\sqrt{1-t^2}} = \int \frac{z\,dz}{(1+1-z^2)\sqrt{z^2}} = \int \frac{z\,dz}{(2-z^2).z} = \int \frac{dz}{2-z^2} \quad \left[\text{use formula, } \int \frac{dx}{a^2-x^2} = \frac{1}{2a}\log\left|\frac{a+x}{a-x}\right|\right]$$

$$I = \int \frac{dz}{2-z^2} = \int \frac{dz}{\left(\sqrt{2}\right)^2 - z^2} = \frac{1}{2.\sqrt{2}}\log\left[\frac{\sqrt{2}+z}{\sqrt{2}-z}\right] + c = \frac{1}{2\sqrt{2}}\log\left[\frac{\sqrt{2}+\sqrt{1-t^2}}{\sqrt{2}-\sqrt{1-t^2}}\right] + c = \frac{1}{2\sqrt{2}}\log\left[\frac{\sqrt{2}+\sqrt{1-\frac{1}{x^2}}}{\sqrt{2}-\sqrt{1-\frac{1}{x^2}}}\right] + c$$

$$I = \frac{1}{2\sqrt{2}}\log\left[\frac{\sqrt{2}+\sqrt{\frac{x^2-1}{x^2}}}{\sqrt{2}-\sqrt{\frac{x^2-1}{x^2}}}\right] + c = \frac{1}{2\sqrt{2}}\log\left[\frac{\frac{\sqrt{2}x+\sqrt{x^2-1}}{x}}{\frac{\sqrt{2}x-\sqrt{x^2-1}}{x}}\right] + c = \frac{1}{2\sqrt{2}}\log\left[\frac{\sqrt{2}x+\sqrt{x^2-1}}{\sqrt{2}x-\sqrt{x^2-1}}\right] + c \text{ Ans.}$$

Integral of the form: – $\int \frac{dx}{1 \pm \sin ax}$, $\int \frac{dx}{\sqrt{1 \pm \sin ax}}$ and $\int \sqrt{1 \pm \sin ax}\,dx$

Working rule: – change $\sin ax$ into $\cos\left(\frac{\pi}{2} \pm ax\right)$ and then use result of integration.

Example: – (a) $I = \int \frac{dx}{1+\sin x} = \int \frac{dx}{1+\cos\left(\frac{\pi}{2}-x\right)} = \int \frac{dx}{2\cos^2\left(\frac{\frac{\pi}{2}-x}{2}\right)} = \frac{1}{2}\int \frac{dx}{\cos^2\left(\frac{\pi}{4}-\frac{x}{2}\right)}$

$$I = \frac{1}{2}\int \sec^2\left(\frac{\pi}{4}-\frac{x}{2}\right)dx = -\frac{1}{2}.\frac{\tan\left(\frac{\pi}{4}-\frac{x}{2}\right)}{-\frac{1}{2}} + c = \tan\left(\frac{\pi}{4}-\frac{x}{2}\right) + c \quad \text{Ans.}$$

Remember: – $\sin\left(\frac{\pi}{2}-x\right) = \sin\frac{\pi}{2}.\cos x - \cos\frac{\pi}{2}.\sin x = \cos x - 0 = \cos x$

$\sin\left(\frac{\pi}{2}+x\right) = \sin\frac{\pi}{2}.\cos x + \cos\frac{\pi}{2}.\sin x = \cos x + 0 = \cos x$

$\cos\left(\frac{\pi}{2}-x\right) = \cos\frac{\pi}{2}.\cos x + \sin\frac{\pi}{2}.\sin x = 0 + \sin x = \sin x$

$\cos\left(\frac{\pi}{2}+x\right) = \cos\frac{\pi}{2}.\cos x - \sin\frac{\pi}{2}.\sin x = 0 - \sin x = -\sin x$

$\tan\left(\frac{\pi}{2}-x\right) = \frac{\sin\left(\frac{\pi}{2}-x\right)}{\cos\left(\frac{\pi}{2}-x\right)} = \frac{\cos x}{\sin x} = \cot x\,,\quad \cot\left(\frac{\pi}{2}-x\right) = \frac{\cos\left(\frac{\pi}{2}-x\right)}{\sin\left(\frac{\pi}{2}-x\right)} = \frac{\sin x}{\cos x} = \tan x$

$\tan\left(\frac{\pi}{2}+x\right) = \frac{\sin\left(\frac{\pi}{2}+x\right)}{\cos\left(\frac{\pi}{2}+x\right)} = \frac{\cos x}{-\sin x} = -\cot x\,,\quad \cot\left(\frac{\pi}{2}+x\right) = \frac{\cos\left(\frac{\pi}{2}+x\right)}{\sin\left(\frac{\pi}{2}+x\right)} = \frac{-\sin x}{\cos x} = -\tan x$

$\sec\left(\frac{\pi}{2}-x\right) = \frac{1}{\cos\left(\frac{\pi}{2}-x\right)} = \frac{1}{\sin x} = \operatorname{cosec} x\,,\quad \sec\left(\frac{\pi}{2}+x\right) = \frac{1}{\cos\left(\frac{\pi}{2}+x\right)} = \frac{1}{-\sin x} = -\operatorname{cosec} x$

$\operatorname{cosec}\left(\frac{\pi}{2}-x\right) = \frac{1}{\sin\left(\frac{\pi}{2}-x\right)} = \frac{1}{\cos x} = \sec x\,,\quad \operatorname{cosec}\left(\frac{\pi}{2}+x\right) = \frac{1}{\sin\left(\frac{\pi}{2}+x\right)} = \frac{1}{\cos x} = \sec x$

$\sin(\pi - x) = \sin\pi.\cos x - \cos\pi.\sin x = 0 - (-\sin x) = \sin x\,,\quad \sin(\pi + x) = -\sin x$

$\cos(\pi - x) = \cos\pi.\cos x + \sin\pi.\sin x = -\cos x\,,\quad \cos(\pi + x) = \cos\pi.\cos x - \sin\pi.\sin x = -\cos x$

IInd Method:– $I = \int \frac{dx}{1+\sin x} = \int \frac{dx}{\sin^2 x/_2 + \cos^2 x/_2 + 2\sin x/_2.\cos x/_2}$

Divide above and below by $\cos^2 x/_2$

$I = \int \frac{\sec^2 x/_2\, dx}{\tan^2 x/_2 + 2\tan x/_2 + 1}$ Put $\tan x/_2 = t \therefore \frac{1}{2}.\sec^2 x/_2\, dx = dt \quad \therefore \sec^2 x/_2\, dx = 2\,dt$

$I = \int \frac{2\,dt}{t^2+2t+1} = 2\int \frac{dt}{(t+1)^2}$ Put $t + 1 = z \therefore dt = dz$

then $I = 2\int \frac{dz}{z^2} = 2.\frac{(z)^{-2+1}}{-2+1} = -2\frac{1}{z} = -\frac{2}{z} + c = -\frac{2}{t+1} + c = -\frac{2}{\tan x/_2 + 1} + c$ Ans.

IIIrd Method:– $I = \int \frac{dx}{1+\sin x} = \int \left(\frac{1}{1+\sin x} \times \frac{1-\sin x}{1-\sin x}\right) dx = \int \frac{1-\sin x}{1-\sin^2 x} dx = \int \frac{1-\sin x}{\cos^2 x} dx = \int \frac{dx}{\cos^2 x} - \int \frac{\sin x}{\cos^2 x} dx$

$= \int \sec^2 x\, dx - \int \frac{\sin x}{\cos^2 x} dx = \tan x - I_1$

or $I_1 = \int \frac{\sin x}{\cos^2 x} dx$ Put $\cos x = t \therefore -\sin x\, dx = dt$

$\therefore I_1 = \int \frac{\sin x}{\cos^2 x} dx = -\int \frac{dt}{t^2} = -\frac{t^{-2+1}}{-2+1} = \frac{1}{t} + c = \frac{1}{\cos x} + c = \sec x + c = \tan x - \sec x + c$ Ans.

(b) $I = \int \frac{dx}{\sqrt{1+\sin x}} = \int \frac{dx}{\sqrt{\sin^2 x/_2 + \cos^2 x/_2 + 2\sin x/_2.\cos x/_2}}$

Divide above and below by $\cos^4 x/_2$

$I = \int \frac{\sec^4 x/_2}{\sqrt{\tan^2 x/_2 + 2\tan x/_2 + 1}} dx = \int \frac{\sec^2 x/_2.\sec^2 x/_2}{\sqrt{\tan^2 x/_2 + 2\tan x/_2 + 1}} dx = \int \frac{(1+\tan^2 x/_2).\sec^2 x/_2}{\sqrt{\tan^2 x/_2 + 2\tan x/_2 + 1}} dx$

Put $\tan x/_2 = t \therefore \frac{1}{2}.\sec^2 x/_2\, dx = dt \quad \therefore \sec^2 x/_2\, dx = 2\,dt$

$$I=\int\frac{1+t^2}{\sqrt{t^2+2t+1}}.2\,dt=2\int\frac{1+t^2}{\sqrt{(1+t)^2}}dt=2\int\frac{1+t^2}{1+t}dt=2\int\frac{dt}{1+t}+2\int\frac{t^2}{1+t}dt=2\int\frac{dt}{1+t}+2\int\left(t-\frac{t}{1+t}\right)dt$$

$$=2\log(1+t)+2\int t\,dt-2\int\frac{t}{1+t}dt=2\log(1+t)+2.\frac{t^2}{2}-2\int\left(1-\frac{1}{1+t}\right)dt$$

$$=2\log(1+t)+t^2-2\int dt+2\int\frac{1}{1+t}dt$$

$I=2\log(1+t)+t^2-2t+2\log(1+t)+c=4\log(1+t)+t^2-2t+c$ Ans. where $t=\tan {}^{x}/_{2}$

IInd Method: – $I=\int\frac{dx}{\sqrt{1+\sin x}}=\int\frac{dx}{\sqrt{1+\cos\left(\frac{\pi}{2}-x\right)}}=\int\frac{dx}{\sqrt{2\cos^2\left(\frac{\frac{\pi}{2}-x}{2}\right)}}=\frac{1}{\sqrt{2}}\int\frac{dx}{\cos\left(\frac{\pi}{4}-\frac{x}{2}\right)}$

Put $\frac{\pi}{4}-\frac{x}{2}=t \quad \therefore -\frac{1}{2}dx=dt \quad \therefore -dx=2dt$

$$I=-\frac{1}{\sqrt{2}}\int\frac{2\,dt}{\cos t}=-\sqrt{2}\int\frac{\sin^2 t+\cos^2 t}{\cos t}dt=-\sqrt{2}\int\frac{\sin^2 t}{\cos t}dt-\sqrt{2}\int\frac{\cos^2 t}{\cos t}dt=-\sqrt{2}\int\frac{\sqrt{1-\cos^2 t}.\sin t}{\cos t}dt-\sqrt{2}\int\cos t\,dt=-\sqrt{2}I_1-\sqrt{2}\sin t$$

$I_1=\int\frac{\sqrt{1-\cos^2 t}.\sin t}{\cos t}dt=\int\frac{\sqrt{1-\cos^2 t}.\sin t.\cos t}{\cos^2 t}dt$ Put $1-\cos^2 t=z^2 \quad \therefore -2\cos t\sin t\,dt=2z\,dz$

$$I_1=-\int\frac{z.z\,dz}{1-z^2}=-\int\frac{z^2}{1-z^2}dz=-\int\left(\frac{1}{1-z^2}-1\right)dz=-\int\frac{1}{1-z^2}dz+\int dz$$

$\left[\text{using formula, } \int\frac{dx}{a^2-x^2}=\frac{1}{2a}\log\left|\frac{a+x}{a-x}\right|\right]$

$$I_1=-\frac{1}{2.1}\log\left|\frac{1+z}{1-z}\right|+z=-\frac{1}{2}\log\left[\frac{1+\sqrt{1-\cos^2 t}}{1-\sqrt{1-\cos^2 t}}\right]+\sqrt{1-\cos^2 t}+c$$

$I=-\sqrt{2}I_1-\sqrt{2}\sin t=\sqrt{2}.\frac{1}{2}\log\left[\frac{1+\sqrt{1-\cos^2 t}}{1-\sqrt{1-\cos^2 t}}\right]-\sqrt{2}.\sqrt{1-\cos^2 t}-\sqrt{2}\sin t+c=\frac{1}{\sqrt{2}}\log\left[\frac{1+\sin t}{1-\sin t}\right]-2\sqrt{2}\sin t+c$ Ans. where $t=\left(\frac{\pi}{4}-\frac{x}{2}\right)$

(c) $I=\int\sqrt{1+\sin x}\,dx=\int\sqrt{1+\cos\left(\frac{\pi}{2}-x\right)}dx=\int\sqrt{2\cos^2\left(\frac{\frac{\pi}{2}-x}{2}\right)}dx=\sqrt{2}\int\cos\left(\frac{\pi}{4}-\frac{x}{2}\right)dx=\sqrt{2}.\sin\left(\frac{\pi}{4}-\frac{x}{2}\right).\frac{1}{-\frac{1}{2}}$

$$=-2\sqrt{2}\sin\left(\frac{\pi}{4}-\frac{x}{2}\right)+c=-2\sqrt{2}\left(\sin\frac{\pi}{4}.\cos\frac{x}{2}-\cos\frac{\pi}{4}.\sin\frac{x}{2}\right)+c$$

$$I=-2\sqrt{2}\left(\frac{1}{\sqrt{2}}.\cos\frac{x}{2}-\frac{1}{\sqrt{2}}.\sin\frac{x}{2}\right)+c=-2\sqrt{2}.\left(\frac{\cos\frac{x}{2}-\sin\frac{x}{2}}{\sqrt{2}}\right)+c=-2\left(\cos\frac{x}{2}-\sin\frac{x}{2}\right)+c \quad \text{Ans.}$$

IInd Method: – $I=\int\sqrt{1+\sin x}\,dx=I=\int\sqrt{1+\sin x\times\frac{1-\sin x}{1-\sin x}}dx=\int\sqrt{\frac{1-\sin^2 x}{1-\sin x}}dx=\int\frac{\cos x}{\sqrt{1-\sin x}}dx$

Put $1-\sin x=t^2 \quad \therefore -\cos x\,dx=2t\,dt$

$$I=\int\frac{\cos x}{\sqrt{1-\sin x}}dx=-\int\frac{2t\,dt}{\sqrt{t^2}}=-2\int\frac{t\,dt}{t}=-2\int dt=-2t+c=-2\sqrt{1-\sin x}+c \quad \text{Ans.}$$

Integral of improper rational function: – Improper rational function: –

A rational function in which degree of $N^r \geq$ degree of D^r is called improper rational function.

e.g. (a) $\frac{x^2+1}{2x+3}$ (b) $\frac{x^2+1}{x^2-1}$ (c) $\frac{3x^3-3x^2+1}{2+3x}$

Remarks: – In improper rational we always divide N^r by D^r and then use result of integration.

Example: – (a) $I = \int \frac{x^6}{x-1} dx$, Divide N^r by D^r and then use result of integration.

Divide

$$\therefore \; x^6 = (x-1)(x^5 + x^4 + x^3 + x^2 + x + 1) + 1$$

Divide both of sides by $(x-1)$ and integrate

$$\int \frac{x^6}{x-1} dx = \int \frac{(x-1)(x^5 + x^4 + x^3 + x^2 + x + 1) + 1}{(x-1)} dx = \int \frac{(x-1)(x^5 + x^4 + x^3 + x^2 + x + 1)}{(x-1)} dx + \int \frac{1}{x-1} dx$$

$$= \int (x^5 + x^4 + x^3 + x^2 + x + 1)\, dx + \int \frac{1}{x-1} dx$$

$$I = \int \frac{x^6}{x-1} dx = \int x^5\, dx + \int x^4\, dx + \int x^3\, dx + \int x^2\, dx + \int x\, dx + \int dx + \int \frac{1}{x-1} dx = \frac{x^6}{6} + \frac{x^5}{5} + \frac{x^4}{4} + \frac{x^3}{3} + \frac{x^2}{2} + x + \log|x-1| + c \quad \text{Ans.}$$

IInd Method: – $\int \frac{x^6}{x-1} dx = \int \frac{(x-1)x^5 + (x-1)x^4 + (x-1)x^3 + (x-1)x^2 + (x-1)x + (x-1) + 1}{x-1} dx$

$$I = \int \frac{(x-1)(x^5 + x^4 + x^3 + x^2 + x + 1) + 1}{(x-1)} dx = \int (x^5 + x^4 + x^3 + x^2 + x + 1)\, dx + \int \frac{1}{x-1} dx$$

$$= \int x^5\, dx + \int x^4\, dx + \int x^3\, dx + \int x^2\, dx + \int x\, dx + \int dx + \int \frac{1}{x-1} dx$$

$$= \frac{x^6}{6} + \frac{x^5}{5} + \frac{x^4}{4} + \frac{x^3}{3} + \frac{x^2}{2} + x + \log|x-1| + c \quad \text{Ans.}$$

(b) $I = \int \frac{x^2+1}{x^2-1} dx = \int \frac{(x^2-1) + 2(x^2-1) + 6}{3(x^2-1)} dx = \frac{1}{3}\int \frac{x^2-1}{x^2-1} dx + \frac{2}{3}\int \frac{x^2-1}{x^2-1} dx + 2\int \frac{1}{x^2-1} dx$

$I = \frac{1}{3}\int dx + \frac{2}{3}\int dx + 2\int \frac{1}{x^2-1} dx$ $\left[\text{use formula, } \int \frac{dx}{x^2-a^2} = \frac{1}{2a}\log\left|\frac{x-a}{x+a}\right|\right]$

$$I = \frac{1}{3}x + \frac{2}{3}x + 2.\frac{1}{2.1}\log\left|\frac{x-1}{x+1}\right| + c = x + \log\left[\frac{x-1}{x+1}\right] + c \quad \text{Ans.}$$

(c) $I = \int \frac{6x^3 + 7x^2 - 4x - 4}{(3x+2)} dx$, Divide N^r by D^r and then use result of integration.

Divide

$$\therefore \; 6x^3 + 7x^2 - 4x - 4 = (3x+2)(2x^2 + x - 2) + 0$$

Divide $(3x+2)$ both of sides and integrate

$$\int \frac{6x^3 + 7x^2 - 4x - 4}{(3x+2)} dx = \int \frac{(3x+2)(2x^2 + x - 2)}{(3x+2)} dx = \int (2x^2 + x - 2)\, dx = \int 2x^2\, dx + \int x\, dx - \int 2\, dx$$

$$\therefore \; I = \int \frac{6x^3 + 7x^2 - 4x - 4}{(3x+2)} dx = 2.\frac{x^3}{3} + \frac{x^2}{2} - 2x + c = \frac{2}{3}x^3 + \frac{1}{2}x^2 - 2x + c \quad \text{Ans.}$$

Integration by substitution: – $\int f(g(x)).g'(x)\, dx$, Putting $g(x) = z \;\; \therefore \; g'(x)\, dx = dz$

$$\boxed{\therefore \int f(z)\, dz}$$

After evaluating this integral we substitute the value of z in term of x, i.e $z = g(x)$.

Example: – (a) $I = \int \frac{\cos x}{\cos(x-3)} dx$, Put $x - 3 = z$ or $x = z + 3$ $\therefore$ $dx = dz$

$$I = \int \frac{\cos(z+3)}{\cos z} dz = \int \frac{\cos z . \cos 3 - \sin z . \sin 3}{\cos z} dz = \int \frac{\cos z . \cos 3}{\cos z} dz - \int \frac{\sin z . \sin 3}{\cos z} dz$$

[formula, $\cos(A+B) = \cos A \cos B - \sin A \sin B$]

$$I = \int \cos 3\, dz - \int \sin 3 . \tan z\, dz = \cos 3 . z - \sin 3 . (-\log(\cos z)) + c \quad \left[\int \tan x\, dx = -\log(\cos x) + c\right]$$

$$I = z \cos 3 + \sin 3 . \log[\cos z] + c = (x-3). \cos 3 + \sin 3 . \log[\cos(x-3)] + c \quad \text{Ans.}$$

(b) $I = \int \frac{dx}{(\sin^{-1} x)^3 . \sqrt{1-x^2}}$ Putting $\sin^{-1} x = z$ $\therefore$ $\frac{1}{\sqrt{1-x^2}} dx = dz$ $\left[\frac{d}{dx}(\sin^{-1} x) = \frac{1}{\sqrt{1-x^2}}\right]$

$$\therefore \ I = \int \frac{dz}{z^3} = \int z^{-3} dz = \frac{z^{-3+1}}{-3+1} + c = -\frac{1}{2} z^{-2} + c = -\frac{1}{2z^2} + c = -\frac{1}{2(\sin^{-1} x)^2} + c \quad \text{Ans.}$$

(c) $I = \int \sin^3 x . \sqrt{\cos x}\, dx = \int \sin^2 x . \sqrt{\cos x} . \sin x\, dx = \int (1 - \cos^2 x). \sqrt{\cos x} . \sin x\, dx$

Putting $\cos x = z^2$ $\therefore$ $-\sin x\, dx = 2z\, dz$

$$I = -\int (1-(z^2)^2). \sqrt{z^2} . 2z\, dz = -2\int (1-z^4). z^2\, dz = -2\int (z^2 - z^6)\, dz = -2\int z^2\, dz + 2\int z^6\, dz = -2\frac{z^{2+1}}{2+1} + 2\frac{z^{6+1}}{6+1} + c = \frac{2}{7} z^7 - \frac{2}{3} z^3 + c$$
$$= \frac{2}{7}\left(\sqrt{\cos x}\right)^7 - \frac{2}{3}\left(\sqrt{\cos x}\right)^3 + c \quad \text{Ans.}$$

Integral of the form: – $\int e^{ax} \cos bx\, dx$ and $\int e^{ax} \sin bx\, dx$

Let $c = \int e^{ax} \cos bx\, dx$ ………… (i) and $s = \int e^{ax} \sin bx\, dx$ …………… (ii) use [$\cos\theta + i \sin\theta = e^{i\theta}$]

Multiplying equation (i) × 1 and (ii) × i and adding equation (i) & (ii), we have

$$c + is = \int e^{ax} \cos bx\, dx + \int e^{ax} . i \sin bx\, dx = \int e^{ax}(\cos bx + i \sin bx)\, dx = \int e^{ax} . e^{ibx}\, dx$$

$$c + is = \int e^{x(a+ib)}\, dx = \frac{e^{x(a+ib)}}{(a+ib)} = \frac{e^{ax} . e^{ibx}}{a+ib} \times \frac{a-ib}{a-ib} = \frac{e^{ax}}{a^2+b^2}(\cos bx + i \sin bx). (a-ib) = \frac{e^{ax}}{a^2+b^2}[a \cos bx - ib \cos bx + ia \sin bx + b \sin bx]$$
$$= \frac{e^{ax}}{a^2+b^2}[(a \cos bx + b \sin bx) + i(a \sin bx - b \cos bx)]$$

$$c + is = \frac{e^{ax}}{a^2+b^2}(a \cos bx + b \sin bx) + i\frac{e^{ax}}{a^2+b^2}(a \sin bx - b \cos bx)$$

$$c = \frac{e^{ax}}{a^2+b^2}(a \cos bx + b \sin bx) \text{ and } s = \frac{e^{ax}}{a^2+b^2}(a \sin bx - b \cos bx) \quad \text{Ans.}$$

Generalized rule for integration by part: –

$$\int u . v\, dx = u . v_1 - u' . v_2 + u'' . v_3 - u''' . v_4 + \cdots \ldots\ldots\ldots\ldots$$

where $u' = \frac{du}{dx}$, $u'' = \frac{du'}{dx} = \frac{d^2y}{dx^2}$, $u''' = \frac{du''}{dx}$ ………

and $v_1 = \int v\, dx$, $v_2 = \int v_1\, dx$, $v_3 = \int v_2\, dx$ ………

Example: – (a) $I = \int x^2 . \cos x\, dx = x^2 . \int \cos x\, dx - \int \left[\frac{d(x^2)}{dx} . \int \cos x\, dx\right] dx = x^2 . \sin x - \int 2x . \sin x\, dx$

$$I = x^2.\sin x - 2\left\{x.\int \sin x\ dx - \int \left[\frac{dx}{dx}.\int \sin x\ dx\right] dx\right\} = x^2.\sin x - 2x.(-\cos x) + 2\int -\cos x\ dx = x^2.\sin x + 2x.\cos x - 2\sin x + c \quad \text{Ans.}$$

IInd Method: $-\quad I = \int x^2.\cos x\ dx = x^2.\int \cos x\ dx - \frac{d(x^2)}{dx}.\int dx.\int \cos x\ dx + \cdots \ldots \ldots$

$$= x^2.\sin x - 2x.\int \sin x\ dx + \frac{d(2x)}{dx}.\int \cos x\ dx - 0 + 0 \ldots\ldots$$

or $I = x^2.\sin x - 2x(-\cos x) + 2\sin x + c \quad \therefore\ I = x^2 \sin x + 2x\cos x + 2\sin x + c \quad$ Ans.

(b) $I = \int x^3.e^x\,dx = x^3.\int e^x\,dx - \int \left[\frac{d(x^3)}{dx}.\int e^x\,dx\right] dx = x^3.e^x - \int 3x^2.e^x\,dx = x^3.e^x - 3\left\{x^2.\int e^x\,dx - \int \left[\frac{d(x^2)}{dx}.\int e^x\,dx\right] dx\right\}$

$$= x^3.e^x - 3x^2.e^x + 3\int 2x.e^x\,dx$$

$$I = x^3.e^x - 3x^2.e^x + 6\left\{x.\int e^x\,dx - \int \left[\frac{dx}{dx}.\int e^x\,dx\right] dx\right\} = x^3.e^x - 3x^2.e^x + 6x.e^x - 6\int e^x\,dx = x^3.e^x - 3x^2.e^x + 6x.e^x - 6e^x + c \quad \text{Ans.}$$

IInd Method: $-\quad I = \int x^3.e^x\,dx \quad$ [use formula generalized rule for integration by part]

$$I = x^3\int e^x\,dx - \frac{d(x^3)}{dx}.\int e^x\,dx + \frac{d(3x^2)}{dx}.\int e^x\,dx - \frac{d(6x)}{dx}.\int e^x\,dx + \frac{d(6)}{dx}.\int e^x\,dx - \cdots \ldots \ldots = x^3.e^x - 3x^2.e^x + 6x.e^x - 6e^x + 0 - \cdots..+c$$

$$= x^3.e^x - 3x^2.e^x + 6x.e^x - 6e^x + c \quad \text{Ans.}$$

here $u = x^3,\ u' = 3x^2,\ u'' = 6x,\ u''' = 6,\ u'''' = 0$ and $v = e^x,\ v_1 = \int v\,dx = e^x,\ v_2 = \int v_1\,dx = e^x \ldots\ldots$

$$\left[\text{use formula, } \int u.v\ dx = u.v_1 - u'.v_2 + u''.v_3 - u'''.v_4 + \cdots \ldots \ldots \ldots \ldots\right]$$

Integration of proper rational function by partial function: –

If the denominator of proper rational function $\frac{P(x)}{Q(x)}$ is at the form

$Q(x) = (x-a)^m(x-b)^n(x-c)^t(px^2+qx+r)^s(ex^2+f.x+d)^l$ where $q^2-4pr,\ f^2-4ed<0$ *then,*

$$\frac{P(x)}{Q(x)} = \frac{A_1}{(x-a)} + \frac{A_2}{(x-a)^2} + \cdots \ldots \ldots + \frac{A_m}{(x-a)^m} + \frac{B_1}{(x-b)} + \frac{B_2}{(x-b)^2} + \cdots \ldots \ldots + \frac{B_n}{(x-b)^n} + \frac{C_1}{(x-c)} + \frac{C_2}{(x-c)^2} + \cdots \ldots \ldots + \frac{C_t}{(x-c)^t}$$
$$+ \frac{D_1x+E_1}{(px^2+qx+r)} + \frac{D_2x+E_2}{(px^2+qx+r)^2} + \cdots \ldots \ldots + \frac{D_sx+E_s}{(px^2+qx+r)^s} + \frac{F_1x+G_1}{(ex^2+f.x+d)} + \frac{F_2x+G_2}{(ex^2+f.x+d)^2}$$
$$+ \cdots \ldots \ldots \ldots + \frac{F_lx+G_l}{(ex^2+f.x+d)^l}$$

e.g. (a) $\frac{2x+3}{(x+1)(x+2)^2} = \frac{A}{x+1} + \frac{B}{x+2} + \frac{C}{(x+2)^2}$

(b) $\frac{2-x^2}{(x-2)(x+3)(x-5)} = \frac{A}{x-2} + \frac{B}{x+3} + \frac{C}{x-5}$ (c) $\frac{1}{x^2(x^2-2x+3)} = \frac{A}{x} + \frac{B}{x^2} + \frac{Cx+D}{(x^2-2x+3)}$

(d) $\frac{2x^2+2x-15}{(x+5)^2(x^2+2x+3)^2} = \frac{A}{x+5} + \frac{B}{(x+5)^2} + \frac{Cx+D}{x^2+2x+3} + \frac{Ex+F}{(x^2+2x+3)^2}$

	Linear factors	Quadratic factors

Non repeated factor	$\frac{P(x)}{(x-a)(x-b)(x+c)} = \frac{A}{x-a}+\frac{B}{x-b}+\frac{C}{x+c}$ find, $A = \left(\frac{P(x)}{(x-b)(x+c)}\right)_{x=a}$ $B = \left(\frac{P(x)}{(x-a)(x+c)}\right)_{x=b}$ $C = \left(\frac{P(x)}{(x-a)(x-b)}\right)_{x=-c}$	$\frac{P(x)}{(x^2+x-1)(x^2+x-3)(x-2)}$ $= \frac{A}{x-2}+\frac{Bx+C}{x^2+x-1}+\frac{Dx+E}{x^2+x-3}$ A, B, C, D, E are obtained by multiplying by $(x^2+x-1)(x^2+x-3)(x-2)$ on both of sides and then equate coefficient of suitable power of x on both side.
Repeated factor	$\frac{P(x)}{(x+a)^2(x+b)^3} = \frac{A}{x+a}+\frac{B}{(x+a)^2}+\frac{C}{x+b}+\frac{D}{(x+b)^2}+\frac{E}{(x+b)^3}$ $B = \left(\frac{P(x)}{(x+b)^3}\right)_{x=-a}$ $E = \left(\frac{P(x)}{(x+a)^2}\right)_{x=-b}$ A, C, D are obtained by multiplying by $(x+a)^2(x+b)^3$ on both sides and then equate of coefficient of suitable power of x.	$\frac{P(x)}{(x^2-x+1)^2(x^2+x-3)}$ $= \frac{Ax+B}{x^2-x+1}+\frac{Cx+D}{(x^2-x+1)^2}+\frac{Ex+F}{x^2+x-3}$ A, B, C, D, E, F are obtained by multiplying by $(x^2-x+1)^2(x^2+x-3)$on both sides and then equate coefficient of suitable power of x on both sides.

Example: – (a) $I = \int \frac{3x^2-5x+6}{(2x-1)(x+1)}dx$, Let $\frac{3x^2-5x+6}{(2x-1)(x+1)} = \frac{A}{2x-1}+\frac{B}{x+1}$ ……………..(A)

$A = \frac{3x^2-5x+6}{(x+1)}$ $\quad\therefore\ 2x-1=0$ or $2x=1$ $\therefore\ x=\frac{1}{2}$

$$A = \left(\frac{3x^2-5x+6}{(x+1)}\right)_{x=\frac{1}{2}} = \frac{3.\left(\frac{1}{2}\right)^2-5.\frac{1}{2}+6}{\frac{1}{2}+1} = \frac{\frac{3}{4}-\frac{5}{2}+6}{\frac{1+2}{2}} = \frac{\frac{3-10+24}{4}}{\frac{3}{2}} = \frac{17}{4}\times\frac{2}{3} = \frac{17}{6}$$

$B = \frac{3x^2-5x+6}{(2x-1)}$ $\quad\therefore\ x+1=0$ $\quad\therefore\ x=-1$

$$B = \left(\frac{3x^2-5x+6}{(2x-1)}\right)_{x=-1} = \frac{3.(-1)^2-5.(-1)+6}{2.(-1)-1} = \frac{3+5+6}{-2-1} = -\frac{14}{3}$$

From equation (A), we have $\Rightarrow \frac{3x^2-5x+6}{(2x-1)(x+1)} = \frac{17}{6}.\frac{1}{2x-1}-\frac{14}{3}.\frac{1}{x+1}$

Integrating, $\int \frac{3x^2-5x+6}{(2x-1)(x+1)}dx = \frac{17}{6}\int\frac{1}{2x-1}dx - \frac{14}{3}\int\frac{1}{x+1}dx = \frac{17}{6}.\frac{\log(2x-1)}{2}-\frac{14}{3}.\log(x+1)+c$

$= \frac{17}{12}\log(2x-1)-\frac{14}{3}\log(x+1)+c$ Ans.

(b) $I = \int \frac{x^2+3x+5}{(x-1)(x+3)(x-4)} dx$, Let $\frac{x^2+3x+5}{(x-1)(x+3)(x-4)} = \frac{A}{x-1} + \frac{B}{x+3} + \frac{C}{x-4}$ .... (A)

See above question, Solve equation (A) and find A, B, C $\therefore A = -\frac{3}{4}$, $B = \frac{5}{28}$ and $C = \frac{33}{21}$

$$I = \int \frac{x^2+3x+5}{(x-1)(x+3)(x-4)} dx = -\frac{3}{4}\log(x-1) + \frac{5}{28}\log(x+3) + \frac{33}{21}\log(x-4) + c \quad \text{Ans.}$$

(c) $I = \int \frac{3x+2}{(x+1)(x-2)^2} dx$, Let $\frac{3x+2}{(x+1)(x-2)^2} = \frac{A}{x+1} + \frac{B}{x-2} + \frac{C}{(x-2)^2}$ .... (A)

$$\therefore A = \left.\frac{3x+2}{(x-2)^2}\right|_{x=-1} = \frac{3.(-1)+2}{(-3)^2} = \frac{-1}{9} = -\frac{1}{9}$$

or $\frac{3x+2}{(x+1)(x-2)^2} = \frac{B}{x-2}$, Multiplying by $(x+1)(x-2)^2$ both of sides, we have

or $\frac{3x+2}{(x+1)(x-2)^2} \times (x+1)(x-2)^2 = \frac{B}{x-2} \times (x+1)(x-2)^2 = B(x+1)(x-2) = B(x^2-2x+x-2)$

$\Rightarrow 3x+2 = B(x^2-x-2) = Bx^2 - Bx - 2B \quad \Rightarrow 0.x^2 + 3x + 2 = Bx^2 - Bx - 2B$

$$\text{or } -Bx = 3x \quad \therefore B = -3$$

or $\frac{3x+2}{(x+1)(x-2)^2} = \frac{C}{(x-2)^2}$, Multiplying by $(x+1)(x-2)^2$ both of sides, we have

$$\Rightarrow 3x+2 = C(x+1) = Cx + C \quad \therefore C = 3$$

Put value of A, B and C in equation (A) and integrate, we have

or $\frac{3x+2}{(x+1)(x-2)^2} = -\frac{1}{9}.\frac{1}{x+1} + (-3).\frac{1}{x-2} + 3.\frac{1}{(x-2)^2}$

$$\therefore \int \frac{3x+2}{(x+1)(x-2)^2} dx = -\frac{1}{9}\int \frac{dx}{x+1} - 3\int \frac{dx}{x-2} + 3\int \frac{dx}{(x-2)^2} = -\frac{1}{9}\log(x+1) - 3\log(x-2) + 3.\frac{(x-2)^{-2+1}}{-2+1}$$

$$\therefore I = \int \frac{3x+2}{(x+1)(x-2)^2} dx = -\frac{1}{9}\log(x+1) - 3\log(x-2) - 3.\frac{1}{x-2} + c \quad \text{Ans.}$$

(d) $I = \int \frac{1-x^2}{x^2(x-3)^2} dx$, Let $\frac{1-x^2}{x^2(x-3)^2} = \frac{A}{x} + \frac{B}{x^2} + \frac{C}{x-3} + \frac{D}{(x-3)^2}$ .... (A)

$\therefore B = \left.\frac{1-x^2}{(x-3)^2}\right|_{x=0} = \frac{1-0}{(0-3)^2} = \frac{1}{9}$ and $D = \left.\frac{1-x^2}{x^2}\right|_{x=3} = \frac{1-(3)^2}{(3)^2} = \frac{1-9}{9} = -\frac{8}{9}$

To be A and C are obtained by multiplying by $x^2(x-3)^2$ on both of sides and then equate

coefficient of suitable power of x on both sides.

or $\frac{1-x^2}{x^2(x-3)^2} \times x^2(x-3)^2 = \frac{A}{x} \times x^2(x-3)^2$

or $1-x^2 = Ax(x-3)^2 = Ax(x^2-6x+9) = Ax^3 - 6Ax^2 + 9Ax \quad \therefore -1 = -6A$ or $6A = 1 \quad \therefore A = \frac{1}{6}$

Again, $\frac{1-x^2}{x^2(x-3)^2} = \frac{C}{x-3}$ or $\frac{1-x^2}{x^2(x-3)^2} \times x^2(x-3)^2 = \frac{C}{x-3} \times x^2(x-3)^2$

$$\text{or } 1-x^2 = Cx^2(x-3) \text{ or } 1-x^2 = Cx^3 - 3Cx^2 \text{ or } -1 = -3C \text{ or } 1 = 3C \quad \therefore C = \frac{1}{3}$$

Put value of A, B, C and D in equation (A) and integrate, we have or $\dfrac{1-x^2}{x^2(x-3)^2}=\dfrac{A}{x}+\dfrac{B}{x^2}+\dfrac{C}{x-3}+\dfrac{D}{(x-3)^2}=\dfrac{1}{6x}+\dfrac{1}{9x^2}+\dfrac{1}{3(x-3)}-\dfrac{8}{9(x-3)^2}$

or $\displaystyle\int\frac{1-x^2}{x^2(x-3)^2}\,dx=\frac{1}{6}\int\frac{1}{x}\,dx+\frac{1}{9}\int\frac{1}{x^2}\,dx+\frac{1}{3}\int\frac{1}{(x-3)}\,dx-\frac{8}{9}\int\frac{1}{(x-3)^2}\,dx$

$$I=\frac{1}{6}\log x+\frac{1}{9}.\frac{x^{-2+1}}{-2+1}+\frac{1}{3}\log(x-3)-\frac{8}{9}.\frac{(x-3)^{-2+1}}{-2+1}+c=\frac{1}{6}\log x-\frac{1}{9}x^{-1}+\frac{1}{3}\log(x-3)+\frac{8}{9}(x-3)^{-1}+c$$

$$=\frac{1}{6}\log x-\frac{1}{9x}+\frac{1}{3}\log(x-3)+\frac{8}{9(x-3)}+c \quad \text{Ans.}$$

Example

(1) Integrate: – (a) $I=\displaystyle\int\frac{\cos x}{\sin^4 x+\cos^4 x}\,dx$ (b) $I=\displaystyle\int\frac{dx}{4x^2+3}$ (c) $I=\displaystyle\int\frac{dx}{\sqrt{x^2-2x+3}}$

(d) $I=\displaystyle\int\frac{dx}{\sqrt{3+(1-2x)^2}}$ (e) $I=\displaystyle\int\frac{dx}{\sqrt{4x^2+4x-5}}$ (f) $I=\displaystyle\int\frac{dx}{\sqrt{3-(1-x)^2}}$ (g) $I=\displaystyle\int\frac{dx}{x^2-4x+3}$

(2) (a) $I=\displaystyle\int\frac{dx}{2-(3-x)^2}$ (b) $I=\displaystyle\int\frac{dx}{9-6x^2}$ (c) $I=\displaystyle\int\sqrt{x^2-8}\,dx$ (d) $I=\displaystyle\int\sqrt{x^2+3}\,dx$

(e) $I=\displaystyle\int\sqrt{2-x^2}\,dx$ (f) $I=\displaystyle\int\sqrt{(5x+2)^2-3}\,dx$ (g) $I=\displaystyle\int\sqrt{(2x-1)^2+4}\,dx$ (h) $I=\displaystyle\int\sqrt{5-(2-x)^2}\,dx$

(3) (a) $I=\displaystyle\int\frac{x\,dx}{(2x^2-3)^4}$ (b) $I=\displaystyle\int\frac{2ax+b}{\sqrt{2ax^2+2bx+c}}\,dx$ (c) $I=\displaystyle\int\frac{x^2}{\sqrt{2-3x^3}}\,dx$ (d) $I=\displaystyle\int\frac{\cos x}{\sqrt{1+\sin x}}\,dx$

(e) $I=\displaystyle\int\frac{dx}{e^x\sqrt{1+e^{-x}}}$ (f) $I=\displaystyle\int\frac{x^x(1+\log x)+\frac{1}{x}}{\sqrt{x^x+\log x}}\,dx$ (g) $I=\displaystyle\int\frac{dx}{(2x-1)^{3/2}}$ (h) $I=\displaystyle\int\frac{(1+\cos\theta)}{(\theta+\sin\theta)^{5/2}}\,d\theta$

(4) (a) $I=\displaystyle\int\frac{1+\sin\theta}{\cos\theta}\,d\theta$ (b) $I=\displaystyle\int\frac{1+\cos\theta}{\sin\theta}\,d\theta$ (c) $I=\displaystyle\int\frac{\sec^2\theta}{1+\tan\theta}\,d\theta$ (d) $I=\displaystyle\int\frac{d}{dx}\left(\frac{\log(1+x)}{x}\right)dx$

(e) $I=\dfrac{d}{dx}\left(\displaystyle\int\frac{\log x+1}{2x}\,dx\right)$ (f) $I=\displaystyle\int\frac{1+\log x}{x\log x}\,dx$ (g) $I=\displaystyle\int\frac{x\cos x}{x\sin x+\cos x}\,dx$ (h) $I=\displaystyle\int\frac{x}{2x^2-5}\,dx$

(5) (a) $I=\displaystyle\int\frac{(2x-1)}{2x^2-2x+5}\,dx$ (b) $I=\displaystyle\int\frac{\cos x}{\sin^3 x}\,dx$ (c) $I=\displaystyle\int\frac{\sec^2 x}{\tan^4 x}\,dx$ (d) $I=\displaystyle\int\frac{dx}{x^2+x+3}$

(e) $I=\displaystyle\int\frac{dx}{2x^2+4x-5}$ (f) $I=\displaystyle\int\frac{dx}{\sqrt{1+4x-2x^2}}$ (g) $I=\displaystyle\int\frac{dx}{\sqrt{x^2+2x+3}}$ (h) $I=\displaystyle\int\frac{dx}{\sqrt{3+4x-2x^2}}$

(6) (a) $I=\displaystyle\int\sqrt{-x^2+3x+5}\,dx$ (b) $I=\displaystyle\int\frac{\sec^2 x}{(\sec x+\tan x)^{9/2}}\,dx$ (c) $I=\displaystyle\int\frac{\sin^2(\log x)+\cos(\log x)}{x\tan(\log x)}\,dx$

(d) $I=\displaystyle\int\frac{dx}{x\sin(\log x).\cos(\log x)}$ (e) $I=\displaystyle\int\frac{\log x}{x^2}\,dx$ (f) $I=\displaystyle\int\frac{dx}{x\sqrt{1+\log x}}$

(7) (a) $I=\displaystyle\int\frac{x^3}{(x-1)^2(x^2+2x+3)}\,dx$ (b) $I=\displaystyle\int\frac{x}{x^3+1}\,dx$ (c) $I=\displaystyle\int\frac{x^4}{(x-1)(x^2+1)}\,dx$

(d) $I=\displaystyle\int\frac{x-1}{x+1}.\frac{1}{\sqrt{x^3+x^2+x}}\,dx$

Solution

(1) (a) $I=\displaystyle\int\frac{\cos x}{\sin^4 x+\cos^4 x}\,dx$, Put $\sin x=t$ $\therefore$ $\cos x\,dx=dt$

then $\sin^4 x=t^4$ and $\cos^4 x=(1-\sin^2 x)^2=(1-t^2)^2$

$$I=\int\frac{dt}{t^4+(1-t^2)^2}=\int\frac{dt}{t^4+1-2t^2+t^4}=\int\frac{dt}{2t^4-2t^2+1}=\int\frac{dt}{\left(\sqrt{2}t^2-\frac{1}{\sqrt{2}}\right)^2+\left(\frac{1}{\sqrt{2}}\right)^2}=\frac{1}{\frac{1}{\sqrt{2}}}\tan^{-1}\left(\frac{\sqrt{2}t^2-\frac{1}{\sqrt{2}}}{\frac{1}{\sqrt{2}}}\right).\frac{1}{\text{d.c of}\left(\sqrt{2}t^2-\frac{1}{\sqrt{2}}\right)}$$

$$=\sqrt{2}\tan^{-1}(2t^2-1).\frac{1}{2\sqrt{2}t}=\frac{1}{2t}\tan^{-1}(2t^2-1)$$

$$I=\frac{1}{2\sin x}\tan^{-1}(2\sin^2x-1)+c=\frac{1}{2}\sec x.\tan^{-1}(2\sin^2x-1)+c\quad \text{Ans.}$$

(b) $I=\int\frac{dx}{4x^2+3}=\int\frac{dx}{(2x)^2+(\sqrt{3})^2}$ $\left[\text{use formula, }\int\frac{dx}{x^2+a^2}=\frac{1}{a}\tan^{-1}\frac{x}{a}\right]$

$$\therefore\ I=\int\frac{dx}{(2x)^2+(\sqrt{3})^2}=\frac{1}{\sqrt{3}}\tan^{-1}\left(\frac{2x}{\sqrt{3}}\right).\frac{1}{2}+c=\frac{1}{2\sqrt{3}}\tan^{-1}\left(\frac{2x}{\sqrt{3}}\right)+c\quad \text{Ans.}$$

(c) $I=\int\frac{dx}{\sqrt{x^2-2x+3}}=\int\frac{dx}{\sqrt{(x-1)^2+2}}=\int\frac{dx}{\sqrt{(x-1)^2+(\sqrt{2})^2}}$ formula, $\int\frac{dx}{\sqrt{x^2+a^2}}=\log\left|x+\sqrt{x^2+a^2}\right|$

$$\therefore\ I=\int\frac{dx}{\sqrt{(x-1)^2+(\sqrt{2})^2}}=\log\left[(x-1)+\sqrt{(x-1)^2+(\sqrt{2})^2}\right]+c=\log\left[(x-1)+\sqrt{x^2-2x+3}\right]+c\quad \text{Ans.}$$

(d) $I=\int\frac{dx}{\sqrt{3+(1-2x)^2}}=\int\frac{dx}{\sqrt{(\sqrt{3})^2+(1-2x)^2}}$ $\left[\text{use formula, }\int\frac{dx}{\sqrt{x^2+a^2}}=\log\left|x+\sqrt{x^2+a^2}\right|\right]$

$$\therefore\ I=\int\frac{dx}{\sqrt{(\sqrt{3})^2+(1-2x)^2}}=\log\left|(1-2x)+\sqrt{(\sqrt{3})^2+(1-2x)^2}\right|.\frac{1}{\frac{d}{dx}(1-2x)}=\log\left[(1-2x)+\sqrt{3+(1-2x)^2}\right].\frac{1}{-2}+c$$

$$=-\frac{1}{2}\log\left[(1-2x)+\sqrt{3+(1-2x)^2}\right]+c\quad \text{Ans.}$$

(e) $I=\int\frac{dx}{\sqrt{4x^2+4x-5}}=\int\frac{dx}{\sqrt{(2x+1)^2-6}}=\int\frac{dx}{\sqrt{(2x+1)^2-(\sqrt{6})^2}}$

$\left[\text{use formula, }\int\frac{dx}{\sqrt{x^2-a^2}}=\log\left|x+\sqrt{x^2-a^2}\right|\right]$

$$\therefore\ I=\int\frac{dx}{\sqrt{(2x+1)^2-(\sqrt{6})^2}}=\log\left|(2x+1)+\sqrt{(2x+1)^2-(\sqrt{6})^2}\right|.\frac{1}{2}+c=\frac{1}{2}\log\left|(2x+1)+\sqrt{4x^2+4x-5}\right|+c\quad \text{Ans.}$$

(f) $I=\int\frac{dx}{\sqrt{3-(1-x)^2}}=\int\frac{dx}{\sqrt{(\sqrt{3})^2-(1-x)^2}}$ $\left[\text{use formula, }\int\frac{dx}{\sqrt{a^2-x^2}}=\sin^{-1}\frac{x}{a}\right]$

$$\therefore\ I=\int\frac{dx}{\sqrt{(\sqrt{3})^2-(1-x)^2}}=\sin^{-1}\left[\frac{1-x}{\sqrt{3}}\right].\frac{1}{-1}+c=-\sin^{-1}\left[\frac{1-x}{\sqrt{3}}\right]+c\quad \text{Ans.}$$

(g) $I=\int\frac{dx}{x^2-4x+3}=\int\frac{dx}{(x-2)^2-1}$ $\left[\text{use formula, }\int\frac{dx}{x^2-a^2}=\frac{1}{2a}\log\left|\frac{x-a}{x+a}\right|\right]$

$$\therefore\ I=\int\frac{dx}{(x-2)^2-1}=\frac{1}{2.1}\log\left|\frac{(x-2)-1}{(x-2)+1}\right|+c=\frac{1}{2}\log\left|\frac{x-3}{x-1}\right|+c\quad \text{Ans.}$$

(2) (a) $I=\int\frac{dx}{2-(3-x)^2}=\int\frac{dx}{(\sqrt{2})^2-(3-x)^2}$ $\left[\text{use formula, }\int\frac{dx}{a^2-x^2}=\frac{1}{2a}\log\left|\frac{a+x}{a-x}\right|\right]$

$$\therefore\ I=\int\frac{dx}{(\sqrt{2})^2-(3-x)^2}=\frac{1}{2.\sqrt{2}}\log\left|\frac{\sqrt{2}+(3-x)}{\sqrt{2}-(3-x)}\right|.\frac{1}{-1}+c=-\frac{1}{2\sqrt{2}}\log\left|\frac{\sqrt{2}+(3-x)}{\sqrt{2}-(3-x)}\right|+c\quad \text{Ans.}$$

(b) $I=\int\frac{dx}{9-6x^2}=\int\frac{dx}{3^2-(\sqrt{6}x)^2}=\frac{1}{6\sqrt{6}}\log\left|\frac{3+\sqrt{6}x}{3-\sqrt{6}x}\right|+c$ Ans. (see above question)

(c) $I=\int\sqrt{x^2-8}\,dx=\int\sqrt{x^2-(2\sqrt{2})^2}\,dx$ $\left[\text{use formula, } \int\sqrt{x^2-a^2}\,dx=\frac{x}{2}\sqrt{x^2-a^2}-\frac{a^2}{2}\log\left|x+\sqrt{x^2-a^2}\right|\right]$

$$\therefore\ I=\int\sqrt{x^2-(2\sqrt{2})^2}\,dx=\frac{x}{2}\sqrt{x^2-8}-4\log\left|x+\sqrt{x^2-8}\right|+c\quad \text{Ans.}$$

(d) $I=\int\sqrt{x^2+3}\,dx=\int\sqrt{x^2+(\sqrt{3})^2}\,dx$ $\left[\text{use formula, } \int\sqrt{x^2+a^2}\,dx=\frac{x}{2}\sqrt{x^2+a^2}+\frac{a^2}{2}\log\left|x+\sqrt{x^2+a^2}\right|\right]$

$$\therefore\ I=\int\sqrt{x^2+(\sqrt{3})^2}\,dx=\frac{x}{2}\sqrt{x^2+3}+\frac{3}{2}\log\left|x+\sqrt{x^2+3}\right|+c\quad \text{Ans.}$$

(e) $I=\int\sqrt{2-x^2}\,dx=\int\sqrt{(\sqrt{2})^2-x^2}\,dx$ $\left[\text{use formula, } \int\sqrt{a^2-x^2}\,dx=\frac{x}{2}\sqrt{a^2-x^2}+\frac{a^2}{2}\sin^{-1}\frac{x}{a}\right]$

$$\therefore\ I=\int\sqrt{(\sqrt{2})^2-x^2}\,dx=\frac{x}{2}\sqrt{2-x^2}+\sin^{-1}\left(\frac{x}{\sqrt{2}}\right)+c\quad \text{Ans.}$$

(f) $I=\int\sqrt{(5x+2)^2-3}\,dx=\int\sqrt{(5x+2)^2-(\sqrt{3})^2}\,dx$ [see question no. $-(2)(c)$]

$$\therefore\ I=\frac{(5x+2)}{10}\sqrt{(5x+2)^2-3}-\frac{3}{10}\log\left|(5x+2)+\sqrt{(5x+2)^2-3}\right|+c\quad \text{Ans.}$$

(g) $I=\int\sqrt{(2x-1)^2+4}\,dx=\int\sqrt{(2x-1)^2+2^2}\,dx$ [see question no. $-(2)(d)$]

$$\therefore\ I=\frac{1}{2}\left\{\frac{(2x-1)}{2}\sqrt{(2x-1)^2+4}+2\log\left|(2x-1)+\sqrt{(2x-1)^2+4}\right|+c\right\}$$
$$=\frac{(2x-1)}{4}\sqrt{(2x-1)^2+4}+\log\left|(2x-1)+\sqrt{(2x-1)^2+4}\right|+c\quad \text{Ans.}$$

(h) $I=\int\sqrt{5-(2-x)^2}\,dx=\int\sqrt{(\sqrt{5})^2-(2-x)^2}\,dx$ [see question no. $-(2)(e)$]

$$\therefore\ I=\left[\frac{(2-x)}{2}\sqrt{5-(2-x)^2}+\frac{5}{2}\sin^{-1}\left(\frac{(2-x)}{\sqrt{5}}\right)\right].\frac{1}{-1}+c=-\left[\frac{(2-x)}{2}\sqrt{5-(2-x)^2}+\frac{5}{2}\sin^{-1}\left(\frac{(2-x)}{\sqrt{5}}\right)\right]+c\ \text{Ans.}$$

(3) (a) $I=\int\frac{x\,dx}{(2x^2-3)^4}=\int(2x^2-3)^{-4}.x\,dx$, Let $f(x)=2x^2-3$ $\therefore f'(x)=4x$

$$\therefore\ I=\int\{f(x)\}^{-4}.\frac{f'(x)}{4}\,dx\qquad\left[\text{using formula, } I=\int\{f(x)\}^n.f'(x)\,dx=\frac{\{f(x)\}^{n+1}}{n+1},\ n\neq-1\right]$$

$$\therefore\ I=\frac{\{f(x)\}^{-4+1}}{-4+1}.\frac{1}{4}=-\frac{(2x^2-3)^{-3}}{12}+c=-\frac{1}{12}(2x^2-3)^{-3}+c\quad \text{Ans.}$$

(b) $I=\int\frac{2ax+b}{\sqrt{2ax^2+2bx+c}}\,dx$, Let $f(x)=2ax^2+2bx+c$ $\therefore f'(x)=4ax+2b=2(2ax+b)$ or $2ax+b=\frac{f'(x)}{2}$

$\left[\text{use formula, } \int\frac{f'(x)}{\sqrt{f(x)}}\,dx=2\sqrt{f(x)}\right]$ $\therefore\ I=\int\frac{f'(x)}{2.\sqrt{f(x)}}\,dx=\frac{2}{2}.\sqrt{f(x)}=\sqrt{2ax^2+2bx+c}+d$ Ans.

(c) $I=\int\frac{x^2}{\sqrt{2-3x^3}}\,dx$, Let $f(x)=2-3x^3$ $\therefore f'(x)=-9x^2$ or $x^2=-\frac{f'(x)}{9}$

or $I=\int \frac{-\frac{f'(x)}{9}}{\sqrt{f(x)}}dx = -\frac{1}{9}\int \frac{f'(x)}{\sqrt{f'(x)}}dx$ (see above question) $\therefore\ I = -\frac{1}{9}.2\sqrt{f(x)} = -\frac{2}{9}\sqrt{2-3x^3}+c$ Ans.

(d) $I=\int \frac{\cos x}{\sqrt{1+\sin x}}dx$, Let $1+\sin x = f(x)$ $\therefore\ f'(x)=\cos x$

$\therefore\ I=\int \frac{f'(x)}{\sqrt{f(x)}}dx = 2\sqrt{f(x)}$ $\therefore\ I=\int \frac{\cos x}{\sqrt{1+\sin x}}dx = 2\sqrt{1+\sin x}+c$ Ans. $\left[\text{formula}, \int \frac{f'(x)}{\sqrt{f(x)}}dx = 2\sqrt{f(x)}\right]$

(e) $I=\int \frac{dx}{e^x\sqrt{1+e^{-x}}}$, Let $f(x)=1+e^{-x}$ $\therefore\ f'(x)=-e^{-x}=-\frac{1}{e^x}$

or $I=-\int \frac{1}{\sqrt{1+e^{-x}}}.\left(-\frac{1}{e^x}\right)dx = -\int \frac{f'(x)}{\sqrt{f(x)}}dx$ $\left[\text{formula}, \int \frac{f'(x)}{\sqrt{f(x)}}dx = 2\sqrt{f(x)}\right]$

$\therefore\ I=-\int \frac{f'(x)}{\sqrt{f(x)}}dx = -2\sqrt{f(x)} = -2\sqrt{1+e^{-x}}+c$ Ans.

(f) $I=\int \frac{x^x(1+\log x)+\frac{1}{x}}{\sqrt{x^x+\log x}}dx$, Let $f(x)=x^x+\log x$ put $y=x^x$ or $\log y = x\log x$ $\therefore\ \frac{1}{y}.\frac{dy}{dx}=x.\frac{1}{x}+\log x.1$

or $\frac{dy}{dx}=y(1+\log x)=x^x(1+\log x)$ then $f'(x)=x^x(1+\log x)+\frac{1}{x}$

$\therefore\ I=\int \frac{x^x(1+\log x)+\frac{1}{x}}{\sqrt{x^x+\log x}}dx = \int \frac{f'(x)}{\sqrt{f(x)}}dx = 2\sqrt{f(x)}+c = 2\sqrt{x^x+\log x}+c$ Ans.

(g) $I=\int \frac{dx}{(2x-1)^{3/2}}$, Let $2x-1=z$ or $2\,dx=dz$ or $dx=\frac{dz}{2}$

$\therefore\ I=\int \frac{dz}{2.z^{3/2}}=\frac{1}{2}\int z^{-3/2}dz=\frac{1}{2}.\frac{z^{-\frac{3}{2}+1}}{-\frac{3}{2}+1}+c=-z^{-\frac{1}{2}}+c=-\frac{1}{\sqrt{z}}+c=-\frac{1}{\sqrt{2x-1}}+c$ Ans.

IInd Method: $-\ I=\int \frac{dx}{(2x-1)^{3/2}}=\int (2x-1)^{-3/2}dx$, Let $f(x)=2x-1$ $\therefore f'(x)=2$

or $I=\frac{1}{2}\int (2x-1)^{-3/2}.2\,dx=\frac{1}{2}\int [f(x)]^{-3/2}.f'(x)\,dx$ $\left[\text{using formula, } I=\int \{f(x)\}^n.f'(x)\,dx=\frac{\{f(x)\}^{n+1}}{n+1}\right]$

$\therefore\ I=\frac{1}{2}\int [f(x)]^{-3/2}.f'(x)\,dx=\frac{1}{2}.\frac{[f(x)]^{-3/2+1}}{-\frac{3}{2}+1}+c=-[f(x)]^{-1/2}+c=-\frac{1}{\sqrt{f(x)}}+c=-\frac{1}{\sqrt{2x-1}}+c$ Ans.

(h) $I=\int \frac{(1+\cos\theta)}{(\theta+\sin\theta)^{5/2}}d\theta=\int (\theta+\sin\theta)^{-5/2}.(1+\cos\theta)\,d\theta$, Let $f(\theta)=\theta+\sin\theta$ $\therefore f'(\theta)=1+\cos\theta$

or $I=\int (\theta+\sin\theta)^{-5/2}.(1+\cos\theta)\,d\theta=\int [f(\theta)]^{-5/2}.f'(\theta)d\theta$ $\left[\text{formula, } \int \{f(x)\}^n.f'(x)\,dx=\frac{\{f(x)\}^{n+1}}{n+1}\right]$

or $I=\int [f(\theta)]^{-5/2}.f'(\theta)\,d\theta=\frac{[f(\theta)]^{-5/2+1}}{-\frac{5}{2}+1}+c=-\frac{2}{3}.[f(\theta)]^{-3/2}+c=-\frac{2}{3}.\frac{1}{[f(\theta)]^{3/2}}+c=-\frac{2}{3}.\frac{1}{[\theta+\sin\theta]^{3/2}}+c$

$=-\frac{2}{3}.\frac{1}{(\theta+\sin\theta).\sqrt{\theta+\sin\theta}}+c$ Ans.

(4) (a) $I=\int \frac{1+\sin\theta}{\cos\theta}d\theta=\int \frac{\sin^2{}^{\theta}/_2+\cos^2{}^{\theta}/_2+2\sin{}^{\theta}/_2\cos{}^{\theta}/_2}{\cos^2{}^{\theta}/_2-\sin^2{}^{\theta}/_2}d\theta$, Divide above and below by $\cos^2{}^{\theta}/_2$

$$I = \int \frac{\tan^2 \theta/_2 + 1 + 2\tan \theta/_2}{1 - \tan^2 \theta/_2}\, d\theta = \int \frac{(1+\tan \theta/_2)^2}{(1+\tan \theta/_2)(1-\tan \theta/_2)}\, d\theta = \int \frac{(1+\tan \theta/_2)}{(1-\tan \theta/_2)}\, d\theta = \int \frac{\sec^2 \theta/_2}{(1-\tan \theta/_2)}\, d\theta$$

Put $\tan \theta/_2 = t \quad \therefore \quad \frac{1}{2}\sec^2 \theta/_2 \; d\theta = dt \quad \text{or} \quad \sec^2 \theta/_2 \; d\theta = 2\, dt$

$$\therefore \quad I = \int \frac{2\, dt}{1-t} = 2\log(1-t)\,.\frac{1}{\text{d.c of }(1-t)} + c = -2\log(1-\tan \theta/_2) + c \qquad \text{Ans.}$$

(b) $I = \int \frac{1+\cos\theta}{\sin\theta}\, d\theta = \int \frac{\sin^2 \theta/_2 + \cos^2 \theta/_2 + \cos^2 \theta/_2 - \sin^2 \theta/_2}{2\sin \theta/_2 \cos \theta/_2}\, d\theta = \int \frac{2\cos^2 \theta/_2}{2\sin \theta/_2 \cos \theta/_2}\, d\theta = \int \frac{\cos \theta/_2}{\sin \theta/_2}\, d\theta$

$\therefore \quad I = \int \frac{\cos \theta/_2}{\sin \theta/_2}\, d\theta \quad$ Put $\sin \theta/_2 = t \quad \text{or} \quad \frac{1}{2}\cos \theta/_2 \; d\theta = dt \quad \text{or} \quad \cos \theta/_2 \; d\theta = 2\, dt$

$$\therefore \quad I = \int \frac{2\, dt}{t} = 2\log t + c = 2\log(\sin \theta/_2) + c \quad \text{Ans.}$$

(c) $I = \int \frac{\sec^2\theta}{1+\tan\theta}\, d\theta \qquad$ Put $1 + \tan\theta = t \quad \therefore \sec^2\theta \; d\theta = dt$

$$\therefore \quad I = \int \frac{dt}{t} = \log t + c = \log(1+\tan\theta) + c \quad \text{Ans.}$$

(d) $I = \int \frac{d}{dx}\left(\frac{\log(1+x)}{x}\right) dx \qquad \left[\text{use formula, } \frac{d}{dx}\int f(x)\, dx = f(x) \text{ or } \int \frac{d}{dx}[f(x)]\, dx = f(x)\right]$

$$\therefore \quad I = \int \frac{d}{dx}\left(\frac{\log(1+x)}{x}\right) dx = \frac{\log(1+x)}{x} + c \quad \text{Ans.}$$

(e) $I = \frac{d}{dx}\left(\int \frac{\log x + 1}{2x}\, dx\right) = \frac{\log x + 1}{2x} + c \quad$ Ans. (see above formula and solve)

(f) $I = \int \frac{1+\log x}{x\log x}\, dx \quad$ Let $f(x) = x\log x \quad \therefore \quad f'(x) = x\frac{1}{x} + \log x\,.1 = 1 + \log x$

or $I = \int \frac{f'(x)}{f(x)}\, dx = \log|f(x)| + c = \log|x\log x| + c = \log|x| + \log|\log x| + c \quad$ Ans.

IInd Method: $-\quad I = \int \frac{1+\log x}{x\log x}\, dx,\quad$ Let $x\log x = t$ or $\left(x.\frac{1}{x} + \log x\,.1\right) dx = dt$ or $(1+\log x)\, dx = dt$

or $I = \int \frac{dt}{t} = \log t + c = \log(x\log x) + c = \log x + \log(\log x) + c \quad$ Ans. $[\log(m.n) = \log m + \log n]$

(g) $I = \int \frac{x\cos x}{x\sin x + \cos x}\, dx \quad$ Let $f(x) = x\sin x + \cos x$ or $f'(x) = x\cos x + \sin x - \sin x = x\cos x$

$$\therefore \quad I = \int \frac{f'(x)}{f(x)}\, dx = \log|f(x)| + c = \log|x\sin x + \cos x| + c \quad \text{Ans.}$$

(h) $I = \int \frac{x}{2x^2 - 5}\, dx = \frac{1}{4}\log|2x^2 - 5| + c \qquad$ Ans. (solve, same as above question)

(5) (a) $I = \int \frac{(2x-1)}{2x^2 - 2x + 5}\, dx \quad$ Let $2x^2 - 2x + 5 = z \quad \therefore \quad (4x-2)\, dx = dz$ or $(2x-1)\, dx = \frac{dz}{2}$

$$\text{or } I = \int \frac{(2x-1)}{2x^2 - 2x + 5}\, dx = \int \frac{dz}{2z} = \frac{1}{2}\log z + c = \frac{1}{2}\log(2x^2 - 2x + 5) + c = \log\left(\sqrt{2x^2 - 2x + 5}\right) + c \quad \text{Ans.}$$

(b) $I = \int \frac{\cos x}{\sin^3 x}\, dx = \int \sin^{-3} x\,.\cos x\, dx \qquad \left[\text{formula, } \int [f(x)]^n.\, f'(x)\, dx = \frac{[f(x)]^{n+1}}{n+1}\right]$

or $I = \int \sin^{-3} x . \cos x \ dx$, Let $f(x) = \sin x \ \therefore \ f'(x) = \cos x$

or $I = \int [f(x)]^{-3}. f'(x) \ dx = \frac{[f(x)]^{-3+1}}{-3+1} = -\frac{1}{2}[f(x)]^{-2} + c$ where $f(x) = \sin x$ and $f'(x) = \cos x$

$$\therefore \ I = -\frac{1}{2}[f(x)]^{-2} + c = -\frac{1}{2}(\sin x)^{-2} + c = -\frac{1}{2\sin^2 x} + c = -\frac{1}{2}\operatorname{cosec}^2 x + c \quad \text{Ans.}$$

IInd Method: – $I = \int \frac{\cos x}{\sin^3 x} dx$ Let $\sin x = z \ \therefore \ \cos x \ dx = dz$

or $I = \int \frac{\cos x}{\sin^3 x} dx = \int \frac{dz}{z^3} = \int z^{-3} \ dz = \frac{z^{-3+1}}{-3+1} + c = -\frac{1}{2}z^{-2} + c = -\frac{1}{2z^2} + c = -\frac{1}{2\sin^2 x} + c = -\frac{1}{2}\operatorname{cosec}^2 x + c$ Ans.

(c) $I = \int \frac{\sec^2 x}{\tan^4 x} dx = \int \tan^{-4} x . \sec^2 x \ dx$ Let $f(x) = \tan x \ \therefore \ f'(x) = \sec^2 x$

$$\left[\text{using formula, } \int [f(x)]^n . f'(x) \ dx = \frac{[f(x)]^{n+1}}{n+1}\right]$$

or $I = \int [f(x)]^{-4}. f'(x) \ dx = \frac{[f(x)]^{-4+1}}{-4+1} + c = -\frac{1}{3}[f(x)]^{-3} + c = -\frac{1}{3[f(x)]^3} + c = -\frac{1}{3(\tan x)^3} + c$

$= -\frac{1}{3}\cot^3 x + c$ Ans. or $-\frac{\cos^3 x}{3\sin^3 x} + c$ Ans.

(d) $I = \int \frac{dx}{x^2 + x + 3}$ or $x^2 + x + 3 = a\left\{\left(x + \frac{b}{2a}\right)^2 - \frac{D}{4a^2}\right\}$

here $a = 1, \ b = 1$ and $c = 3$ then $D = b^2 - 4ac = 1 - 12 = -11$

or $x^2 + x + 3 = 1\left\{\left(x + \frac{1}{2}\right)^2 - \frac{-11}{4.1}\right\} = \left(x + \frac{1}{2}\right)^2 + \frac{11}{4} = \left(x + \frac{1}{2}\right)^2 + \left(\frac{\sqrt{11}}{2}\right)^2$

$I = \int \frac{dx}{x^2 + x + 3} = \int \frac{dx}{\left(x + \frac{1}{2}\right)^2 + \left(\frac{\sqrt{11}}{2}\right)^2}$ using formula, $\int \frac{dx}{x^2 + a^2} = \frac{1}{a}\tan^{-1}\left(\frac{x}{a}\right)$

$$I = \int \frac{dx}{\left(x + \frac{1}{2}\right)^2 + \left(\frac{\sqrt{11}}{2}\right)^2} = \frac{1}{\frac{\sqrt{11}}{2}}\tan^{-1}\left(\frac{x + \frac{1}{2}}{\frac{\sqrt{11}}{2}}\right) + c = \frac{2}{\sqrt{11}}\tan^{-1}\left(\frac{2x+1}{\sqrt{11}}\right) + c \quad \text{Ans.}$$

(e) $I = \int \frac{dx}{2x^2 + 4x - 5} = \frac{1}{2}\int \frac{dx}{(x+1)^2 - \left(\frac{\sqrt{7}}{\sqrt{2}}\right)^2} = \frac{1}{2\sqrt{14}}\log\left|\frac{\sqrt{2}x + \sqrt{2} - \sqrt{7}}{\sqrt{2}x + \sqrt{2} + \sqrt{7}}\right| + c$ Ans.

(f) $I = \int \frac{dx}{\sqrt{1 + 4x - 2x^2}}$ $\therefore \ 1 + 4x - 2x^2 = a\left\{\left(x + \frac{b}{2a}\right)^2 - \frac{D}{4a^2}\right\}$ here $a = -2, b = 4, c = 1$ and $D = 24$

or $1 + 4x - 2x^2 = -2\left\{\left(x + \frac{4}{-4}\right)^2 - \frac{24}{16}\right\} = -2\left\{(x-1)^2 - \frac{3}{2}\right\} = 2\left\{\frac{3}{2} - (x-1)^2\right\} = 2\left\{\left(\sqrt{\frac{3}{2}}\right)^2 - (x-1)^2\right\}$

or $I = \int \frac{dx}{\sqrt{2\left[\left(\frac{\sqrt{3}}{\sqrt{2}}\right)^2 - (x-1)^2\right]}} = \frac{1}{\sqrt{2}}\int \frac{dx}{\sqrt{\left(\frac{\sqrt{3}}{\sqrt{2}}\right)^2 - (x-1)^2}}$ using formula, $\int \frac{dx}{\sqrt{a^2 - x^2}} = \sin^{-1}\frac{x}{a}$

$$\therefore\ I=\frac{1}{\sqrt{2}}\int\frac{dx}{\sqrt{\left(\frac{\sqrt{3}}{\sqrt{2}}\right)^2-(x-1)^2}}=\frac{1}{\sqrt{2}}\sin^{-1}\left(\frac{x-1}{\frac{\sqrt{3}}{\sqrt{2}}}\right)+c=\frac{1}{\sqrt{2}}\sin^{-1}\left(\frac{\sqrt{2}(x-1)}{\sqrt{3}}\right)+c \quad \text{Ans.}$$

(g) $I=\int\frac{dx}{\sqrt{x^2+2x+3}}=\int\frac{dx}{\sqrt{(x+1)^2+2}}=\int\frac{dx}{\sqrt{(x+1)^2+\left(\sqrt{2}\right)^2}}$

use formula, $\int\frac{dx}{\sqrt{x^2+a^2}}=\log\left|x+\sqrt{x^2+a^2}\right|$

or $I=\int\frac{dx}{\sqrt{(x+1)^2+\left(\sqrt{2}\right)^2}}=\log\left|(x+1)+\sqrt{(x+1)^2+\left(\sqrt{2}\right)^2}\right|+c=\log\left|(x+1)+\sqrt{x^2+2x+3}\right|+c$ Ans.

(h) $I=\int\frac{dx}{\sqrt{3+4x-2x^2}}=\int\frac{dx}{\sqrt{a\left\{\left(x+\frac{b}{2a}\right)^2-\frac{D}{4a^2}\right\}}}=\int\frac{dx}{\sqrt{-2\left\{\left(x+\frac{4}{-4}\right)^2-\frac{40}{16}\right\}}}=\int\frac{dx}{\sqrt{-2\left\{(x-1)^2-\frac{5}{2}\right\}}}=\int\frac{dx}{\sqrt{2\left\{\frac{5}{2}-(x-1)^2\right\}}}$

$=\frac{1}{\sqrt{2}}\int\frac{dx}{\sqrt{\left(\frac{\sqrt{5}}{\sqrt{2}}\right)^2-(x-1)^2}}$ using formula, $\int\frac{dx}{\sqrt{a^2-x^2}}=\sin^{-1}\frac{x}{a}$

or $I=\frac{1}{\sqrt{2}}\int\frac{dx}{\sqrt{\left(\frac{\sqrt{5}}{\sqrt{2}}\right)^2-(x-1)^2}}=\frac{1}{\sqrt{2}}\sin^{-1}\left(\frac{x-1}{\frac{\sqrt{5}}{\sqrt{2}}}\right)+c=\frac{1}{\sqrt{2}}\sin^{-1}\left[\frac{\sqrt{2}(x-1)}{\sqrt{5}}\right]+c$ Ans.

(6) (a) $I=\int\sqrt{-x^2+3x+5}\,dx=\int\sqrt{a\left\{\left(x+\frac{b}{2a}\right)^2-\frac{D}{4a^2}\right\}}\,dx=\int\sqrt{-1\left\{\left(x+\frac{3}{-2}\right)^2-\frac{29}{4}\right\}}\,dx=\int\sqrt{\frac{29}{4}-\left(x-\frac{3}{2}\right)^2}\,dx$

$=\int\sqrt{\left(\frac{\sqrt{29}}{2}\right)^2-\left(x-\frac{3}{2}\right)^2}\,dx$

using formula, $\int\sqrt{a^2-x^2}\,dx=\frac{x}{2}\sqrt{a^2-x^2}+\frac{a^2}{2}\sin^{-1}\frac{x}{a}$

or $I=\frac{x-\frac{3}{2}}{2}\sqrt{\left(\frac{\sqrt{29}}{2}\right)^2-\left(x-\frac{3}{2}\right)^2}+\frac{29}{8}\sin^{-1}\left(\frac{x-\frac{3}{2}}{\frac{\sqrt{29}}{2}}\right)+c=\frac{2x-3}{4}\sqrt{-x^2+3x+5}+\frac{29}{8}\sin^{-1}\left(\frac{2x-3}{\sqrt{29}}\right)+c$ Ans.

(b) $I=\int\frac{\sec^2 x}{(\sec x+\tan x)^{9/2}}\,dx=-\frac{1}{(\sec x+\tan x)^{11/2}}\left\{\frac{1}{11}+\frac{1}{7}(\sec x+\tan x)^2\right\}+k$ Ans.

(c) $I=\int\frac{\sin^2(\log x)+\cos(\log x)}{x\tan(\log x)}\,dx$ Let $\log x=t$ $\therefore\ \frac{1}{x}dx=dt$

or $I=\int\frac{\sin^2(\log x)+\cos(\log x)}{\tan(\log x)}.\frac{1}{x}\,dx=\int\frac{\sin^2 t+\cos t}{\tan t}\,dt=\int\frac{\sin^2 t+\cos t}{\frac{\sin t}{\cos t}}\,dt=\int\frac{\cos t\,(\sin^2 t+\cos t)}{\sin t}\,dt=\int\frac{\sin^2 t.\cos t}{\sin t}\,dt+\int\frac{\cos^2 t}{\sin t}\,dt$

$=\int\sin t.\cos t\ dt+\int\frac{1-\sin^2 t}{\sin t}\,dt=I_1+I_2$ (say)

or $I_1=\int\sin t.\cos t\ dt$ Let $\sin t=u$ $\therefore\ \cos t\ dt=du$

$\therefore\ I_1=\int u.du=\frac{u^2}{2}+c=\frac{1}{2}.\sin^2 t+c=\frac{\sin^2(\log x)}{2}+c$

Now, $I_2 = \int \frac{1-\sin^2 t}{\sin t}\, dt = \int \frac{1}{\sin t}\, dt - \int \frac{\sin^2 t}{\sin t}\, dt = \int \operatorname{cosec} t\, dt - \int \sin t\, dt = \log(\tan {}^{t}/_{2}) + \cos t + c = \log\left[\tan\left(\frac{\log x}{2}\right)\right] + \cos(\log x) + c$

or $I = I_1 + I_2 = \frac{\sin^2(\log x)}{2} + c + \log\left[\tan\left(\frac{\log x}{2}\right)\right] + \cos(\log x) + c = \frac{\sin^2(\log x)}{2} + \log\left[\tan\left(\frac{\log x}{2}\right)\right] + \cos(\log x) + k$ Ans.

(d) $I = \int \frac{dx}{x \sin(\log x).\cos(\log x)}$ Let $\log x = t$ $\therefore \frac{1}{x}\, dx = dt$

or $I = \int \frac{1}{\sin(\log x).\cos(\log x)}.\frac{1}{x}\, dx = \int \frac{dt}{\sin t.\cos t} = \int \frac{\sin^2 t + \cos^2 t}{\sin t.\cos t}\, dt = \int \frac{\sin^2 t}{\sin t.\cos t}\, dt + \int \frac{\cos^2 t}{\sin t.\cos t}\, dt$

$\therefore\ I = \int \tan t\, dt + \int \cot t\, dt = -\log(\cos t) + \log(\sin t) + c = \log(\sin t) - \log(\cos t) + c = \log\left(\frac{\sin t}{\cos t}\right) + c = \log(\tan t) + c$

$= \log[\tan(\log x)] + c$ Ans.

(e) $I = \int \frac{\log x}{x^2}\, dx$, use integration by part formula $\int f(x).g(x)\, dx = f(x).\int g(x)\, dx - \int \left[\frac{d(f(x))}{dx}.\int g(x)\, dx\right] dx$

or $I = \int x^{-2}.\log x\, dx = \log x.\int x^{-2}\, dx - \int \left[\frac{d(\log x)}{dx}.\int x^{-2}\, dx\right] dx = \log x.\frac{x^{-2+1}}{-2+1} - \int \frac{1}{x}.\frac{x^{-2+1}}{-2+1}\, dx = -\log x.\frac{1}{x} + \int \frac{1}{x^2}\, dx$

$= -\frac{\log x}{x} + \int x^{-2}\, dx = -\frac{\log x}{x} + \frac{x^{-2+1}}{-2+1} + c = -\frac{\log x}{x} - \frac{1}{x} + c$

$\therefore\ I = \frac{-\log x - 1}{x} + c = \frac{-(\log x + 1)}{x} + c$ Ans.

(f) $I = \int \frac{dx}{x\sqrt{1+\log x}}$ Let $\log x = t$ $\therefore \frac{1}{x}\, dx = dt$

$\therefore\ I = \int \frac{1}{\sqrt{1+\log x}}.\frac{1}{x} dx = \int \frac{dt}{\sqrt{1+t}} = \int (1+t)^{-\frac{1}{2}}\, dt = \frac{(1+t)^{-\frac{1}{2}+1}}{-\frac{1}{2}+1} + c = 2\sqrt{1+t} + c = 2\sqrt{1+\log x} + c$ Ans.

(7) (a) $I = \int \frac{x^3}{(x-1)^2(x^2+2x+3)}\, dx$, Let $\frac{x^3}{(x-1)^2(x^2+2x+3)} = \frac{A}{x-1} + \frac{B}{(x-1)^2} + \frac{Cx+D}{x^2+2x+3}$ ………..(A)

$\therefore\ (x-1)^2 = 0$ or $x = 1$ then $B = \left.\frac{x^3}{(x^2+2x+3)}\right|_{x=1} = \frac{1}{1+2+3} = \frac{1}{6}$

To be find A, C and D are obtained by multiplying by $(x-1)^2(x^2+2x+3)$ on both of sides and then equate coefficient of suitable power of x on both sides.

or $\frac{x^3}{(x-1)^2(x^2+2x+3)} \times (x-1)^2(x^2+2x+3) = \frac{A}{x-1} \times (x-1)^2(x^2+2x+3)$

or $x^3 = A(x-1)(x^2+2x+3) = A[x^3+2x^2+3x-x^2-2x-3] = A[x^3+x^2+x-3] = Ax^3+Ax^2+Ax-3A$

$\therefore\ Ax^3 = x^3$ or $A = 1$

similarly, $x^3 = A(x+1)(x^2+2x+3) + B(x^2+2x+3) + (Cx+D)(x-1)^2$

solve and find value of A, C & *D and put value of A, B, C and D in equation* (A)and solve it.

(b) $I = \int \frac{x}{x^3+1}\, dx$, Let $\frac{x}{x^3+1} = \frac{x}{(x+1)(x^2-x+1)} = \frac{A}{x+1} + \frac{Bx+C}{x^2-x+1}$ ………………..(A)

$\therefore\ x+1 = 0$ or $x = -1$ then $A = \left.\frac{x}{(x^2-x+1)}\right|_{x=-1} = \frac{-1}{1+1+1} = -\frac{1}{3}$

Now, $\frac{x}{(x+1)(x^2-x+1)} = \frac{Bx+C}{x^2-x+1}$ Multiplying both of side by $(x+1)(x^2-x+1)$

and then equate coefficient of suitable power of x on both sides.

or $x = (Bx + C)(x + 1) = Bx^2 + Bx + Cx + C = Bx^2 + x(B + C) + C$

$$\therefore\ B + C = 1 \text{ and } C = 0 \text{ then } B = 1$$

Put value of A, B and C in equation (A) and integrate, we get

or $\dfrac{x}{(x + 1)(x^2 - x + 1)} = \dfrac{A}{x + 1} + \dfrac{Bx + C}{x^2 - x + 1} = -\dfrac{1}{3}.\dfrac{1}{x + 1} + \dfrac{x}{x^2 - x + 1}$

or $I = \displaystyle\int \frac{x}{(x + 1)(x^2 - x + 1)}\,dx = -\frac{1}{3}\int \frac{dx}{x + 1} + \int \frac{x}{x^2 - x + 1}\,dx = -\frac{1}{3}\log(x + 1) + \int \frac{\frac{1}{2}[2x - 1] + \frac{1}{2}}{x^2 - x + 1}\,dx$

$$I = -\frac{1}{3}\log(x + 1) + \int \frac{\frac{1}{2}[2x - 1]}{x^2 - x + 1}\,dx + \frac{1}{2}\int \frac{dx}{x^2 - x + 1}\,dx = -\frac{1}{3}\log(x + 1) + \frac{1}{2}\int \frac{f'(x)}{f(x)}\,dx + \frac{1}{2}\int \frac{dx}{\left(x - \frac{1}{2}\right)^2 + \frac{3}{4}}$$

use formula, $\displaystyle\int \frac{f'(x)}{f(x)}\,dx = \log[f(x)]$ and $\displaystyle\int \frac{dx}{x^2 + a^2} = \frac{1}{a}\tan^{-1}\frac{x}{a}$

$$I = -\frac{1}{3}\log(x + 1) + \frac{1}{2}\int \frac{f'(x)}{f(x)}\,dx + \frac{1}{2}\int \frac{dx}{\left(x - \frac{1}{2}\right)^2 + \left(\frac{\sqrt{3}}{2}\right)^2} = -\frac{1}{3}\log(x + 1) + \frac{1}{2}\log f(x) + \frac{1}{2}.\frac{1}{\frac{\sqrt{3}}{2}}\tan^{-1}\left(\frac{\left(x - \frac{1}{2}\right)}{\frac{\sqrt{3}}{2}}\right)$$

$$= -\frac{1}{3}\log(x + 1) + \log\left(\sqrt{x^2 - x + 1}\right) + \frac{1}{\sqrt{3}}\tan^{-1}\left(\frac{2x - 1}{\sqrt{3}}\right) + c \quad \text{Ans.}$$

(c) $I = \displaystyle\int \frac{x^4}{(x - 1)(x^2 + 1)}\,dx$, Let $\dfrac{x^4}{(x - 1)(x^2 + 1)} = \dfrac{A}{x - 1} + \dfrac{Bx + C}{x^2 + 1}$ ………………….(A)

$$\therefore\ x - 1 = 0 \text{ or } x = 1 \text{ then } A = \left.\frac{x^4}{(x^2 + 1)}\right|_{x=1} = \frac{1}{1 + 1} = \frac{1}{2}$$

Now, $\dfrac{x^4}{(x - 1)(x^2 + 1)} = \dfrac{Bx + C}{x^2 + 1}$

Multiplying both of side by $(x - 1)(x^2 + 1)$ and then equate coefficient of suitable power of x on both sides.

or $x^4 = (Bx + C)(x - 1) = Bx^2 - Bx + Cx - C$ $\quad\therefore\ C = 0$ or $x(C - B) = 0.x$ or $B = 0$

(d) $I = \displaystyle\int \frac{x - 1}{x + 1}.\frac{1}{\sqrt{x^3 + x^2 + x}}\,dx = \int \frac{(x - 1)(x + 1)}{(x + 1)^2.\sqrt{x^3 + x^2 + x}}\,dx = \int \frac{(x^2 - 1)\,dx}{(x^2 + 2x + 1).x.\sqrt{x + 1 + \frac{1}{x}}} = \int \frac{\left(1 - \frac{1}{x^2}\right)}{\left(x + \frac{1}{x} + 2\right).\sqrt{x + 1 + \frac{1}{x}}}\,dx$

Putting $x + \dfrac{1}{x} + 1 = z^2$ $\quad\therefore \left(1 - \dfrac{1}{x^2}\right)dx = 2z\,dz$ or $x + \dfrac{1}{x} = z^2 - 1$

or $I = \displaystyle\int \frac{2z\,dz}{(z^2 - 1 + 2).\sqrt{z^2}} = \int \frac{2z\,dz}{(z^2 + 1).z} = 2\int \frac{dz}{z^2 + 1} = 2.\frac{1}{1}\tan^{-1}\left(\frac{z}{1}\right) = 2\tan^{-1}\left(\sqrt{x + \frac{1}{x} + 1}\right) + c$ Ans.

Integral of the form: – $\displaystyle\int x^m(a + bx^n)^p\,dx$ where m, n and p are rational numbers.

Case I: – when $p \in I$

subcase: – (a) when $p \in I_+$, Expand $(a + bx^n)^p$ using binomal theorem.

(b) when $p \in I_-$, Putting $x = t^\alpha$, where α = L. C. M of D^r of fractions m and n.

Case II: – when $p \notin I$

subcase: – (a) when $\frac{m+1}{n}$ = integer, then putting $a + bx^n = t^\alpha$ where α is the D^r of fractions p.

(b) when $\frac{m+1}{n} + p \neq$ integer, then putting $a + bx^n = x^n t^\alpha$ where α is the D^r of fractions p.

Example: – (1) (a) $I = \int \sqrt{x}.(3+\sqrt[3]{x})^2\, dx = \int x^{\frac{1}{2}}.\left(3+x^{\frac{1}{3}}\right)^2 dx$

Here $m = \frac{1}{2}$, $n = \frac{1}{3}$ and $p = 2$ where p is positive integer.

Putting $x = t^\alpha$ where α = L. C. M of D^r of fractions m and n. $\therefore \alpha = \{2,3\} = 6$

$$\therefore\ x = t^6 \text{ or } t = x^{\frac{1}{6}} \text{ or } dx = 6t^5\, dt \text{ and } x^{\frac{1}{2}} = (t^6)^{\frac{1}{2}} = t^3,\ x^{\frac{1}{3}} = (t^6)^{\frac{1}{3}} = t^2$$

$$I = \int t^3.(3+t^2)^2.6t^5\, dt = 6\int t^8.(3+t^2)^2\, dt = 6\int t^8.(9+6t^2+t^4)\, dt = 6\int (9t^8+6t^{10}+t^{12})\, dt = 54\int t^8\, dt + 36\int t^{10}\, dt + 6\int t^{12}\, dt$$

$$= 54.\frac{t^{8+1}}{8+1} + 36.\frac{t^{10+1}}{10+1} + 6.\frac{t^{12+1}}{12+1} + c$$

$$I = 54.\frac{t^9}{9} + 36.\frac{t^{11}}{11} + 6.\frac{t^{13}}{13} + c = 6t^9 + \frac{36}{11}t^{11} + \frac{6}{13}t^{13} + c$$

put $t = x^{\frac{1}{6}}$ then $t^9 = x^{\frac{3}{2}}$, $t^{11} = x^{\frac{11}{6}}$, $t^{13} = x^{\frac{13}{6}}$ $\therefore\ I = 6x^{\frac{3}{2}} + \frac{36}{11}x^{\frac{11}{6}} + \frac{6}{13}x^{\frac{13}{6}} + c$ Ans.

IInd Method: – $I = \int \sqrt{x}.(3+\sqrt[3]{x})^2\, dx = \int x^{\frac{1}{2}}.\left(3+x^{\frac{1}{3}}\right)^2 dx = \int x^{\frac{1}{2}}.\left(9+6x^{\frac{1}{3}}+x^{\frac{2}{3}}\right) dx = \int \left(9x^{\frac{1}{2}} + 6x^{\frac{1}{2}}.x^{\frac{1}{3}} + x^{\frac{1}{2}}.x^{\frac{2}{3}}\right) dx$

$$= 9\int x^{\frac{1}{2}}\, dx + 6\int x^{\frac{5}{6}}\, dx + \int x^{\frac{7}{6}}\, dx$$

$$I = 9.\frac{x^{\frac{1}{2}+1}}{\frac{1}{2}+1} + 6.\frac{x^{\frac{5}{6}+1}}{\frac{5}{6}+1} + \frac{x^{\frac{7}{6}+1}}{\frac{7}{6}+1} + c = 6x^{\frac{3}{2}} + \frac{36}{11}x^{\frac{11}{6}} + \frac{6}{13}x^{\frac{13}{6}} + c \quad \text{Ans.}$$

(b) $I = \int x^{-\frac{1}{2}}.\left(2+x^{\frac{1}{2}}\right)^{-1} dx$ $\therefore$ here $m = -\frac{1}{2}$, $n = \frac{1}{2}$ and $p = -1$ (p is a negative integer)

Putting $x = t^\alpha$ where α = L. C. M of D^r of fractions m and n. $\therefore \alpha = \{2,2\} = 2$

$$\therefore\ x = t^2 \text{ or } t = \sqrt{x} \quad \therefore\ dx = 2t\, dt$$

$$\therefore\ I = \int x^{-\frac{1}{2}}.\left(2+x^{\frac{1}{2}}\right)^{-1} dx = \int (t^2)^{\frac{-1}{2}}.\left[2+(t^2)^{\frac{1}{2}}\right]^{-1}.2t\, dt = 2\int t^{-1}.(2+t)^{-1}.t\, dt = 2\int \frac{dt}{2+t} = 2\log(2+t) + c = \log(2+t)^2 + c$$

$$= \log(2+\sqrt{x})^2 + c = \log\left[2+x^{\frac{1}{2}}\right]^2 + c \quad \text{Ans.}$$

(c) $I = \int \frac{\sqrt{3+\sqrt[3]{x}}}{\sqrt[3]{x^2}}\, dx = \int x^{-\frac{2}{3}}.\left(3+x^{\frac{1}{3}}\right)^{\frac{1}{2}} dx$ Here $m = -\frac{2}{3}$, $n = \frac{1}{3}$ and $p = \frac{1}{2} \neq$ integer

when, $\frac{m+1}{n} = \frac{-\frac{2}{3}+1}{\frac{1}{3}} = 1$ = integer. Putting, $3 + x^{\frac{1}{3}} = t^\alpha$ where α is the D^r of fractions p $\therefore \alpha = 2$

$$\therefore\ 3 + x^{\frac{1}{3}} = t^2 \text{ or } \frac{1}{3}x^{\frac{1}{3}-1}dx = 2t\, dt \quad \therefore\ x^{\frac{-2}{3}}\, dx = 6t\, dt$$

$$I = \int x^{-\frac{2}{3}}.\left(3+x^{\frac{1}{3}}\right)^{\frac{1}{2}} dx = \int (t^2)^{\frac{1}{2}}.6t\, dt = 6\int t^2\, dt = 6.\frac{t^{2+1}}{2+1} + c = 2t^3 + c$$

put $3 + x^{\frac{1}{3}} = t^2$ $\therefore\ t = \left(3+x^{\frac{1}{3}}\right)^{\frac{1}{2}}$ then $I = 2t^3 + c = 2\left[\left(3+x^{\frac{1}{3}}\right)^{\frac{1}{2}}\right]^3 + c = 2\left(3+x^{\frac{1}{3}}\right)^{\frac{3}{2}} + c$ Ans.

(d) $I=\int \frac{dx}{x^9(2+x^3)^{\frac{1}{3}}}=\int x^{-9}.(2+x^3)^{\frac{-1}{3}}\,dx$ Here $m=-9,\ n=3$ and $p=-\frac{1}{3}\neq$ integer

when, $\frac{m+1}{n}=\frac{-9+1}{3}=-\frac{8}{3}\neq$ integer $\therefore \frac{m+1}{n}+p=-\frac{8}{3}-\frac{1}{3}=-\frac{9}{3}=-3=$ negative integer

$\therefore$ Putting, $2+x^3=x^3t^\alpha$ where α is D^r of the fractions p $\therefore \alpha=3$

$$\therefore 2+x^3=x^3t^3 \text{ or } x^3t^3-x^3=2 \text{ or } x^3(t^3-1)=2 \text{ or } x^3=\frac{2}{t^3-1} \quad \therefore x=\left(\frac{2}{t^3-1}\right)^{\frac{1}{3}}$$

$$\text{or } 3x^2\,dx=\frac{(t^3-1).0-2.3t^2}{(t^3-1)^2}\,dt=\frac{-6t^2}{(t^3-1)^2}\,dt$$

$$\text{or } dx=\frac{-6t^2}{(t^3-1)^2.3x^2}\,dt=\frac{-2t^2}{(t^3-1)^2.\left[\left(\frac{2}{t^3-1}\right)^{\frac{1}{3}}\right]^2}\,dt$$

$$\text{or } I=\int x^{-9}.(2+x^3)^{\frac{-1}{3}}\,dx=\int\left(\frac{2}{t^3-1}\right)^{-3}.\left(2+\frac{2}{t^3-1}\right)^{-\frac{1}{3}}.\frac{-2t^2}{(t^3-1)^2.\left(\frac{2}{t^3-1}\right)^{\frac{2}{3}}}\,dt=-\int\left(\frac{2}{t^3-1}\right)^{-3}\times\left(\frac{2t^3-2+2}{t^3-1}\right)^{-\frac{1}{3}}\times\frac{2t^2.(t^3-1)^{\frac{2}{3}}}{(t^3-1)^2.2^{\frac{2}{3}}}\,dt$$

$$I=-\int 2^{-3}.(t^3-1)^3.\frac{(2t^3)^{-\frac{1}{3}}}{(t^3-1)^{\frac{-1}{3}}}.\frac{2t^2.(t^3-1)^{\frac{2}{3}}}{(t^3-1)^2.2^{\frac{2}{3}}}\,dt=-\int\frac{2^{-3}.2^{\frac{-2}{3}}.2^{\frac{-1}{3}}.2.t^2.t^{-1}}{(t^3-1)^{-1}.(t^3-1)^{\frac{-1}{3}}.(t^3-1)^{\frac{-2}{3}}}\,dt$$

$$I=-\int\frac{2^{\left(-3-\frac{1}{3}-\frac{2}{3}+1\right)}.t}{(t^3-1)^{\left(-1-\frac{1}{3}-\frac{2}{3}\right)}}\,dt=-\int\frac{2^{-3}.t}{(t^3-1)^{-2}}\,dt=-\frac{1}{8}\int t(t^3-1)^2\,dt=-\frac{1}{8}\int t(t^6-2t^3+1)\,dt=-\frac{1}{8}\int(t^7-2t^4+t)\,dt$$

$$=-\frac{1}{8}\int t^7\,dt+\frac{1}{4}\int t^4\,dt-\frac{1}{8}\int t\,dt$$

$$I=-\frac{1}{8}.\frac{t^{7+1}}{7+1}+\frac{1}{4}.\frac{t^{4+1}}{4+1}-\frac{1}{8}.\frac{t^{1+1}}{1+1}+c=-\frac{1}{64}t^8+\frac{1}{20}t^5-\frac{1}{16}t^2+c$$

Putting, $2+x^3=x^3t^3$ or $t^3=\frac{2+x^3}{x^3}$ $\therefore t=\left(\frac{2+x^3}{x^3}\right)^{\frac{1}{3}}$

$$\therefore I=-\frac{1}{64}t^8+\frac{1}{20}t^5-\frac{1}{16}t^2+c=-\frac{1}{64}\left[\left(\frac{2+x^3}{x^3}\right)^{\frac{1}{3}}\right]^8+\frac{1}{20}\left[\left(\frac{2+x^3}{x^3}\right)^{\frac{1}{3}}\right]^5-\frac{1}{16}\left[\left(\frac{2+x^3}{x^3}\right)^{\frac{1}{3}}\right]^2+c$$

$$=-\frac{1}{64}\left(\frac{2+x^3}{x^3}\right)^{\frac{8}{3}}+\frac{1}{20}\left(\frac{2+x^3}{x^3}\right)^{\frac{5}{3}}-\frac{1}{16}\left(\frac{2+x^3}{x^3}\right)^{\frac{2}{3}}+c \quad \text{Ans.}$$

(2) Do yourself. find the integral: – $\int x^m(a+bx^n)^p\,dx$

(a) $I=\int \sqrt[3]{x}(2+\sqrt{x})^2.\,dx$, Ans: – $3x^{\frac{4}{3}}+\frac{24}{11}x^{\frac{11}{6}}+\frac{3}{7}x^{\frac{7}{3}}+c$

(b) $I=\int x^{-\frac{2}{3}}\left(1+x^{\frac{2}{3}}\right)^{-1}\,dx$, Ans: – $3\tan^{-1}(x)^{\frac{1}{3}}+c$ (c) $I=\int\frac{\sqrt{1+\sqrt[3]{x}}}{\sqrt[3]{x^2}}\,dx$, Ans: – $2\left(1+x^{\frac{1}{3}}\right)^{\frac{3}{2}}+c$

(d) $I=\int\frac{dx}{x^{11}(1+x^4)^{\frac{1}{2}}}$, Ans: – $-\frac{1}{10}\left(\frac{1+x^4}{x^4}\right)^{\frac{5}{2}}+\frac{1}{3}\left(\frac{1+x^4}{x^4}\right)^{\frac{3}{2}}-\frac{1}{2}\left(\frac{1+x^4}{x^4}\right)^{\frac{1}{2}}+c$

Other method of integrating: – $\int R\left(x,\sqrt{ax^2+bx+c}\right)dx$

Integral of the form: $-\int \frac{P_m(x)}{\sqrt{ax^2+bx+c}}\,dx$, where P_m is polynomial in x of degree m and $x \neq 0$

Method of integration: $-$ Let $\int \frac{P_m(x)}{\sqrt{ax^2+bx+c}}\,dx = P_{m-1}(x).\sqrt{ax^2+bx+c} + k\int \frac{dx}{\sqrt{ax^2+bx+c}}$ ……. (A)

Differentiating both sides with respect to x and then equate coefficient of suitable power of x from this

find the value of constants and put in equation (A).

Example: $-$ $I = \int \frac{x^3+2}{\sqrt{x^2+2x+3}}\,dx$

Let $\int \frac{x^3+2}{\sqrt{x^2+2x+3}}\,dx = (ax^2+bx+c).\sqrt{x^2+2x+3} + k\int \frac{dx}{\sqrt{x^2+2x+3}}$ …………………. (A)

Differentiating both of sides, we get

or $\frac{x^3+2}{\sqrt{x^2+2x+3}} = (ax^2+bx+c).\frac{1}{2\sqrt{x^2+2x+3}}.(2x+2) + \sqrt{x^2+2x+3}.(2ax+b) + k.\frac{1}{\sqrt{x^2+2x+3}}$

Multiplying both of sides by $\sqrt{x^2+2x+3}$, we get

or $x^3+2 = (ax^2+bx+c)(x+1) + (x^2+2x+3)(2ax+b) + k$

or $x^3 + 0.x^2 + 0.x + 2 = ax^3 + bx^2 + cx + ax^2 + bx + c + 2ax^3 + 4ax^2 + 6ax + bx^2 + 2bx + 3b + k$

$\therefore$ Equating coefficient of similar power of x on both sides, we get

$\therefore\ a + 2a = 1$ or $a = \frac{1}{3}$ and $a + b + 4a + b = 0$ or $5a + 2b = 0$ $\therefore\ b = -\frac{5}{6}$

or $c + b + 6a + 2b = 0$ or $6a + 3b + c = 0$ $\therefore\ c = -6.\frac{1}{3} - 3.\frac{-5}{6} = -2 + \frac{5}{2} = \frac{-4+5}{2} = \frac{1}{2}$

or $3b + c + k = 2$ or $3.\frac{-5}{6} + \frac{1}{2} + k = 2$ $\therefore\ k = 2 + \frac{5}{2} - \frac{1}{2} = \frac{4+5-1}{2} = \frac{8}{2} = 4$

Put value a, b, c and k in equation (A), we get

$$\int \frac{x^3+2}{\sqrt{x^2+2x+3}}\,dx = (ax^2+bx+c).\sqrt{x^2+2x+3} + k\int \frac{dx}{\sqrt{x^2+2x+3}} = \left(\frac{1}{3}x^2 - \frac{5}{6}x + \frac{1}{2}\right).\sqrt{x^2+2x+3} + 4\int \frac{dx}{\sqrt{(x+1)^2 + (\sqrt{2})^2}}$$

$$\int \frac{x^3+2}{\sqrt{x^2+2x+3}}\,dx = \left(\frac{1}{3}x^2 - \frac{5}{6}x + \frac{1}{2}\right).\sqrt{x^2+2x+3} + 4\log\left[(x+1) + \sqrt{x^2+2x+3}\right] + c \quad \text{Ans.}$$

Integral of the form: $-$ $\int \frac{dx}{(x \pm x_1)^m.\sqrt{ax^2+bx+c}}$, Method of integral: $-$ Putting $x \pm x_1 = \frac{1}{t}$

Example: $-$ (a) $I = \int \frac{dx}{(x-1)^2.\sqrt{x^2-2x+2}}$, Putting $x - 1 = \frac{1}{t}$ or $x = \frac{1+t}{t}$ $\therefore\ dx = -\frac{1}{t^2}\,dt$

$$I = \int -\frac{1}{t^2}.\frac{dt}{\left(\frac{1}{t}\right)^2.\sqrt{\left(\frac{1+t}{t}\right)^2 - 2.\left(\frac{1+t}{t}\right) + 2}} = -\int \frac{dt}{\sqrt{\frac{(1+t)^2}{t^2} - 2.\frac{1+t}{t} + 2}} = -\int \frac{dt}{\sqrt{\frac{1+2t+t^2-2t-2t^2+2t^2}{t^2}}} = -\int \frac{t\,dt}{\sqrt{t^2+1}}$$

Let $1 + t^2 = z^2$ $\therefore$ $2t\,dt = 2z\,dz$ or $t\,dt = z\,dz$

$$I = -\int \frac{z\,dz}{z} = -\int dz = -z + c = -\sqrt{1+t^2} + c = -\sqrt{1 + \left(\frac{1}{x-1}\right)^2} + c = -\sqrt{\frac{(x-1)^2+1}{(x-1)^2}} + c = -\frac{\sqrt{x^2-2x+1+1}}{x-1} + c$$

$$= -\frac{\sqrt{x^2-2x+2}}{x-1} + c \quad \text{Ans.}$$

(b) $I = \int \frac{x\,dx}{(x^2 - 4x + 3).\sqrt{x^2 - 5x + 4}} = \int \frac{x\,dx}{(x-1)(x-3).\sqrt{x^2 - 5x + 4}}$

Let $\frac{x}{(x-1)(x-3)} = \frac{A}{x-1} + \frac{B}{x-3}$ ……………. (i) then find A and B.

$$\therefore\ A = \frac{x}{(x-3)}\bigg|_{x=1} = \frac{1}{1-3} = -\frac{1}{2} \quad \text{and} \quad B = \frac{x}{(x-1)}\bigg|_{x=3} = \frac{3}{3-1} = \frac{3}{2}$$

$$\therefore\ \frac{x}{(x-1)(x-3)} = \frac{A}{x-1} + \frac{B}{x-3} = -\frac{1}{2(x-1)} + \frac{3}{2(x-3)}$$

$$\therefore\ I = \int \frac{x\,dx}{(x-1)(x-3).\sqrt{x^2-5x+4}} = \int \left\{\frac{3}{2(x-3)} - \frac{1}{2(x-1)}\right\}.\frac{dx}{\sqrt{x^2-5x+4}} = \frac{3}{2}\int \frac{dx}{(x-3).\sqrt{x^2-5x+4}} - \frac{1}{2}\int \frac{dx}{(x-1).\sqrt{x^2-5x+4}}$$
$$= \frac{3}{2}I_1 - \frac{1}{2}I_2 \quad \text{……………. (ii)}$$

solve I_1 and I_2 and put in equation (ii)

where $I_1 = \int \frac{dx}{(x-3).\sqrt{x^2-5x+4}}$ Putting $x - 3 = \frac{1}{t}$ or $x = \frac{1+3t}{t}$ or $t = \frac{1}{x-3}$ $\therefore\ dx = -\frac{1}{t^2}\,dt$

$$\text{or } I_1 = -\int \frac{1}{t^2}.\frac{dt}{\frac{1}{t}.\sqrt{\left(\frac{1+3t}{t}\right)^2 - 5.\left(\frac{1+3t}{t}\right) + 4}} = -\int \frac{dt}{t.\sqrt{\frac{1+6t+9t^2-5t-15t^2+4t^2}{t^2}}} = -\int \frac{dt}{\sqrt{1+t-2t^2}} = -\int \frac{dt}{\sqrt{2\left[\frac{9}{16} - \left(t-\frac{1}{4}\right)^2\right]}}$$
$$= -\frac{1}{\sqrt{2}}\int \frac{dt}{\sqrt{\left(\frac{3}{4}\right)^2 - \left(t-\frac{1}{4}\right)^2}}$$

use formula, $ax^2 + bx + c = a\left[\left(x + \frac{b}{2a}\right)^2 - \frac{D}{4a^2}\right]$ and $\int \frac{dx}{\sqrt{a^2 - x^2}} = \sin^{-1}\frac{x}{a}$

$$\therefore\ I_1 = -\frac{1}{\sqrt{2}}.\sin^{-1}\left(\frac{t-\frac{1}{4}}{\frac{3}{4}}\right) + c = -\frac{1}{\sqrt{2}}\sin^{-1}\left(\frac{4t-1}{3}\right) + c = -\frac{1}{\sqrt{2}}\sin^{-1}\left(\frac{4.\frac{1}{x-3} - 1}{3}\right) + c = -\frac{1}{\sqrt{2}}\sin^{-1}\left(\frac{4-x+3}{3x-9}\right) + c$$
$$= -\frac{1}{\sqrt{2}}\sin^{-1}\left(\frac{7-x}{3x-9}\right) + c$$

similarly, $I_2 = \int \frac{dx}{(x-1).\sqrt{x^2-5x+4}}$, Putting $x - 1 = \frac{1}{t}$ or $x = \frac{1+t}{t}$ $\therefore\ dx = -\frac{1}{t^2}\,dt$

$$\text{or } I_2 = -\int \frac{1}{t^2}.\frac{dt}{\frac{1}{t}.\sqrt{\left(\frac{1+t}{t}\right)^2 - 5.\left(\frac{1+t}{t}\right) + 4}} = -\int \frac{dt}{\sqrt{1+2t+t^2-5t-5t^2+4t^2}} = -\int \frac{dt}{\sqrt{1-3t}} = \int \frac{-dt}{\sqrt{1-3t}}$$

$$\text{Put } 1 - 3t = z^2 \quad \therefore\ -3\,dt = 2z\,dz \quad \text{or } -dt = \frac{2}{3}z\,dz$$

$$\therefore\ I_2 = \frac{2}{3}\int \frac{z\,dz}{z} = \frac{2}{3}\int dz = \frac{2}{3}z + c = \frac{2}{3}\sqrt{1-3t} + c = \frac{2}{3}\sqrt{1 - 3.\frac{1}{x-1}} + c = \frac{2}{3}\sqrt{\frac{x-1-3}{x-1}} + c = \frac{2}{3}\sqrt{\frac{x-4}{x-1}} + c$$

Put value of I_1 and I_2 in equation (ii), we get

$$\text{or } I = \frac{3}{2}I_1 - \frac{1}{2}I_2 = -\frac{3}{2}.\frac{1}{\sqrt{2}}\sin^{-1}\left(\frac{7-x}{3x-9}\right) - \frac{1}{2}.\frac{2}{3}\sqrt{\frac{x-4}{x-1}} + k = -\frac{3}{2\sqrt{2}}\sin^{-1}\left(\frac{7-x}{3x-9}\right) - \frac{1}{3}.\sqrt{\frac{x-4}{x-1}} + k \quad \text{Ans.}$$

\# **Irrational function**: − If in $y = f(x)$ operations of addition, subtraction, multiplication, division and raising to a power with rational non integer are performed on the R. H. S , then $y = f(x)$ is called irrational function.

e.g. (i) $y=\dfrac{\sqrt[3]{x}+3.\sqrt{x}}{2.\sqrt[5]{x}+4}=\dfrac{x^{\frac{1}{3}}+3.x^{\frac{1}{2}}}{2.x^{\frac{1}{5}}+4}=R\left(x,x^{\frac{1}{3}},x^{\frac{1}{2}},x^{\frac{1}{5}}\right)$ (ii) $y=-\frac{1}{3}.x^{\frac{1}{4}}=R\left(x,x^{\frac{1}{4}}\right)$

(iii) $y=\dfrac{\sqrt{x+3}-\sqrt[3]{x-2}}{(x-5)^{\frac{1}{4}}}=R\left(x,(x+3)^{\frac{1}{2}},(x-2)^{\frac{1}{3}},(x-5)^{\frac{1}{4}}\right)$

Integral of the form: – $\displaystyle\int R\left(x,x^{\frac{p_1}{q_1}},x^{\frac{p_2}{q_2}},\ldots\ldots\ldots.x^{\frac{p_n}{q_n}}\right)dx$

Putting $x=t^{\alpha}$, where $\alpha=$ L.C.M of $\{q_1,q_2,q_3,\ldots\ldots\ldots\ldots\ldots..q_n\}$

Integral of the form: – $\displaystyle\int R\left[x,\left(\frac{ax+b}{cx+d}\right)^{\frac{1}{n}}\right]dx$, Putting $\dfrac{ax+b}{cx+d}=t^n$

Example: – find the integral $I=\displaystyle\int\frac{x-\sqrt[3]{x}+\sqrt[5]{x^2}}{x(1+\sqrt[4]{x})}dx$, Here integral is of the form $\displaystyle\int R\left(x,x^{\frac{1}{3}},x^{\frac{2}{5}},x^{\frac{1}{4}}\right)dx$

Putting $x=t^{\alpha}$, where $\alpha=$ L.C.M of $\{3,5,4\}=60$ i.e $x=t^{60}$ $\therefore$ $dx=60t^{59}\,dt$

$$I=\int\frac{x-\sqrt[3]{x}+\sqrt[5]{x^2}}{x(1+\sqrt[4]{x})}dx=\int\frac{t^{60}-t^{\frac{60}{3}}+t^{\frac{2\times60}{5}}}{t^{60}\left(1+t^{\frac{60}{4}}\right)}.60t^{59}\,dt=60\int\frac{t^{60}-t^{20}+t^{24}}{t(1+t^{15})}dt=60\int\frac{t^{59}-t^{19}+t^{23}}{(1+t^{15})}dt$$

$$I=60\int\frac{t^{59}}{(1+t^{15})}dt-60\int\frac{t^{19}}{(1+t^{15})}dt+60\int\frac{t^{23}}{(1+t^{15})}dt=60I_1-60I_2+60I_3\ \text{(say)}\ldots\ldots(A)$$

solve I_1, I_2 and I_3 and put equation (A),

where $I_1=\displaystyle\int\frac{t^{59}}{(1+t^{15})}dt$, Let $1+t^{15}=z$ or $t^{15}=z-1$ $\therefore$ $15t^{14}\,dt=dz$ or $t^{14}\,dt=\dfrac{dz}{15}$

where $I_1=\displaystyle\int\frac{t^{45}.t^{14}}{(1+t^{15})}dt=\int\frac{(z-1)^3}{z}.\frac{dz}{15}=\frac{1}{15}\int\frac{z^3-3z^2+3z-1}{z}dz$ $\quad[(a-b)^3=a^3-b^3-3a^2b+3ab^2]$

$$\text{or } I_1=\frac{1}{15}\int\frac{z^3}{z}dz-\frac{3}{15}\int\frac{z^2}{z}dz+\frac{3}{15}\int\frac{z}{z}dz-\frac{1}{15}\int\frac{dz}{z}=\frac{1}{15}\int z^2\,dz-\frac{3}{15}\int z\,dz+\frac{3}{15}\int dz-\frac{1}{15}\int\frac{dz}{z}$$
$$=\frac{1}{15}.\frac{z^{2+1}}{2+1}-\frac{3}{15}.\frac{z^{1+1}}{1+1}+\frac{3}{15}.z-\frac{1}{15}.\log z+c=\frac{z^3}{45}-\frac{z^2}{10}+\frac{3z}{15}-\frac{1}{15}.\log z+c$$

$$\therefore\ I_1=\frac{(1+t^{15})^3}{45}-\frac{(1+t^{15})^2}{10}+\frac{3(1+t^{15})}{15}-\frac{1}{15}.\log(1+t^{15})+c$$

similarly, $I_2=\displaystyle\int\frac{t^{19}}{(1+t^{15})}dt=\int\frac{t^5.t^{14}}{(1+t^{15})}dt=\int\frac{(z-1)^{\frac{1}{3}}}{z}.\frac{dz}{15}$

Integral of the form: – Euler'ssubstitution: – $\displaystyle\int R\left(x,\sqrt{ax^2+bx+c}\right)dx$

Case I: – when root of equation $ax^2+bx+c=0$ are imaginary.

Subcase: – (a) when $a>0$, *Putting* $\sqrt{ax^2+bx+c}=t-x\sqrt{a}$

(b) when $c>0$, *Putting* $\sqrt{ax^2+bx+c}=tx-\sqrt{c}$

Case II: – when root of equation $ax^2+bx+c=0$ are real.

Putting $\sqrt{ax^2+bx+c}=(x-\alpha)t$, where α is a root of equation $ax^2+bx+c=0$.

Example: – $I=\displaystyle\int\frac{dx}{x+\sqrt{x^2+4x+8}}$ Integral of the form $\displaystyle\int R\left(x,\sqrt{x^2+4x+8}\right)dx$

when $a = 1 > 0$, *Putting* $\sqrt{x^2+4x+8} = t - x\sqrt{1} = t - x$ squaring both of sides,

or $x^2+4x+8 = (t-x)^2 = t^2 - 2tx + x^2$ or $4x + 8 = t^2 - 2tx$ or $4x + 2tx = t^2 - 8$

$$\text{or } 2x(2+t) = t^2 - 8 \quad \therefore\ 2x = \frac{t^2-8}{2+t} \quad \text{or } x = \frac{t^2-8}{4+2t}$$

$$\therefore\ dx = \frac{(4+2t).2t - (t^2-8).2}{[2(2+t)]^2}\,dt = \frac{8t + 4t^2 - 2t^2 + 16}{4(2+t)^2}\,dt = \frac{t^2+4t+8}{2(2+t)^2}\,dt$$

$$I = \int \frac{1}{x+t-x}.\frac{t^2+4t+8}{2(2+t)^2}\,dt = \frac{1}{2}\int \frac{t^2+4t+8}{t(2+t)^2}\,dt = \frac{1}{2}\int \frac{(2+t)^2+4}{t(2+t)^2}\,dt = \frac{1}{2}\int \frac{(2+t)^2}{t(2+t)^2}\,dt + \int \frac{2}{t(2+t)^2}\,dt = \frac{1}{2}\int \frac{dt}{t} + 2\int\left(\frac{1}{t} - \frac{t+4}{(2+t)^2}\right)dt$$
$$= \frac{1}{2}\int \frac{dt}{t} + 2\int \frac{dt}{t} - \int \frac{2t+8}{(2+t)^2}\,dt = \frac{5}{2}\int \frac{dt}{t} - \int \frac{2t+8}{4+4t+t^2}\,dt$$

$$I = \frac{5}{2}\int \frac{dt}{t} - \int \frac{2(t+2)+4}{(2+t)^2}\,dt = \frac{5}{2}\int \frac{dt}{t} - \int \frac{2(t+2)}{(2+t)^2}\,dt - \int \frac{4}{(2+t)^2}\,dt = \frac{5}{2}.\log t - 2\int \frac{dt}{t+2} - 4\int \frac{dt}{(2+t)^2}$$
$$= \frac{5}{2}.\log t - 2\log(t+2) - 4.\frac{(2+t)^{-2+1}}{-2+1} + c = \frac{5}{2}.\log t - 2\log(t+2) + \frac{4}{2+t} + c \qquad \text{Ans.}$$

Put $\sqrt{x^2+4x+8} = t - x$ or $t = x + \sqrt{x^2+4x+8}$

$$I = \frac{5}{2}.\log\left(x + \sqrt{x^2+4x+8}\right) - 2.\log\left[(x+2) + \sqrt{x^2+4x+8}\right] + \frac{4}{(x+2)+\sqrt{x^2+4x+8}} + c \qquad \text{Ans.}$$

Integral of the form: – $\int \sin^m x \cos^n x\ dx$

Case I: – when $m \in$ odd positive integer then putting $z = \cos x$

Case II: – when $n \in$ odd positive integer then putting $z = \sin x$

Case III: – when m and n both are odd positive integer then putting $z = \sin x$ or $\cos x$

Case IV: – when neither m nor n is odd integer,

(a) when $m + n =$ even negative integer then putting $z = \tan x$

(b) when $m + n =$ odd negative integer, multiplying suitable power of $\sin^2 x + \cos^2 x$

(c) when $m + n =$ even positive integer, Let $z = \cos x + i \sin x$ or $\frac{1}{z} = \cos x - i\sin x$

$$\text{or } \cos x = \frac{1}{2}\left(z + \frac{1}{z}\right) \text{ and } \sin x = \frac{1}{2i}\left(z - \frac{1}{z}\right) \quad \text{or } z^n + \frac{1}{z^n} = 2\cos nx \text{ and } z^n - \frac{1}{z^n} = 2i\sin nx$$

Example: – (a) $I = \int \sin^5 x.\cos^2 x\ dx$ Here $m = 5$ and $n = 2$ when m is odd positive integer

$$\text{Then putting } z = \cos x \quad \therefore\ dz = -\sin x\ dx$$

$$I = \int \sin^5 x.\cos^2 x\ dx = \int \cos^2 x.\sin^4 x.\sin x\ dx = \int \cos^2 x.(1-\cos^2 x)^2.\sin x\ dx = -\int z^2.(1-z^2)^2\ dz = -\int z^2(1 - 2z^2 + z^4)\,dz$$
$$= -\int (z^2 - 2z^4 + z^6)\ dz = -\int z^2\,dz + 2\int z^4\,dz - \int z^6\,dz$$

$$I = -\frac{z^{2+1}}{2+1} + 2.\frac{z^{4+1}}{4+1} - \frac{z^{6+1}}{6+1} + c = -\frac{1}{3}z^3 + \frac{2}{5}z^5 - \frac{1}{7}z^7 + c = -\frac{1}{3}\cos^3 x + \frac{2}{5}\cos^5 x - \frac{1}{7}\cos^7 x + c \qquad \text{Ans.}$$

(b) $I = \int \sin^4 x.\cos^3 x\ dx$ Here $m = 4$ and $n = 3$ when n is odd positive integer

$$\text{Putting, } z = \sin x \quad \therefore\ dz = \cos x\ dx$$

$$I = \int \sin^4 x.\cos^2 x.\cos x\,dx = \int \sin^4 x\,(1-\sin^2 x).\cos x\,dx = \int z^4(1-z^2)\,dz = \int (z^4 - z^6)\,dz = \int z^4\,dz - \int z^6\,dz = \frac{z^{4+1}}{4+1} - \frac{z^{6+1}}{6+1} + c$$

$$= \frac{z^5}{5} - \frac{z^7}{7} + c = \frac{1}{5}\sin^5 x - \frac{1}{7}\sin^7 x + c \quad \text{Ans.}$$

(c) $I = \int \frac{\cos^3 x}{\sin^5 x}\,dx = \int \cos^3 x.\sin^{-5} x\,dx$ Here $m = -5$ and $n = 3$ when n is odd positive integer

Putting $z = \sin x \quad \therefore\ dz = \cos x\,dx$

$$I = \int \cos^2 x.\sin^{-5} x.\cos x\,dx = \int (1-\sin^2 x).\sin^{-5} x.\cos x\,dx = \int (1-z^2).z^{-5}\,dz = \int (z^{-5} - z^{-3})\,dz = \int z^{-5}\,dz - \int z^{-3}\,dz$$

$$= \frac{z^{-5+1}}{-5+1} - \frac{z^{-3+1}}{-3+1} + c = -\frac{z^{-4}}{4} + \frac{z^{-2}}{2} + c = \frac{1}{2z^2} - \frac{1}{4z^4} + c$$

$$I = \frac{1}{2.\sin^2 x} - \frac{1}{4.\sin^4 x} + c = \frac{1}{2\sin^2 x}\left[1 - \frac{1}{2\sin^2 x}\right] + c = \frac{1}{2\sin^2 x}\left[\frac{2\sin^2 x - 1}{2\sin^2 x}\right] + c = \frac{1}{2\sin^2 x}\left[\frac{-\cos 2x}{2\sin^2 x}\right] + c = -\frac{1}{4}\operatorname{cosec}^4 x.\cos 2x + c \quad \text{Ans.}$$

(d) $I = \int \frac{dx}{\sqrt[4]{\sin^3 x.\cos^5 x}} = \int \sin^{-\frac{3}{4}} x.\cos^{-\frac{5}{4}} x\,dx$ Here $m = -\frac{3}{4}$ and $n = -\frac{5}{4}$

when $m + n = -\frac{3}{4} - \frac{5}{4} = \frac{-3-5}{4} = -\frac{8}{4} = -2 =$ even negative integer

Putting $z = \tan x \quad \therefore\ dz = \sec^2 x\,dx$ or $dx = \frac{dz}{\sec^2 x} = \frac{dz}{1+\tan^2 x} = \frac{dz}{1+z^2}$

or $z = \tan x \quad \therefore\ \sin x = \frac{\tan x}{\sqrt{1+\tan^2 x}} = \frac{z}{\sqrt{1+z^2}}$ and $\cos x = \frac{1}{\sqrt{1+\tan^2 x}} = \frac{1}{\sqrt{1+z^2}}$

$$I = \int \sin^{\frac{-3}{4}} x.\cos^{\frac{-5}{4}} x\,dx = \int \left(\frac{z}{\sqrt{1+z^2}}\right)^{\frac{-3}{4}}.\left(\frac{1}{\sqrt{1+z^2}}\right)^{\frac{-5}{4}}.\frac{dz}{1+z^2} = \int \left(\frac{\sqrt{1+z^2}}{z}\right)^{\frac{3}{4}}.\left(\sqrt{1+z^2}\right)^{\frac{5}{4}}.\frac{dz}{1+z^2}$$

$$I = \int \frac{(\sqrt{1+z^2})^{\frac{3}{4}}}{z^{\frac{3}{4}}}.\left(\sqrt{1+z^2}\right)^{\frac{5}{4}}.\frac{dz}{1+z^2} = \int \frac{(\sqrt{1+z^2})^{\frac{3}{4}+\frac{5}{4}}}{z^{\frac{3}{4}}(1+z^2)}dz = \int \frac{(\sqrt{1+z^2})^2}{z^{\frac{3}{4}}(1+z^2)}dz = \int \frac{1+z^2}{z^{\frac{3}{4}}(1+z^2)}dz$$

$$I = \int z^{\frac{-3}{4}}\,dz = \frac{z^{\frac{-3}{4}+1}}{\frac{-3}{4}+1} + c = \frac{z^{\frac{1}{4}}}{\frac{1}{4}} + c = 4.z^{\frac{1}{4}} + c = 4(\tan x)^{\frac{1}{4}} + c \quad \text{Ans.}$$

(e) Do yourself. $I = \int \frac{dx}{\sqrt{\sin^7 x.\cos^5 x}} = \int \sin^{\frac{-7}{2}} x.\cos^{\frac{-5}{2}} x\,dx$ Here $m = -\frac{7}{2}$ and $n = -\frac{5}{2}$

when $m + n = -\frac{7}{2} - \frac{5}{2} = -6 =$ even negative integer, putting $z = \tan x \quad \therefore\ dz = \sec^2 x\,dx$

Ans:– $-\frac{2}{5}(\tan x)^{\frac{-5}{2}} + 2(\tan x)^{\frac{5}{2}} + \frac{20}{3}(\tan x)^{\frac{3}{2}} + \frac{20}{7}(\tan x)^{\frac{7}{2}} + \frac{10}{11}(\tan x)^{\frac{11}{2}} + \frac{2}{15}(\tan x)^{\frac{15}{2}} + c$

(f) $I = \int \sin^4 x.\cos^6 x\,dx$, Here $m = 4$ and $n = 6$, when $m + n = 4 + 6 = 10 =$ even positive integer

Let $z = \cos x + i\sin x$, $\frac{1}{z} = \cos x - i\sin x$ and $\cos x = \frac{1}{2}\left(z + \frac{1}{z}\right)$ and $\sin x = \frac{1}{2i}\left(z - \frac{1}{z}\right)$

Also, $z^n + \frac{1}{z^n} = 2\cos nx$

Now, $\sin^4 x.\cos^6 x = \left\{\frac{1}{2i}\left(z - \frac{1}{z}\right)\right\}^4.\left\{\frac{1}{2}\left(z + \frac{1}{z}\right)\right\}^6 = \frac{1}{1024}.\left(z - \frac{1}{z}\right)^4.\left(z + \frac{1}{z}\right)^6 = \frac{1}{1024}.\left(z - \frac{1}{z}\right)^2.\left(z - \frac{1}{z}\right)^2.\left(z + \frac{1}{z}\right)^2.\left(z + \frac{1}{z}\right)^2.\left(z + \frac{1}{z}\right)^2$

$\sin^4 x.\cos^6 x = \frac{1}{1024}.\left(z^2 - \frac{1}{z^2}\right)^2.\left(z^2 - \frac{1}{z^2}\right)^2.\left(z + \frac{1}{z}\right)^2 = \frac{1}{1024}.\left(z^4 - 2.z^2.\frac{1}{z^2} + \frac{1}{z^4}\right).\left(z^4 - 2.z^2.\frac{1}{z^2} + \frac{1}{z^4}\right).\left(z^2 - 2.z.\frac{1}{z} + \frac{1}{z^2}\right)$

$$\sin^4 x.\cos^6 x = \frac{1}{1024}.\left(z^4+\frac{1}{z^4}-2\right).\left(z^4+\frac{1}{z^4}-2\right).\left(z^2+\frac{1}{z^2}+2\right)$$
$$= \frac{1}{1024}.\left[z^8+z^4.\frac{1}{z^4}-2z^4+\frac{1}{z^4}.z^4+\frac{1}{z^4}.\frac{1}{z^4}-\frac{2}{z^4}-2z^4-\frac{2}{z^4}+4\right]\left(z^2+\frac{1}{z^2}+2\right)$$

$$\sin^4 x.\cos^6 x = \frac{1}{1024}.\left[\left(z^8+\frac{1}{z^8}\right)-2\left(z^4+\frac{1}{z^4}\right)-2\left(z^4+\frac{1}{z^4}\right)+6\right]\left(z^2+\frac{1}{z^2}+2\right) = \frac{1}{1024}.(2\cos 8x-8\cos 4x+6)(2\cos 2x+2)$$
$$= \frac{4}{1024}.(\cos 8x-4\cos 4x+3)(\cos 2x+1)$$

$$\sin^4 x.\cos^6 x = \frac{1}{256}.(\cos 8x-4\cos 4x+3+\cos 8x.\cos 2x-4\cos 4x.\cos 2x+3\cos 2x)$$

Integrating both of sides, we get

$$\int \sin^4 x.\cos^6 x\ dx = \frac{1}{256}\int(\cos 8x-4\cos 4x+3+\cos 8x.\cos 2x-4\cos 4x.\cos 2x+3\cos 2x)\,dx$$

$$I = \frac{1}{256}\int\cos 8x\,dx-\frac{4}{256}\int\cos 4x\,dx+\frac{3}{256}\int dx+\frac{1}{256}\int\cos 8x.\cos 2x\,dx-\frac{2}{256}\int 2\cos 4x.\cos 2x\,dx+\frac{3}{256}\int\cos 2x\,dx$$

$$I = \frac{1}{256}.\frac{\sin 8x}{8}-\frac{4}{256}.\frac{\sin 4x}{4}+\frac{3}{256}x+\frac{3}{256}.\frac{\sin 2x}{2}+\frac{1}{512}\int 2\cos 8x.\cos 2x\,dx-\frac{2}{256}\int 2\cos 4x.\cos 2x\,dx$$

Let $I_1 = \int 2\cos 8x.\cos 2x\,dx$ [use formula, $2\cos A\cos B = \cos(A+B)+\cos(A-B)$]

$$I_1 = \int(\cos 10x+\cos 6x)\,dx = \int\cos 10x\,dx+\int\cos 6x\,dx = \frac{\sin 10x}{10}+\frac{\sin 6x}{6}+c$$

Again, $I_2 = \int 2\cos 4x.\cos 2x\,dx = \int\cos 6x\ dx+\int\cos 2x\ dx = \frac{\sin 6x}{6}+\frac{\sin 2x}{2}+c$

$$I = \frac{1}{256}.\frac{\sin 8x}{8}-\frac{4}{256}.\frac{\sin 4x}{4}+\frac{3}{256}x+\frac{3}{256}.\frac{\sin 2x}{2}+\frac{1}{512}\left(\frac{\sin 10x}{10}+\frac{\sin 6x}{6}\right)-\frac{2}{256}\left(\frac{\sin 6x}{6}+\frac{\sin 2x}{2}\right)+c$$

$$I = \frac{1}{2048}\sin 8x-\frac{1}{256}\sin 4x+\frac{3}{256}x+\frac{3}{512}\sin 2x+\frac{1}{5120}\sin 10x+\frac{1}{3072}\sin 6x-\frac{1}{768}\sin 6x-\frac{1}{256}\sin 2x+c$$

$$I = \frac{1}{5120}\sin 10x+\frac{1}{2048}\sin 8x-\frac{1}{1024}\sin 6x-\frac{1}{256}\sin 4x+\frac{1}{512}\sin 2x+\frac{3}{256}x+c \quad \text{Ans.}$$

Integral of the form: – $\int\sqrt{\frac{ax+b}{cx+d}}\,dx$, Multiplying, N^r and D^r by $\sqrt{ax+b}$

Example: – (a) $I = \int\sqrt{\frac{x+3}{x-5}}\,dx$, Multiplying, N^r and D^r by $\sqrt{x+3}$, we get

$$I = \int\frac{\sqrt{x+3}}{\sqrt{x-5}}\times\frac{\sqrt{x+3}}{\sqrt{x+3}}\,dx = \int\frac{x+3}{\sqrt{(x-5)(x+3)}}\,dx = \int\frac{x+3}{\sqrt{x^2+3x-5x-15}}\,dx = \int\frac{x+3}{\sqrt{x^2-2x-15}}\,dx = \int\frac{(x-1)+4}{\sqrt{x^2-2x-15}}\,dx$$
$$= \int\frac{x-1}{\sqrt{x^2-2x-15}}\,dx+\int\frac{4}{\sqrt{x^2-2x-15}}\,dx = I_1+I_2 \quad \text{(say)}$$

Now, $I_1 = \int\frac{x-1}{\sqrt{x^2-2x-15}}\,dx$, Let $x^2-2x-15 = z^2$ $\therefore$ $(2x-2)dx = 2z\,dz$ or $(x-1)dx = z\,dz$

or $I_1 = \int\frac{z\,dz}{\sqrt{z^2}} = \int dz = z+c = \sqrt{x^2-2x-15}+c$ $\left[\text{Direct use formula, } \int\frac{f'(x)}{f(x)}dx = 2\sqrt{f(x)}+c\right]$

Now, $I_2 = \int\frac{4}{\sqrt{x^2-2x-15}}\,dx = 4\int\frac{dx}{\sqrt{(x-1)^2-16}}$ use formula, $\int\frac{dx}{\sqrt{x^2-a^2}} = \log\left|x+\sqrt{x^2-a^2}\right|$

or $I_2 = 4\int\frac{dx}{\sqrt{(x-1)^2-(4)^2}} = 4.\log\left[(x-1)+\sqrt{(x-1)^2-(4)^2}\right]+c = 4\log\left[(x-1)+\sqrt{x^2-2x-15}\right]+c$

or $I = I_1 + I_2 = \sqrt{x^2 - 2x - 15} + c + 4\log\left[(x-1) + \sqrt{x^2 - 2x - 15}\right] + c = \sqrt{x^2 - 2x - 15} + 4\log\left[(x-1) + \sqrt{x^2 - 2x - 15}\right] + k$ Ans.

(b) $I = \int \sqrt{\frac{3-x}{x+4}}\, dx$ Ans: $-\sqrt{-x^2 - x + 12} + \frac{7}{2}\sin^{-1}\left(\frac{2x+1}{7}\right) + c$ (solve same as above question)

Integral of the form: – $\int f(e^x)\, dx$, Putting $e^x = z$

Example: – (a) $I = \int \frac{dx}{\sqrt{1 - e^{2x}}}$

Let $1 - e^{2x} = z^2$ or $e^{2x} = 1 - z^2$ $\therefore -e^{2x}.2\,dx = 2z\,dz$ $\therefore dx = -\frac{z\,dz}{e^{2x}} = -\frac{z\,dz}{1 - z^2}$

or $I = -\int \frac{1}{z}.\frac{z\,dz}{(1 - z^2)} = -\int \frac{dz}{1 - z^2}$ use formula, $\int \frac{dx}{a^2 - x^2} = \frac{1}{2a}\log\left|\frac{a+x}{a-x}\right|$, when $x < a$

or $I = -\frac{1}{2.1}\log\left[\frac{1+z}{1-z}\right] + c = -\frac{1}{2}\log\left[\frac{1+\sqrt{1-e^{2x}}}{1-\sqrt{1-e^{2x}}}\right] + c = -\frac{1}{2}\log\left[\frac{(1+\sqrt{1-e^{2x}})(1+\sqrt{1-e^{2x}})}{(1-\sqrt{1-e^{2x}})(1+\sqrt{1-e^{2x}})}\right] + c = -\frac{1}{2}\log\left[\frac{(1+\sqrt{1-e^{2x}})^2}{1 - 1 + e^{2x}}\right] + c$

$$= -\frac{1}{2}\log\left[\frac{1+\sqrt{1-e^{2x}}}{e^x}\right]^2 + c = -\log\left(\frac{1+\sqrt{1-e^{2x}}}{e^x}\right) + c$$

or $I = \log\left(\frac{1+\sqrt{1-e^{2x}}}{e^x}\right)^{-1} + c = \log\left(\frac{e^x}{1+\sqrt{1-e^{2x}}}\right) + c$ Ans. or $I = -\log\left(\frac{1}{e^x} + \sqrt{\frac{1}{e^{2x}} - 1}\right) + c$ Ans.

IInd Method: – $I = \int \frac{dx}{\sqrt{1 - e^{2x}}}$, Putting $e^x = z$ $\therefore e^x\,dx = dz$ or $dx = \frac{dz}{e^x} = \frac{dz}{z}$

$I = \int \frac{1}{\sqrt{1 - z^2}}.\frac{dz}{z} = \int \frac{dz}{z.\sqrt{1 - z^2}}$

use formula, $\int \frac{dx}{P\sqrt{Q}}$ where P is linear and Q is quadratic, putting $P = \frac{1}{z}$

Putting, $z = \frac{1}{t}$ or $t = \frac{1}{z}$ $\therefore dz = -\frac{1}{t^2}dt$

$\therefore I = \int \frac{-\frac{1}{t^2}dt}{\frac{1}{t}.\sqrt{1 - \frac{1}{t^2}}} = -\int \frac{dt}{\sqrt{t^2 - 1}}$ use formula, $\int \frac{dx}{\sqrt{x^2 - a^2}} = \log\left|x + \sqrt{x^2 - a^2}\right|$

$$\therefore I = -\log\left(t + \sqrt{t^2 - 1}\right) + c = -\log\left(\frac{1}{z} + \sqrt{\frac{1}{z^2} - 1}\right) + c = -\log\left[\frac{1}{e^x} + \frac{\sqrt{1 - e^{2x}}}{e^x}\right] + c \quad \text{Ans.}$$

(b) $I = \int \frac{e^x}{e^{2x} + 1}\, dx$, Putting $e^x = z$ $\therefore e^x dx = dz$ $\therefore I = \int \frac{dz}{z^2 + 1} = \tan^{-1} z + c = \tan^{-1} e^x + c$ Ans.

(c) $I = \int \frac{e^x.\log(e^x + 1)}{e^x + 1}\, dx$, Putting $1 + e^x = z$ $\therefore e^x\,dx = dz$

$\therefore I = \int \frac{\log z}{z}\, dz$, Let $\log z = t$ $\therefore \frac{1}{z}dz = dt$ $\therefore I = \int t\, dt = \frac{t^2}{2} + c = \frac{(\log z)^2}{2} + c = \frac{[\log(1 + e^x)]^2}{2} + c$ Ans.

Integral of the form: – $\int e^x[f(x) + f'(x)]\, dx = e^x f(x) + c$

Proof: – Let $I = \int e^x[f(x) + f'(x)]\, dx = \int e^x f(x)\, dx + \int e^x f'(x)\, dx$

$$I = f(x)\int e^x\,dx - \int f'(x).dx\int e^x\,dx + \int e^x f'(x)\,dx = f(x).e^x - \int e^x f'(x)\,dx + \int e^x f'(x)\,dx = e^x f(x) + c \text{ Proved.}$$

Example: – (a) $I = \int e^x(\sin x + \cos x)\,dx,\qquad f(x) = \sin x \;\therefore f'(x) = \cos x$

$$\therefore I = \int e^x[f(x) + f'(x)]\,dx = e^x f(x) + c = e^x \sin x + c \quad \text{Ans.}$$

(b) $I = \int e^x\left(\log x + \frac{1}{x}\right)dx = e^x \log x + c$ Ans. $\left(\text{Let } f(x) = \log x\,, f'(x) = \frac{1}{x}\right)$

(c) $I = \int e^x(x^2 + 2x)\,dx = e^x.x^2 + c = x^2e^x + c$ Ans.

$\left(\text{Let } f(x) = x^2, f'(x) = 2x \text{ then } \int e^x[f(x) + f'(x)]\,dx = e^x f(x) + c\right)$

(d) $I = \int \frac{e^{\sin x}}{\cos^2 x}(x\cos^3 x - \sin x)\,dx$, Putting $\sin x = z$ or $x = \sin^{-1} z \;\therefore dx = \frac{1}{\sqrt{1-z^2}}dz$

$$I = \int \frac{e^z}{1-z^2}\left(\sin^{-1} z.(1-z^2)^{\frac{3}{2}} - z\right).\frac{1}{\sqrt{1-z^2}}dz = \int e^z\left(\sin^{-1} z - \frac{z}{(1-z^2)^{\frac{3}{2}}}\right)dz = \int e^z\left\{\left(\sin^{-1} z - \frac{1}{\sqrt{1-z^2}}\right) + \left(\frac{1}{\sqrt{1-z^2}} - \frac{z}{(1-z^2)^{\frac{3}{2}}}\right)\right\}dz$$

$$= \int e^z[f(z) + f'(z)]dz$$

$$I = e^z\left(\sin^{-1} z - \frac{1}{\sqrt{1-z^2}}\right) + c = e^{\sin x}\left(x - \frac{1}{\cos x}\right) + c = e^{\sin x}(x - \sec x) + c \quad \text{Ans.}$$

(e) $I = \int \left[\frac{1}{\log x} - \frac{1}{(\log x)^2}\right]dx$, Putting $\log x = z$ or $x = e^z \;\therefore dx = e^z dz$

$$I = \int \left(\frac{1}{z} - \frac{1}{z^2}\right).e^z dz = \int e^z\left(\frac{1}{z} + \frac{-1}{z^2}\right)dz, \qquad \text{Let } f(x) = \frac{1}{z} \;\therefore f'(x) = -\frac{1}{z^2}$$

use formula, $\int e^x[f(x) + f'(x)]\,dx = e^x f(x) + c$ or $I = e^z.\frac{1}{z} + c = \frac{e^{\log x}}{\log x} + c = \frac{x}{\log x} + c$ Ans.

Integral of the form: –

Expression	Substitution
$\sqrt{a^2 - x^2}$ or $a^2 - x^2$	$x = a\sin\theta$
$\sqrt{x^2 - a^2}$ or $x^2 - a^2$	$x = a\sec\theta$
$\sqrt{a^2 + x^2}$ or $a^2 + x^2$	$x = \tan\theta$
$\sqrt{\frac{a-x}{a+x}}$ or $\sqrt{\frac{a+x}{a-x}}$	$x = \cos 2\theta$
$\sqrt{\frac{x}{a-x}}$ or $\sqrt{\frac{a-x}{x}}$	$x = a\sin^2\theta$
$\sqrt{\frac{x}{a+x}}$ or $\sqrt{\frac{a+x}{x}}$ or $\frac{x}{\sqrt{x+a}}$	$x = a\tan\theta$ or $x = a\cot\theta$
$\sqrt{(a-x)(x-b)}$	$x = a\cos^2\theta + b\sin^2\theta$

Example: – (1) (a) $I = \int \frac{x}{\sqrt{1+x^4}}dx$

Putting $x = a\tan\theta \therefore a = 1$ or $x = \tan\theta \quad \therefore dx = \sec^2\theta\, d\theta$

$$I = \int \frac{\tan\theta}{\sqrt{1+\tan^4\theta}}.\sec^2\theta\, d\theta = \int \frac{\tan\theta.\sec^2\theta\ d\theta}{\sqrt{1+(\tan^2\theta)^2}}, \qquad \text{Let } \tan^2\theta = z \ \therefore\ 2\tan\theta.\sec^2\theta\ d\theta = dz$$

$$I = \int \frac{dz}{2\sqrt{1+z^2}} = \frac{1}{2}.\log\left[z+\sqrt{1+z^2}\right]+c \qquad \text{use formula, } \int \frac{dx}{\sqrt{x^2+a^2}} = \log\left|x+\sqrt{x^2+a^2}\right|$$

$$I = \frac{1}{2}\log\left[\tan^2\theta+\sqrt{1+\tan^4\theta}\right]+c = \frac{1}{2}\log\left[x^2+\sqrt{1+x^4}\right]+c \qquad \text{Ans.}$$

(b) $I = \int \sqrt{1-x^2}\, dx$, Putting $x = a\sin\theta \quad \therefore a = 1$ or $x = \sin\theta \quad \therefore dx = \cos\theta\, d\theta$

$$I = \int \sqrt{1-\sin^2\theta}.\cos\theta\, d\theta = \int \cos^2\theta\, d\theta = \int \frac{1+\cos 2\theta}{2} d\theta = \frac{1}{2}\int d\theta + \frac{1}{2}\int \cos 2\theta\, d\theta = \frac{1}{2}\theta + \frac{1}{2}\frac{\sin 2\theta}{2} + c = \frac{1}{2}\sin^{-1}x + \frac{1}{4}.2\sin\theta\cos\theta + c$$
$$= \frac{1}{2}\sin^{-1}x + \frac{1}{2}x\sqrt{1-x^2}+c \quad \text{Ans.}$$

(c) $I = \int \sqrt{\frac{1+\sqrt{x}}{1-\sqrt{x}}}dx$, Putting $\sqrt{x} = a\cos 2\theta \quad \therefore a = 1$ or $\sqrt{x} = \cos 2\theta$

or $\frac{1}{2\sqrt{x}}dx = -2\sin 2\theta\, d\theta$ or $dx = -4\cos 2\theta \sin 2\theta\, d\theta$

$$I = \int \sqrt{\frac{1+\cos 2\theta}{1-\cos 2\theta}}.-4\cos 2\theta\sin 2\theta\, d\theta = -4\int \sqrt{\frac{1+\cos^2\theta-\sin^2\theta}{1-\cos^2\theta+\sin^2\theta}}.\cos 2\theta\sin 2\theta\, d\theta = -4\int \frac{\cos\theta}{\sin\theta}.2\sin\theta\cos\theta\cos 2\theta\, d\theta$$
$$= -8\int \cos 2\theta.\cos^2\theta\, d\theta = -8\int \cos 2\theta.\left(\frac{1+\cos 2\theta}{2}\right)d\theta$$

$$\left[\cos 2\theta = 2\cos^2\theta - 1 \text{ or } \cos^2\theta = \frac{1+\cos 2\theta}{2} \text{ or } \cos^2\theta-\sin^2\theta = \cos 2\theta \text{ or } \cos 2\theta = 1-2\sin^2\theta\right]$$

$$I = -4\int(\cos 2\theta+\cos^2 2\theta)\, d\theta = -4\int \cos 2\theta\, d\theta - 4\int \cos^2 2\theta\, d\theta = -4.\frac{\sin 2\theta}{2} - 4\int \frac{1+\cos 4\theta}{2}d\theta = -2\sin 2\theta - 2\int d\theta - 2\int \cos 4\theta\, d\theta$$
$$= -2\sin 2\theta - 2\theta - 2.\frac{\sin 4\theta}{4}+c$$

$\therefore$ Put $\sqrt{x} = \cos 2\theta$ or $2\theta = \cos^{-1}\sqrt{x}$ or $\theta = \frac{\cos^{-1}\sqrt{x}}{2}$

$$I = -2\sin 2\theta - 2\theta - \frac{\sin 4\theta}{2} + c = -2\sin\left(\cos^{-1}\sqrt{x}\right) - 2.\frac{\cos^{-1}\sqrt{x}}{2} - \frac{\sin\left(2\cos^{-1}\sqrt{x}\right)}{2}+c$$
$$= -2\sin\left(\cos^{-1}\sqrt{x}\right) - \frac{\sin\left(2\cos^{-1}\sqrt{x}\right)}{2} - \cos^{-1}\sqrt{x}+c \quad \text{Ans.}$$

(d) $I = \int \cos^{-1}\left(\frac{x+1}{\sqrt{x^2+2x+5}}\right)dx = \int \cos^{-1}\left(\frac{x+1}{\sqrt{(x+1)^2+2^2}}\right)$, Putting $x+1 = a\cot\theta \quad \therefore a = 2$

or $x+1 = 2\cot\theta \quad \therefore dx = -2\,\text{cosec}^2\theta\, d\theta$ or $\cot\theta = \frac{x+1}{2}$ or $\theta = \cot^{-1}\left(\frac{x+1}{2}\right)$

$$I = \int \cos^{-1}\left(\frac{2\cot\theta}{\sqrt{(2\cot\theta)^2+4}}\right).-2\,\text{cosec}^2\theta\, d\theta = -2\int \cos^{-1}\left(\frac{2\cot\theta}{\sqrt{4(\cot^2\theta+1)}}\right).\text{cosec}^2\theta\, d\theta$$

$$I = -2\int \cos^{-1}\left(\frac{\cot\theta}{\sqrt{\text{cosec}^2\theta}}\right).\text{cosec}^2\theta\, d\theta = -2\int \cos^{-1}\left(\frac{\cot\theta}{\text{cosec}\,\theta}\right).\text{cosec}^2\theta\, d\theta \quad [\cot^2\theta+1 = \text{cosec}^2\theta]$$

$$I = -2\int \cos^{-1}\left(\frac{\frac{\cos\theta}{\sin\theta}}{\frac{1}{\sin\theta}}\right).\text{cosec}^2\theta\, d\theta = -2\int \theta.\text{cosec}^2\theta\, d\theta$$

formula, $\int f(x).g(x)\,dx = f(x).\int g(x)dx - \int f'(x)dx\int g(x)dx$ and $\int \operatorname{cosec}^2\theta\,d\theta = -\cot\theta + c$

and $\int \cot\theta\,d\theta = \log(\sin\theta) + c$

$$I = -2\left[\theta.\int \operatorname{cosec}^2\theta\,d\theta - \int\frac{d(\theta)}{d\theta}.d\theta.\int \operatorname{cosec}^2\theta\,d\theta\right] = -2\left[\theta.(-\cot\theta) + \int\cot\theta\,d\theta\right] = -2[\theta\,(-\cot\theta)\ + \log(\sin\theta)] + c$$
$$= 2\theta\cot\theta - 2\log(\sin\theta) + c$$

$$I = 2\cot^{-1}\left(\frac{x+1}{2}\right).\cot\left[\cot^{-1}\left(\frac{x+1}{2}\right)\right] - 2\log\left\{\sin\left[\cot^{-1}\left(\frac{x+1}{2}\right)\right]\right\} + c = 2\cot^{-1}\left(\frac{x+1}{2}\right).\left(\frac{x+1}{2}\right) - 2\log\left\{\sin\left[\cot^{-1}\left(\frac{x+1}{2}\right)\right]\right\} + c \quad \text{Ans.}$$

(2) find the following integral: – (a) $I = \int\frac{x^3}{\sqrt{1+x^8}}dx$, Putting $x^4 = \tan\theta \quad \therefore 4x^3dx = \sec^2\theta\,d\theta$

$I = \frac{1}{4}\int\frac{\sec^2\theta\,d\theta}{\sqrt{1+\tan^2\theta}}$ Let $\tan\theta = z \quad \therefore \sec^2\theta\,d\theta = dz$ using formula, $\int\frac{dx}{\sqrt{a^2+x^2}} = \log\left|x + \sqrt{x^2+a^2}\right|$

$$I = \frac{1}{4}\int\frac{dz}{\sqrt{1+z^2}} = \frac{1}{4}\log\left(z + \sqrt{1+z^2}\right) + c = \frac{1}{4}\log\left[\tan\theta + \sqrt{1+\tan^2\theta}\right] + c = \frac{1}{4}\log\left[x^4 + \sqrt{1+x^8}\right] + c \quad \text{Ans.}$$

(b) $I = \int\sin^{-1}\left(\frac{2x+2}{\sqrt{(2x)^2 + 2.2.2x + 4 + 9}}\right)dx$

Putting $2x+2 = a\tan\theta$ or $2x+2 = 3\tan\theta \quad \therefore 2dx = 3\sec^2\theta\,d\theta$ or $dx = \frac{3}{2}\sec^2\theta\,d\theta$ or $\tan\theta = \frac{2x+2}{3}$

$$I = \int\sin^{-1}\left(\frac{2x+2}{\sqrt{(2x+2)^2+9}}\right)dx = \int\sin^{-1}\left(\frac{3\tan\theta}{\sqrt{9\tan^2\theta+9}}\right).\frac{3}{2}\sec^2\theta\,d\theta = \frac{3}{2}\int\sin^{-1}\left(\frac{3\tan\theta}{3\sqrt{1+\tan^2\theta}}\right).\sec^2\theta\,d\theta$$
$$= \frac{3}{2}\int\sin^{-1}\left(\frac{\sin\theta/\cos\theta}{1/\cos\theta}\right).\sec^2\theta\,d\theta = \frac{3}{2}\int\sin^{-1}(\sin\theta).\sec^2\theta\,d\theta = \frac{3}{2}\int\theta.\sec^2\theta\,d\theta$$

$$I = \frac{3}{2}\left[\theta.\int\sec^2\theta\,d\theta - \int\frac{d(\theta)}{d\theta}.d\theta\int\sec^2\theta\,d\theta\right] = \frac{3}{2}\left[\theta.\tan\theta - \int\tan\theta\,d\theta\right] = \frac{3}{2}[\theta.\tan\theta + \log(\cos\theta)] + c$$

formula, $\int f(x).g(x)\,dx = f(x).\int g(x)dx - \int f'(x)dx\int g(x)dx$ and $\int\sec^2\theta\,d\theta = \tan\theta + c$

and $\int\tan\theta\,d\theta = -\log(\cos\theta) + c$ Put $\tan\theta = \frac{2x+2}{3} \quad \therefore \theta = \tan^{-1}\left(\frac{2x+2}{3}\right)$

$$I = \frac{3}{2}[\theta.\tan\theta + \log(\cos\theta)] + c = \frac{3}{2}\left\{\tan^{-1}\left(\frac{2x+2}{3}\right).\tan\left[\tan^{-1}\left(\frac{2x+2}{3}\right)\right] + \log\left[\cos\tan^{-1}\left(\frac{2x+2}{3}\right)\right]\right\} + c$$
$$= \frac{3}{2}.\left(\frac{2x+2}{3}\right).\tan^{-1}\left(\frac{2x+2}{3}\right) + \frac{3}{2}.\log\left[\cos\tan^{-1}\left(\frac{2x+2}{3}\right)\right] + c$$

$$I = \frac{3}{2}.\frac{2}{3}(x+1).\tan^{-1}\left(\frac{2x+2}{3}\right) + \frac{3}{2}.\log\left[\cos\tan^{-1}\left(\frac{2x+2}{3}\right)\right] + c = (x+1)\tan^{-1}\left(\frac{2x+2}{3}\right) + \frac{3}{2}.\log\left[\cos\tan^{-1}\left(\frac{2x+2}{3}\right)\right] + c \quad \text{Ans.}$$

(c) $I = \int\sin^{-1}\left(\frac{x+1}{\sqrt{x^2+2x+2}}\right)dx = \int\sin^{-1}\left(\frac{x+1}{\sqrt{(x+1)^2+1}}\right)$

Putting $x+1 = a\tan\theta$ or $\theta = \tan^{-1}(x+1)$ or $x+1 = \tan\theta \quad \therefore dx = \sec^2\theta\,d\theta$

$$I = \int\sin^{-1}\left(\frac{\tan\theta}{\sqrt{\tan^2\theta+1}}\right).\sec^2\theta\,d\theta = \int\sin^{-1}\left(\frac{\sin\theta/\cos\theta}{1/\cos\theta}\right).\sec^2\theta\,d\theta = \int\sin^{-1}(\sin\theta).\sec^2\theta\,d\theta$$
$$= \int\theta.\sec^2\theta\,d\theta \quad \text{(use integrating by part formula)}$$

$$I = \theta.\int\sec^2\theta\,d\theta - \int\frac{d(\theta)}{d\theta}.d\theta\int\sec^2\theta\,d\theta = \theta\tan\theta - \int\tan\theta\,d\theta = \theta\tan\theta + \log(\cos\theta) + c$$

$$I = \tan^{-1}(x+1).\tan[\tan^{-1}(x+1)] + \log\{\cos[\tan^{-1}(x+1)]\} + c = (x+1)\tan^{-1}(x+1) + \log\{\cos[\tan^{-1}(x+1)]\} + c \quad \text{Ans.}$$

Exercise – A8

(1) (a) $I = \int \frac{dx}{\cos x + \tan x}$ (b) $I = \int \frac{dx}{\sin x + \cot x}$ (c) $I = \int \frac{dx}{1 + \sin 3x}$ (d) $I = \int \frac{dx}{\sqrt{1 + \sin 5x}}$

(2) (a) $I = \int \sqrt{1 - \sin 2x}\, dx$ (b) $I = \int \frac{x^4}{x - 1} dx$ (c) $I = \int \frac{4x^3 + 12x^2 + 9x - 1}{2x + 3} dx$ (d) $I = \int \frac{dx}{(\sin^{-1} x)^3 . \sqrt{1 - x^2}}$

(3) (a) $I = \int \frac{x + 1}{x^2 + x \log x} dx$ (b) $I = \int \frac{x^3}{\sqrt{1 - x^2}} dx$ (c) $I = \int \sin^5 x . \sqrt{\cos x}\, dx$ (d) $I = \int \frac{x^2}{(x \sin x + \cos x)^2} dx$

(4) (a) $I = \int \sin^{-1} \sqrt{x}\, dx$ (b) $I = \int \cos^{-1} \sqrt{x}\, dx$ (c) $I = \int \tan^{-1} \sqrt{x}\, dx$ (d) $I = \int \cot^{-1} \sqrt{x}\, dx$

(5) (a) $I = \int \sqrt{e^{2x} + 1}\, dx$ (b) $I = \int \frac{dx}{(x - 4)\sqrt{x - 2}}$ (c) $I = \int e^x \log x\, dx$ (d) $I = \int x^2 e^x\, dx$

(6) (a) $I = \int x^3 \log x\, dx$ (b) $I = \int e^x \sin x\, dx$ (c) $I = \int \frac{\cos^{-1} \sqrt{x} - \sin^{-1} \sqrt{x}}{\sin^{-1} \sqrt{x} + \cos^{-1} \sqrt{x}} dx$ (d) $I = \int \cos x . \log(\sin x)\, dx$

(7) (a) $I = \int \log\left(\sqrt{1 - x^2}\right) dx$ (b) $I = \int \sin x . \log(\cot x)\, dx$ (c) $I = \int \frac{x \tan x}{\sec^3 x} dx$ (d) $I = \int x(e^x + 1)\, dx$

(8) (a) $I = \int \frac{\log\{\log(1 + \sqrt{x})\}}{(x + \sqrt{x})} dx$ (b) $I = \int 2^{3 \log_2 \sqrt[3]{x}}\, dx$ (c) $I = \int x^3 \cot^{-1} x\, dx$ (d) $I = \int \frac{(2^x - 3^x)^2}{2^x . 3^x} dx$

(9) (a) $I = \int \frac{\sin 2x . \cos x + 2 \cos^2 x}{(1 + 2 \sin {}^{x}/_{2} \cos {}^{x}/_{2})} dx$ (b) $I = \int \frac{x^2 + 1}{x(x + 3)(x - 4)} dx$ (c) $I = \int \frac{(x^3 + 64)(x - 2)}{x^2 - 4x + 16} dx$

(d) $I = \int \frac{dx}{\sqrt{x - 1} + \sqrt{x + 2}}$ (e) $I = \int \cos^{-1}\left(\frac{2 \cot x}{1 + \cot^2 x}\right) dx$ (f) $I = \int \frac{x}{x^4 + 4} dx$

(10) (a) $I = \int \frac{dx}{\sqrt{x} - \sqrt{x - 2}}$ (b) $I = \int \frac{x^8}{\sqrt{1 + x^3}} dx$ (c) $I = \int \frac{x + 3}{\sqrt{4x + 5}} dx$ (d) $I = \int \frac{x}{2x^2 - 3x + 5} dx$

(11) (a) $I = \int \frac{3x + 2}{x^2 + 2x + 3} dx$ (b) $I = \int \frac{4x + 1}{x^2 + 3x + 2} dx$ (c) $I = \int \frac{2x - 3}{x^2 + 4x - 5} dx$ (d) $I = \int \frac{x^5 + x^2}{x^6 + 16} dx$

(12) (a) $I = \int \frac{\sin 2\theta - 5 \cos \theta}{7 - \cos^2 \theta + 4 \sin \theta} d\theta$ (b) $I = \int \sin(\log x)\, dx + \int \cos(\log x)\, dx$ (c) $I = \int \cos^{-1}\left(\frac{2x}{1 + x^2}\right) dx$

(13) (a) $I = \int \sin^{-1}\left(\frac{x}{\sqrt{1 + x^2}}\right) dx$ (b) $I = \int \tan^{-1}\left(\frac{x}{\sqrt{1 - x^2}}\right) dx$ (c) $I = \int \frac{\cos^{-1} x}{\sqrt{1 - x^2}} dx$ (d) $I = \int \frac{\cos^{-1} x}{(1 - x^2)^{\frac{3}{2}}} dx$

(14) (a) $I = \int \sqrt{\sec \theta - 1}\, d\theta$ (b) $I = \int \frac{x - \sqrt[3]{x^2}}{x(\sqrt[4]{x} + \sqrt[6]{x^2})} dx$ (c) $I = \int e^{3x}(3x^2 + 4x + 1)\, dx$ (d) $I = \int \frac{x^2 - 9}{x + 2} dx$

(15) (a) $I = \int e^{\sin 2x}(1 + \cos 2x)\, dx$ (b) $I = \int \sin^2 x \cot^2 x\, dx$ (c) $I = \int \frac{\cot x \operatorname{cosec} x}{3 + \operatorname{cosec} x} dx$ (d) $I = \int \tan^4 x\, dx$

(16) (a) $I = \int (\sin^4 x + \cos^4 x)\, dx$ (b) $I = \int \frac{dx}{1 - \sqrt{1 + x^2}}$ (c) $I = \int \frac{(\sin x - \sqrt[3]{\sin^2 x}) . \cos x}{\sqrt{\sin^3 x} + \sqrt[4]{\sin^6 x}} dx$

(d) $I = \int \frac{x - 1}{(x - 2)\sqrt{x + 3}} dx$

(17) (a) $I = \int \sin x . \sin 3x \ dx$ (b) $I = \int \cos x . \cos 3x . \cos 5x \ dx$ (c) $I = \int \tan 2x . \tan 3x . \tan 5x \ dx$

(d) $I = \int \sin x . \sin 3x . \sin 5x . \sin 7x \ dx$ (e) $I = \int \sin x . \cos 3x \ dx$

(18) (a) $I = \int \cos 2x . \sin 3x \ dx$ (b) $I = \int \sin x . \sin 2x . \cos x . \cos 2x \ dx$

(c) $I = \int \frac{\sin 2x}{\cos(x+\theta) . \cos(x-\theta)} dx$ (d) $I = \int \frac{dx}{\sin x . \sin(x+\theta)}$

Answer

(1) (a) $I = \int \frac{dx}{\cos x + \tan x} = \int \frac{dx}{\cos x + \frac{\sin x}{\cos x}} = \int \frac{\cos x}{\cos^2 x + \sin x} dx = \int \frac{\cos x \ dx}{1 - \sin^2 x + \sin x}$

Let $\sin x = t \quad \therefore \cos x \, dx = dt$

$$I = \int \frac{dt}{1 - t^2 + t} = \int \frac{dt}{1 + t - t^2} = \int \frac{dt}{\frac{5}{4} - \left(t - \frac{1}{2}\right)^2} = \int \frac{dt}{\left(\frac{\sqrt{5}}{2}\right)^2 - \left(t - \frac{1}{2}\right)^2}$$

using formula, $\int \frac{dx}{a^2 - x^2} = \frac{1}{2a} \log \left| \frac{a+x}{a-x} \right|$ or $I = \frac{1}{2 . \frac{\sqrt{5}}{2}} \log \left(\frac{\frac{\sqrt{5}}{2} + \left(t - \frac{1}{2}\right)}{\frac{\sqrt{5}}{2} - \left(t - \frac{1}{2}\right)} \right) + c = \frac{1}{\sqrt{5}} \log \left(\frac{\sqrt{5} + 2t - 1}{\sqrt{5} - 2t + 1} \right) + c$

Put $t = \sin x \quad \therefore I = \frac{1}{\sqrt{5}} \log \left(\frac{\sqrt{5} + 2 \sin x - 1}{\sqrt{5} - 2 \sin x + 1} \right) + c$ Ans.

(b) $I = \int \frac{dx}{\sin x + \cot x} = \int \frac{\sin x}{\sin^2 x + \cos x} dx = \int \frac{\sin x}{1 - \cos^2 x + \cos x} dx$

Put $\cos x = t \ \therefore -\sin x \, dx = dt$ use formula, $\int \frac{dx}{x^2 - a^2} = \frac{1}{2a} \log \left| \frac{x-a}{x+a} \right|$, when $x > a$

$$I = \int \frac{dt}{t^2 - t - 1} = \int \frac{dt}{\left(t - \frac{1}{2}\right)^2 - \left(\frac{\sqrt{5}}{2}\right)^2} = \frac{1}{\sqrt{5}} \log \left| \frac{2t - 1 - \sqrt{5}}{2t - 1 + \sqrt{5}} \right| + c = \frac{1}{\sqrt{5}} \log \left| \frac{2 \cos x - 1 - \sqrt{5}}{2 \cos x - 1 + \sqrt{5}} \right| + c \quad \text{Ans.}$$

(c) $I = \int \frac{dx}{1 + \sin 3x} = \int \frac{dx}{1 + 2 \sin \frac{3x}{2} \cos \frac{3x}{2}}$ divide above and below by $\cos^2 \frac{3x}{2}$

Do yourself Ans: $- -\frac{2}{3} . \frac{\cos \frac{3x}{2}}{\sin \frac{3x}{2} + \cos \frac{3x}{2}} + c$

IInd Method: $- \ I = \int \frac{dx}{1 + \sin 3x} = \int \frac{\cos 3x \, dx}{(1 + \sin 3x) \cos 3x} = \int \frac{\cos 3x \, dx}{(1 + \sin 3x) . \sqrt{1 - \sin^2 3x}}$

Let $\sin 3x = t \ \therefore 3 \cos 3x \ dx = dt$ or $\cos 3x \ dx = \frac{dt}{3}$

$I = \int \frac{dt}{3(1+t) . \sqrt{1 - t^2}}$, Put $1 + t = z \ \therefore dt = dz$ or $t = z - 1$

$I = \frac{1}{3} \int \frac{dz}{z . \sqrt{1 - (z-1)^2}} = \frac{1}{3} \int \frac{dz}{z . \sqrt{2z - z^2}} = \frac{1}{3} \int \frac{dz}{z^2 . \sqrt{\frac{2}{z} - 1}}$, Let $\frac{2}{z} - 1 = u^2 \ \therefore -\frac{2}{z^2} dz = 2u \, du$

$$I=-\frac{1}{3}\int\frac{u\,du}{u}=-\frac{1}{3}\int du=-\frac{1}{3}u+c=-\frac{1}{3}\sqrt{\frac{2}{z}-1}+c=-\frac{1}{3}\sqrt{\frac{2-z}{z}}+c=-\frac{1}{3}\sqrt{\frac{2-1-t}{1+t}}+c=-\frac{1}{3}\sqrt{\frac{1-t}{1+t}}+c=-\frac{1}{3}\sqrt{\frac{1-t}{1+t}\times\frac{1-t}{1-t}}+c$$

$$=-\frac{1}{3}.\frac{1-t}{\sqrt{1-t^2}}+c=-\frac{1}{3}.\frac{1-\sin 3x}{\sqrt{1-\sin^2 3x}}+c=-\frac{1}{3}.\frac{1-\sin 3x}{\cos 3x}+c \quad \text{Ans.}$$

IIIrd Method: – $I=\int\frac{dx}{1+\sin 3x}=\int\frac{dx}{\sin^2\left(\frac{3x}{2}\right)+\cos^2\left(\frac{3x}{2}\right)+2\sin\left(\frac{3x}{2}\right)\cos\left(\frac{3x}{2}\right)}=\int\frac{dx}{\left(\sin\left(\frac{3x}{2}\right)+\cos\left(\frac{3x}{2}\right)\right)^2}$

$$=\int\frac{dx}{\left[\sqrt{2}\left\{\sin\left(\frac{3x}{2}\right).\cos\left(\frac{\pi}{4}\right)+\cos\left(\frac{3x}{2}\right).\sin\left(\frac{\pi}{4}\right)\right\}\right]^2}=\int\frac{dx}{2\left[\sin\left(\frac{3x}{2}+\frac{\pi}{4}\right)\right]^2}=\frac{1}{2}\int\operatorname{cosec}^2\left(\frac{3x}{2}+\frac{\pi}{4}\right)dx$$

$$I=-\frac{1}{2}\cot\left(\frac{3x}{2}+\frac{\pi}{4}\right).\frac{1}{\frac{3}{2}}+c=-\frac{1}{3}\cot\left(\frac{\pi}{4}+\frac{3x}{2}\right)+c \qquad \text{Ans.} \qquad \left[\int\operatorname{cosec}^2 x\,dx=-\cot x\right]$$

(d) $I=\int\frac{dx}{\sqrt{1+\sin 5x}}=\int\frac{dx}{\sqrt{1+2\sin\left(\frac{5x}{2}\right)\cos\left(\frac{5x}{2}\right)}}$, Divide above and below by $\cos^4\left(\frac{5x}{2}\right)$

$I=\int\frac{\sec^4\left(\frac{5x}{2}\right)}{\sqrt{\sec^2\left(\frac{5x}{2}\right)+2\tan\left(\frac{5x}{2}\right)}}dx$, Let $\tan\left(\frac{5x}{2}\right)=t \therefore \sec^2\left(\frac{5x}{2}\right).\frac{5}{2}dx=dt$ or $\sec^2\left(\frac{5x}{2}\right)dx=\frac{2}{5}dt$

$$I=\int\frac{\sec^2\left(\frac{5x}{2}\right).\sec^2\left(\frac{5x}{2}\right)}{\sqrt{\sec^2\left(\frac{5x}{2}\right)+2\tan\left(\frac{5x}{2}\right)}}dx=\int\frac{\left[1+\tan^2\left(\frac{5x}{2}\right)\right].\sec^2\left(\frac{5x}{2}\right)}{\sqrt{1+\tan^2\left(\frac{5x}{2}\right)+2\tan\left(\frac{5x}{2}\right)}}dx=\frac{2}{5}\int\frac{(1+t^2)}{\sqrt{1+t^2+2t}}dt=\frac{2}{5}\int\frac{(1+t^2)}{\sqrt{(1+t)^2}}dt=\frac{2}{5}\int\frac{1+t^2}{1+t}dt$$

$$=\frac{2}{5}\int\frac{1}{1+t}dt+\frac{2}{5}\int\frac{t^2}{1+t}dt=\frac{2}{5}\int\frac{1}{1+t}dt+\frac{2}{5}\int\left(t-\frac{t}{1+t}\right)dt=\frac{2}{5}\int\frac{1}{1+t}dt+\frac{2}{5}\int t\,dt-\frac{2}{5}\int\frac{t}{1+t}dt$$

$$=\frac{2}{5}\int\frac{1}{1+t}dt+\frac{2}{5}\int t\,dt-\frac{2}{5}\int\left(1-\frac{1}{1+t}\right)dt$$

$$I=\frac{2}{5}\int\frac{1}{1+t}dt+\frac{2}{5}\int t\,dt-\frac{2}{5}\int dt+\frac{2}{5}\int\frac{1}{1+t}dt=\frac{2}{5}\log(1+t)+\frac{2}{5}.\frac{t^2}{2}-\frac{2}{5}.t+\frac{2}{5}\log(1+t)+c=\frac{4}{5}\log(1+t)+\frac{1}{5}t^2-\frac{2}{5}t+c$$

$$=\frac{4}{5}\log\left[1+\tan\left(\frac{5x}{2}\right)\right]+\frac{\tan^2\left(\frac{5x}{2}\right)}{2}-\frac{2\tan\left(\frac{5x}{2}\right)}{5}+c \quad \text{Ans.}$$

IInd Method: – $I=\int\frac{dx}{\sqrt{1+\sin 5x}}=\int\frac{dx}{\sqrt{\sin^2\left(5x/2\right)+\cos^2\left(5x/2\right)+2\sin\left(5x/2\right)\cos\left(5x/2\right)}}=\int\frac{dx}{\sqrt{\left[\sin\left(5x/2\right)+\cos\left(5x/2\right)\right]^2}}$

$$=\int\frac{dx}{\sin\left(5x/2\right)+\cos\left(5x/2\right)}$$

$$I=\int\frac{dx}{\sqrt{2}\left[\sin\left(\frac{5x}{2}\right).\cos\left(\frac{\pi}{4}\right)+\cos\left(\frac{5x}{2}\right).\sin\left(\frac{\pi}{4}\right)\right]}=\frac{1}{\sqrt{2}}\int\frac{dx}{\sin\left[\frac{5x}{2}+\frac{\pi}{4}\right]}=\frac{1}{\sqrt{2}}\int\operatorname{cosec}\left[\frac{5x}{2}+\frac{\pi}{4}\right]dx$$

$$I=\frac{1}{\sqrt{2}}.\log\left\{\tan\left[\frac{\left(\frac{5x}{2}+\frac{\pi}{4}\right)}{2}\right]\right\}.\frac{1}{\frac{5}{2}}+c=\frac{2}{5\sqrt{2}}\log\left\{\tan\left[\frac{\left(\frac{5x}{2}+\frac{\pi}{4}\right)}{2}\right]\right\}+c=\frac{\sqrt{2}}{5}\log\left\{\tan\left[\frac{\left(\frac{5x}{2}+\frac{\pi}{4}\right)}{2}\right]\right\}+c \quad \text{Ans.}$$

$$\left[\text{use formula, } \int\operatorname{cosec}x\,dx=\log\left[\tan\left(\frac{x}{2}\right)\right]\right]$$

(2) (a) $I=\int\sqrt{1-\sin 2x}\,dx=\int\sqrt{\sin^2 x+\cos^2 x-2\sin x\cos x}\,dx=\int\sqrt{(\sin x-\cos x)^2}\,dx=\int(\sin x-\cos x)\,dx$

$$I=\int\sin x\,dx-\int\cos x\,dx=-\cos x-\sin x+c=-(\sin x+\cos x)+c \quad \text{Ans.}$$

(b) $I=\int\frac{x^4}{x-1}dx=\int\frac{(x^3+x^2+x+1)(x-1)+1}{(x-1)}dx$, factorised of x^4 is $(x^3+x^2+x+1)(x-1)+1$

$$I = \int \frac{(x^3 + x^2 + x + 1)(x - 1)}{(x - 1)} dx + \int \frac{1}{x - 1} dx = \int (x^3 + x^2 + x + 1)\, dx + \int \frac{1}{x - 1} dx = \int x^3\, dx + \int x^2\, dx + \int x\, dx + \int dx + \int \frac{1}{x - 1} dx$$

$$= \frac{x^4}{4} + \frac{x^3}{3} + \frac{x^2}{2} + x + \log(x - 1) + c \quad \text{Ans.}$$

(c) $I = \int \frac{4x^3 + 12x^2 + 9x - 1}{2x + 3} dx = \int \frac{(2x + 3)(2x^2 + 3x) - 1}{2x + 3} dx = \int \frac{(2x + 3)(2x^2 + 3x)}{(2x + 3)} dx - \int \frac{dx}{2x + 3}$

$$I = \int (2x^2 + 3x)\, dx - \int \frac{dx}{2x + 3} = \int 2x^2\, dx + \int 3x\, dx - \int \frac{dx}{2x + 3} = 2.\frac{x^3}{3} + 3.\frac{x^2}{2} - \log(2x + 3).\frac{1}{\text{d. c of } 2x + 3} + c$$

$$\therefore\ I = \frac{2x^3}{3} + \frac{3x^2}{2} - \frac{\log(2x + 3)}{2} + c \quad \text{Ans.}$$

(d) $I = \int \frac{dx}{(\sin^{-1} x)^3 . \sqrt{1 - x^2}}$, Put $\sin^{-1} x = z \quad \therefore \quad \frac{1}{\sqrt{1 - x^2}} dx = dz$

$$I = \int \frac{dz}{z^3} = \int z^{-3}\, dz = \frac{z^{-3+1}}{-3 + 1} + c = -\frac{1}{2} z^{-2} + c = -\frac{1}{2z^2} + c = -\frac{1}{2(\sin^{-1} x)^2} + c \quad \text{Ans.}$$

(3) (a) $I = \int \frac{x + 1}{x^2 + x \log x} dx = \int \frac{x + 1}{x(x + \log x)} dx$, Put $x + \log x = z \quad \therefore \quad \left(1 + \frac{1}{x}\right) dx = dz$ or $\left(\frac{x + 1}{x}\right) dx = dz$

$$I = \int \frac{1}{(x + \log x)} . \frac{x + 1}{x} dx = \int \frac{dz}{z} = \log z + c = \log[x + \log x] + c \quad \text{Ans.}$$

(b) $I = \int \frac{x^3}{\sqrt{1 - x^2}} dx$, Put $1 - x^2 = z^2$ or $x^2 = 1 - z^2 \quad \therefore\ -2x\, dx = 2z\, dz$ or $x\, dx = -z\, dz$

$$I = \int \frac{x^2}{\sqrt{1 - x^2}} . x\, dx = \int \frac{1 - z^2}{z} . (-z)\, dz = \int (z^2 - 1)\, dz = \int z^2\, dz - \int dz = \frac{z^{2+1}}{2 + 1} - z + c = \frac{z^3}{3} - z + c$$

$$I = \frac{z^2 . z}{3} - z + c = \frac{(1 - x^2)\sqrt{1 - x^2}}{3} - \sqrt{1 - x^2} + c = \frac{1}{3}(1 - x^2)^{\frac{3}{2}} - \sqrt{1 - x^2} + c \quad \text{Ans.}$$

(c) $I = \int \sin^5 x . \sqrt{\cos x}\, dx$, Put $\cos x = t \quad \therefore\ -\sin x\, dx = dt$

$$I = \int \sin^4 x . \sqrt{\cos x} . \sin x\ dx = \int (1 - \cos^2 x)^2 . \sqrt{\cos x} . \sin x\ dx = -\int (1 - t^2)^2 . \sqrt{t}\ dt = -\int (1 - 2t^2 + t^4) . \sqrt{t}\ dt$$

$$I = -\int \sqrt{t}\ dt + 2\int t^2 . \sqrt{t}\ dt - \int t^4 . \sqrt{t}\ dt = -\int t^{\frac{1}{2}}\ dt + 2\int t^{\frac{5}{2}}\ dt - \int t^{\frac{9}{2}}\ dt = -\frac{t^{\frac{3}{2}}}{\frac{3}{2}} + 2.\frac{t^{\frac{7}{2}}}{\frac{7}{2}} - \frac{t^{\frac{11}{2}}}{\frac{11}{2}} + c$$

$$I = -\frac{2}{3} t^{\frac{3}{2}} + \frac{4}{7} t^{\frac{7}{2}} - \frac{2}{11} t^{\frac{11}{2}} + c = -\frac{2}{3}(\cos x)^{\frac{3}{2}} + \frac{4}{7}(\cos x)^{\frac{7}{2}} - \frac{2}{11}(\cos x)^{\frac{11}{2}} + c \quad \text{Ans.}$$

(d) $I = \int \frac{x^2}{(x \sin x + \cos x)^2} dx = \int \frac{x^2}{x \cos x} . \frac{x \cos x}{(x \sin x + \cos x)^2} dx = \int u.v\ dx$ (integrating by part)

where $u = \frac{x^2}{x \cos x} \quad \therefore \quad \frac{du}{dx} = \frac{x \cos x . 2x - x^2(x. - \sin x + \cos x)}{(x \cos x)^2} = \frac{2x^2 \cos x + x^3 \sin x - x^2 \cos x}{x^2 \cos^2 x}$

or $\frac{du}{dx} = \frac{x^2 \cos x + x^3 \sin x}{x^2 \cos^2 x} = \frac{x^2(\cos x + x \sin x)}{x^2 \cos^2 x} = \frac{(\cos x + x \sin x)}{\cos^2 x}$

Now, $v = \int \frac{x \cos x}{(x \sin x + \cos x)^2} dx$, Put $x \sin x + \cos x = t \ \therefore (x \cos x + \sin x - \sin x)\, dx = dt$ or $x \cos x\ dx = dt$

$$v = \int \frac{dt}{t^2} = \int t^{-2}\ dt = \frac{t^{-2+1}}{-2 + 1} = -\frac{1}{t} + c = -\frac{1}{(x \sin x + \cos x)} + c$$

$$I = \int u.v\ dx = u.\int v\, dx - \int \left[\frac{du}{dx} . \int v . dx\right] dx = \frac{x^2}{x \cos x} . \left[-\frac{1}{(x \sin x + \cos x)}\right] - \int \left[-\frac{1}{(x \sin x + \cos x)} . \frac{(\cos x + x \sin x)}{\cos^2 x}\right] dx$$

$$I=-\frac{x}{\cos x}.\left[\frac{1}{(x\sin x+\cos x)}\right]+\int \sec^2 x\ dx=-\frac{x}{\cos x}.\left[\frac{1}{(x\sin x+\cos x)}\right]+\tan x+c \quad \text{Ans.}$$

(4) (a) $I=\int \sin^{-1}\sqrt{x}\ dx$, Put $\sqrt{x}=t$ $\therefore$ $\frac{1}{2\sqrt{x}}dx=dt$ or $dx=2t\ dt$

$I=\int \sin^{-1}t.2t\ dt=2\int t.\sin^{-1}t\ dt$ [use integrating by part formula]

$$I=2\left\{\sin^{-1}t.\int t\ dt-\int\left[\frac{d(\sin^{-1}t)}{dt}.\int t\ dt\right]dt\right\}=2\sin^{-1}t.\frac{t^2}{2}-2\int\frac{1}{\sqrt{1-t^2}}.\frac{t^2}{2}.dt=t^2\sin^{-1}t-\int\frac{t^2}{\sqrt{1-t^2}}dt$$

Let $1-t^2=z^2$ or $t=\sqrt{1-z^2}$ $\therefore$ $-2t\,dt=2z\,dz$ or $t\,dt=-z\,dz$

$$I=t^2\sin^{-1}t+\int\frac{z.\sqrt{1-z^2}\,dz}{z}=t^2\sin^{-1}t+\int\sqrt{1-z^2}\ dz$$

using formula, $\int\sqrt{a^2-x^2}\,dx=\frac{x}{2}\sqrt{a^2-x^2}+\frac{a^2}{2}\sin^{-1}\frac{x}{a}$

$I=t^2\sin^{-1}t+\frac{z}{2}\sqrt{1-z^2}+\frac{1}{2}\sin^{-1}z+c$ put $t=\sqrt{x}$ and $z=\sqrt{1-t^2}=\sqrt{1-x}$

$$I=x\sin^{-1}\sqrt{x}+\frac{\sqrt{1-x}}{2}\sqrt{1-1+x}+\frac{1}{2}\sin^{-1}\sqrt{1-x}+c=x\sin^{-1}\sqrt{x}+\frac{\sqrt{x}}{2}.\sqrt{1-x}+\frac{1}{2}\sin^{-1}\sqrt{1-x}+c \quad \text{Ans.}$$

(b) $I=\int\cos^{-1}\sqrt{x}\ dx=x\cos^{-1}\sqrt{x}-\frac{\sqrt{x}}{2}.\sqrt{1-x}-\frac{1}{2}\sin^{-1}\sqrt{1-x}+c$ Ans. (see above question)

(c) $I=\int\tan^{-1}\sqrt{x}\ dx=(x+1)\tan^{-1}\sqrt{x}-\sqrt{x}+c$ Ans. (Do yourself, see above question)

(d) $I=\int\cot^{-1}\sqrt{x}\ dx=x\cot^{-1}\sqrt{x}-\tan^{-1}\sqrt{x}+\sqrt{x}+c$ Ans. (Do yourself, see above question)

(5) (a) $I=\int\sqrt{e^{2x}+1}\,dx=\int\frac{\sqrt{e^{2x}+1}.e^{2x}}{e^{2x}}dx$, Put $1+e^{2x}=z^2$ $\therefore$ $2.e^{2x}\,dx=2z\,dz$

or $e^{2x}\,dx=z\,dz$ $\therefore$ $1+e^{2x}=z^2$ or $e^{2x}=z^2-1$

$$I=\int\frac{\sqrt{z^2}.z\,dz}{z^2-1}=\int\frac{z^2}{z^2-1}dz=\int\left(1+\frac{1}{z^2-1}\right)dz=\int dz+\int\frac{dz}{z^2-1}$$

use formula, $\int\frac{dx}{x^2-a^2}=\frac{1}{2a}\log\left|\frac{x-a}{x+a}\right|$ or $I=z+\frac{1}{2.1}\log\left|\frac{z-1}{z+1}\right|+c=\sqrt{e^{2x}+1}+\frac{1}{2}\log\left|\frac{\sqrt{e^{2x}+1}-1}{\sqrt{e^{2x}+1}+1}\right|+c$ Ans.

(b) $I=\int\frac{dx}{(x-4)\sqrt{x-2}}$, Let $x-2=t^2$ or $t=\sqrt{x-2}$ or $x=t^2+2$ $\therefore$ $dx=2t\,dt$

$I=\int\frac{2t}{(t^2+2-4).t}dt=2\int\frac{dt}{t^2-2}=2\int\frac{dt}{t^2-(\sqrt{2})^2}$ use formula, $\int\frac{dx}{x^2-a^2}=\frac{1}{2a}\log\left|\frac{x-a}{x+a}\right|$

$$I=2.\frac{1}{2.\sqrt{2}}\log\left|\frac{t-\sqrt{2}}{t+\sqrt{2}}\right|+c=\frac{1}{\sqrt{2}}\log\left|\frac{\sqrt{x-2}-\sqrt{2}}{\sqrt{x-2}+\sqrt{2}}\right|+c \quad \text{Ans.}$$

(c) $I=\int e^x\log x\,dx$ (use integrating by part formula)

(d) $I=\int x^2e^x\,dx=\int u.v\,dx$ (use integrating by part formula)

$I=x^2.e^x-\int 2x.e^x\ dx$ Again use integrating by part formula

$$I = x^2.e^x - 2\left[x.e^x - \int e^x\, dx\right] = x^2.e^x - 2xe^x + 2e^x + c = e^x[x^2 - 2x + 2] + c = e^x[(x-1)^2 + 1] + c \quad \text{Ans.}$$

(6) (a) $I = \int x^3 \log x\, dx = \frac{x^4 \log x}{4} - \frac{x^4}{16} + c$ Ans. [use integrating by part formula]

(b) $I = \int e^x \sin x\, dx = \frac{1}{2} e^x[\sin x - \cos x] + c$ Ans. [use integrating by part formula]

(c) $I = \int \frac{\cos^{-1}\sqrt{x} - \sin^{-1}\sqrt{x}}{\sin^{-1}\sqrt{x} + \cos^{-1}\sqrt{x}} dx = \frac{2}{\pi}\int \left(\cos^{-1}\sqrt{x} - \sin^{-1}\sqrt{x}\right) dx \quad \left[\sin^{-1}\sqrt{x} + \cos^{-1}\sqrt{x} = \frac{\pi}{2}\right]$

$I = \frac{2}{\pi}\int \left(\cos^{-1}\sqrt{x}\right) dx - \frac{2}{\pi}\int \left(\sin^{-1}\sqrt{x}\right) dx$ (see question no. –(4)(a & b)

$$I = \frac{2}{\pi}\left[x\cos^{-1}\sqrt{x} - \frac{\sqrt{x}}{2}.\sqrt{1-x} - \frac{1}{2}\sin^{-1}\sqrt{1-x}\right] - \frac{2}{\pi}\left[x\sin^{-1}\sqrt{x} + \frac{\sqrt{x}}{2}.\sqrt{1-x} + \frac{1}{2}\sin^{-1}\sqrt{1-x}\right] + c$$

(d) $I = \int \cos x.\log(\sin x)\, dx = \sin x\,[\log(\sin x) - 1] + c$ Ans. [use integrating by part formula]

(7) (a) $I = \int \log\left(\sqrt{1-x^2}\right) dx = x\log\left(\sqrt{1-x^2}\right) + \frac{1}{2}\log\left|\frac{1+x}{1-x}\right| - x + c$ Ans.

$\left[\text{use integrating by part formula and } \int \frac{dx}{a^2 - x^2} = \frac{1}{2a}\log\left|\frac{a+x}{a-x}\right|\right]$

(b) $I = \int \sin x.\log(\cot x)\, dx = -\cos x.\log(\cot x) + \log\left[\tan\left(\frac{x}{2}\right)\right] + c$ Ans.

$\left[\text{use integrating by part formula and } \int \operatorname{cosec} x\, dx = \log\left[\tan\left(\frac{x}{2}\right)\right]\right]$

(c) $I = \int \frac{x\tan x}{\sec^3 x} dx = \int \frac{x.\frac{\sin x}{\cos x}}{\frac{1}{\cos^3 x}} dx = \int x\sin x.\cos^2 x\, dx$

Let $\cos x = t$ or $x = \cos^{-1} t \quad \therefore -\sin x\, dx = dt$

$$I = -\int \cos^{-1} t.t^2\, dt = -\left[\cos^{-1} t.\frac{t^3}{3} + \int \frac{1}{\sqrt{1-t^2}}.\frac{t^3}{3}.dt\right] = -\frac{t^3}{3}.\cos^{-1} t - \frac{1}{3}\int \frac{t^3}{\sqrt{1-t^2}} dt$$

Put $1 - t^2 = z^2$ or $t = \sqrt{1-z^2} \quad \therefore -2t\,dt = 2z\,dz$ or $t\,dt = -z\,dz$

$$I = -\frac{t^3}{3}.\cos^{-1} t + \frac{1}{3}\int \frac{z.(1-z^2)}{z} dz = -\frac{t^3}{3}.\cos^{-1} t + \frac{1}{3}\int (1-z^2)\, dz = -\frac{t^3}{3}.\cos^{-1} t + \frac{1}{3}\int dz - \frac{1}{3}\int z^2\, dz$$

$I = -\frac{t^3}{3}.\cos^{-1} t + \frac{1}{3}z - \frac{1}{3}.\frac{z^3}{3} + c$ Put $t = \cos x$ and $z = \sqrt{1-t^2} = \sqrt{1-\cos^2 x} = \sin x$

$$\therefore\ I = \frac{1}{3}\sin x - \frac{1}{9}\sin^3 x - \frac{\cos^3 x}{3}.\cos^{-1}(\cos x) + c = \frac{1}{3}\sin x - \frac{1}{9}\sin^3 x - \frac{x\cos^3 x}{3} + c \quad \text{Ans.}$$

(d) $I = \int x(e^x + 1)\, dx = e^x(x-1) + \frac{x^2}{2} + c$ Ans. [use integrating by part formula]

(8) (a) $I = \int \frac{\log\{\log(1+\sqrt{x})\}}{(x+\sqrt{x})} dx$

Let $\log(1+\sqrt{x}) = z \quad \therefore \frac{1}{1+\sqrt{x}}.\frac{1}{2\sqrt{x}} dx = dz$ or $\frac{1}{(x+\sqrt{x})} dx = 2\, dz$

$I = \int \log z.2\, dz = 2\int \log z\, dz = 2\left[z\log z - \int \frac{1}{z}.z\, dz\right] = 2\left[z\log z - \int dz\right]$ [use integrating by part formula]

$I = 2z\log z - 2z + c = 2z(\log z - 1) = 2\log(1+\sqrt{x}).\left[\log\log(1+\sqrt{x}) - 1\right] + c$ Ans.

(b) $I = \int 2^{3\log_2 \sqrt[3]{x}}\,dx = \int 2^{\log_2(\sqrt[3]{x})^3}\,dx = \int 2^{\log_2 x}\,dx = \int x\,dx = \frac{x^2}{2} + c$ Ans. $\left[e^{\log_e x} = x\right]$

(c) $I = \int x^3\cot^{-1}x\,dx = \frac{x^4}{4}\cot^{-1}x + \frac{x^3}{12} - \frac{x}{4} + \frac{1}{4}\tan^{-1}x + c$ Ans. [integrating by part]

(d) $I = \int \frac{(2^x - 3^x)^2}{2^x.3^x}\,dx = \int \frac{2^{2x} + 3^{2x} - 2.2^x.3^x}{2^x.3^x}\,dx = \int \frac{2^{2x}}{2^x.3^x}\,dx + \int \frac{3^{2x}}{2^x.3^x}\,dx - 2\int \frac{2^x.3^x}{2^x.3^x}\,dx = \int \frac{2^x}{3^x}\,dx + \int \frac{3^x}{2^x}\,dx - 2\int dx$

$= \int \left(\frac{2}{3}\right)^x dx + \int \left(\frac{3}{2}\right)^x dx - 2\int dx$

$I = \frac{\left(\frac{2}{3}\right)^x}{\log\left(\frac{2}{3}\right)} + \frac{\left(\frac{3}{2}\right)^x}{\log\left(\frac{3}{2}\right)} - 2x + c$ Ans. $\left[\text{use} \int a^x\,dx = \frac{a^x}{\log a}\right]$

(9) (a) $I = \int \frac{\sin 2x.\cos x + 2\cos^2 x}{(1 + 2\sin{}^x\!/_2\cos{}^x\!/_2)}dx = x + \frac{\sin 2x}{2} + c$ Ans. (Do yourself)

(b) $I = \int \frac{x^2+1}{x(x+3)(x-4)}dx$, Let $\frac{x^2+1}{x(x+3)(x-4)} = \frac{A}{x} + \frac{B}{x+3} + \frac{C}{x-4}$ ……………… (i)

$\therefore\ A|_{x=0} = \frac{1}{3.(-4)} = -\frac{1}{12}$ and $B|_{x=-3} = \frac{10}{21}$ and $C|_{x=4} = \frac{17}{28}$

Put value of A, B and C in equation (i) and integrating, we get

$I = \int \frac{x^2+1}{x(x+3)(x-4)}dx = -\frac{1}{12}\int\frac{dx}{x} + \frac{10}{21}\int\frac{dx}{x+3} + \frac{17}{28}\int\frac{dx}{x-4} = -\frac{1}{12}\log x + \frac{10}{21}\log(x+3) + \frac{17}{28}\log(x-4) + c$ Ans.

(c) $I = \int \frac{(x^3+64)(x-2)}{x^2-4x+16}dx = \int \frac{(x^3+4^3)(x-2)}{x^2-4x+16}dx$ $[a^3+b^3 = (a+b)(a^2-ab+b^2)]$

$I = \int \frac{(x+4)(x^2-4x+16)(x-2)}{(x^2-4x+16)}dx = \int (x+4)(x-2)\,dx = \int (x^2+2x-8)\,dx = \int x^2\,dx + 2\int x\,dx - 8\int dx = \frac{x^3}{3} + 2\frac{x^2}{2} - 8x + c$

$= \frac{x^3}{3} + x^2 - 8x + c$ Ans.

(d) $I = \int \frac{dx}{\sqrt{x-1}+\sqrt{x+2}} = \int \frac{dx}{\sqrt{x-1}+\sqrt{x+2}} \times \frac{\sqrt{x-1}-\sqrt{x+2}}{\sqrt{x-1}-\sqrt{x+2}} = -\frac{1}{3}\int\left(\sqrt{x-1}-\sqrt{x+2}\right)dx = \frac{2}{9}(x+2)^{\frac{3}{2}} - \frac{2}{9}(x-1)^{\frac{3}{2}} + c$ Ans.

(e) $I = \int \cos^{-1}\left(\frac{2\cot x}{1+\cot^2 x}\right)dx = \int \cos^{-1}\left(\frac{2\cot x}{\operatorname{cosec}^2 x}\right)dx = \int \cos^{-1}\left(\frac{2\frac{\cos x}{\sin x}}{\frac{1}{\sin^2 x}}\right)dx = \int \cos^{-1}(2\sin x\cos x)\,dx = \int \cos^{-1}(\sin 2x)\,dx$

$= \int \cos^{-1}\left[\cos\left(\frac{\pi}{2} - 2x\right)\right]dx = \int\left(\frac{\pi}{2} - 2x\right)dx$ $[\cos^{-1}(\cos\theta) = \theta]$

$I = \frac{\pi}{2}\int dx - 2\int x\,dx = \frac{\pi}{2}x - 2.\frac{x^2}{2} + c = -x^2 + \frac{\pi}{2}x + c$ Ans.

(f) $I = \int \frac{x}{x^4+4}dx = \int \frac{x}{(x^2)^2+2^2}dx$, Let $x^2 = t$ $\therefore$ $2x\,dx = dt$ or $x\,dx = \frac{dt}{2}$

$I = \frac{1}{2}\int\frac{dt}{t^2+2^2} = \frac{1}{2}.\frac{1}{2}\tan^{-1}\left(\frac{t}{2}\right) + c = \frac{1}{4}\tan^{-1}\left(\frac{x^2}{2}\right) + c$ Ans. $\left[\text{formula, } \int\frac{dx}{x^2+a^2} = \frac{1}{a}\tan^{-1}\frac{x}{a}\right]$

(10) (a) $I = \int \frac{dx}{\sqrt{x}-\sqrt{x-2}} = \int \frac{dx}{\sqrt{x}-\sqrt{x-2}} \times \frac{\sqrt{x}+\sqrt{x-2}}{\sqrt{x}+\sqrt{x-2}} = \frac{1}{2}\int\left(\sqrt{x}+\sqrt{x-2}\right)dx = \frac{1}{2}\int\sqrt{x}\,dx + \frac{1}{2}\int\sqrt{x-2}\,dx = \frac{1}{2}.\frac{x^{\frac{3}{2}}}{\frac{3}{2}} + \frac{1}{2}.\frac{(x-2)^{\frac{3}{2}}}{\frac{3}{2}} + c$

$= \frac{1}{3}(x)^{\frac{3}{2}} + \frac{1}{3}(x-2)^{\frac{3}{2}} + c$ Ans.

(b) $I = \int \frac{x^8}{\sqrt{1+x^3}} dx$, Let $1 + x^3 = t^2$ or $x^3 = t^2 - 1$ or $x^6 = (t^2 - 1)^2$ $\therefore$ $3x^2\,dx = 2t\,dt$ or $x^2\,dx = \frac{2t\,dt}{3}$

$$I = \int \frac{x^6 . x^2\,dx}{\sqrt{1+x^3}} = \frac{2}{3}\int \frac{(t^2-1)^2 . t}{\sqrt{t^2}} dt = \frac{2}{3}\int (t^2-1)^2\,dt = \frac{2}{3}\int (t^4 - 2t^2 + 1)\,dt = \frac{2}{3}\int t^4\,dt - \frac{4}{3}\int t^2\,dt + \frac{2}{3}\int dt = \frac{2}{3}.\frac{t^5}{5} - \frac{4}{3}.\frac{t^3}{3} + \frac{2}{3}t + c$$

$$= \frac{2}{15}\left(\sqrt{1+x^3}\right)^5 - \frac{4}{9}\left(\sqrt{1+x^3}\right)^3 + \frac{2}{3}\sqrt{1+x^3} + c \quad \text{Ans.}$$

(c) $I = \int \frac{x+3}{\sqrt{4x+5}} dx$, Let $4x + 5 = t^2$ or $t = \sqrt{4x+5}$ or $x = \frac{t^2-5}{4}$ $\therefore$ $4\,dx = 2t\,dt$ or $dx = \frac{t\,dt}{2}$

$$I = \frac{1}{2}\int \frac{\frac{t^2-5}{4} + 3}{\sqrt{t^2}} . t\,dt = \frac{1}{2}\int \frac{t^2 - 5 + 12}{4}\,dt = \frac{1}{8}\int (t^2 + 7)\,dt = \frac{1}{8}\int t^2\,dt + \frac{7}{8}\int dt = \frac{1}{8}.\frac{t^3}{3} + \frac{7}{8}t + c = \frac{1}{24}t^3 + \frac{7}{8}t + c$$

$$= \frac{1}{24}\left(\sqrt{4x+5}\right)^3 + \frac{7}{8}.\sqrt{4x+5} + c = \frac{1}{24}(4x+5)^{\frac{3}{2}} + \frac{7}{8}(4x+5)^{\frac{1}{2}} + c \quad \text{Ans.}$$

(d) $I = \int \frac{x}{2x^2 - 3x + 5} dx = \int \left(\frac{\frac{1}{4}.(4x-3) + \frac{3}{4}}{2x^2 - 3x + 5}\right) dx = \frac{1}{4}\int \frac{4x-3}{2x^2 - 3x + 5}\,dx + \frac{3}{4}\int \frac{dx}{2x^2 - 3x + 5}$

$$I = \frac{1}{4}\int \frac{f'(x)}{f(x)} dx + \frac{3}{4}\int \frac{dx}{\left(\sqrt{2}x - \frac{3}{2\sqrt{2}}\right)^2 + \left(\frac{\sqrt{31}}{2\sqrt{2}}\right)^2}$$

using formula $\left[\int \frac{f'(x)}{f(x)} dx = \log f(x) \text{ and } \int \frac{dx}{x^2 + a^2} = \frac{1}{a}\tan^{-1}\left(\frac{x}{a}\right).\frac{1}{\text{d.c of x}}\right]$

$$I = \frac{1}{4}\log(2x^2 - 3x + 5) + \frac{3}{4}.\frac{1}{\frac{\sqrt{31}}{2\sqrt{2}}}\tan^{-1}\left(\frac{\sqrt{2}x - \frac{3}{2\sqrt{2}}}{\frac{\sqrt{31}}{2\sqrt{2}}}\right).\frac{1}{\sqrt{2}} + c = \frac{1}{4}\log(2x^2 - 3x + 5) + \frac{3}{2\sqrt{31}}\tan^{-1}\left(\frac{4x-3}{\sqrt{31}}\right) + c \quad \text{Ans.}$$

(11) (a) $I = \int \frac{3x+2}{x^2 + 2x + 3} dx$, Let $3x + 2 = l(\text{d.c of } x^2 + 2x + 3) + m = l(2x + 2) + m$ ……….. (A)

or $2l = 3$ $\therefore$ $l = \frac{3}{2}$ and $2l + m = 2$ $\therefore$ $m = 2 - 2.\frac{3}{2} = 2 - 3 = -1$

Put value of l and m in equation (A), we get $\therefore$ $3x + 2 = \frac{3}{2}(2x + 2) - 1 = 3(x + 1) - 1$

Divide by $x^2 + 2x + 3$ both of sides, we get $\therefore$ $\frac{3x+2}{x^2 + 2x + 3} = \frac{3}{2}.\frac{(2x+2)}{x^2 + 2x + 3} - \frac{1}{x^2 + 2x + 3}$

Integrating both of sides, $\int \frac{3x+2}{x^2 + 2x + 3}\,dx = \frac{3}{2}\int \frac{(2x+2)}{x^2 + 2x + 3}\,dx - \int \frac{1}{x^2 + 2x + 3}\,dx$

$I = \frac{3}{2}\int \frac{f'(x)}{f(x)}\,dx - \int \frac{dx}{(x+1)^2 + \left(\sqrt{2}\right)^2}$ using $\int \frac{f'(x)}{f(x)}\,dx = \log f(x)$ and $\int \frac{dx}{x^2 + a^2} = \frac{1}{a}\tan^{-1}\left(\frac{x}{a}\right)$

where $f(x) = x^2 + 2x + 3$, $f'(x) = 2x + 2$ $\therefore$ $I = \frac{3}{2}\log(x^2 + 2x + 3) - \frac{1}{\sqrt{2}}\tan^{-1}\left(\frac{x+1}{\sqrt{2}}\right) + c$ Ans.

(b) Do yourself. $I = \int \frac{4x+1}{x^2 + 3x + 2} dx = 2\log|x^2 + 3x + 2| - 5\log\left|\frac{x+1}{x+2}\right| + c$ Ans.

(c) $I = \int \frac{2x-3}{x^2 + 4x - 5} dx = \log|x^2 + 4x - 5| - \frac{7}{6}\log\left|\frac{x-1}{x+5}\right| + c$ Ans. [Do yourself, see question no. –(11) (a)]

(d) $I = \int \frac{x^5 + x^2}{x^6 + 16}\,dx = \int \frac{x^5}{x^6 + 16}\,dx + \int \frac{x^2}{x^6 + 16}\,dx = I_1 + I_2$ (say)

Now, $I_1 = \int \frac{x^5}{x^6 + 16} dx$, Let $x^6 + 16 = t$ $\therefore$ $6x^5 dx = dt$ or $x^5 dx = \frac{dt}{6}$

$$\therefore \ I_1 = \frac{1}{6}\int \frac{dt}{t} = \frac{1}{6}\log|t| + c = \frac{1}{6}\log|x^6 + 16| + c$$

Now, $I_2 = \int \frac{x^2}{x^6 + 16} dx = \int \frac{x^2}{(x^3)^2 + 4^2} dx$, Let $x^3 = z$ $\therefore$ $3x^2 dx = dz$ or $x^2 dx = \frac{dz}{3}$

or $I_2 = \frac{1}{3}\int \frac{dz}{z^2 + 4^2} = \frac{1}{3}.\frac{1}{4}\tan^{-1}\left(\frac{z}{4}\right) + c = \frac{1}{12}\tan^{-1}\left(\frac{x^3}{4}\right) + c$ $\left[\int \frac{dx}{x^2 + a^2} = \frac{1}{a}\tan^{-1}\left(\frac{x}{a}\right)\right]$

Then, $I = I_1 + I_2 = \frac{1}{6}\log|x^6 + 16| + \frac{1}{12}\tan^{-1}\left(\frac{x^3}{4}\right) + k$ Ans.

(12) (a) $I = \int \frac{\sin 2\theta - 5\cos\theta}{7 - \cos^2\theta + 4\sin\theta} d\theta$, Let $\sin 2\theta - 5\cos\theta = l(\text{d.c of } 7 - \cos^2\theta + 4\sin\theta) + m$

or $\sin 2\theta - 5\cos\theta = l(2\sin\theta\cos\theta + 4\cos\theta) + m = l(\sin 2\theta + 4\cos\theta) + m$ ....(A)

$$\therefore \ l = 1 \ \text{ and } \ 4l + m = -5 \qquad \therefore \ m = -5 - 4 = -9$$

Put value of l and m in equation (A), we get or $\sin 2\theta - 5\cos\theta = (\sin 2\theta + 4\cos\theta) - 9$

Divide by $7 - \cos^2\theta + 4\sin\theta$ both of sides and integrating, we get

$$\int \frac{\sin 2\theta - 5\cos\theta}{7 - \cos^2\theta + 4\sin\theta} d\theta = \int \frac{\sin 2\theta + 4\cos\theta}{7 - \cos^2\theta + 4\sin\theta} d\theta - 9\int \frac{d\theta}{7 - \cos^2\theta + 4\sin\theta} = I_1 - I_2 \ \text{(say)}$$

Now, $I_1 = \int \frac{\sin 2\theta + 4\cos\theta}{7 - \cos^2\theta + 4\sin\theta} d\theta$, Let $f(\theta) = 7 - \cos^2\theta + 4\sin\theta$, $f'(\theta) = \sin 2\theta + 4\cos\theta$

$$\therefore \ I_1 = \int \frac{f'(\theta)}{f(\theta)} d\theta = \log[f(\theta)] + c = \log[7 - \cos^2\theta + 4\sin\theta] + c$$

Now, $I_2 = 9\int \frac{d\theta}{7 - \cos^2\theta + 4\sin\theta} = 9\int \frac{d\theta}{7 - (1 - \sin^2\theta) + 4\sin\theta} = 9\int \frac{d\theta}{\sin^2\theta + 4\sin\theta + 6}$

$I_2 = 9\int \frac{d\theta}{(\sin\theta + 2)^2 + \left(\sqrt{2}\right)^2}$ $\left[\text{formula, } \int \frac{dx}{(bx)^2 + a^2} = \frac{1}{a}\tan^{-1}\left(\frac{bx}{a}\right).\frac{1}{\text{d.c of bx}}\right]$

$$\therefore \ I_2 = 9.\frac{1}{\sqrt{2}}\tan^{-1}\left(\frac{\sin\theta + 2}{\sqrt{2}}\right).\frac{1}{\text{d.c of }(\sin\theta + 2)} + c = \frac{9}{\sqrt{2}}\sec\theta\tan^{-1}\left(\frac{\sin\theta + 2}{\sqrt{2}}\right) + c$$

Then, $I = I_1 - I_2 = \log[7 - \cos^2\theta + 4\sin\theta] - \frac{9}{\sqrt{2}}\sec\theta\tan^{-1}\left(\frac{\sin\theta + 2}{\sqrt{2}}\right) + k$ Ans.

(b) $I = \int \sin(\log x)\, dx + \int \cos(\log x)\, dx = x\sin(\log x) + c$ Ans. [use integrating by part formula]

(c) $I = \int \cos^{-1}\left(\frac{2x}{1 + x^2}\right) dx$, Put $x = \tan\theta$ $\therefore$ $dx = \sec^2\theta\, d\theta = (1 + \tan^2\theta)\, d\theta$

$$I = \int \cos^{-1}\left(\frac{2\tan\theta}{1 + \tan^2\theta}\right).\sec^2\theta\, d\theta = \int \cos^{-1}(\sin 2\theta).\sec^2\theta\, d\theta = \int \cos^{-1}\left[\cos\left(\frac{\pi}{2} - 2\theta\right)\right].\sec^2\theta\, d\theta = \int \left(\frac{\pi}{2} - 2\theta\right).\sec^2\theta\, d\theta$$
$$= \int \frac{\pi}{2}.\sec^2\theta\, d\theta - 2\int \theta.\sec^2\theta\, d\theta$$

$$I = \frac{\pi}{2}.\tan\theta - 2\left[\theta.\int \sec^2\theta\, d\theta - \int \frac{d(\theta)}{d\theta}.d\theta.\int \sec^2\theta\, d\theta\right] = \frac{\pi}{2}.\tan\theta - 2\theta\tan\theta + 2\int \tan\theta\, d\theta = \frac{\pi}{2}.\tan\theta - 2\theta\tan\theta - 2\log|\cos\theta| + c$$

Put $x = \tan\theta$ $\left[\int \tan\theta\, d\theta = -\log|\cos\theta|\right]$

$I = \frac{\pi}{2}x - 2x\tan^{-1}x - 2\log\left|\frac{1}{\sqrt{1+x^2}}\right| + c$ Ans. $\left[\cos\theta = \frac{1}{\sqrt{1+\tan^2\theta}} = \frac{1}{\sqrt{1+x^2}}\right]$

(13) (a) $I = \int \sin^{-1}\left(\frac{x}{\sqrt{1+x^2}}\right)dx$, Put $x = \tan\theta$ or $\theta = \tan^{-1}x$ $\therefore$ $dx = \sec^2\theta\ d\theta$

$$I = \int \sin^{-1}\left(\frac{\tan\theta}{\sqrt{1+\tan^2\theta}}\right).\sec^2\theta\ d\theta = \int \sin^{-1}\left(\frac{\tan\theta}{\sec\theta}\right).\sec^2\theta\ d\theta = \int \sin^{-1}(\sin\theta).\sec^2\theta\ d\theta = \int \theta.\sec^2\theta\ d\theta$$

$$I = \theta.\int \sec^2\theta\ d\theta - \int \frac{d(\theta)}{d\theta}.d\theta.\int \sec^2\theta\ d\theta = \theta.\tan\theta - \int \tan\theta\, d\theta = \theta\tan\theta + \log|\cos\theta| + c$$

Put $\theta = \tan^{-1}x$ then, $I = \tan^{-1}x.\tan(\tan^{-1}x) + \log\left|\frac{1}{\sqrt{1+x^2}}\right| + c = x\tan^{-1}x + \log\left|\frac{1}{\sqrt{1+x^2}}\right| + c$ Ans.

Put $\cos\theta = \frac{1}{\sqrt{1+\tan^2\theta}} = \frac{1}{\sqrt{1+x^2}}$ $\left[\int \sec^2\theta\ d\theta = \tan\theta \text{ and } \int \tan\theta\, d\theta = -\log|\cos\theta|\right]$

(b) $I = \int \tan^{-1}\left(\frac{x}{\sqrt{1-x^2}}\right)dx = x\sin^{-1}x + \sqrt{1-x^2} + c$ Ans. [Put $x = \sin\theta$ $\therefore$ $dx = \cos\theta\, d\theta$]

(c) $I = \int \frac{\cos^{-1}x}{\sqrt{1-x^2}}dx$, Put $\cos^{-1}x = z$ $\therefore$ $-\frac{1}{\sqrt{1-x^2}}dx = dz$

$$\therefore\ I = -\int z\,dz = -\frac{z^2}{2} + c = -\frac{(\cos^{-1}x)^2}{2} + c \quad \text{Ans.}$$

(d) $I = \int \frac{\cos^{-1}x}{(1-x^2)^{\frac{3}{2}}}dx = \int \frac{\cos^{-1}x}{(1-x^2).\sqrt{1-x^2}}dx$, Put $\cos^{-1}x = z$ or $x = \cos z$ $\therefore$ $-\frac{1}{\sqrt{1-x^2}}dx = dz$

$I = -\int \frac{z}{1-\cos^2 z}dz = -\int \frac{z}{\sin^2 z}dz = -\int z.\mathrm{cosec}^2 z\ dz$ [use integrating by part formula]

$$I = -\left[z.\int \mathrm{cosec}^2 z\ dz - \int \frac{d(z)}{dz}.dz.\int \mathrm{cosec}^2 z\ dz\right] = -\left[z.(-\cot z) + \int \cot z\ dz\right] = z.\cot z - \log|\sin z| + c$$

$I = z.\frac{\cos z}{\sqrt{1-\cos^2 z}} - \log\left|\sqrt{1-\cos^2 z}\right| + c = \cos^{-1}x.\frac{x}{\sqrt{1-x^2}} - \log\left|\sqrt{1-x^2}\right| + c$ Ans.

(14) (a) $I = \int \sqrt{\sec\theta - 1}\ d\theta = -\log\left|\frac{1}{2} + \cos\theta + \sqrt{\cos\theta + \cos^2\theta}\right| + c$ Ans.

$\left[\text{Do yourself, use formula } \int \frac{dx}{\sqrt{x^2-a^2}} = \log\left|x + \sqrt{x^2-a^2}\right|\right]$

(b) $I = \int \frac{x - \sqrt[3]{x^2}}{x(\sqrt[4]{x} + \sqrt[6]{x^2})}dx$, Integral of the form $\int R\left(x, x^{\frac{2}{3}}, x^{\frac{1}{4}}, x^{\frac{1}{3}}\right)dx$

Put $x = t^{\alpha}$ where $\alpha =$ L. C. M of (3,4,3) = 12 or $x = t^{12}$ $\therefore$ $dx = 12t^{11}\,dt$

Ans:– $\frac{12}{5}(x)^{\frac{5}{12}} + \frac{12}{7}(x)^{\frac{7}{12}} - 2(x)^{\frac{1}{2}} - \frac{3}{2}(x)^{\frac{2}{3}} + c$

(c) $I = \int e^{3x}(3x^2 + 4x + 1)\,dx = e^{3x}\left[x^2 + \frac{2}{3}x + \frac{7}{9}\right] + c$ Ans. [use integrating by part formula]

(d) $I = \int \frac{x^2 - 9}{x + 2}dx = \frac{x^2}{2} - 2x - 5\log|x + 2| + c$ Ans.

(15) (a) $I = \int e^{\sin 2x}(1 + \cos 2x)\,dx = \frac{1}{2}e^{\sin 2x}(1 + \sec 2x) + c$ Ans.

(b) $I = \int \sin^2 x\cot^2 x\ dx = \frac{x}{2} + \frac{\sin 2x}{4} + c$ Ans. (c) $I = \int \frac{\cot x\,\mathrm{cosec}\,x}{3 + \mathrm{cosec}\,x}dx = -\log|3 + \mathrm{cosec}\,x| + c$ Ans.

(d) $I = \int \tan^4 x \, dx = \frac{\tan^3 x}{3} - \tan x + x + c$ Ans.

(16) (a) $I = \int (\sin^4 x + \cos^4 x) \, dx = \frac{3x}{4} + \frac{\sin 4x}{16} + c$ Ans.

(b) $I = \int \frac{dx}{1 - \sqrt{1 + x^2}}$ Integral of the form $\int R\left(x, (1 + x^2)^{\frac{1}{2}}\right)$, Put $\sqrt{1 + x^2} = t - x\sqrt{a}$ $\therefore$ $a = 1$

or $\sqrt{1 + x^2} = t - x$ $\therefore$ $\frac{1}{2\sqrt{1 + x^2}}.(2x) \, dx = dt - 1$ or $\left(\frac{x}{\sqrt{1 - x^2}} + 1\right) dx = dt$

(c) $I = \int \frac{(\sin x - \sqrt[3]{\sin^2 x}).\cos x}{\sqrt{\sin^3 x} + \sqrt[4]{\sin^6 x}} dx$, Put $\sin x = z$ $\therefore$ $\cos x \, dx = dz$

$I = \int \frac{(z - \sqrt[3]{z^2})}{\sqrt{z^3} + \sqrt[4]{z^6}} \, dz$ Integral of the form $\int R\left(z, z^{\frac{2}{3}}, z^{\frac{3}{2}}, z^{\frac{6}{4}}\right) dz$, Put $z = t^\alpha$ where $\alpha =$ L. C. M of $(3,2,4) = 12$

or $z = t^{12}$ $\therefore$ $dz = 12t^{11} \, dt$ then, $I = \int \frac{(t^{12} - t^8)}{(t^{18} + t^{18})}.12t^{11} \, dt = 6 \int \frac{(t^{12} - t^8)}{t^7}.dt$

$I = 6 \int t^5 \, dt - 6 \int t \, dt = 6.\frac{t^6}{6} - 6.\frac{t^2}{2} + c = t^6 - 3t^2 + c = \sqrt{z} - 3\sqrt[6]{z} + c = \sqrt{\sin x} - 3.\sqrt[6]{\sin x} + c$ Ans.

(d) $I = \int \frac{x - 1}{(x - 2)\sqrt{x + 3}} dx$, Put $x + 3 = z^2$ or $x = z^2 - 3$ $\therefore$ $dx = 2z \, dz$

$I = \int \frac{z^2 - 3 - 1}{(z^2 - 3 - 2).\sqrt{z^2}}.2z \, dz = 2\int \frac{z^2 - 4}{z^2 - 5} \, dz = 2\int \left(1 + \frac{1}{z^2 - 5}\right) dz = 2\int dz + 2\int \frac{1}{z^2 - 5} \, dz = 2z + 2.\frac{1}{2.\sqrt{5}}\log\left|\frac{z - \sqrt{5}}{z + \sqrt{5}}\right| + c$

put $z = \sqrt{x + 3}$ use formula, $\int \frac{dx}{x^2 - a^2} = \frac{1}{2a}\log\left|\frac{x - a}{x + a}\right|$ or $I = 2\sqrt{x + 3} + \frac{1}{\sqrt{5}}\log\left|\frac{\sqrt{x + 3} - \sqrt{5}}{\sqrt{x + 3} + \sqrt{5}}\right| + c$ Ans.

(17) (a) $I = \int \sin x.\sin 3x \, dx = \frac{1}{2}\int 2\sin x.\sin 3x \, dx$, using formula, $2\sin A \sin B = \cos(A - B) - \cos(A + B)$

$I = \frac{1}{2}\int [\cos(-2x) - \cos 4x] \, dx = \frac{1}{2}\int \cos 2x \, dx - \frac{1}{2}\int \cos 4x \, dx = \frac{1}{2}.\frac{\sin 2x}{2} - \frac{1}{2}.\frac{\sin 4x}{4} + c$

$\therefore I = \frac{\sin 2x}{4} - \frac{\sin 4x}{8} + c$ Ans. $[\cos(-\theta) = \cos\theta]$

(b) $I = \int \cos x.\cos 3x.\cos 5x \, dx = \frac{1}{2}\int (2\cos x.\cos 3x).\cos 5x \, dx$

using formula $[\, 2\cos A.\cos B = \cos(A + B) + \cos(A - B) \,]$

$I = \frac{1}{2}\int [\cos 4x + \cos(-2x)].\cos 5x \, dx = \frac{1}{2}\int \cos 4x \cos 5x \, dx + \frac{1}{2}\int \cos 2x \cos 5x \, dx$

$= \frac{1}{4}\int 2\cos 4x \cos 5x \, dx + \frac{1}{4}\int 2\cos 2x \cos 5x \, dx$ Again use above formula

$I = \frac{1}{4}\int [\cos 9x + \cos x] \, dx + \frac{1}{4}\int [\cos 7x + \cos 3x] \, dx = \frac{1}{4}\int \cos 9x \, dx + \frac{1}{4}\int \cos x \, dx + \frac{1}{4}\int \cos 7x \, dx + \frac{1}{4}\int \cos 3x \, dx$

$= \frac{1}{4}.\frac{\sin 9x}{9} + \frac{1}{4}.\sin x + \frac{1}{4}.\frac{\sin 7x}{7} + \frac{1}{4}.\frac{\sin 3x}{3} + c = \frac{\sin 9x}{36} + \frac{\sin x}{4} + \frac{\sin 7x}{28} + \frac{\sin 3x}{12} + c$ Ans.

(c) $I = \int \tan 2x.\tan 3x.\tan 5x \, dx$, Let $\tan 5x = \tan(3x + 2x) = \frac{\tan 3x + \tan 2x}{1 - \tan 3x \tan 2x}$

or $\tan 5x\,(1 - \tan 3x \tan 2x) = \tan 3x + \tan 2x$ or $\tan 5x - \tan 2x.\tan 3x.\tan 5x = \tan 3x + \tan 2x$

or $\tan 2x.\tan 3x.\tan 5x = \tan 5x - \tan 3x - \tan 2x$ Integrating both of sides

$$\int \tan 2x.\tan 3x.\tan 5x\ dx = \int (\tan 5x - \tan 3x - \tan 2x)\ dx = \int \tan 5x\ dx - \int \tan 3x\ dx - \int \tan 2x\ dx$$

$$I = -\frac{\log(\cos 5x)}{5} + \frac{\log(\cos 3x)}{3} + \frac{\log(\cos 2x)}{2} + c \qquad \left[\int \tan x\, dx = -\log\cos x = \log\sec x\right]$$

$$I = \frac{\log(\cos 3x)}{3} + \frac{\log(\cos 2x)}{2} - \frac{\log(\cos 5x)}{5} + c \text{ Ans. or } I = \frac{\log(\sec 5x)}{5} - \frac{\log(\sec 3x)}{3} - \frac{\log(\sec 2x)}{2} + c \text{ Ans.}$$

(d) $I = \int \sin x.\sin 3x.\sin 5x.\sin 7x\ dx$ use formula, $2\sin A\sin B = \cos(A-B) - \cos(A+B)$

$$I = \frac{1}{8}\left[x + \frac{\sin 4x}{4} - \frac{\sin 14x}{14} - \frac{\sin 10x}{10} - \frac{\sin 6x}{6} - \frac{\sin 2x}{2} + \frac{\sin 16x}{16} + \frac{\sin 8x}{8}\right] + c \quad \text{Ans.}$$

(e) $I = \int \sin x.\cos 3x\ dx$ use formula, $2\sin A\cos B = \sin(A+B) + \sin(A-B)$

$$I = \frac{1}{2}\int 2\sin x.\cos 3x\ dx = \frac{1}{2}\int [\sin 4x + \sin(-2x)]\, dx = \frac{1}{2}\int \sin 4x\ dx - \frac{1}{2}\int \sin 2x\ dx \quad [\sin(-\theta) = -\sin\theta]$$

$$I = -\frac{1}{2}.\frac{\cos 4x}{4} + \frac{1}{2}.\frac{\cos 2x}{2} + c = \frac{\cos 2x}{4} - \frac{\cos 4x}{8} + c \quad \text{Ans.}$$

(18) (a) $I = \int \cos 2x.\sin 3x\ dx$ use formula, $2\cos A\sin B = \sin(A+B) - \sin(A-B)$

$$I = \frac{1}{2}\int 2\cos 2x.\sin 3x\ dx = \frac{1}{2}\int [\sin 5x + \sin x]\, dx = \frac{1}{2}\int \sin 5x\ dx + \frac{1}{2}\int \sin x\ dx \quad [\sin(-\theta) = \sin\theta]$$

$$I = -\frac{1}{2}.\frac{\cos 5x}{5} - \frac{1}{2}\cos x + c = -\left(\cos x + \frac{\cos 5x}{5}\right) + c \quad \text{Ans.}$$

(b) $I = \int \sin x.\sin 2x.\cos x.\cos 2x\ dx = \frac{\sin^3 2x}{12} + c$ Ans.

(c) $$I = \int \frac{\sin 2x}{\cos(x+\theta).\cos(x-\theta)}\, dx = \int \frac{\sin[(x+\theta)+(x-\theta)]}{\cos(x+\theta).\cos(x-\theta)}\, dx = \int \frac{\sin(x+\theta).\cos(x-\theta) + \cos(x+\theta).\sin(x-\theta)}{\cos(x+\theta).\cos(x-\theta)}\, dx$$

$$I = \int \frac{\sin(x+\theta).\cos(x-\theta)}{\cos(x+\theta).\cos(x-\theta)}\, dx + \int \frac{\cos(x+\theta).\sin(x-\theta)}{\cos(x+\theta).\cos(x-\theta)}\, dx = \int \frac{\sin(x+\theta)}{\cos(x+\theta)}\, dx + \int \frac{\sin(x-\theta)}{\cos(x-\theta)}\, dx = \int \tan(x+\theta)\ dx + \int \tan(x-\theta)\ dx$$
$$= -\log[\cos(x+\theta)] - \log[\cos(x-\theta)] + c = -\{\log[\cos(x+\theta)] + \log[\cos(x-\theta)]\} + c$$
$$= -\{\log[\cos(x+\theta).\cos(x-\theta)]\} + c \quad \text{Ans.}$$

(d) $$I = \int \frac{dx}{\sin x.\sin(x+\theta)} = \frac{1}{\sin\theta}\int \frac{\sin(x+\theta-x)}{\sin x.\sin(x+\theta)}\, dx = \frac{1}{\sin\theta}\int \frac{\sin(x+\theta).\cos x - \cos(x+\theta).\sin x}{\sin x.\sin(x+\theta)}\, dx$$

$$I = \frac{1}{\sin\theta}\int \frac{\sin(x+\theta).\cos x}{\sin x.\sin(x+\theta)}\, dx - \frac{1}{\sin\theta}\int \frac{\cos(x+\theta).\sin x}{\sin x.\sin(x+\theta)}\, dx = \frac{1}{\sin\theta}\int \frac{\cos x}{\sin x}\, dx - \frac{1}{\sin\theta}\int \frac{\cos(x+\theta)}{\sin(x+\theta)}\, dx$$
$$= \frac{1}{\sin\theta}\int \cot x\ dx - \frac{1}{\sin\theta}\int \cot(x+\theta)\ dx = \frac{1}{\sin\theta}\{\log(\sin x) - \log[\sin(x+\theta)]\} + c = \frac{1}{\sin\theta}\left\{\log\left[\frac{\sin x}{\sin(x+\theta)}\right]\right\} + c$$
$$= \frac{1}{\sin\theta}.\log\left[\frac{\sin x}{\sin(x+\theta)}\right] + c \quad \text{Ans.}$$

Definite Integral

Definition: – If f(x) be a single valued continuous function in the interval (a, b), where $b > a$

and if the interval (a, b) be divided into n equal part of length h by the points

$$a + h, a + 2h, a + 3h, \ldots\ldots\ldots\ldots\ldots\ldots\ldots, a + (\overline{n-1})h$$

so that, $a + nh = b$ or $nh = b - a$ then sum of limit: –

$$\lim_{h\to 0} h[f(a) + f(a+h) + f(a+2h) + f(a+3h) + \cdots \ldots\ldots\ldots + f(a + (\overline{n-1})h)]$$

written, $\lim\limits_{h\to 0} h \sum\limits_{r=0}^{n-1} f(a+rh)$ or $\lim\limits_{n\to\infty} h \sum\limits_{r=0}^{n-1} f(a+rh)$ where $h = \frac{b-a}{n}$ and when $h \to 0, n \to \infty$

f(x) is defined is a functions whose differentiation is f(x) or F(x) is the integral of f(x) with respect to x.

Summation of series: – (i) $\sum\limits_{r=1}^{n} r = 1 + 2 + 3 + \cdots \ldots\ldots\ldots\ldots\ldots\ldots + n = \frac{n(n+1)}{2}$

(ii) $\sum\limits_{r=1}^{n} r^2 = 1 + 4 + 9 + \cdots \ldots\ldots\ldots\ldots\ldots\ldots\ldots\ldots + n^2 = \frac{n(n+1)(2n+1)}{6}$

(iii) $\sum\limits_{r=1}^{n} r^3 = 1 + 2^3 + 3^3 + \cdots \ldots\ldots\ldots\ldots\ldots\ldots\ldots\ldots + n^3 = \left[\frac{n(n+1)}{2}\right]^2 = \frac{n^2(n+1)^2}{4}$

(iv) $\sum 1 = n$

(v) $\sum\limits_{r=1}^{n} x^r = x + x^2 + x^3 + \cdots \ldots\ldots\ldots\ldots\ldots\ldots\ldots + x^n =$ sum of a G. P $= \frac{x(x^n - 1)}{x-1}$

(vi) $\sum\limits_{r=1}^{n} ax^{r-1} = a + ax + ax^2 + \cdots \ldots\ldots\ldots\ldots\ldots\ldots + ax^{n-1} = \frac{a(x^n - 1)}{x-1}$

(vii) we know that $\int_a^b f(x)\,dx = \lim\limits_{n\to\infty} h \sum\limits_{r=1}^{n} f(a+rh)$, where $nh = b - a$ or $h = \frac{b-a}{n}$

Now, Put $a = 0, b = 1, nh = 1 - 0 = 1$ $\therefore$ $h = \frac{1}{n}$

$\int_0^1 f(x)\,dx = \lim\limits_{n\to\infty} \frac{1}{n} \sum f\left(\frac{r}{n}\right)$ Replace $\frac{r}{n}$ by x and $\frac{1}{n}$ by dx and the limit of the sum is $\int_0^1 f(x)\,dx$

Example: – $I = \int_a^b x^2\,dx = \frac{1}{3}(b^3 - a^3)$

Solution: – $\int_a^b x^2\,dx$, Here $f(x) = x^2, f(a) = a^2, f(a+h) = (a+h)^2, f(a+2h) = (a+2h)^2$ etc.

Now, $\int_a^b f(x)\,dx = \lim\limits_{n\to 0} \sum\limits_{r=0}^{n-1} h.f(a+rh) = \lim\limits_{n\to\infty} \sum\limits_{r=0}^{n-1} h.f(a+rh)$ where $nh = b - a$

$$I = \int_a^b x^2\,dx = \lim_{h\to\infty} h[f(a) + f(a+h) + f(a+2h) + f(a+3h) + \cdots \ldots\ldots\ldots + f(a + (\overline{n-1})h)]$$

$$= \lim_{h\to\infty} h\left[a^2 + (a+h)^2 + (a+2h)^2 + \cdots \ldots\ldots\ldots + \left[a + (\overline{n-1})h\right]^2\right]$$

Grouping the terms of $a^2, 2a^2h$ and h^2 to the above, we get

$$I = \lim_{n\to\infty} h\left[a^2 \sum 1 + 2ah.\{1 + 2 + 3 + \cdots \ldots\ldots + (n-1)\}\right] + h^2[1^2 + 2^2 + \cdots \ldots\ldots + (n-1)^2]$$

Now, $\sum 1 = n$, $\sum\limits_{r=1}^{n} r = \frac{n(n+1)}{2}$ and $\sum\limits_{r=1}^{n} r^2 = \frac{n(n+1)(2n+1)}{6}$

Replacing n by n – 1 in the above $\sum\limits_{r=1}^{n-1} r = \frac{n(n-1)}{2}$ and $\sum\limits_{r=1}^{n-1} r^2 = \frac{n(n-1)(2n-1)}{6}$

$$I = \lim_{n\to\infty} h\left\{a^2.n + 2ah.\frac{(n-1).n}{2} + h^2.\frac{(n-1)n(2n-1)}{6}\right\}$$

we have to take the limit $n \to \infty$ or $h \to 0$ and $nh = b - a$

$$I = \lim_{n\to\infty}\left\{a^2(nh) + a(nh)^2\left(1-\frac{1}{n}\right) + \frac{1}{6}(nh)^3\left(1-\frac{1}{n}\right)\left(2-\frac{1}{n}\right)\right\} \text{ when } n\to\infty,\ \frac{1}{n}\to 0 \text{ and } nh = b-a$$

$$I = \left\{a^2(b-a) + a(b-a)^2.1 + \frac{1}{6}(b-a)^3.1.2\right\} = (b-a)\left[a^2 + a(b-a) + \frac{1}{3}(b-a)^2\right] = \frac{(b-a)}{3}[3a^2 + 3(ab-a^2) + (b^2 - 2ab + a^2)]$$

$$= \frac{(b-a)}{3}[b^2 + ab + a^2] = \frac{1}{3}(b^3 - a^3) \quad \text{Proved.}$$

Note: – $\int_a^b f(x)\,dx = \lim_{n\to\infty}\sum_{r=1}^{n} f(a+rh)$, where $nh = b - a$ or $h = \frac{b-a}{n}$

Exercise – A9

(1) (a) $\lim_{n\to\infty}\left[\frac{1}{n+1} + \frac{1}{n+2} + \frac{1}{n+3} + \cdots \ldots \ldots + \frac{1}{2n}\right]$

(b) $\lim_{n\to\infty}\left[\frac{1}{n+1} + \frac{1}{n+2} + \frac{1}{n+3} + \cdots \ldots \ldots + \frac{1}{8n}\right]$

(c) $\lim_{n\to\infty}\left[\frac{1}{\sqrt{1+n^2}} + \frac{1}{\sqrt{4+n^2}} + \frac{1}{\sqrt{9+n^2}} + \cdots \ldots \ldots + \frac{1}{\sqrt{49+n^2}}\right]$

(2) (a) $\lim_{n\to\infty}\frac{1}{n}\sum_{r=1}^{n}\cos\left(\frac{r\pi}{n}\right)$ (b) $\lim_{n\to\infty}\frac{1}{n}\sum_{r=1}^{2n}\cos^2\left(\frac{r\pi}{2n}\right)$

(c) $\lim_{n\to\infty}\frac{1}{n}\left[\sec^2\left(\frac{\pi}{3n}\right) + \sec^2\left(\frac{2\pi}{3n}\right) + \cdots \ldots + \sec^2\left(\frac{n\pi}{3n}\right)\right]$

(3) (a) If $S_n = 1 + \frac{n-1}{n+1} + \frac{n-2}{n+2} + \cdots \ldots + 0$ then find $\lim_{n\to\infty}\frac{S_n}{n}$.

(b) If $S_n = \left[\left(1+\frac{1}{n}\right)\left(1+\frac{2}{n}\right)\left(1+\frac{3}{n}\right) \ldots \ldots \ldots \left(1+\frac{n}{n}\right)\right]^{\frac{1}{n}}$ then find $\lim_{n\to\infty} S_n$.

(4) (a) $\lim_{n\to\infty}\frac{1}{n}\sum_{r=1}^{n}\frac{r}{\sqrt{n^2+r^2}}$ (b) $\lim_{n\to\infty}\frac{1}{n^{m+1}}[1^m + 2^m + 3^m + \cdots \ldots \ldots + n^m]$, $m > -1$

(c) $\lim_{n\to\infty}\left[\frac{n!}{n^n}\right]^{\frac{1}{n}}$ (d) $\lim_{n\to\infty}\left[\frac{(2n)!}{n^n.n!}\right]^{\frac{1}{n}}$

(5) (a) $\lim_{n\to\infty}\frac{1}{n}\left[e^{\left(1+\frac{1}{n}\right)+\left(1+\frac{2}{n}\right)+\left(1+\frac{3}{n}\right)+\cdots\ldots+\left(1+\frac{n}{n}\right)}\right]$

(b) $\lim_{n\to\infty}\frac{1}{n}\left[\sin\left(\frac{n+1}{n}\right) + \sin\left(\frac{n+2}{n}\right) + \cdots \ldots + \sin\left(\frac{n+n}{n}\right)\right]$

(6) use $\int_a^b f(x)\,dx = \lim_{n\to\infty}\sum_{r=1}^{n} f(a+rh)$, where $nh = b - a$ and find the following integrals: –

(a) $\int_1^2 x^3\,dx = \frac{15}{4}$ (b) $\int_a^b \sin x\,dx = \cos a - \cos b$ (c) $\int_a^b e^{2x}\,dx = \frac{e^{2b} - e^{2a}}{2}$ (d) $\int_a^b \frac{1}{x}\,dx = \log\left(\frac{b}{a}\right)$

(7) (a) $\int_0^3 (x-3)\,dx = -4$ (b) $\int_1^3 (2x^2 - 3x)\,dx = \frac{16}{3}$ (c) $\int_0^2 (x^3+3)\,dx = 10$ (d) $\int_0^1 x\,dx$ (e) $\int_0^{\frac{\pi}{2}} \cos x\,dx = 1$

(8) (a) $\int_a^b \cos^2 x\,dx = \frac{1}{2}\left[(b-a) + \frac{\sin 2b - \sin 2a}{2}\right]$ (b) $\int_0^1 \sqrt{x}\,dx$

(c) $\int_2^3 e^{x+1}\,dx = e^4 - e^3$ (d) $\int_1^2 \frac{2}{\sqrt{x}}\,dx = 4(\sqrt{2}-1)$

Answer

(1) (a) $\lim_{n\to\infty}\left[\frac{1}{n+1}+\frac{1}{n+2}+\frac{1}{n+3}+\cdots\ldots\ldots\ldots\ldots+\frac{1}{2n}\right] = \lim_{n\to\infty}\frac{1}{n}\left[\frac{1}{1+\frac{1}{n}}+\frac{1}{1+\frac{2}{n}}+\cdots\ldots\ldots+\frac{1}{1+\frac{n}{n}}\right] = \lim_{n\to\infty}\frac{1}{n}\sum_{r=1}^{n}\frac{1}{\left(1+\frac{r}{n}\right)}$ standard form

Put $\frac{r}{n} = x$ and $\frac{1}{n} = dx$ when $r = 1, x = \frac{1}{n} \to 0$ when $r = n, x = \frac{n}{n} = 1$ as $n \to \infty$

$I = \int_0^1 \frac{1}{1+x}\,dx = [\log(1+x)]_0^1 = \log(1+1) - \log 1 = \log 2 - 0 = \log 2$ Ans.

(b) Do yourself. $\lim_{n\to\infty}\left[\frac{1}{n+1}+\frac{1}{n+2}+\frac{1}{n+3}+\cdots\ldots\ldots\ldots\ldots+\frac{1}{8n}\right] = \log 8 = 3\log 2$ Ans. (solve same as above question)

(c) Do yourself. $\lim_{n\to\infty}\left[\frac{1}{\sqrt{1+n^2}}+\frac{1}{\sqrt{4+n^2}}+\frac{1}{\sqrt{9+n^2}}+\cdots\ldots\ldots\ldots\ldots+\frac{1}{\sqrt{49+n^2}}\right] = \log(1+\sqrt{2})$ or $\sinh^{-1}(1)$ Ans.

(2) (a) $\lim_{n\to\infty}\frac{1}{n}\sum_{r=1}^{n}\cos\left(\frac{r\pi}{n}\right) = 0$ Ans. (Do yourself)

(b) $\lim_{n\to\infty}\frac{1}{n}\sum_{r=1}^{2n}\cos^2\left(\frac{r\pi}{2n}\right) = \frac{\pi^2}{4}$ Ans. (c) $\lim_{n\to\infty}\frac{1}{n}\left[\sec^2\left(\frac{\pi}{3n}\right)+\sec^2\left(\frac{2\pi}{3n}\right)+\cdots\ldots\ldots+\sec^2\left(\frac{n\pi}{3n}\right)\right] = \frac{3\sqrt{3}}{\pi}$ Ans.

(3) (a) $2\log 2 - 1$ or $\log 4 - 1$ Ans. (b) $A = \frac{4}{e}$ Let $A = S_n$

(4) (a) $\sqrt{2}-1$ Ans. (b) $\frac{1}{m+1}$ Ans. (c) $\frac{1}{e}$ Ans. (d) $\frac{4}{e}$ Ans.

(5) (a) $e^2 - e$ or $e(e-1)$ Ans. (b) $\cos 1 - \cos 2$ Ans.

Properties of Definite Integrals: – Property I: – $\int_a^b f(x)\,dx = \int_a^b f(t)\,dt$

change of variable does not make any difference

Proof: – $L.H.S = \int_a^b f(x)\,dx$, Put $x = t$ $\therefore$ $dx = dt$ or $x = a, x = b$ then $t = a, t = b$

$L.H.S = \int_a^b f(x)\,dx = \int_a^b f(t)\,dt = R.H.S$ Proved

Property II: – $\int_a^b f(x)\,dx = -\int_b^a f(x)\,dx$ Interchanging the limit to change of sign.

Property III: – $\int_a^b f(x)\,dx = \int_a^c f(x)\,dx + \int_c^b f(x)\,dx$, If $\int f(x)\,dx = F(x) + c$

Proof: – $L.H.S = \int_a^b f(x)\,dx = [F(x)+c]_a^b = F(b) - F(a) + c - c = F(b) - F(a)$

$R.H.S = \int_a^c f(x)\,dx + \int_c^b f(x)\,dx = [F(x)+c]_a^c + [F(x)+c]_c^b = F(c) + c - F(a) - c + F(b) + c - F(c) - c = F(b) - F(a)$

$\therefore$ $L.H.S = F(b) - F(a) = R.H.S$ Proved.

Property IV: – $\int_0^a f(x)\,dx = \int_0^a f(a-x)\,dx$ $\left\{\int_0^a f(x)\,dx,\ \text{Put } t = a + 0 - x \text{ or } t = a - x \text{ or } x = a - t\right\}$

Proof: – $\text{L.H.S} = \int_0^a f(x)\,dx$, Put $x = a - t \quad \therefore \quad dx = -dt$

If $x = 0$ then $x = a - t$ or $t = a - x = a$ and If $x = a$ then $t = a - x = a - a = 0$

$$\therefore \int_0^a f(x)\,dx = \int_a^0 f(a-t)\,(-dt) = -\int_a^0 f(a-t)\,dt = \int_0^a f(a-t)\,dt \quad \text{[form property I]}$$

$$\int_0^a f(a-t)\,dt = \int_0^a f(a-x)\,dx \qquad \text{Proved.} \quad \text{[from property I]}$$

Example: – (a) If $f(x) = \int_0^x \sin^2 t\,dt$ then find $f(x+\pi)$.

Solution: – $f(x) = \int_0^x \sin^2 t\,dt$

$$f(x+\pi) = \int_0^{x+\pi} \sin^2 t\,dt = \int_0^{\pi} \sin^2 t\,dt + \int_{\pi}^{x+\pi} \sin^2 t\,dt = f(\pi) + I_2 \quad \text{by prop. I}$$

or $I_2 = \int_{\pi}^{x+\pi} \sin^2 t\,dt$, Put $t = \theta + \pi$ or $t = \pi,\ \theta = 0$ and $t = x + \pi,\ \theta = x$

$$\therefore\ I_2 = \int_{\pi}^{x+\pi} \sin^2 t\,dt = \int_0^x \sin^2\theta\,d\theta = f(x) \quad \text{by property I.}$$

$$\therefore\ f(x+\pi) = f(\pi) + I_2 = f(\pi) + f(x) \qquad \text{Ans.}$$

$$\text{(b) } I = \int_0^{\frac{\pi}{2}} \frac{\cos x}{\sin x + \cos x}\,dx = \int_0^{\frac{\pi}{2}} \frac{\cos\left(\frac{\pi}{2} - x\right)}{\sin\left(\frac{\pi}{2} - x\right) + \cos\left(\frac{\pi}{2} - x\right)}\,dx = \int_0^{\frac{\pi}{2}} \frac{\sin x}{\sin x + \cos x}\,dx = I \text{ (say)} \quad \text{by property IV.}$$

$$\text{Adding, } 2I = \int_0^{\frac{\pi}{2}} \left(\frac{\cos x}{\sin x + \cos x} + \frac{\sin x}{\sin x + \cos x}\right)dx = \int_0^{\frac{\pi}{2}} \left(\frac{\sin x + \cos x}{\sin x + \cos x}\right)dx = \int_0^{\frac{\pi}{2}} dx = [x]_0^{\frac{\pi}{2}} = \frac{\pi}{2}$$

$$\therefore\ 2I = \frac{\pi}{2} \quad \text{or} \quad I = \frac{\pi}{4} \qquad \text{Ans.}$$

$$\text{(c) } I = \int_0^{\frac{\pi}{2}} \frac{e^{\sin x}}{e^{\sin x} + e^{\cos x}}\,dx = \int_0^{\frac{\pi}{2}} \frac{e^{\sin\left(\frac{\pi}{2} - x\right)}}{e^{\sin\left(\frac{\pi}{2} - x\right)} + e^{\cos\left(\frac{\pi}{2} - x\right)}}\,dx = \int_0^{\frac{\pi}{2}} \frac{e^{\cos x}}{e^{\cos x} + e^{\sin x}}\,dx = I \quad \text{(say)}$$

$$\left\{\text{by property IV. } \int_0^a f(x)\,dx = \int_0^a f(a-x)\,dx\right\}$$

$$\text{Adding, } 2I = \int_0^{\frac{\pi}{2}} \left(\frac{e^{\sin x}}{e^{\sin x} + e^{\cos x}} + \frac{e^{\cos x}}{e^{\cos x} + e^{\sin x}}\right)dx = \int_0^{\frac{\pi}{2}} \left(\frac{e^{\sin x} + e^{\cos x}}{e^{\sin x} + e^{\cos x}}\right)dx = \int_0^{\frac{\pi}{2}} dx = [x]_0^{\frac{\pi}{2}} = \frac{\pi}{2}$$

$$\therefore\ 2I = \frac{\pi}{2} \quad \text{or} \quad I = \frac{\pi}{4} \qquad \text{Ans.}$$

Property V: –
$$\int_{-a}^{a} f(x)\,dx = \begin{cases} 2\int_0^a f(x)\,dx, & \text{if } f(-x) = f(x) \text{ [Even function]} \\ 0, & \text{if } f(-x) = -f(x) \text{ [Odd function]} \end{cases}$$

Proof: – $\int_{-a}^{a} f(x)\,dx = \int_{-a}^{0} f(x)\,dx + \int_0^a f(x)\,dx$, Put $x = -t \quad \therefore\ dx = -dt$

$$\int_{-a}^{a} f(x)\,dx = -\int_a^0 f(-t)\,dt + \int_0^a f(x)\,dx = \int_0^a f(-x)\,dx + \int_0^a f(x)\,dx \quad \text{[by property I and II]}$$

$$\therefore \quad I = \int_{-a}^{a} f(x)\,dx = \begin{cases} 2\int_{0}^{a} f(x)\,dx, & \text{if } f(-x) = f(x) \text{ [Even function]} \\ 0, & \text{if } f(-x) = -f(x) \text{ [Odd function]} \end{cases} \qquad \text{Proved.}$$

Example: – (a) Evaluate $\int_{-\frac{\pi}{3}}^{\frac{\pi}{3}} \sin^3 x\,dx$

Let $f(x) = \sin^3 x$, $f(-x) = [\sin(-x)]^3 = [-\sin x]^3 = -\sin^3 x = -f(x)$

$\therefore$ f(x) is an odd function then $\int_{-\frac{\pi}{3}}^{\frac{\pi}{3}} \sin^3 x\,dx = 0$ Ans. [by property V.]

(b) $I = \int_{-\pi}^{\pi} \cos^2 x\,dx$, Let $f(x) = \cos^2 x$, $f(-x) = \cos^2(-x) = \cos^2 x = f(x)$ $\therefore$ f(x) is an even function

$$I = \int_{-\pi}^{\pi} \cos^2 x\,dx = 2\int_{0}^{\pi} \cos^2 x\,dx = 2\int_{0}^{\pi} \frac{1-\cos 2x}{2}\,dx = 2\int_{0}^{\pi} \frac{1}{2}\,dx - 2\int_{0}^{\pi} \frac{\cos 2x}{2}\,dx \quad \text{[by property V.]}$$

$$I = 2.\frac{1}{2}[x]_0^{\pi} - 2.\frac{1}{2}\left[\frac{\sin 2x}{2}\right]_0^{\pi} = \pi - \frac{1}{2}[0-0] = \pi - 0 = \pi \quad \text{Ans.}$$

(c) $I = \int_{-\frac{\pi}{4}}^{\frac{\pi}{4}} \cos^3 x\,dx$, Let $f(x) = \cos^3 x$, $f(-x) = \cos^3(-x) = \cos^3 x = f(x)$ [$\cos(-\theta) = \cos\theta$]

$\therefore$ f(x) is an even function

$$I = \int_{-\frac{\pi}{4}}^{\frac{\pi}{4}} \cos^3 x\,dx = 2\int_{0}^{\frac{\pi}{4}} \frac{\cos 3x + 3\cos x}{4}\,dx = \frac{1}{2}\int_{0}^{\frac{\pi}{4}} (\cos 3x + 3\cos x)\,dx = \frac{1}{2}\int_{0}^{\frac{\pi}{4}} \cos 3x\,dx + \frac{3}{2}\int_{0}^{\frac{\pi}{4}} \cos x\,dx \quad \text{[by prop. V]}$$

$$I = \frac{1}{2}\left[\frac{\sin 3x}{3}\right]_0^{\frac{\pi}{4}} + \frac{3}{2}[\sin x]_0^{\frac{\pi}{4}} = \frac{1}{6}\left[\sin\left(\frac{3\pi}{4}\right) - \sin 0\right] + \frac{3}{2}\left[\sin\left(\frac{\pi}{4}\right) - \sin 0\right] = \frac{1}{6}.\frac{1}{\sqrt{2}} + \frac{3}{2}.\frac{1}{\sqrt{2}} = \frac{5}{3\sqrt{2}} \quad \text{Ans.}$$

Property VI: – $\int_{a}^{b} f(x)\,dx = \int_{a}^{b} f(a+b-x)\,dx$

Example: – $I = \int_{-\frac{\pi}{4}}^{\frac{\pi}{2}} \cos x\,dx = \int_{-\frac{\pi}{4}}^{\frac{\pi}{2}} \cos\left(-\frac{\pi}{4} + \frac{\pi}{2} - x\right)dx = \int_{-\frac{\pi}{4}}^{\frac{\pi}{2}} \cos\left(\frac{\pi}{4} - x\right)dx$ [$\cos(A-B) = \cos A\cos B + \sin A \sin B$]

$$I = \int_{-\frac{\pi}{4}}^{\frac{\pi}{2}} \left[\cos\left(\frac{\pi}{4}\right)\cos x + \sin\left(\frac{\pi}{4}\right)\sin x\right]dx = \frac{1}{\sqrt{2}}\int_{-\frac{\pi}{4}}^{\frac{\pi}{2}} [\cos x + \sin x]\,dx = \frac{1}{\sqrt{2}}[\sin x - \cos x]_{-\frac{\pi}{4}}^{\frac{\pi}{2}} = \frac{1}{\sqrt{2}}\left[(1-0) - \left(-\frac{1}{\sqrt{2}} - \frac{1}{\sqrt{2}}\right)\right] = \frac{1}{\sqrt{2}}\left[1 + \frac{1}{\sqrt{2}} + \frac{1}{\sqrt{2}}\right]$$

$$= \frac{1}{\sqrt{2}}\left[\frac{\sqrt{2}+1+1}{\sqrt{2}}\right] = \frac{1}{\sqrt{2}}.\frac{\sqrt{2}+2}{\sqrt{2}}$$

$$I = \frac{\sqrt{2}+2}{2} = \frac{\sqrt{2}}{2} + \frac{2}{2} = \frac{\sqrt{2}}{\sqrt{2}\sqrt{2}} + 1 = \frac{1}{\sqrt{2}} + 1 = \frac{1+\sqrt{2}}{\sqrt{2}} \quad \text{Ans.}$$

Property VII: – $\int_{0}^{2a} f(x)\,dx = \begin{cases} 2\int_{0}^{a} f(x)\,dx, & \text{if } f(2a-x) = f(x) \\ 0, & \text{if } f(2a-x) = -f(x) \end{cases}$

Proof: – $\int_{0}^{2a} f(x)\,dx = \int_{0}^{a} f(x)\,dx + \int_{a}^{2a} f(x)\,dx$ ……………………..(A)

Putting, $x = 2a - t$ in last integral $\therefore$ $dx = -dt$ If $x = a$ then $t = 2a - x = 2a - a = a$

and if $x = 2a$ then $t = 2a - x = 2a - 2a = 0$

From equation (A), $\int_{0}^{2a} f(x)\,dx = \int_{0}^{a} f(x)\,dx + \int_{a}^{0} f(2a-t)(-dt) = \int_{0}^{a} f(x)\,dx + \int_{0}^{a} f(2a-t)\,dt$

$$\Rightarrow \int_0^{2a} f(x)\,dx = \int_0^{a} f(x)\,dx + \int_0^{a} f(2a-x)\,dx \quad \text{by property I.}$$

$$\left[\int_a^0 f(x)\,dx = -\int_0^a f(x)\,dx \quad \text{and} \quad \int_a^b f(x)\,dx = \int_a^b f(t)\,dt\right]$$

$$\Rightarrow \int_0^{2a} f(x)\,dx = \begin{cases} \int_0^a f(x)\,dx + \int_0^a f(x)\,dx, & \text{if } f(2a-x) = f(x) \\ \int_0^a f(x)\,dx + \int_0^a -f(x)\,dx, & \text{if } f(2a-x) = -f(x) \end{cases} = \begin{cases} 2\int_0^a f(x)\,dx, & \text{if } f(2a-x) = f(x) \\ 0, & \text{if } f(2a-x) = -f(x) \end{cases} \quad \text{Proved.}$$

Example: – (i) Evaluate: – $I = \int_{-\frac{\pi}{2}}^{\frac{\pi}{2}} \cos^3 x\,dx$

Solution: – $I = \int_{-\frac{\pi}{2}}^{\frac{\pi}{2}} \cos^3 x\,dx$, Let $f(x) = \cos^3 x$, $f(-x) = \cos^3(-x) = \cos^3 x = f(x)$

$\therefore$ f(x) is an even function

$$I = \int_{-\frac{\pi}{2}}^{\frac{\pi}{2}} \cos^3 x\,dx = 2\int_0^{\frac{\pi}{2}} \cos^3 x\,dx = 2\int_0^{\frac{\pi}{2}} \left(\frac{\cos 3x + 3\cos x}{4}\right) dx = \frac{1}{2}\int_0^{\frac{\pi}{2}} (\cos 3x + 3\cos x)\,dx = \frac{1}{2}\left[\frac{\sin 3x}{3} + 3\sin x\right]_0^{\frac{\pi}{2}}$$

$$= \frac{1}{2}\left[\frac{\sin\left(\frac{3\pi}{2}\right)}{3} + 3\sin\left(\frac{\pi}{2}\right) - \frac{\sin 0}{3} - 3\sin 0\right] = \frac{1}{2}\left[-\frac{1}{3} + 3\right] = \frac{1}{2}.\left[\frac{-1+9}{3}\right] = \frac{1}{2}.\frac{8}{3} = \frac{4}{3} \quad \text{Ans.}$$

(ii) Evaluate: – $I = \int_{-\frac{\pi}{4}}^{\frac{\pi}{4}} \sin^3 x\,dx$

Solution: – Let $f(x) = \sin^3 x$, $f(-x) = \sin^3(-x) = -\sin^3 x = -f(x)$

$\therefore$ f(x) is an odd function then $I = \int_{-\frac{\pi}{4}}^{\frac{\pi}{4}} \sin^3 x\,dx = 0$ Ans.

(iii) Evaluate: – $I = \int_{-\frac{\pi}{3}}^{\frac{\pi}{3}} \cos^2 x\,dx$,

Solution: – Let $f(x) = \cos^2 x$, $f(-x) = \cos^2(-x) = \cos^2 x = f(x)$ $\therefore$ f(x) is an even function.

$\left[\cos 2x = \cos^2 x - \sin^2 x,\ \cos 2x = \cos^2 x - 1 + \cos^2 x = 2\cos^2 x - 1,\ 2\cos^2 x = 1 + \cos 2x \text{ or } \cos^2 x = \frac{1+\cos 2x}{2}\right]$

$$I = \int_{-\frac{\pi}{3}}^{\frac{\pi}{3}} \cos^2 x\,dx = 2\int_0^{\frac{\pi}{3}} \cos^2 x\,dx = 2\int_0^{\frac{\pi}{3}} \left(\frac{1+\cos 2x}{2}\right) dx = 2.\frac{1}{2}\int_0^{\frac{\pi}{3}} dx + 2.\frac{1}{2}\int_0^{\frac{\pi}{3}} \cos 2x\,dx = (x)_0^{\frac{\pi}{3}} + \left(\frac{\sin 2x}{2}\right)_0^{\frac{\pi}{3}} = \frac{\pi}{3} - 0 + \frac{\sqrt{3}}{4} - 0 = \frac{\pi}{3} + \frac{\sqrt{3}}{4}$$

$$= \frac{4\pi + 3\sqrt{3}}{12} \quad \text{Ans.}$$

(iv) Evaluate: – $I = \int_{-\frac{\pi}{6}}^{\frac{\pi}{6}} \sin^2 x\,dx$

Solution: – Let $f(x) = \sin^2 x$, $f(-x) = \sin^2(-x) = (-\sin x)^2 = \sin^2 x = f(x)$ $\therefore$ f(x) is an even function.

$\left[\cos 2x = \cos^2 x - \sin^2 x,\ \cos 2x = 1 - \sin^2 x - \sin^2 x = 1 - 2\sin^2 x,\ 2\sin^2 x = 1 - \cos 2x \text{ or } \sin^2 x = \frac{1-\cos 2x}{2}\right]$

$$I = \int_{-\frac{\pi}{6}}^{\frac{\pi}{6}} \sin^2 x\,dx = 2\int_0^{\frac{\pi}{6}} \sin^2 x\,dx = 2\int_0^{\frac{\pi}{6}} \left(\frac{1-\cos 2x}{2}\right) dx = \int_0^{\frac{\pi}{6}} dx - \int_0^{\frac{\pi}{6}} \cos 2x\,dx = (x)_0^{\frac{\pi}{6}} - \left(\frac{\sin 2x}{2}\right)_0^{\frac{\pi}{6}} = \frac{\pi}{6} - \frac{\sqrt{3}}{4} = \frac{2\pi - 3\sqrt{3}}{12} \quad \text{Ans.}$$

(v) Evaluate: – $I = \int_0^{\pi} \cos^2 x \, dx$

Solution: – Let $f(x) = \cos^2 x$, $f(\pi - x) = \cos^2(\pi - x) = [\cos(\pi - x)]^2 = \cos^2 x = f(x)$ [use property VII.]

$$\int_0^{\pi} \cos^2 x \, dx = 2\int_0^{\frac{\pi}{2}} \cos^2 x \, dx = 2\int_0^{\frac{\pi}{2}} \left(\frac{1+\cos 2x}{2}\right) dx = 2.\frac{1}{2}\int_0^{\frac{\pi}{2}} dx + 2.\frac{1}{2}\int_0^{\frac{\pi}{2}} \cos 2x \, dx \quad \left[\therefore 2a = \pi \quad \therefore a = \frac{\pi}{2}\right]$$

$$I = (x)_0^{\frac{\pi}{2}} + \left(\frac{\sin 2x}{2}\right)_0^{\frac{\pi}{2}} = \frac{\pi}{2} - 0 + 0 - 0 = \frac{\pi}{2} \quad \text{Ans.}$$

(vi) Evaluate: – $I = \int_{-\pi}^{\pi} \sin^4 x \, dx$

Solution: – Let $f(x) = \sin^4 x$, $f(-x) = [\sin(-x)]^4 = [-\sin x]^4 = \sin^4 x = f(x)$

$\therefore$ f(x) is an even function [use property V.]

$$I = \int_{-\pi}^{\pi} \sin^4 x \, dx = 2\int_0^{\pi} \sin^4 x \, dx = 2\int_0^{\pi} [\sin^2 x]^2 \, dx = 2\int_0^{\pi} \left[\frac{1-\cos 2x}{2}\right]^2 dx = \frac{2}{4}\int_0^{\pi} [1 - 2\cos 2x + \cos^2 2x] \, dx$$

$$= \frac{1}{2}\int_0^{\pi} dx - \int_0^{\pi} \cos 2x \, dx + \frac{1}{2}\int_0^{\pi} \cos^2 2x \, dx = \frac{1}{2}\int_0^{\pi} dx - \int_0^{\pi} \cos 2x \, dx + \frac{1}{2}\int_0^{\pi} \left(\frac{1+\cos 4x}{2}\right) dx$$

$$I = \frac{1}{2}\int_0^{\pi} dx - \int_0^{\pi} \cos 2x \, dx + \frac{1}{4}\int_0^{\pi} dx + \frac{1}{4}\int_0^{\pi} \cos 4x \, dx = \frac{1}{2}(x)_0^{\pi} - \left(\frac{\sin 2x}{2}\right)_0^{\pi} + \frac{1}{4}(x)_0^{\pi} + \frac{1}{4}\left(\frac{\sin 4x}{4}\right)_0^{\pi}$$

$$= \frac{1}{2}(\pi - 0) - \frac{1}{2}(0 - 0) + \frac{1}{4}(\pi - 0) + \frac{1}{16}(0 - 0) = \frac{\pi}{2} + \frac{\pi}{4} = \frac{2\pi + \pi}{4} = \frac{3\pi}{4} \quad \text{Ans.}$$

(vii) Evaluate: – $I = \int_0^{\pi} \sin^5 x \, dx$

Solution: – Let $f(x) = \sin^5 x$, $f(\pi - x) = [\sin(\pi - x)]^5 = \sin^5 x = f(x)$ [use property VII.]

$$I = \int_0^{\pi} \sin^5 x \, dx = 2\int_0^{\frac{\pi}{2}} \sin^5 x \, dx = 2\int_0^{\frac{\pi}{2}} \sin^4 x . \sin x \, dx = 2\int_0^{\frac{\pi}{2}} (\sin^2 x)^2 . \sin x \, dx = 2\int_0^{\frac{\pi}{2}} (1 - \cos^2 x)^2 . \sin x \, dx$$

Put $\cos x = t \; \therefore \; -\sin x \, dx = dt$ when $x = \frac{\pi}{2}$, $t = 0$ and $x = 0$, $t = 1$ upper and lower limit is $t = (0,1)$

$$I = -2\int_1^0 [1 - t^2]^2 \, dt = 2\int_0^1 (1 - 2t^2 + t^4) \, dt = 2\int_0^1 dt - 4\int_0^1 t^2 \, dt + 2\int_0^1 t^4 \, dt = 2(t)_0^1 - 4\left(\frac{t^3}{3}\right)_0^1 + 2\left(\frac{t^5}{5}\right)_0^1$$

$$= 2(1 - 0) - \frac{4}{3}(1 - 0) + \frac{2}{5}(1 - 0) = 2 - \frac{4}{3} + \frac{2}{5} = \frac{30 - 20 + 6}{15} = \frac{16}{15} \quad \text{Ans.}$$

(viii) Evaluate: – $\int_{-\frac{\pi}{2}}^{\frac{\pi}{2}} \frac{\cos^2 x}{1 - a^x} dx$

Solution: – Let $I = \int_{-\frac{\pi}{2}}^{\frac{\pi}{2}} \frac{\cos^2 x}{1 - a^x} dx$ ……………….(A)

$$I = \int_{-\frac{\pi}{2}}^{\frac{\pi}{2}} \frac{\cos^2\left(-\frac{\pi}{2} + \frac{\pi}{2} - x\right)}{1 - a^{\left(-\frac{\pi}{2}+\frac{\pi}{2}-x\right)}} dx = \int_{-\frac{\pi}{2}}^{\frac{\pi}{2}} \frac{\cos^2(-x)}{1 - a^{-x}} dx = \int_{-\frac{\pi}{2}}^{\frac{\pi}{2}} \frac{\cos^2(-x)}{\frac{a^x - 1}{a^x}} dx = \int_{-\frac{\pi}{2}}^{\frac{\pi}{2}} \frac{a^x \cos^2 x}{a^x - 1} dx \text{ ……………….(B)}$$

Adding equation (A) and (B), we get

$$2I = \int_{-\frac{\pi}{2}}^{\frac{\pi}{2}} \left[\frac{\cos^2 x}{1 - a^x} + \frac{a^x \cos^2 x}{a^x - 1}\right] dx = \int_{-\frac{\pi}{2}}^{\frac{\pi}{2}} \left[\frac{\cos^2 x}{1 - a^x} - \frac{a^x \cos^2 x}{1 - a^x}\right] dx = \int_{-\frac{\pi}{2}}^{\frac{\pi}{2}} \left[\frac{\cos^2 x - a^x \cos^2 x}{1 - a^x}\right] dx = \int_{-\frac{\pi}{2}}^{\frac{\pi}{2}} \cos^2 x . \left[\frac{1 - a^x}{1 - a^x}\right] dx = \int_{-\frac{\pi}{2}}^{\frac{\pi}{2}} \cos^2 x \, dx$$

$$= \int_{-\frac{\pi}{2}}^{\frac{\pi}{2}} \left(\frac{1 + \cos 2x}{2}\right) dx = \frac{1}{2}\int_{-\frac{\pi}{2}}^{\frac{\pi}{2}} dx + \frac{1}{2}\int_{-\frac{\pi}{2}}^{\frac{\pi}{2}} \cos 2x \, dx = \frac{1}{2}(x)_{-\frac{\pi}{2}}^{\frac{\pi}{2}} + \frac{1}{2}\left(\frac{\sin 2x}{2}\right)_{-\frac{\pi}{2}}^{\frac{\pi}{2}}$$

$$2I=\frac{1}{2}\left(\frac{\pi}{2}-\frac{-\pi}{2}\right)+\frac{1}{4}(\sin\pi-\sin(-\pi))=\frac{1}{2}\left(\frac{\pi}{2}+\frac{\pi}{2}\right)+\frac{1}{4}(0+0)=\frac{1}{2}\cdot\frac{2\pi}{2}=\frac{\pi}{2} \quad \text{or} \quad \boxed{I=\frac{\pi}{4} \quad \text{Ans.}}$$

(ix) Evaluate: – $\displaystyle\int_{-\pi}^{\pi}\frac{\sin^2 x}{1+a^x}\,dx$

Solution: – Let $\displaystyle I=\int_{-\pi}^{\pi}\frac{\sin^2 x}{1+a^x}\,dx \ldots\ldots\ldots\ldots (A)$

$$I=\int_{-\pi}^{\pi}\frac{\sin^2(-\pi+\pi-x)}{1+a^{(-\pi+\pi-x)}}\,dx=\int_{-\pi}^{\pi}\frac{\sin^2(-x)}{1+a^{(-x)}}\,dx=\int_{-\pi}^{\pi}\frac{\sin^2 x}{1+\frac{1}{a^x}}\,dx=\int_{-\pi}^{\pi}\frac{\sin^2 x}{\frac{1+a^x}{a^x}}\,dx=\int_{-\pi}^{\pi}\frac{a^x\sin^2 x}{1+a^x}\,dx \ldots\ldots\ldots\ldots (B)$$

Adding equation (A) and (B), we get

$$2I=\int_{-\pi}^{\pi}\left(\frac{\sin^2 x}{1+a^x}+\frac{a^x\sin^2 x}{1+a^x}\right)dx=\int_{-\pi}^{\pi}\left(\frac{\sin^2 x+a^x\sin^2 x}{1+a^x}\right)dx=\int_{-\pi}^{\pi}\sin^2 x.\left(\frac{1+a^x}{1+a^x}\right)dx=\int_{-\pi}^{\pi}\sin^2 x\,dx=\int_{-\pi}^{\pi}\left(\frac{1-\cos 2x}{2}\right)dx$$

$$=\frac{1}{2}\int_{-\pi}^{\pi}dx-\frac{1}{2}\int_{-\pi}^{\pi}\cos 2x\,dx=\frac{1}{2}(x)_{-\pi}^{\pi}-\frac{1}{2}\left(\frac{\sin 2x}{2}\right)_{-\pi}^{\pi}=\frac{1}{2}[\pi+\pi]-\frac{1}{4}[0-0]$$

$$2I=\frac{1}{2}.2\pi=\pi \qquad \therefore\ I=\frac{\pi}{2} \qquad \text{Ans.}$$

(x) Evaluate: – $\displaystyle\int_{-\frac{\pi}{4}}^{\frac{\pi}{4}}(\cos^4 x+\sin^2 x)\,dx$

Solution: – Let $\displaystyle I=\int_{-\frac{\pi}{4}}^{\frac{\pi}{4}}(\cos^4 x+\sin^2 x)\,dx=\int_{-\frac{\pi}{4}}^{\frac{\pi}{4}}\cos^4 x\,dx+\int_{-\frac{\pi}{4}}^{\frac{\pi}{4}}\sin^2 x\,dx=I_1+I_2$ (say)

$\displaystyle I_1=\int_{-\frac{\pi}{4}}^{\frac{\pi}{4}}\cos^4 x\,dx=\frac{3\pi+8}{16}$ (solve do yourself) and $\displaystyle I_2=\int_{-\frac{\pi}{4}}^{\frac{\pi}{4}}\sin^2 x\,dx=\frac{\pi-2}{4}$ (solve do yourself)

$$I=I_1+I_2=\frac{3\pi+8}{16}+\frac{\pi-2}{4}=\frac{7\pi}{16} \qquad \text{Ans.}$$

Property VIII: – If f(x) is a periodic function with period T then $\displaystyle\int_a^{a+nT}f(x)\,dx=n\int_0^T f(x)\,dx$

If f(x) is a periodic function with period T then, $f(x+nT)=f(x)$

Proof: – $f(x+T)=f(x)$, Replace x by $x+T$ $\Rightarrow f(x+2T)=f(x+T)=f(x)$

$$f(x+3T)=f(x+2T)=f(x)$$

………………………………………

………………………………………

$$f(x+nT)=f(x) \qquad \text{Proved.}$$

Again, Proof: – $\displaystyle\int_a^{a+nT}f(x)\,dx=\int_a^{nT}f(x)\,dx+\int_{nT}^{a+nT}f(x)\,dx$, Putting $x=nT+y$ in last integral.

$\therefore\ dx=dy$ if $x=nT+y$ when $x=nT$, $y=0$ and $x=a+nT$, $y=a$

$$\int_a^{a+nT}f(x)\,dx=\int_a^{nT}f(x)\,dx+\int_0^a f(nT+y)\,dy=\int_0^a f(nT+x)\,dx+\int_a^{nT}f(x)\,dx=\int_0^a f(x)\,dx+\int_a^{nT}f(x)\,dx \ \text{ by prop. I}$$

$$\int_a^{a+nT}f(x)\,dx=\int_0^{nT}f(x)\,dx=\int_0^{T}f(x)\,dx+\int_T^{2T}f(x)\,dx+\int_{2T}^{3T}f(x)\,dx+\cdots\ldots\ldots\ldots+\int_{(n-1)T}^{nT}f(x)\,dx$$

$$\therefore\ \int_a^{a+nT}f(x)\,dx=I_1+I_2+I_3+\cdots\ldots\ldots\ldots\ldots\ldots+I_n \quad \ldots\ldots\ldots\ldots\ldots\ldots (A)$$

where $I_1 = \int_0^T f(x)\,dx$ and $I_2 = \int_T^{2T} f(x)\,dx$, putting $x = T + y \ \therefore\ dx = dy$

when $x = T,\ y = 0$ and $x = 2T,\ y = T$

$$I_2 = \int_T^{2T} f(x)\,dx = \int_0^T f(T+y)\,dy = \int_0^T f(T+x)\,dx = \int_0^T f(x)\,dx = I_1 \quad \text{by property I.}$$

$$I_3 = \int_{2T}^{3T} f(x)\,dx = I_1 \quad \text{and} \quad I_4 = \int_{3T}^{4T} f(x)\,dx = I_1, \ldots\ldots\ldots \text{and } I_n = \int_{(n-1)T}^{nT} f(x)\,dx = I_1$$

From equation (A), $\Rightarrow \int_a^{a+nT} f(x)\,dx = nI_1 = n\int_0^T f(x)\,dx$ Proved.

Particular case: – (i) $\int_0^{nT} f(x)\,dx = n\int_0^T f(x)\,dx$ (ii) $\int_a^{a+T} f(x)\,dx = \int_0^T f(x)\,dx$

Property IX: – If f(x) is a periodic function with period T then, $\int_{mT}^{nT} f(x)\,dx = (n-m)\int_0^T f(x)\,dx$

Proof: – $L.H.S = \int_{mT}^{nT} f(x)\,dx = \int_{mT}^{0} f(x)\,dx + \int_0^{nT} f(x)\,dx = -\int_0^{mT} f(x)\,dx + \int_0^{nT} f(x)\,dx = -m\int_0^T f(x)\,dx + n\int_0^T f(x)\,dx$

$= (n-m)\int_0^T f(x)\,dx$ Proved.

Property X: – If f(x) is a periodic function with period T then, $\int_{a+nT}^{b+nT} f(x)\,dx = \int_a^b f(x)\,dx$

Proof: – $L.H.S = \int_{a+nT}^{b+nT} f(x)\,dx = \int_{a+nT}^{a} f(x)\,dx + \int_a^b f(x)\,dx + \int_b^{b+nT} f(x)\,dx = -\int_a^{a+nT} f(x)\,dx + \int_a^b f(x)\,dx + \int_b^{b+nT} f(x)\,dx$

$= -n\int_0^T f(x)\,dx + \int_a^b f(x)\,dx + n\int_0^T f(x)\,dx = \int_a^b f(x)\,dx$ Proved. (by property VIII.)

Example: – (a) $\int_0^{200\pi} \sqrt{1+\cos 2x}\,dx$

Solution: – Let $I = \int_0^{200\pi} \sqrt{1+\cos 2x}\,dx = \int_0^{200\pi} \sqrt{2\cos^2 x}\,dx = \int_0^{200\pi} \sqrt{2}|\cos x|\,dx = 200.\sqrt{2}\int_0^{\pi} \cos x\,dx = 200.\sqrt{2}(\sin x)_0^{\pi}$

$= 200.\sqrt{2}(0-0) = 0$ Ans. (b) $\int_0^{101} e^{x-[x]}\,dx$

Solution: – Let $I = \int_0^{101} e^{x-[x]}\,dx = 101\int_0^1 e^{x-[x]}\,dx = 101\int_0^1 e^x\,dx = 101(x)_0^1 = 101(e^1 - e^0) = 101(e-1)$ Ans.

(c) $\int_0^{50\pi} (\sin x + \cos x)\,dx$

Solution: – Let $I = \int_0^{50\pi} (\sin x + \cos x)\,dx = 50\int_0^{\pi} (\sin x + \cos x)\,dx = 50\int_0^{\pi} \sin x\,dx + 50\int_0^{\pi} \cos x\,dx = 50(-\cos x)_0^{\pi} + 50(\sin x)_0^{\pi}$

$= -50(-1-1) + 50.0 = 100$ Ans.

(d) show that $\int_0^{n\pi+v} |\cos x|\,dx = \sin v$ where n is a positive integer and $0 \le v < \pi$.

Proof: – $L.H.S = \int_0^{n\pi+v} |\cos x|\,dx = \int_0^{v} |\cos x|\,dx + \int_v^{n\pi+v} |\cos x|\,dx = \int_0^{v} |\cos x|\,dx + \int_0^{n\pi} |\cos x|\,dx = \int_0^{v} |\cos x|\,dx + n\int_0^{v} |\cos x|\,dx$

$= (\sin x)_0^{v} + n(\sin x)_0^{\pi} = \sin v$ Proved.

(e) Let $T > 0$ *be a fixed number suppose* f *is a continuous function for all* $x \in R, f(x+T) = f(x)$.

if $I=\int_0^T f(x)\,dx$ then the value of $\int_4^{4+4T} f(3x)\,dx$.

Solution: $-\quad \int_4^{4+4T} f(3x)\,dx=\int_4^4 f(3x)\,dx+\int_4^{4+4T} f(3x)\,dx=\int_4^4 f(3x)\,dx+\int_4^{4+12.\frac{T}{3}} f(3x)\,dx=12\int_0^{\frac{T}{3}} f(3x)\,dx$

Putting, $3x=y\quad \therefore\ dx=\frac{dy}{3}$ when $x=0,\ y=0$ and $x=\frac{T}{3},\ y=T$

$$\Rightarrow\quad \int_4^{4+4T} f(3x)\,dx=\frac{12}{3}\int_0^T f(y)\,dy=4\int_0^T f(x)\,dx=4I\qquad \text{Ans.}$$

(f) If f(x) is an odd function in $\left[-\frac{T}{2},\frac{T}{2}\right]$ and has period equal to T. Prove that $\int_a^x f(y)\,dy$ is also periodic function with period T.

Solution: $-\quad \int_{-\frac{T}{2}}^{\frac{T}{2}} f(x)\,dx=0,\ f(x+T)=f(x)$ Let $g(x)=\int_a^x f(y)\,dy$ to prove $g(x+T)=g(x)$

or $\int_a^{x+T} f(y)\,dy=\int_a^x f(y)\,dy$ or $\int_a^{x+T} f(y)\,dy-\int_a^x f(y)\,dy=0$

$$\text{L.H.S}=\int_a^{x+T} f(y)\,dy-\int_a^x f(y)\,dy=\int_a^x f(y)\,dy+\int_x^{x+T} f(y)\,dy-\int_a^x f(y)\,dy=\int_x^{x+T} f(y)\,dy=\int_x^{-\frac{T}{2}} f(y)\,dy+\int_{-\frac{T}{2}}^{\frac{T}{2}} f(y)\,dy+\int_{\frac{T}{2}}^{x+T} f(y)\,dy$$

$$=\int_x^{-\frac{T}{2}} f(y)\,dy+\int_{\frac{T}{2}}^{x+T} f(y)\,dy$$

Putting, $y=z+T$ in last integral $\quad\therefore\ dy=dz$

y	$\frac{T}{2}$	$x+T$
z	$-\frac{T}{2}$	x

$$\int_a^{x+T} f(y)\,dy=\int_x^{-\frac{T}{2}} f(y)\,dy+\int_{-\frac{T}{2}}^{x} f(z+T)\,dz=\int_x^{-\frac{T}{2}} f(y)\,dy+\int_{-\frac{T}{2}}^{x} f(z)\,dz=\int_x^{-\frac{T}{2}} f(z)\,dz+\int_{-\frac{T}{2}}^{x} f(z)\,dz=-\int_{-\frac{T}{2}}^{x} f(z)\,dz+\int_{-\frac{T}{2}}^{x} f(z)\,dz=0$$

$=\text{R.H.S}$ Proved.

IInd Method: $-\quad \text{L.H.S}=\int_a^{x+T} f(y)\,dy-\int_a^x f(y)\,dy=\int_a^x f(y)\,dy+\int_x^{x+T} f(y)\,dy-\int_a^x f(y)\,dy=\int_x^{x+T} f(y)\,dy$

$\left\{\because\ \int_x^{x+T} f(y)\,dy \text{ is independent of } x\right\},\ \int_x^{x+T} f(y)\,dy=\int_{-\frac{T}{2}}^{\frac{T}{2}} f(y)\,dy=0=\text{R.H.S}$ Proved.

Property XI: $-$ If $I(t)=\int_a^b f(x,t)\,dx$ then, $\frac{d[I(t)]}{dt}=\int_a^b \frac{\partial[f(x,t)]}{\partial t}.dx$

Example: $-$ (a) Evaluate: $-\quad \int_0^1 \frac{x^b+1}{\log x}\,dx\quad (b\ge 0)$

Solution: $-$ Let $I(b)=\int_0^1 \frac{x^b+1}{\log x}\,dx$, Differentiating both of sides with respect to b, we get

$$\therefore\ \frac{d[I(b)]}{db}=\int_0^1 \frac{\partial}{\partial b}\left(\frac{x^b+1}{\log x}\right).dx=\int_0^1\left[\frac{1}{\log x}.x^b.\log x\right]dx=\int_0^1 x^b\,dx=\left(\frac{x^{b+1}}{b+1}\right)_0^1=\frac{1}{b+1}-0=\frac{1}{b+1}$$

$\therefore\ I'(b)=\frac{1}{b+1}$, Integrating $\int I'(b)\,db=\int\frac{db}{b+1}\quad \therefore\ I(b)=\log|1+b|+c$ ……… (i)

Putting, $b = 0$, $I(0) = c$ $\therefore$ $c = 0$ from equation (i) $\Rightarrow I(b) = \log|1 + b| + 0 = \log|1 + b|$

$$\therefore \int_0^1 \frac{x^b + 1}{\log x} dx = \log|1 + b| \quad \text{Ans.}$$

(b) $\int_{\frac{\pi}{6}}^{\frac{\pi}{3}} (\tan^2\theta + b^2\sec^2\theta)\, d\theta$

Solution: – Let $I(b) = \int_{\frac{\pi}{6}}^{\frac{\pi}{3}} (\tan^2\theta + b^2\sec^2\theta)\, d\theta$

Differentiating both sides w. r. t b, we get

$$\frac{d[I(b)]}{db} = \int_{\frac{\pi}{6}}^{\frac{\pi}{3}} \frac{\partial}{\partial b}(\tan^2\theta + b^2\sec^2\theta)\, d\theta = \int_{\frac{\pi}{6}}^{\frac{\pi}{3}} 2b\sec^2\theta\, d\theta = 2b\int_{\frac{\pi}{6}}^{\frac{\pi}{3}} \sec^2\theta\, d\theta = 2b(\tan\theta)_{\frac{\pi}{6}}^{\frac{\pi}{3}} = 2b\left(\tan\frac{\pi}{3} - \tan\frac{\pi}{6}\right)$$

$$I'(b) = 2b\left(\sqrt{3} - \frac{1}{\sqrt{3}}\right) = \frac{2b(3-1)}{\sqrt{3}} = \frac{4b}{\sqrt{3}}$$

Integrating, $\int I'(b)\, db = \int \frac{4b}{\sqrt{3}} db = \frac{4}{\sqrt{3}}\int b\, db = \frac{4}{\sqrt{3}}.\frac{b^2}{2} + c = \frac{2b^2}{\sqrt{3}} + c$

$$\therefore I(b) = \frac{2b^2}{\sqrt{3}} + c \ \ldots\ldots\ldots\ldots\ldots \text{(i)}$$

Put $b = 0$, $c = \int_{\frac{\pi}{6}}^{\frac{\pi}{3}} \tan^2\theta\, d\theta = \int_{\frac{\pi}{6}}^{\frac{\pi}{3}} (\sec^2\theta - 1)\, d\theta = \int_{\frac{\pi}{6}}^{\frac{\pi}{3}} \sec^2\theta\, d\theta - \int_{\frac{\pi}{6}}^{\frac{\pi}{3}} d\theta = (\tan\theta)_{\frac{\pi}{6}}^{\frac{\pi}{3}} - (\theta)_{\frac{\pi}{6}}^{\frac{\pi}{3}} = \left(\tan\frac{\pi}{3} - \tan\frac{\pi}{6}\right) - \left(\frac{\pi}{3} - \frac{\pi}{6}\right) = \sqrt{3} - \frac{1}{\sqrt{3}} - \frac{\pi}{6}$

$= \frac{2}{\sqrt{3}} - \frac{\pi}{6}$

Put the value of c in equation (i), we get $\therefore I(b) = \frac{2b^2}{\sqrt{3}} + c = \frac{2b^2}{\sqrt{3}} + \frac{2}{\sqrt{3}} - \frac{\pi}{6}$ Ans.

<u>Exercise – A10</u>

(1) Evaluate the following definite integral: – (a) $I = \int_1^{\sqrt{3}} \frac{x^3}{\sqrt{x^4 - 1}} dx$ (b) $I = \int_0^{\frac{\pi}{6}} \cos 3x\, dx$

(c) $I = \int_0^2 \frac{3}{x+2} dx$ (d) $I = \int_1^e \frac{3}{2\sqrt{1+x}} dx$ (e) $I = \int_0^{\frac{\pi}{2}} \tan^{-1}\left(\frac{x}{\sqrt{1-x^2}}\right) dx$ (f) $I = \int_0^1 \tan^{-1}\left(\frac{x-1}{x+1}\right) dx$

(2) Find the following integral: – (a) $I = \int_0^{\frac{\pi}{2}} \log(\sin x)\, dx$ (b) $I = \int_0^{\frac{\pi}{2}} \log(\cos x)\, dx$ (c) $I = \int_0^{\frac{\pi}{2}} \log(\tan x)\, dx$

(d) $I = \int_0^{\frac{\pi}{2}} \log(\cot x)\, dx$ (e) $I = \int_0^{\frac{\pi}{2}} \log(\sec x)\, dx$ (f) $I = \int_0^{\frac{\pi}{2}} \log(\text{cosec}\, x)\, dx$ (g) $I = \int_0^{\frac{\pi}{2}} \log x\, dx$

(h) $I = \int_0^{\frac{\pi}{2}} (\sin x + \cos x)\, dx$ (i) $I = \int_0^{\frac{\pi}{4}} (\tan x + \cot x)\, dx$

(3) Evaluate: – (a) $I = \int_0^1 \frac{x^2 + 3x + 2}{(x+1)} dx$ (b) $I = \int_1^2 \frac{\log x}{x} dx$ (c) $I = \int_0^{\pi} \log(1 + \cos x).\sin x\, dx$

(d) $I = \int_0^1 \sin^{-1} x\, dx$ (e) $I = \int_0^1 \cos^{-1} x\, dx$ (f) $I = \int_0^1 \tan^{-1} x\, dx$ (g) $I = \int_0^1 \cot^{-1} x\, dx$

(h) $I = \int_0^1 \sec^{-1} x\, dx$ (i) $I = \int_0^1 \text{cosec}^{-1} x\, dx$

(4) Evaluate: – (a) $I = \int_0^{2\pi} 2^{3\log_2 x}\, dx$ (b) $I = \int_0^1 \cos^{-1}\left(\frac{1+\cos x}{2\cos^{x}/_{2}}\right) dx$ (c) $I = \int_1^2 x^2 \log x\, dx$

(d) $I = \int_0^1 \frac{1}{\sqrt{x+1}-\sqrt{x+2}}\, dx$ (e) $I = \int_0^5 |x-3|\, dx$ (f) $I = \int_0^{\frac{\pi}{2}} \sin^5 x\, dx$

(g) $I = \int_1^2 \frac{1}{x\left(\log x + \frac{1}{\log x}\right)}\, dx$ (h) $I = \int_{-\frac{\pi}{2}}^{\frac{\pi}{2}} e^x . \sin x\, dx$

(5) Evaluate: – (a) $I = \int_{-\frac{\pi}{2}}^{\frac{\pi}{2}} \frac{\cos x}{2+\sin x}\, dx$ (b) $I = \int_{\frac{\pi}{4}}^{\frac{\pi}{2}} \sin(2x+3)\, dx$ (c) $I = \int_1^3 (5x+2)^5\, dx$

(d) $I = \int_1^5 \frac{\sqrt{6-x}}{\sqrt{x}+\sqrt{6-x}}\, dx$ (e) $I = \int_0^{\frac{\pi}{2}} \frac{\cos x}{\sqrt{\sin^2 x + 2\sin x + 3}}\, dx$ (f) $I = \int_0^{\frac{\pi}{4}} \tan^2(x+1)\, dx$

(g) $I = \int_0^{\frac{\pi}{2}} \frac{e^{\sqrt{\cos x}}}{e^{\sqrt{\sin x}} + e^{\sqrt{\cos x}}}\, dx$ (h) $I = \int_0^{\frac{\pi}{2}} \sin^2 x . \cos^3 x\, dx$

(6) Evaluate: – (a) $I = \int_0^1 \frac{dx}{x^2+x+1}$ (b) $I = \int_0^2 |x^2+x|\, dx$ (c) $I = \int_0^2 \frac{dx}{(x-1).\sqrt{x+2}}$

(d) $I = \int_1^e \frac{\log x}{x}\, dx$ (e) $I = \int_6^7 \frac{dx}{(x+3).\sqrt{x-5}}$ (f) $I = \int_2^3 \frac{dx}{(x+2)(x+3)}$

(g) $I = \int_{-1}^6 \frac{dx}{(x-1)(x-5)}$ (h) $I = \int_1^3 \frac{dx}{(x+1)\sqrt{x^2+2x+3}}$

(7) Evaluate: – (a) $I = \int_0^1 \frac{2x^2-5x+1}{\sqrt{x}}\, dx$ (b) $I = \int_1^2 \frac{3x^3+4x^2-x-2}{(3x-2)}\, dx$

(c) $I = \int_0^{\frac{\pi}{2}} \frac{\sin 2x}{2-\cos^2 x}\, dx$ (d) $I = \int_0^{\log 2} e^{2x}\, dx$ (e) $I = \int_{\log 2}^{\log 3} (e^x - 1)\, dx$ (f) $I = \int_e^{e^2} |x-4|\, dx$

(g) $I = \int_{-2}^2 \frac{e^x}{e^x+e^{-x}}\, dx$ (h) $I = \int_0^{\frac{\pi}{2}} \frac{\sin x}{\sin x + \cos x}\, dx$ (i) $I = \int_1^{\sin^2\theta} \frac{\sin^{-1}\sqrt{x}}{\sqrt{1-x}}\, dx$

(j) $I = \int_0^{\frac{\pi}{2}} \frac{dx}{1+\cos x}$ (k) $I = \int_0^{\frac{\pi}{2}} \frac{dx}{1+\tan x}$ (l) $I = \int_0^1 \frac{\sqrt{x}+\sqrt[3]{x}}{\sqrt[4]{x}}\, dx$ (m) $I = \int_0^{\frac{\pi}{2}} \frac{\sqrt{\cos x}}{\sqrt{\sin x}+\sqrt{\cos x}}\, dx$

(8) Evaluate: – (a) $I = \int_1^2 \frac{\sin(1+\log x)}{x}\, dx$ (b) $I = \int_{-\frac{1}{3}}^{\frac{1}{3}} e^{\cos x} . \sin x\, dx$

(c) $I = \int_{-\frac{\pi}{4}}^{\frac{\pi}{4}} e^{\sec x} . \sec x \tan x\, dx$ (d) $I = \int_{-5}^5 \frac{3x^5+2x^4+5x^3+4x^2+2x+1}{(x^2+2)}\, dx$

(e) $I = \int_{-3}^3 \frac{5x^3+x}{x^2+1}\, dx$ (f) $I = \int_{-1}^1 \frac{x^3 \cos x}{x^2+1}\, dx$ (g) $I = \int_{\frac{\pi}{4}}^{\frac{3\pi}{4}} \frac{dx}{1-\cos x}$

(9) Evaluate the following integral: – (a) $I = \int_0^{\frac{\pi}{2}} \frac{\sqrt{\sin x}}{\sqrt{\sin x}+\sqrt{\cos x}}\, dx$ (b) $I = \int_0^1 \frac{e^{1-x}}{e^x+e^{1-x}}\, dx$

(c) $I = \int_0^{\pi} \frac{e^{\cos x}}{e^{\cos x}+e^{-\cos x}}\, dx$ (d) $I = \int_0^{\frac{\pi}{2}} \frac{dx}{1-\tan^2 x}$

(10) Evaluate: – (a) $I = \int_{\frac{\pi}{6}}^{\frac{\pi}{3}} \frac{dx}{1+\sqrt{\tan x}}$ (b) $I = \int_{\sqrt{\log 2}}^{\sqrt{\log 3}} \frac{x \sin x^2}{\sin x^2 + \sin(\log 6 - x^2)}\, dx$

(c) $I = \int_0^1 \frac{x^4(1-x)^4}{1+x^2}\,dx$ (d) $I = \int_{e^{-1}}^{e^2} \left|\frac{\log_e x}{x}\right| dx$

(11) Evaluate: – (a) $\int_0^{\pi} [2\sin x]\,dx$ (b) $\int_0^{102} [\tan^{-1} x]\,dx$

(c) $\int_0^{2n\pi} [\sin x + \cos x]\,dx$ (d) $\int_0^{\frac{5\pi}{12}} [\tan x]\,dx$

(e) $\int_0^{n^2} [\sqrt{x}]\,dx,\ n \in N$ where [.] denotes the greatest integer functions (G. I. F).

(f) Prove that $\int_0^x [t]\,dt = \frac{[x]([x]-1)}{2} + [x](x-[x])$ where [.] denotes the G. I. F.

Answer

(1) (a) $I = \int_1^{\sqrt{3}} \frac{x^3}{\sqrt{x^4-1}}dx = \frac{\sqrt{8}}{2} = \frac{2\sqrt{2}}{2} = \sqrt{2}$ or 1.414 Ans. [Do yourself, Put $x^4 - 1 = t^2$]

(b) $I = \int_0^{\frac{\pi}{6}} \cos 3x\,dx = \left(\frac{\sin 3x}{3}\right)_0^{\frac{\pi}{6}} = \frac{\sin 3.\frac{\pi}{6}}{3} = \frac{\sin\frac{\pi}{2}}{3} = \frac{1}{3}$ or 0.33 or 0.334 Ans.

(c) $I = \int_0^2 \frac{3}{x+2}dx = 3\int_0^2 \frac{dx}{x+2} = 3[\log(x+2)]_0^2 = 3[\log 4 - \log 2] = 3[\log 2^2 - \log 2] = 3[2\log 2 - \log 2] = 3\log 2$ Ans.

(d) $I = \int_1^e \frac{3}{2\sqrt{1+x}}dx = 3\left(\sqrt{e+1} - \sqrt{2}\right)$ Ans. (e) $I = \int_0^{\frac{\pi}{2}} \tan^{-1}\left(\frac{x}{\sqrt{1-x^2}}\right)dx = \frac{\pi}{2} - 1$ Ans.

(Do yourself, put $x = \sin\theta$ and use integration by part formula Q. No. –(e))

(f) $I = \int_0^1 \tan^{-1}\left(\frac{x-1}{x+1}\right)dx = \int_0^1 \tan^{-1}\left(\frac{x-1}{1+x.1}\right)dx = \int_0^1 (\tan^{-1}x - \tan^{-1}1)\,dx = \int_0^1 \tan^{-1}x\,dx - \int_0^1 \tan^{-1}1\,dx$

Let $I_1 = \int_0^1 \tan^{-1}x\,dx = \left[\tan^{-1}x.x - \int \frac{x}{1+x^2}dx\right]_0^1$, Put $1 + x^2 = t$ $\therefore$ $2x\,dx = dt$ or $x\,dx = \frac{dt}{2}$

$I_1 = \left[\tan^{-1}x.x - \frac{1}{2}\int\frac{dt}{t}\right]_0^1 = \left[\tan^{-1}x.x - \frac{1}{2}\log t\right]_0^1 = \left[\tan^{-1}x.x - \frac{1}{2}\log(1+x^2)\right]_0^1 = \tan^{-1}1 - \frac{1}{2}\log 2 = \frac{\pi}{4} - \frac{1}{2}\log 2$

Let $I_2 = \int_0^1 \tan^{-1}1\,dx = \tan^{-1}1.(x)_0^1 = \tan^{-1}1 = \frac{\pi}{4}$

$I = I_1 - I_2 = \frac{\pi}{4} - \frac{1}{2}\log 2 - \frac{\pi}{4} = -\frac{1}{2}\log 2$ or $-\log\sqrt{2}$ or $\log\left(\frac{1}{\sqrt{2}}\right)$ Ans.

(2) (a) $I = \int_0^{\frac{\pi}{2}} \log(\sin x)\,dx$ $\left[\text{use by property, } \int_0^a f(x)\,dx = \int_0^a f(a-x)\,dx\right]$

or $I = \int_0^{\frac{\pi}{2}} \log\left[\sin\left(\frac{\pi}{2} - x\right)\right]dx = \int_0^{\frac{\pi}{2}} \log(\cos x)\,dx = I$

Adding, $2I = \int_0^{\frac{\pi}{2}} \log(\sin x)\,dx + \int_0^{\frac{\pi}{2}} \log(\cos x)\,dx = \int_0^{\frac{\pi}{2}} [\log(\sin x) + \log(\cos x)]\,dx = \int_0^{\frac{\pi}{2}} \log(\sin x.\cos x)\,dx$

or $2I = \int_0^{\frac{\pi}{2}} \log\left(\frac{2\sin x.\cos x}{2}\right)dx = \int_0^{\frac{\pi}{2}} \log\left(\frac{\sin 2x}{2}\right)dx = \int_0^{\frac{\pi}{2}} \log(\sin 2x)\,dx - \int_0^{\frac{\pi}{2}} \log 2\,dx = I_1 - I_2$ (say) ……….. (i)

where $I_1 = \int_0^{\frac{\pi}{2}} \log(\sin 2x)\,dx$ and $I_2 = \int_0^{\frac{\pi}{2}} \log 2\,dx$

Solve, $I_1 = \int_0^{\frac{\pi}{2}} \log(\sin 2x)\ dx$ Put $2x = t$ $\therefore$ $2dx = dt$ or $dx = \frac{dt}{2}$ and limit become 0 to π.

$$I_1 = \int_0^{\pi} \log(\sin t) . \frac{dt}{2} = \frac{1}{2}\int_0^{\pi} \log(\sin t)\ dt$$

use property, $\int_0^{2a} f(x)\,dx = \begin{cases} 2\int_0^{a} f(x)\,dx, & \text{if } f(2a-x) = f(x) \\ 0, & \text{if } f(2a-x) = -f(x) \end{cases}$

$$\therefore\ a = \pi,\ f(t) = \log(\sin t),\ f(2a - t) = f(2\pi - t) = \log[\sin(2\pi - t)] = \log(\sin t) = f(t)$$

$$\therefore\ I_1 = \frac{1}{2}\int_0^{\pi} \log(\sin t)\ dt = \frac{1}{2}.2\int_0^{\frac{\pi}{2}} \log(\sin t)\ dt = \int_0^{\frac{\pi}{2}} \log(\sin t)\ dt = I$$

Now, solve $I_2 = \int_0^{\frac{\pi}{2}} \log 2\ dx = \log 2 \int_0^{\frac{\pi}{2}} dx = \log 2\,(x)_0^{\frac{\pi}{2}} = \log 2\left(\frac{\pi}{2} - 0\right) = \frac{\pi}{2}\log 2$

Put value I_1 and I_2 in equation (i), we get

$$\therefore\ 2I = I_1 - I_2 = I - \frac{\pi}{2}\log 2 \quad \text{or} \quad 2I - I = -\frac{\pi}{2}\log 2 \quad \therefore\ I = \frac{\pi}{2}\log\frac{1}{2} \quad \text{Ans.}$$

(b) $I = \int_0^{\frac{\pi}{2}} \log(\cos x)\,dx = \frac{\pi}{2}\log\frac{1}{2}$ Ans. (Do yourself, to be solve same as above question)

(c) $I = \int_0^{\frac{\pi}{2}} \log(\tan x)\,dx = \int_0^{\frac{\pi}{2}} \log\left(\frac{\sin x}{\cos x}\right) dx = \int_0^{\frac{\pi}{2}} \log(\sin x)\,dx - \int_0^{\frac{\pi}{2}} \log(\cos x)\,dx = \frac{\pi}{2}\log\frac{1}{2} - \frac{\pi}{2}\log\frac{1}{2} = 0$ Ans.

(d) $I = \int_0^{\frac{\pi}{2}} \log(\cot x)\,dx = 0$ Ans.

$$\boxed{\therefore\ \int_0^{\frac{\pi}{2}} \log(\tan x)\,dx = \int_0^{\frac{\pi}{2}} \log(\cot x)\,dx = 0 \quad \text{Ans.}}$$

(e) $I = \int_0^{\frac{\pi}{2}} \log(\sec x)\,dx = \int_0^{\frac{\pi}{2}} \log\left(\frac{1}{\cos x}\right) dx = \int_0^{\frac{\pi}{2}} \log 1\,dx - \int_0^{\frac{\pi}{2}} \log(\cos x)\,dx = 0 - \int_0^{\frac{\pi}{2}} \log(\cos x)\,dx = -\int_0^{\frac{\pi}{2}} \log(\cos x)\,dx = -\frac{\pi}{2}\log\frac{1}{2}$

$= \frac{\pi}{2}\log 2$ Ans. [see question no. –(1)(b)]

(f) $I = \int_0^{\frac{\pi}{2}} \log(\operatorname{cosec} x)\,dx = \int_0^{\frac{\pi}{2}} \log\left(\frac{1}{\sin x}\right) dx = \int_0^{\frac{\pi}{2}} \log 1\,dx - \int_0^{\frac{\pi}{2}} \log(\sin x)\,dx = 0 - \int_0^{\frac{\pi}{2}} \log(\sin x)\,dx = -\int_0^{\frac{\pi}{2}} \log(\sin x)\,dx = -\frac{\pi}{2}\log\frac{1}{2}$

$= \frac{\pi}{2}\log 2$ Ans. [see question no. –(1)(a)]

$$\boxed{\therefore\ \int_0^{\frac{\pi}{2}} \log(\sec x)\,dx = \int_0^{\frac{\pi}{2}} \log(\operatorname{cosec} x)\,dx = \frac{\pi}{2}\log 2 \quad \text{Ans.}}$$

(g) $I = \int_0^{\frac{\pi}{2}} \log x\,dx = \int_0^{\frac{\pi}{2}} \log x . 1\ dx$ use integration by part formula, $\int u.v\ dx = u.\int v\ dx - \int\left[\frac{du}{dx}.\int v\ dx\right] dx$

$$I = \log x \int 1\,dx - \int\left\{\frac{d(\log x)}{dx}.\int 1\ dx\right\} dx = \log x . x - \int \frac{1}{x}.x\ dx = x\log x - \int dx = x\log x - x = x(\log x - 1)$$

Put limit 0 to $\frac{\pi}{2}$ then $I = [x(\log x - 1)]_0^{\frac{\pi}{2}} = \left[\frac{\pi}{2}\left(\log\frac{\pi}{2} - 1\right) - 0\right] = \frac{\pi}{2}\left(\log\frac{\pi}{2} - 1\right)$ Ans.

(h) $I = \int_0^{\frac{\pi}{2}} (\sin x + \cos x)\,dx = \int_0^{\frac{\pi}{2}} \sin x\ dx + \int_0^{\frac{\pi}{2}} \cos x\ dx = (-\cos x)_0^{\frac{\pi}{2}} + (\sin x)_0^{\frac{\pi}{2}} = -\cos\left(\frac{\pi}{2}\right) + \cos 0 + \sin\left(\frac{\pi}{2}\right) - \sin 0 = -0 + 1 + 1 - 0$

$= 1 + 1 = 2$ Ans.

(i) $I = \int_0^{\frac{\pi}{4}} (\tan x + \cot x)\,dx = \int_0^{\frac{\pi}{4}} \left(\tan x + \frac{1}{\tan x}\right) dx = \int_0^{\frac{\pi}{4}} \left(\frac{\tan^2 x + 1}{\tan x}\right) dx = \int_0^{\frac{\pi}{4}} \frac{\sec^2 x}{\tan x}\ dx$

Put $\tan x = t \quad \therefore \sec^2 x \, dx = dt$

$$I = \int_0^{\frac{\pi}{4}} \frac{dt}{t} = (\log t)_0^{\frac{\pi}{4}} = [\log(\tan x)]_0^{\frac{\pi}{4}} = \left[\log\left(\tan\frac{\pi}{4}\right) - \log(\tan 0)\right] = \log 1 - \log 0 = 0 \qquad \text{Ans.}$$

(3) (a) $I = \int_0^1 \frac{x^2+3x+2}{(x+1)} dx = \int_0^1 \frac{x^2+2x+x+2}{(x+1)} dx = \int_0^1 \frac{x(x+2)+1(x+2)}{(x+1)} dx = \int_0^1 \frac{(x+1)(x+2)}{(x+1)} dx = \int_0^1 (x+2)\, dx = \left[\frac{x^2}{2}+2x\right]_0^1$

$$= \left[\frac{1}{2}+2-0\right] = \frac{1}{2}+2 = \frac{5}{2} \qquad \text{Ans.}$$

(b) $I = \int_1^2 \frac{\log x}{x} dx$, Put $\log x = t \quad \therefore \frac{1}{x} dx = dt$ and limit become 0 to log 2

$$I = \int_0^{\log 2} t\, dt = \left(\frac{t^2}{2}\right)_0^{\log 2} = \frac{(\log 2)^2}{2} - 0 = \frac{[\log 2]^2}{2} \qquad \text{Ans.}$$

x	1	2
t	0	log 2

(c) $I = \int_0^{\pi} \log(1+\cos x) . \sin x \, dx$

Put $1+\cos x = t \quad \therefore -\sin x\, dx = dt \quad \therefore \sin x\, dx = -dt \qquad$ Limit $1+\cos x = t$

x	0	π
t	2	0

$I = -\int_2^0 \log t\, dt = \int_0^2 \log t\, dt \qquad$ use integration by part formula,

$I = [t(\log t - 1)]_0^2 = 2(\log 2 - 1) - 0 = 2(\log 2 - 1) \quad$ Ans. [see question (1)(g).]

(d) $I = \int_0^1 \sin^{-1} x\, dx$

$\left\{\text{use integration by part formula, } \int u.v\, dx = u.\int v\, dx - \int \left[\frac{du}{dx}.\int v\, dx\right] dx\right\}$

$$I = \sin^{-1} x . \int dx - \int \left[\frac{d(\sin^{-1} x)}{dx}.\int dx\right] dx = \sin^{-1} x . x - \int \frac{x}{\sqrt{1-x^2}} dx$$

Put $1-x^2 = t^2 \quad \therefore -2x\, dx = 2t\, dt \quad \therefore x\, dx = -t\, dt$

$$I = x\sin^{-1} x + \int \frac{t\, dt}{t} = x\sin^{-1} x + \int dt = x\sin^{-1} x + t = x\sin^{-1} x + \sqrt{1-x^2}$$

Put limit 0 to 1 then, $I = \left(x\sin^{-1} x + \sqrt{1-x^2}\right)_0^1 = (1.\sin^{-1} 1 + 0 - 0 - 1) = \frac{\pi}{2} - 1 \qquad$ Ans.

(e) $I = \int_0^1 \cos^{-1} x\, dx = 1 \qquad$ Ans. (see above question)

(f) $I = \int_0^1 \tan^{-1} x\, dx \quad \left\{\text{use integration by part formula, } \int u.v\, dx = u.\int v\, dx - \int \left[\frac{du}{dx}.\int v\, dx\right] dx\right\}$

$$I = \left\{\tan^{-1} x . \int dx - \int \left[\frac{d(\tan^{-1} x)}{dx}.\int dx\right] dx\right\}_0^1 = \left[x\tan^{-1} x - \int \frac{x}{1+x^2} dx\right]_0^1$$

Put $1+x^2 = t$ or $2x\, dx = dt$

$$I = \left[x\tan^{-1} x - \frac{1}{2}\int \frac{dt}{t}\right]_0^1 = \left[x\tan^{-1} x - \frac{1}{2}\log t\right]_0^1 = \left[x\tan^{-1} x - \frac{1}{2}\log(1+x^2)\right]_0^1 = \left[1.\tan^{-1} 1 - \frac{1}{2}\log(1+1)\right] - [0-0] = \tan^{-1} 1 - \frac{1}{2}\log 2$$

$$= \frac{\pi}{4} - \frac{1}{2}\log 2 \quad \text{or} \quad \frac{\pi}{4} - \log\sqrt{2} \qquad \text{Ans.}$$

(g) $I = \int_0^1 \cot^{-1} x\, dx = \frac{\pi}{4} + \frac{1}{2}\log 2$ or $\frac{\pi}{4} + \log\sqrt{2}$ Ans. (Do yourself, To be solve same as above question)

(h) $I = \int_0^1 \sec^{-1} x\ dx$, doyourself $\left[x \sec^{-1} x - \log\left(x + \sqrt{x^2 - 1}\right)\right]_0^1$

(4) (a) $I = \int_0^{2\pi} 2^{3\log_2 x}\ dx = \int_0^{2\pi} 2^{\log_2 x^3}\ dx = \int_0^{2\pi} x^3\ dx = \left(\frac{x^4}{4}\right)_0^{2\pi} = \frac{(2\pi)^4}{4} - 0 = \frac{16\pi^4}{4} = 4\pi^4$ Ans.

(b) $I = \int_0^1 \cos^{-1}\left(\frac{1 + \cos x}{2\cos {}^{x}/_{2}}\right) dx = \frac{1}{4}$ Ans. (Do yourself)

$\left[\text{Hint:} -\ 1 + \cos x = 2\cos^2 {}^{x}/_{2}\ ,\ I = \int_0^1 \cos^{-1}\left(\frac{2\cos^2 {}^{x}/_{2}}{2\cos {}^{x}/_{2}}\right) dx = \int_0^1 \cos^{-1}(\cos {}^{x}/_{2})\, dx = \int_0^1 \frac{x}{2} dx\right]$

(c) $I = \int_1^2 x^2 \log x\ dx = \frac{8}{3}\log 2 - \frac{7}{9}$ Ans. (Do yourself, use integration by part formula)

(d) $I = \int_0^1 \frac{1}{\sqrt{x+1} - \sqrt{x+2}}\ dx = \int_0^1 \left(\frac{1}{\sqrt{x+1} - \sqrt{x+2}} \times \frac{\sqrt{x+1} + \sqrt{x+2}}{\sqrt{x+1} + \sqrt{x+2}}\right) dx = \int_0^1 \frac{\sqrt{x+1} + \sqrt{x+2}}{x + 1 - x - 2}\ dx$

$I = -\int_0^1 \left(\sqrt{x+1} + \sqrt{x+2}\right) dx = -\int_0^1 \sqrt{x+1}\, dx - \int_0^1 \sqrt{x+2}\, dx = -I_1 - I_2$ ……………… (A)

solve, $I_1 = \int_0^1 \sqrt{x+1}\, dx$, Put $x + 1 = t^2$ or $t = \sqrt{x+1}$ $\therefore$ $dx = 2t\, dt$

$\therefore\ I_1 = \int t.2t\ dt = 2\int t^2\ dt = 2.\frac{t^3}{3} = \frac{2}{3}.(x+1).\sqrt{x+1}$, Put limit 0 to 1

$\therefore\ I_1 = \frac{2}{3}.\left((x+1).\sqrt{x+1}\right)_0^1 = \frac{2}{3}(2\sqrt{2} - 1)$

x	0	1
z	2	3

Now, solve $I_2 = \int_0^1 \sqrt{x+2}\, dx$, Put $x + 2 = z$ $\therefore$ $dx = dz$ and limit become 2 to3

$$\therefore\ I_2 = \int_2^3 \sqrt{z}\ dz = \int_2^3 z^{\frac{1}{2}}\ dz = \left(\frac{z^{\frac{3}{2}}}{\frac{3}{2}}\right)_2^3 = \frac{2}{3}\left[z\sqrt{z}\right]_2^3 = \frac{2}{3}(3\sqrt{3} - 2\sqrt{2})$$

Put value of I_1 and I_2 in equation (A), we have $\therefore\ I = -I_1 - I_2 = -\frac{2}{3}(2\sqrt{2} - 1) - \frac{2}{3}(3\sqrt{3} - 2\sqrt{2})$

$$\therefore\ I = -\frac{2}{3}(2\sqrt{2} - 1 + 3\sqrt{3} - 2\sqrt{2}) = -\frac{2}{3}(3\sqrt{3} - 1) = \frac{2}{3}(1 - 3\sqrt{3})\ \text{ or }\ \left(\frac{2}{3} - 2\sqrt{3}\right)\quad \text{Ans.}$$

(e) $I = \int_0^5 |x - 3|\ dx$, Put $f(x) = x - 3$, $f(x) = 0$ or $x - 3 = 0$ or $x = 3$

case I: − If $x > 3$ *then* $f(x)$ is positive in the interval $3 < x < 5$

case II: − If $x < 3$ *then* $f(x)$ is negative in the interval $0 < x < 3$

$$\therefore\ f(x) = \begin{cases} -(x-3), & \text{if } x < 3 \\ (x-3), & \text{if } x > 3 \end{cases}$$

$I = \int_0^3 |x - 3|\, dx + \int_3^5 |x - 3|\, dx = -\int_0^3 (x - 3)\, dx + \int_3^5 (x - 3)\, dx = -\int_0^3 x\, dx + 3\int_0^3 dx + \int_3^5 x\, dx - 3\int_3^5 dx$

$= -\left(\frac{x^2}{2}\right)_0^3 + 3(x)_0^3 + \left(\frac{x^2}{2}\right)_3^5 - 3(x)_3^5 = -\frac{9}{2} + 9 + \frac{25}{2} - \frac{9}{2} - 3(5 - 3) = -\frac{9}{2} + 9 + \frac{25}{2} - \frac{9}{2} - 6 = \frac{-9 + 18 + 25 - 9 - 12}{2}$

$= \frac{13}{2}$ Ans.

(f) $I = \int_0^{\frac{\pi}{2}} \sin^5 x\ dx = \int_0^{\frac{\pi}{2}} \sin^4 x.\sin x\ dx = \int_0^{\frac{\pi}{2}} (\sin^2 x)^2.\sin x\ dx = \int_0^{\frac{\pi}{2}} (1 - \cos^2 x)^2.\sin x\ dx$

Put $\cos x = t \quad \therefore -\sin x\, dx = dt$ or $\sin x\, dx = -dt$ and limit becomes

x	0	$\frac{\pi}{2}$
t	1	0

$$\therefore\ I = -\int_1^0 (1-t^2)^2\, dt = \int_0^1 (1-t^2)^2\, dt = \int_0^1 (1-2t^2+t^4)\, dt$$

$$\therefore\ I = \int_0^1 dt - 2\int_0^1 t^2\, dt + \int_0^1 t^4\, dt = (t)_0^1 - 2\left(\frac{t^3}{3}\right)_0^1 + \left(\frac{t^5}{5}\right)_0^1 = (1-0) - 2\left(\frac{1}{3}-0\right) + \left(\frac{1}{5}-0\right) = 1 - \frac{2}{3} + \frac{1}{5} = \frac{15-10+3}{15} = \frac{18-10}{15}$$

$$= \frac{8}{15} \quad \text{Ans.}$$

(g) $I = \int_1^2 \frac{1}{x\left(\log x + \frac{1}{\log x}\right)}\, dx$

Put $\log x = t \quad \therefore \frac{1}{x}\, dx = dt$ and limit becomes

x	1	2
t	0	log 2

$$\therefore\ I = \int_1^2 \frac{1}{\left(\log x + \frac{1}{\log x}\right)} \cdot \frac{1}{x}\, dx = \int_0^{\log 2} \frac{dt}{\left(t + \frac{1}{t}\right)} = \int_0^{\log 2} \frac{t}{t^2+1}\, dt$$

Put $1+t^2 = z \quad \therefore 2t\, dt = dz$ or $t\, dt = \frac{dz}{2}$

$1+t^2=z$		
t	0	log 2
z	1	$1+(\log 2)^2$

$$\therefore\ I = \frac{1}{2}\int_1^{1+(\log 2)^2} \frac{dz}{z} = \frac{1}{2}(\log z)_1^{1+(\log 2)^2} = \frac{1}{2}\{\log[1+(\log 2)^2] - 0\}$$

$$\therefore\ I = \frac{1}{2}\log[1+(\log 2)^2] \text{ or } \log[1+(\log 2)^2]^{\frac{1}{2}} \text{ or } \log\left(\sqrt{1+(\log 2)^2}\right) \quad \text{Ans.}$$

(h) $I = \int_{-\frac{\pi}{2}}^{\frac{\pi}{2}} e^x . \sin x\, dx$

(5) (a) $I = \int_{-\frac{\pi}{2}}^{\frac{\pi}{2}} \frac{\cos x}{2+\sin x}\, dx$Put $2+\sin x = t \quad \therefore \cos x\, dx = dt$ and limit becomes 1 to 3

x	$-\frac{\pi}{2}$	$\frac{\pi}{2}$
t	1	3

$$\therefore\ I = \int_1^3 \frac{dt}{t} = (\log t)_1^3 = \log 3 - \log 1 = \log 3 - 0 = \log 3 \quad \text{Ans.}$$

(b) $I = \int_{\frac{\pi}{4}}^{\frac{\pi}{2}} \sin(2x+3)\, dx$, Put $2x+3 = t \quad \therefore 2\, dx = dt$ or $dx = \frac{dt}{2}$

$$\therefore\ I = \frac{1}{2}\int_{\frac{\pi}{4}}^{\frac{\pi}{2}} \sin t\, dt = -\frac{1}{2}.(\cos t)_{\frac{\pi}{4}}^{\frac{\pi}{2}} = -\frac{1}{2}[\cos(2x+3)]_{\frac{\pi}{4}}^{\frac{\pi}{2}} = -\frac{1}{2}\left[\cos(\pi+3) - \cos\left(\frac{\pi}{2}+3\right)\right] = -\frac{1}{2}[-\cos 3 + \sin 3]$$

$$= \frac{1}{2}[\cos 3 - \sin 3] \quad \text{Ans.} \quad \text{or} \quad \frac{1}{\sqrt{2}}\left[\sin\left(\frac{3\pi}{4}+3\right)\right] \quad \text{Ans.}$$

(c) $I = \int_1^3 (5x+2)^5\, dx$, Put $5x+2 = t \quad \therefore 5\, dx = dt$ or $dx = \frac{dt}{5}$ and limit become 7 to 17 .

$$\therefore\ I = \int_7^{17} t^5 . \frac{dt}{5} = \frac{1}{5}\int_7^{17} t^5\, dt = \frac{1}{5}\left(\frac{t^6}{6}\right)_7^{17} = \frac{1}{30}[17^6 - 7^6] \text{ or } \frac{1}{30}[24137569 - 117649] = \frac{24019920}{30} = 800664 \quad \text{Ans.}$$

(d) $I = \int_1^5 \frac{\sqrt{6-x}}{\sqrt{x}+\sqrt{6-x}}\, dx \qquad \left[\text{use property VI, } \int_b^a f(x)\, dx = \int_b^a f(a+b-x)\, dx\right]$

$$\therefore\ I = \int_1^5 \frac{\sqrt{6-x}}{\sqrt{x}+\sqrt{6-x}}\, dx = \int_1^5 \frac{\sqrt{6-(6-x)}}{\sqrt{6-x}+\sqrt{6-(6-x)}}\, dx = \int_1^5 \frac{\sqrt{x}}{\sqrt{6-x}+\sqrt{x}}\, dx = I \text{ (say)}$$

Adding, $2I = \int_1^5 \left(\frac{\sqrt{6-x}}{\sqrt{x}+\sqrt{6-x}} + \frac{\sqrt{x}}{\sqrt{6-x}+\sqrt{x}}\right) dx = \int_1^5 \left(\frac{\sqrt{x}+\sqrt{6-x}}{\sqrt{x}+\sqrt{6-x}}\right) dx = \int_1^5 dx = (x)_1^5 = 5-1 = 4$

$\therefore\ 2I = 4 \quad \therefore\ I = 2$ Ans.

(e) $I = \int_0^{\frac{\pi}{2}} \frac{\cos x}{\sqrt{\sin^2 x + 2\sin x + 3}}\, dx$, Put $\sin x = t$ $\therefore\ \cos x\ dx = dt$ and limit become 0 to1

or $I = \int_0^1 \frac{dt}{\sqrt{t^2+2t+3}} = \int_0^1 \frac{dt}{\sqrt{(t+1)^2+2}} = \int_0^1 \frac{dt}{\sqrt{(t+1)^2+(\sqrt{2})^2}}$

x	0	$\frac{\pi}{2}$
t	0	1

use formula, $\int \frac{dx}{\sqrt{x^2+a^2}} = \log\left|x+\sqrt{x^2+a^2}\right|$

or $I = \left[\log\left|(t+1)+\sqrt{(t+1)^2+2}\right|\right]_0^1 = \log|2+\sqrt{6}| - \log|1+\sqrt{3}| = \log\left|\frac{2+\sqrt{6}}{1+\sqrt{3}}\right|$ Ans.

(f) $I = \int_0^{\frac{\pi}{4}} \tan^2(x+1)\, dx$, Put $x+1 = t$ $\therefore\ dx = dt$

$\therefore\ I = \int_0^{\frac{\pi}{4}} \tan^2 t\ dt = \int_0^{\frac{\pi}{4}} (\sec^2 t - 1)\, dt = \int_0^{\frac{\pi}{4}} \sec^2 t\ dt - \int_0^{\frac{\pi}{4}} dt = [\tan t - t]_0^{\frac{\pi}{4}}$ Put $t = x+1$

$\therefore\ I = [\tan(x+1) - (x+1)]_0^{\frac{\pi}{4}} = \tan\left(\frac{\pi}{4}+1\right) - \left(\frac{\pi}{4}+1\right) - \tan 1 + 1 = \tan\left(\frac{\pi}{4}+1\right) - \tan 1 - \frac{\pi}{4}$ Ans.

(g) $I = \int_0^{\frac{\pi}{2}} \frac{e^{\sqrt{\cos x}}}{e^{\sqrt{\sin x}} + e^{\sqrt{\cos x}}}\, dx = \frac{\pi}{4}$ Ans. [Do yourself, see question no. –(4)(d) and use property IV.]

(h) $I = \int_0^{\frac{\pi}{2}} \sin^2 x . \cos^3 x\ dx = \int_0^{\frac{\pi}{2}} \sin^2 x . \cos^2 x . \cos x\ dx = \int_0^{\frac{\pi}{2}} \sin^2 x\,(1-\sin^2 x). \cos x\ dx$

Put $\sin x = t$ $\therefore\ \cos x\ dx = dt$ and limit become 0 to 1

$\therefore I = \int_0^1 t^2(1-t^2)\, dt = \int_0^1 (t^2 - t^4)\, dt = \left[\frac{t^3}{3} - \frac{t^5}{5}\right]_0^1 = \left(\frac{1}{3} - \frac{1}{5} - 0\right) = \frac{5-3}{15} = \frac{2}{15}$ Ans.

x	0	$\frac{\pi}{2}$
t	0	1

(6) (a) $I = \int_0^1 \frac{dx}{x^2+x+1} = \int_0^1 \frac{dx}{\left(x+\frac{1}{2}\right)^2 + \frac{3}{4}} = \int_0^1 \frac{dx}{\left(x+\frac{1}{2}\right)^2 + \left(\frac{\sqrt{3}}{2}\right)^2}$ [use formula, $\int \frac{dx}{x^2+a^2} = \frac{1}{a}\tan^{-1}\left(\frac{x}{a}\right)$]

$\therefore\ I = \frac{1}{\frac{\sqrt{3}}{2}}\tan^{-1}\left(\frac{x+\frac{1}{2}}{\frac{\sqrt{3}}{2}}\right) = \frac{2}{\sqrt{3}}\tan^{-1}\left(\frac{2x+1}{\sqrt{3}}\right)$, Put limit 0 to 1 and $\tan^{-1}\sqrt{3} = \frac{\pi}{3}$, $\tan^{-1}\frac{1}{\sqrt{3}} = \frac{\pi}{6}$

$\therefore\ I = \left[\frac{2}{\sqrt{3}}\tan^{-1}\left(\frac{2x+1}{\sqrt{3}}\right)\right]_0^1 = \frac{2}{\sqrt{3}}\tan^{-1}\left(\frac{3}{\sqrt{3}}\right) - \frac{2}{\sqrt{3}}\tan^{-1}\left(\frac{1}{\sqrt{3}}\right) = \frac{2}{\sqrt{3}}\left[\tan^{-1}(\sqrt{3}) - \tan^{-1}\left(\frac{1}{\sqrt{3}}\right)\right] = \frac{2}{\sqrt{3}}\left[\frac{\pi}{3} - \frac{\pi}{6}\right] = \frac{2}{\sqrt{3}}\left[\frac{2\pi-\pi}{6}\right] = \frac{2}{\sqrt{3}}.\frac{\pi}{6}$

$= \frac{\pi}{3\sqrt{3}}$ Ans.

(b) $I = \int_0^2 |x^2+x|\ dx$ $\therefore\ x^2+x = 0$ or $x(x+1) = 0$ $\therefore\ x = 0, -1$

If limit between $0 < x < 2$ *then* $|x^2+x|$ is positive.

If limit between $-1 < x < 0$ *then* $|x^2+x|$ is negative.

$$I = \int_0^2 |x^2 + x|\, dx = \int_{-1}^0 -(x^2 + x)\, dx + \int_0^2 (x^2 + x)\, dx = -\int_{-1}^0 x^2\, dx - \int_{-1}^0 x\, dx + \int_0^2 x^2\, dx + \int_0^2 x\, dx = -\left(\frac{x^3}{3}\right)_{-1}^0 - \left(\frac{x^2}{2}\right)_{-1}^0 + \left(\frac{x^3}{3}\right)_0^2 + \left(\frac{x^2}{2}\right)_0^2$$

$$= -\left(0 + \frac{1}{3}\right) - \left(0 - \frac{1}{2}\right) + \left(\frac{8}{3} - 0\right) + \left(\frac{4}{2} - 0\right) = -\frac{1}{3} + \frac{1}{2} + \frac{8}{3} + 2 = \frac{-2 + 3 + 16 + 12}{6} = \frac{29}{6} \quad \text{Ans.}$$

(c) $I = \int_0^2 \frac{dx}{(x-1).\sqrt{x+2}}$, Put $x + 2 = t^2$ or $x = t^2 - 2$ $\therefore$ $dx = 2t\, dt$

$$I = \int_0^2 \frac{2t}{(t^2 - 2 - 1)\sqrt{t^2}}\, dt = \int_0^2 \frac{2t}{(t^2 - 3).t}\, dt = 2\int_0^2 \frac{dt}{(t^2 - 3)} = 2\int_0^2 \frac{dt}{\left[t^2 - (\sqrt{3})^2\right]} \quad \text{formula} \int \frac{dx}{x^2 - a^2} = \frac{1}{2a}\log\left|\frac{x-a}{x+a}\right|$$

$$I = \left[2.\frac{1}{2\sqrt{3}}\log\left|\frac{t - \sqrt{3}}{t + \sqrt{3}}\right|\right]_0^2 = \left[\frac{1}{\sqrt{3}}\log\left|\frac{\sqrt{x+2} - \sqrt{3}}{\sqrt{x+2} + \sqrt{3}}\right|\right]_0^2 = \frac{1}{\sqrt{3}}\left[\log\left|\frac{\sqrt{2+2} - \sqrt{3}}{\sqrt{2+2} + \sqrt{3}}\right| - \log\left|\frac{\sqrt{0+2} - \sqrt{3}}{\sqrt{0+2} + \sqrt{3}}\right|\right] = \frac{1}{\sqrt{3}}\left[\log\left|\frac{2 - \sqrt{3}}{2 + \sqrt{3}}\right| - \log\left|\frac{\sqrt{2} - \sqrt{3}}{\sqrt{2} + \sqrt{3}}\right|\right]$$

$$= \frac{1}{\sqrt{3}}\log\left|\frac{2 - \sqrt{3}}{2 + \sqrt{3}} \times \frac{2 - \sqrt{3}}{2 - \sqrt{3}}\right| - \frac{1}{\sqrt{3}}\log\left|\frac{\sqrt{2} - \sqrt{3}}{\sqrt{2} + \sqrt{3}} \times \frac{\sqrt{2} - \sqrt{3}}{\sqrt{2} - \sqrt{3}}\right|$$

$$\therefore\ I = \frac{1}{\sqrt{3}}\log\left|\frac{(2 - \sqrt{3})^2}{4 - 3}\right| - \frac{1}{\sqrt{3}}\log\left|\frac{(\sqrt{2} - \sqrt{3})^2}{2 - 3}\right| = \frac{1}{\sqrt{3}}\log\left|(2 - \sqrt{3})^2\right| - \frac{1}{\sqrt{3}}\log\left|(\sqrt{2} - \sqrt{3})^2\right| \quad \text{Ans.}$$

(d) $I = \int_1^e \frac{\log x}{x}\, dx$, Put $\log x = t$ $\therefore$ $\frac{1}{x}\, dx = dt$ and limit become 0 to1.

$$\text{or } I = \int_0^1 t\, dt = \left(\frac{t^2}{2}\right)_0^1 = \frac{1}{2} - 0 = \frac{1}{2} \quad \text{Ans.}$$

(e) $I = \int_6^7 \frac{dx}{(x+3).\sqrt{x-5}} = \frac{1}{\sqrt{2}}\left[\tan^{-1}\left(\frac{1}{2}\right) - \tan^{-1}\left(\frac{1}{2\sqrt{2}}\right)\right]$ Ans. [Do yourself, see question no. –(5)(c)]

(f) $I = \int_2^3 \frac{dx}{(x+2)(x+3)} = \int_2^3 \frac{dx}{x^2 + 5x + 6} = \int_2^3 \frac{dx}{\left(x + \frac{5}{2}\right)^2 - \frac{1}{4}} = \int_2^3 \frac{dx}{\left(x + \frac{5}{2}\right)^2 - \left(\frac{1}{2}\right)^2}$

use formula $\int \frac{dx}{x^2 - a^2} = \frac{1}{2a}\log\left|\frac{x-a}{x+a}\right|$

$$I = \left[\frac{1}{2.\frac{1}{2}}.\log\left|\frac{x + \frac{5}{2} - \frac{1}{2}}{x + \frac{5}{2} + \frac{1}{2}}\right|\right]_2^3 = \left[\log\left|\frac{2x + 5 - 1}{2x + 5 + 1}\right|\right]_2^3 = \log\left(\frac{2.3 + 5 - 1}{2.3 + 5 + 1}\right) - \log\left(\frac{2.2 + 5 - 1}{2.2 + 5 + 1}\right) = \log\left(\frac{10}{12}\right) - \log\left(\frac{8}{10}\right) = \log\left(\frac{5}{6}\right) - \log\left(\frac{4}{5}\right) = \log\left[\frac{\frac{5}{6}}{\frac{4}{5}}\right]$$

$$= \log\left[\frac{5}{6} \times \frac{5}{4}\right] = \log\left[\frac{25}{24}\right] \quad \text{Ans.}$$

(g) $I = \int_{-1}^6 \frac{dx}{(x-1)(x-5)} = \frac{1}{4}\log\left(\frac{1}{15}\right)$ Ans. (Do yourself, see above question)

(h) $I = \int_1^3 \frac{dx}{(x+1)\sqrt{x^2 + 2x + 3}}$, Put $x + 1 = t$ $\therefore$ $dx = dt$ and limit become tend to 2 to 4 .

$$I = \int_1^3 \frac{dx}{(x+1)\sqrt{(x+1)^2 + 2}} = \int_2^4 \frac{dt}{t\sqrt{t^2 + 2}}$$

Again put $t^2 + 2 = z^2$ $\therefore$ $2t\, dt = 2z\, dz$ or $t\, dt = z\, dz$ or $dt = \frac{z\, dz}{\sqrt{z^2 - 2}}$

$$I = \int_2^4 \frac{z\, dz}{z.\sqrt{z^2 - 2}.\sqrt{z^2 - 2}} = \int_2^4 \frac{dz}{z^2 - 2} = \int_2^4 \frac{dz}{z^2 - (\sqrt{2})^2} \quad \text{use formula, } \int \frac{dx}{x^2 - a^2} = \frac{1}{2a}\log\left|\frac{x-a}{x+a}\right|$$

$$I = \left[\frac{1}{2.\sqrt{2}}\log\left|\frac{z-\sqrt{2}}{z+\sqrt{2}}\right|\right]_2^4 = \frac{1}{2\sqrt{2}}\left[\log\left|\frac{\sqrt{t^2+2}-\sqrt{2}}{\sqrt{t^2+2}+\sqrt{2}}\right|\right]_2^4 = \frac{1}{2\sqrt{2}}\left[\log\left|\frac{\sqrt{16+2}-\sqrt{2}}{\sqrt{16+2}+\sqrt{2}}\right| - \log\left|\frac{\sqrt{4+2}-\sqrt{2}}{\sqrt{4+2}+\sqrt{2}}\right|\right] = \frac{1}{2\sqrt{2}}\left[\log\left|\frac{\sqrt{18}-\sqrt{2}}{\sqrt{18}+\sqrt{2}}\right| - \log\left|\frac{\sqrt{6}-\sqrt{2}}{\sqrt{6}+\sqrt{2}}\right|\right]$$

$$= \frac{1}{2\sqrt{2}}\log\left|\frac{(\sqrt{18}-\sqrt{2})(\sqrt{6}+\sqrt{2})}{(\sqrt{18}+\sqrt{2})(\sqrt{6}-\sqrt{2})}\right| \quad \text{Ans.}$$

(7) (a) $I = \int_0^1 \frac{2x^2-5x+1}{\sqrt{x}}\,dx = \int_0^1\left(\frac{2x^2}{\sqrt{x}} - \frac{5x}{\sqrt{x}} + \frac{1}{\sqrt{x}}\right)dx = \int_0^1 \frac{2x^2}{\sqrt{x}}dx - \int_0^1 \frac{5x}{\sqrt{x}}dx + \int_0^1 \frac{1}{\sqrt{x}}dx$

$$I = 2\int_0^1 x^{3/2}\,dx - 5\int_0^1 x^{1/2}\,dx + \int_0^1 x^{-1/2}\,dx = 2\left(\frac{x^{5/2}}{5/2}\right)_0^1 - 5\left(\frac{x^{3/2}}{3/2}\right)_0^1 + \left(\frac{x^{1/2}}{1/2}\right)_0^1 = \frac{4}{5}(1-0) - \frac{10}{3}(1-0) + 2(1-0) = \frac{4}{5} - \frac{10}{3} + 2$$

$$= \frac{12-50+30}{15} = \frac{42-50}{15} = \frac{-8}{50} = -\frac{8}{15} \quad \text{Ans.}$$

(b) $I = \int_1^2 \frac{3x^3+4x^2-x-2}{(3x-2)}\,dx = \int_1^2 \frac{(3x-2)(x^2+2x+1)}{(3x-2)}\,dx = \int_1^2 (x+1)^2\,dx = \left[\frac{(x+1)^3}{3}\right]_1^2 = \frac{(2+1)^3}{3} - \frac{(1+1)^3}{3} = \frac{(3)^3}{3} - \frac{(2)^3}{3}$

$$= \frac{27}{3} - \frac{8}{3} = \frac{27-8}{3} = \frac{19}{3} \quad \text{Ans.}$$

(c) $I = \int_0^{\frac{\pi}{2}} \frac{\sin 2x}{2-\cos^2 x}\,dx = \int_0^{\frac{\pi}{2}} \frac{2\sin x.\cos x}{2-(1-\sin^2 x)}\,dx = \int_0^{\frac{\pi}{2}} \frac{2\sin x.\cos x}{2-1+\sin^2 x}\,dx = \int_0^{\frac{\pi}{2}} \frac{2\sin x.\cos x}{1+\sin^2 x}\,dx$

Put $1+\sin^2 x = t \quad \therefore \quad 2\sin x.\cos x\,dx = dt$

$$I = \int_0^{\frac{\pi}{2}} \frac{dt}{t} = (\log t)_0^{\frac{\pi}{2}} = [\log(1+\sin^2 x)]_0^{\frac{\pi}{2}} = \log(1+1) - \log(1+0) = \log 2 - \log 1 = \log 2 \quad \text{Ans.}$$

(d) $I = \int_0^{\log 2} e^{2x}\,dx = \left(\frac{e^{2x}}{2}\right)_0^{\log 2} = \frac{1}{2}\left(e^{2\log 2} - e^0\right) = \frac{1}{2}\left(e^{\log 4} - e^0\right) = \frac{1}{2}(4-1) = \frac{3}{2}$ Ans.

(e) $I = \int_{\log 2}^{\log 3} (e^x - 1)\,dx = (e^x - x)_{\log 2}^{\log 3} = \left(e^{\log 3} - \log 3\right) - \left(e^{\log 2} - \log 2\right) = 3 - \log 3 - 2 + \log 2$

$$= 1 + \log 2 - \log 3 \quad \text{or} \quad 1 + \log\left(\frac{2}{3}\right) \quad \text{Ans.}$$

(f) $I = \int_e^{e^2} |x-4|\,dx$, Put $x - 4 = 0 \quad \therefore \quad x = 4$ then $|x-4| = \begin{cases} -(x-4), & \text{if } e < x < 4 \\ (x-4), & \text{if } 4 < x < e^2 \end{cases}$

$$I = \int_e^{e^2} |x-4|\,dx = \int_e^4 -(x-4)\,dx + \int_4^{e^2} (x-4)\,dx = -\int_e^4 x\,dx + \int_e^4 4\,dx + \int_4^{e^2} x\,dx - \int_4^{e^2} 4\,dx = (4x)_e^4 - \left(\frac{x^2}{2}\right)_e^4 + \left(\frac{x^2}{2}\right)_4^{e^2} - (4x)_4^{e^2}$$

$$= (16-4e) - \left(\frac{16}{2} - \frac{e^2}{2}\right) + \left(\frac{e^4}{2} - \frac{16}{2}\right) - (4e^2 - 16) = 16 - 4e - 8 + \frac{e^2}{2} + \frac{e^4}{2} - 8 - 4e^2 + 16 = \frac{e^4 + e^2 - 8e^2 - 8e + 32}{2}$$

$$= \frac{e^4 - 7e^2 - 8e + 32}{2} \quad \text{Ans.}$$

(g) $I = \int_{-2}^{2} \frac{e^x}{e^x + e^{-x}}\,dx = 2$ Ans. $\left[\text{Do yourself, use property VI} : - \int_b^a f(x)\,dx = \int_b^a f(a+b-x)\,dx\right]$

(h) $I = \int_0^{\frac{\pi}{2}} \frac{\sin x}{\sin x + \cos x}\,dx = \frac{\pi}{4}$ Ans. $\left[\text{Do yourself, use property IV} : - \int_0^a f(x)\,dx = \int_0^a f(a-x)\,dx\right]$

(i) $I = \int_1^{\sin^2\theta} \frac{\sin^{-1}\sqrt{x}}{\sqrt{1-x}}\,dx$, Put $\sin^{-1}\sqrt{x} = t \quad \therefore \quad \frac{1}{\sqrt{1-x}}\,dx = dt$

$$I = \int_1^{\sin^2\theta} t\,dt = \left(\frac{t^2}{2}\right)_1^{\sin^2\theta} = \left(\frac{\left(\sin^{-1}\sqrt{x}\right)^2}{2}\right)_1^{\sin^2\theta} = \frac{\left(\sin^{-1}\sqrt{\sin^2\theta}\right)^2}{2} - \frac{\left(\sin^{-1}\sqrt{1}\right)^2}{2} = \frac{(\sin^{-1}(\sin\theta))^2}{2} - \frac{(\sin^{-1}1)^2}{2} = \frac{\theta^2}{2} - \frac{\left(\frac{\pi}{2}\right)^2}{2}$$

$$= \frac{\theta^2}{2} - \frac{\pi^2}{8} \quad \text{or} \quad \frac{1}{2}\left[\theta^2 - \frac{\pi^2}{4}\right] \quad \text{Ans.}$$

(j) $I = \int_0^{\frac{\pi}{2}} \frac{dx}{1+\cos x} = \int_0^{\frac{\pi}{2}} \frac{dx}{1+\cos^2 x/2 - \sin^2 x/2} = \int_0^{\frac{\pi}{2}} \frac{dx}{\cos^2 x/2 + \cos^2 x/2} = \int_0^{\frac{\pi}{2}} \frac{dx}{2\cos^2 x/2} = \frac{1}{2}\int_0^{\frac{\pi}{2}} \sec^2 x/2 \, dx$

Put $\frac{x}{2} = t \quad \therefore\ dx = 2\,dt$

$\therefore\ I = \frac{1}{2}\int_0^{\frac{\pi}{2}} 2\sec^2 t \, dt = \int_0^{\frac{\pi}{2}} \sec^2 t \, dt = (\tan t)_0^{\frac{\pi}{2}} = (\tan x/2)_0^{\frac{\pi}{2}} = \tan\frac{\pi}{4} - \tan 0 = 1 - 0 = 1$ Ans.

(k) $I = \int_0^{\frac{\pi}{2}} \frac{dx}{1+\tan x} = \int_0^{\frac{\pi}{2}} \frac{\cos x\, dx}{\cos x + \sin x} = \frac{\pi}{4}$ Ans. $\left[\text{Do yourself, use property IV}: - \int_0^a f(x)\,dx = \int_0^a f(a-x)\,dx\right]$

(l) $I = \int_0^1 \frac{\sqrt{x}+\sqrt[3]{x}}{\sqrt[4]{x}} dx = \int_0^1 \left(\frac{\sqrt{x}}{\sqrt[4]{x}} + \frac{\sqrt[3]{x}}{\sqrt[4]{x}}\right) dx = \int_0^1 \left(x^{\frac{1}{4}} + x^{\frac{1}{12}}\right) dx = \left[\frac{x^{5/4}}{5/4} + \frac{x^{13/12}}{13/12}\right]_0^1 = \left[\frac{4}{5}x^{5/4} + \frac{12}{13}x^{13/12}\right]_0^1 = \frac{4}{5} + \frac{12}{13} - 0 = \frac{4}{5} + \frac{12}{13} = \frac{52+60}{65}$

$= \frac{112}{65}$ Ans.

(m) $I = \int_0^{\frac{\pi}{2}} \frac{\sqrt{\cos x}}{\sqrt{\sin x} + \sqrt{\cos x}} dx = \frac{\pi}{4}$ Ans. $\left[\text{Do yourself, use property IV}: - \int_0^a f(x)\,dx = \int_0^a f(a-x)\,dx\right]$

(8) (a) $I = \int_1^2 \frac{\sin(1+\log x)}{x} dx$, Put $1 + \log x = t \quad \therefore\ \frac{1}{x}dx = dt$

$I = \int_1^2 \sin t \, dt = (-\cos t)_1^2 = -[\cos(1+\log x)]_1^2 = -[\cos(1+\log 2) - \cos(1+\log 1)] = \cos 1 - \cos(1+\log 2)$ Ans.

(b) $I = \int_{-\frac{1}{3}}^{\frac{1}{3}} e^{\cos x}.\sin x \, dx$, Put $\cos x = t \quad \therefore\ -\sin x\, dx = dt$

$\therefore\ I = -\int_{-\frac{1}{3}}^{\frac{1}{3}} e^t\, dt = -(e^t)_{-\frac{1}{3}}^{\frac{1}{3}} = -[e^{\cos x}]_{-\frac{1}{3}}^{\frac{1}{3}} = -\left[e^{\cos\left(\frac{1}{3}\right)} - e^{\cos\left(-\frac{1}{3}\right)}\right] = e^{\cos\left(\frac{1}{3}\right)} - e^{\cos\left(\frac{1}{3}\right)} = 0$ Ans.

IInd Method: $-\ I = \int_{-\frac{1}{3}}^{\frac{1}{3}} e^{\cos x}.\sin x\, dx = \int_{-\frac{1}{3}}^{\frac{1}{3}} e^{\cos\left(-\frac{1}{3}+\frac{1}{3}-x\right)}.\sin\left(-\frac{1}{3}+\frac{1}{3}-x\right) dx = -\int_{-\frac{1}{3}}^{\frac{1}{3}} e^{\cos x}.\sin x\, dx = -I$

$\therefore\ I + I = 0$ or $2I = 0 \quad \therefore\ I = 0$ Ans.

IIIrd Method: $-\ f(x) = e^{\cos x}.\sin x,\ f(-x) = -e^{\cos x}.\sin x = -f(x)$

$\therefore$ f(x) is odd function, $I = \int_{-\frac{1}{3}}^{\frac{1}{3}} e^{\cos x}.\sin x\, dx = 0$ Ans.

(c) $I = \int_{-\frac{\pi}{4}}^{\frac{\pi}{4}} e^{\sec x}.\sec x \tan x\, dx = 0$ Ans. [Do yourself, $f(x) = e^{\sec x}.\sec x \tan x,\ f(-x) = -f(x)$ it is odd function]

(d) $I = \int_{-5}^{5} \frac{3x^5+2x^4+5x^3+4x^2+2x+1}{(x^2+2)} dx = \int_{-5}^{5} \frac{-3x^5+2x^4-5x^3+4x^2-2x+1}{(x^2+2)} dx = I$

$\left[\text{use property VI}: - \int_b^a f(x)\,dx = \int_b^a f(a+b-x)\,dx\right]$

Adding, $2I = \int_{-5}^{5}\left[\frac{3x^5+2x^4+5x^3+4x^2+2x+1}{(x^2+2)} + \frac{-3x^5+2x^4-5x^3+4x^2-2x+1}{(x^2+2)}\right] dx = \int_{-5}^{5} \frac{4x^4+8x^2+2}{x^2+2} dx = \int_{-5}^{5} \frac{4x^2(x^2+2)+2}{x^2+2} dx$

$= \int_{-5}^{5} \frac{4x^2(x^2+2)}{x^2+2} dx + \int_{-5}^{5} \frac{2}{x^2+2} dx = \int_{-5}^{5} 4x^2\, dx + 2\int_{-5}^{5} \frac{1}{x^2+(\sqrt{2})^2} dx$

$$2I = 4\left(\frac{x^3}{3}\right)_{-5}^{5} + 2\left[\frac{1}{\sqrt{2}}\tan^{-1}\left(\frac{x}{\sqrt{2}}\right)\right]_{-5}^{5} = \frac{4(125+125)}{3} + \sqrt{2}\left[\tan^{-1}\left(\frac{5}{\sqrt{2}}\right) - \tan^{-1}\left(\frac{-5}{\sqrt{2}}\right)\right],$$

use formula $\int \frac{dx}{x^2+a^2} = \frac{1}{a}\tan^{-1}\left(\frac{x}{a}\right)$

$$\therefore\ I = \frac{1}{2}\left\{\frac{1000}{3} + \sqrt{2}\left[\tan^{-1}\left(\frac{5}{\sqrt{2}}\right) - \tan^{-1}\left(\frac{-5}{\sqrt{2}}\right)\right]\right\} = \frac{500}{3} + \frac{1}{\sqrt{2}}\cdot\left[\tan^{-1}\left(\frac{5}{\sqrt{2}}\right) - \tan^{-1}\left(\frac{-5}{\sqrt{2}}\right)\right] \quad \text{Ans.}$$

(e) $I = \int_{-3}^{3} \frac{5x^3+x}{x^2+1}\,dx = 0$ Ans. (Do yourself)

$$\left[\text{use property V :} -\ \int_{-a}^{a} f(x)\,dx = \begin{cases} 2\int_0^a f(x)\,dx, & \text{if } f(-x) = f(x) \text{ [Even function]} \\ 0, & \text{if } f(-x) = -f(x) \text{ [Odd function]} \end{cases}\right]$$

(f) $I = \int_{-1}^{1} \frac{x^3\cos x}{x^2+1}\,dx$, Let $f(x) = \frac{x^3\cos x}{x^2+1}$, $f(-x) = -\frac{x^3\cos x}{x^2+1} = -f(x)$ use property V.

or $f(-x) = -f(x)$ $\therefore$ f(x) is odd function. $\therefore\ I = \int_{-1}^{1}\frac{x^3\cos x}{x^2+1}\,dx = 0$ Ans.

(g) $I = \int_{\frac{\pi}{4}}^{\frac{3\pi}{4}} \frac{dx}{1-\cos x} = \int_{\frac{\pi}{4}}^{\frac{3\pi}{4}} \frac{dx}{1-\cos\left(\frac{3\pi}{4}+\frac{\pi}{4}-x\right)} = \int_{\frac{\pi}{4}}^{\frac{3\pi}{4}} \frac{dx}{1-\cos(\pi - x)} = \int_{\frac{\pi}{4}}^{\frac{3\pi}{4}} \frac{dx}{1+\cos x} = I$

$$\left[\text{use property VI :} -\ \int_b^a f(x)\,dx = \int_b^a f(a+b-x)\,dx\right]$$

Adding, $2I = \int_{\frac{\pi}{4}}^{\frac{3\pi}{4}}\left(\frac{1}{1-\cos x}+\frac{1}{1+\cos x}\right)dx = \int_{\frac{\pi}{4}}^{\frac{3\pi}{4}}\left(\frac{1+\cos x+1-\cos x}{(1-\cos x)(1+\cos x)}\right)dx = \int_{\frac{\pi}{4}}^{\frac{3\pi}{4}}\left(\frac{2}{1-\cos^2 x}\right)dx = \int_{\frac{\pi}{4}}^{\frac{3\pi}{4}}\left(\frac{2}{\sin^2 x}\right)dx = 2\int_{\frac{\pi}{4}}^{\frac{3\pi}{4}}\operatorname{cosec}^2 x\,dx$

$$= 2[-\cot x]_{\frac{\pi}{4}}^{\frac{3\pi}{4}} = -2\left(\cot\frac{3\pi}{4} - \cot\frac{\pi}{4}\right) = -2[-1-1] = 4$$

or $I = \frac{4}{2} = 2$ Ans.

(9) (a) $I = \int_0^{\frac{\pi}{2}} \frac{\sqrt{\sin x}}{\sqrt{\sin x}+\sqrt{\cos x}}\,dx = \frac{\pi}{4}$ Ans. (Do yourself, use property IV.)

(b) $I = \int_0^1 \frac{e^{1-x}}{e^x + e^{1-x}}\,dx = \frac{1}{2}$ Ans. (Do yourself, use property IV.)

(c) $I = \int_0^{\pi} \frac{e^{\cos x}}{e^{\cos x}+e^{-\cos x}}\,dx = \frac{\pi}{2}$ Ans. (Do yourself, use property IV.)

(d) $I = \int_0^{\frac{\pi}{2}} \frac{dx}{1-\tan^2 x} = \int_0^{\frac{\pi}{2}} \frac{\cos^2 x\,dx}{\cos^2 x - \sin^2 x} = \frac{\pi}{4}$ Ans. (Do yourself, use property IV.)

(10) (a) $I = \int_{\frac{\pi}{6}}^{\frac{\pi}{3}} \frac{dx}{1+\sqrt{\tan x}} = \frac{\pi}{12}$ Ans. (Do yourself.) $\left[\text{use property VI :} -\ \int_b^a f(x)\,dx = \int_b^a f(a+b-x)\,dx\right]$

(b) $I = \int_{\sqrt{\log 2}}^{\sqrt{\log 3}} \frac{x\sin x^2}{\sin x^2 + \sin(\log 6 - x^2)}\,dx = \frac{1}{4}\log\left(\frac{3}{2}\right)$ Ans.

(c) $I = \int_0^1 \frac{x^4(1-x)^4}{1+x^2}\,dx = \frac{22}{7} - \pi$ Ans. (d) $I = \int_{e^{-1}}^{e^2}\left|\frac{\log_e x}{x}\right|dx = \frac{5}{2}$ Ans.

(11) (a) $\int_0^{\pi} [2\sin x]\,dx$, Let $y = [2\sin x]$ or $y = [f(x)]$ where $f(x) = 2\sin x$

$$y = \begin{cases} 0, & 0 \le x < \frac{\pi}{6} \\ 1, & \frac{\pi}{6} \le x < \frac{\pi}{2} \\ 1, & \frac{\pi}{2} \le x < \frac{5\pi}{6} \\ 0, & \frac{5\pi}{6} < x < \pi \\ 2, & x = \frac{\pi}{2} \end{cases}$$

Draw Graph

or $2\sin x = 1$ or $\sin x = \frac{1}{2}$ $\therefore$ $x = n\pi \pm (-1)^n.\frac{\pi}{6}$ $\therefore$ $x = \frac{\pi}{6}, \frac{5\pi}{6}$

$$\int_0^{\pi} [2\sin x]\, dx = A_1 + A_2 = \left[\frac{\pi}{2} - \frac{\pi}{6}\right] + \left[\frac{5\pi}{6} - \frac{\pi}{2}\right] = \frac{2\pi}{6} + \frac{2\pi}{6} = \frac{4\pi}{6} = \frac{2\pi}{3} \quad \text{Ans.}$$

(b) $\int_0^{102} [\tan^{-1} x]\, dx$, Let $y = [\tan^{-1} x]$ or $y = [f(x)]$ where $f(x) = \tan^{-1} x$

or $f(x) = 1$ or $\tan^{-1} x = 1$ $\therefore$ $x = \tan 1$

$$y = \begin{cases} 0, & 0 \le x < \tan 1 \\ 1, & \tan 1 \le x \le 102 \end{cases}$$

Draw Graph

$$\int_0^{102} [\tan^{-1} x]\, dx = \int_0^{\tan 1} [\tan^{-1} x]\, dx + \int_{\tan 1}^{102} [\tan^{-1} x]\, dx = 0 + [x]_{\tan 1}^{102} = [102 - \tan 1] = 102 - \tan 1 \quad \text{Ans.}$$

(c) $\int_0^{2n\pi} [\sin x + \cos x]\, dx$ (Do yourself)

(d) $\int_0^{\frac{5\pi}{12}} [\tan x]\, dx$, Let $y = [\tan x]$ or $y = [f(x)]$ where $f(x) = \tan x$

$f(x) = 0,1,2,3$ then $x = 0, \tan^{-1} 1, \tan^{-1} 2, \tan^{-1} 3$ or $y = \begin{cases} 0, & 0 \le x < \tan^{-1} 1 \\ 1, & \tan^{-1} 1 \le x < \tan^{-1} 2 \\ 2, & \tan^{-1} 2 \le x < \tan^{-1} 3 \\ 3, & \tan^{-1} 3 \le x \le \frac{5\pi}{12} \end{cases}$

$$\int_0^{\frac{5\pi}{12}} [\tan x]\, dx = \int_0^{\tan^{-1} 1} 0\, dx + \int_{\tan^{-1} 1}^{\tan^{-1} 2} dx + \int_{\tan^{-1} 2}^{\tan^{-1} 3} 2\, dx + \int_{\tan^{-1} 3}^{\frac{5\pi}{12}} 3\, dx = (\tan^{-1} 2 - \tan^{-1} 1) + 2(\tan^{-1} 3 - \tan^{-1} 2) + 3\left(\frac{5\pi}{12} - \tan^{-1} 3\right)$$

$$\int_0^{\frac{5\pi}{12}} [\tan x]\, dx = \tan^{-1} 2 - \tan^{-1} 1 + 2\tan^{-1} 3 - 2\tan^{-1} 2 + 3.\frac{5\pi}{12} - 3\tan^{-1} 3 = \frac{5\pi}{4} - \tan^{-1} 1 - \tan^{-1} 2 - \tan^{-1} 3 \quad \text{Ans.}$$

(e) $\int_0^{n^2} [\sqrt{x}]\, dx, n \in N$

$$I = \int_0^{1^2} 0\, dx + \int_{1^2}^{2^2} dx + \int_{2^2}^{3^2} 2\, dx + \cdots \ldots\ldots\ldots\ldots\ldots\ldots\ldots\ldots\ldots\ldots + \int_{(n-1)^2}^{n^2} (n-1)\, dx$$
$$= (2^2 - 1^2) + 2(3^2 - 2^2) + 3(4^2 - 3^2) + \cdots \ldots\ldots\ldots\ldots\ldots\ldots + (n-1)[n^2 - (n-1)^2]$$

$$I = (2+1) + 2(3+2) + 3(4+3) + \cdots \ldots\ldots + (n-1)[n^2 - (n-1)^2] = \{1^2 + 2^2 + 3^2 + \cdots \ldots + (n-1)^2\} + \sum_2^n (n-1)n$$
$$= \frac{1}{6}.N(N+1)(2N+1) + \sum_1^n (n-1)n = \frac{1}{6}(n-1)n(2n-1) + \frac{(n-1)n(n+1)}{3}$$

$$I = \frac{1}{6}n(n-1)[2n - 1 + 2n + 2] = \frac{1}{6}n(n-1)(4n+1) \quad \text{Ans.}$$

(f) Prove that $\int_0^x [t]\, dt = \frac{[x]([x]-1)}{2} + [x](x - [x])$

$L.H.S = \int_0^x [t]\,dt$, Let $I = \int_0^x [t]\,dt$ $\therefore \{x\} = x - [x],\ 0 \le \{x\} < 1$ or $x = [x] + \{x\}$ or $x = n + f,\ 0 \le f < 1\ n \in I$

$I = \int_0^{n+f} [x]\,dx = 0 + 1 + 2 + 3 + \cdots \ldots\ldots\ldots. + (n-1) + nf = \frac{n(n-1)}{2} + nf$ Draw Graph

$I = \frac{[x]([x]-1)}{2} + [x](x - [x]) = R.H.S$ Proved.

Area

The area between the curve $y = f(x)$, $x -$ axis and two ordinates at the points $x = a$ and $x = b\ (b > a)$

$$A = \int_a^b f(x)\,dx = \int_a^b y\,dx$$

It represents the shaded and non – shaded area in given below figure –

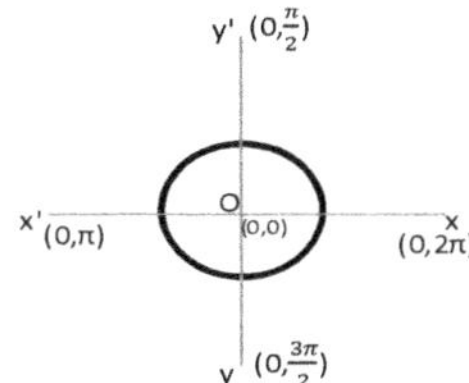

Similarly, if the area be between the curve and y – axis and two abscissas drawn at the

points $y = a$ and $y = b\ (b > a)$ is given by the formula, $\int_a^b x\,dy$

Area between two curves $y = f(x)$ and $y = g(x)$ and the two ordinates drawn at the points

$x = a$ and $x = b$ then, $A = \int_a^b y_1\,dx - \int_a^b y_2\,dx = \text{upper} - \text{lower} = \int_a^b [f(x) - g(x)]\,dx$

where y_1 and y_2 are ordinates y_1 is the ordinate of $y = f(x)$ which is upper curve

and y_2 is the ordinate of $y = g(x)$ which is lower curve.

In the above result we have taken the area to be lying above x – axis and the area to be lying

below x – axis then the value of area will be – ve and hence we take mod. $\therefore\ A = \left|\int_a^b y\,dx\right|$

similarly, if the area be lying to right and left of y – axis then $\therefore\ A = \left|\int_a^b x\,dy\right|$

Consider the following graph: – draw

we shall divide $[a,b]$ into $[a,x_1] + [x_1,x_2] + [x_2,x_3] + [x_3,x_4] + [x_4,x_5] + [x_5,x_6] + [x_6,x_7] + [x_7,b]$

or $\int_a^b f(x)\,dx = \left|\int_a^{x_1} y\,dx\right| + \left|\int_{x_1}^{x_2} y\,dx\right| + \left|\int_{x_2}^{x_3} y\,dx\right| + \left|\int_{x_3}^{x_4} y\,dx\right| + \left|\int_{x_4}^{x_5} y\,dx\right| + \left|\int_{x_5}^{x_6} y\,dx\right| + \left|\int_{x_6}^{x_7} y\,dx\right| + \left|\int_{x_7}^{b} y\,dx\right|$

where $\left|\int_a^{x_1} y\,dx\right| = -ve$, $\left|\int_{x_1}^{x_2} y\,dx\right| = +ve$, $\left|\int_{x_2}^{x_3} y\,dx\right| = -ve$, $\left|\int_{x_3}^{x_4} y\,dx\right| = +ve$, $\left|\int_{x_4}^{x_5} y\,dx\right| = -ve$,

$$\left|\int_{x_5}^{x_6} y\,dx\right| = +ve,\ \left|\int_{x_6}^{x_7} y\,dx\right| = -ve,\ \left|\int_{x_7}^{b} y\,dx\right| = +ve$$

From above figure, the area is lying above x – axis as well as below x – axis in the interval [a, b] then we divide the interval [a, b] into various sub – interval. which the area is lying either above x – axis or below x – axis for below x – axis we take mod. Area of a sub – interval $[a, x_1], [x_2, x_3], [x_4, x_5]$ and $[x_6, x_7]$ are lying below x – axis then we take mod.

Area bounded by two curves: –

we have shown in (P – 1)that $A = \int_a^b (y_1 - y_2)\,dx = \int_a^b [f(x) - g(x)]\,dx$ (fig P – II)

where $y_1 \to$ Upper and $y_2 \to$ Lower

The above formula does not change whether the area is above or below x – axis or both.

Draw graph

$$A = \int_a^b (x_1 - x_2)\,dy = \int_a^b [f(y) - g(y)]\,dy \quad \text{where } x_1 \to \text{Right and } x_2 \to \text{Left}$$

Asymptotes

An asymptote is essentially, a line that a graph approaches, but does not intersect.

An asymptote is a line or curve that approaches a given curve arbitrarily closely, as illustrated in the diagram. For example in the following graph of $y = \frac{1}{x}$, the line approaches the x – axis (y = 0) but never touches it.

Draw

The curve has a vertical asymptote at x = 0 and a horizontal asymptote at y = 0.

Vertical Asymptote

Vertical asymptotes correspond to the 0 of the denominator of a rational function.

Let $f(x) = \frac{N(x)}{D(x)}$ be a rational function. the line x = c is a vertical asymptote of the graph of f(x)

If D(x) = 0 and N(c) ≠ 0.

Example: – find the vertical asymptote of $f(x) = \frac{x^2 + 3x - 2}{x^2 - 1}$.

Solution: – $f(x) = \frac{x^2 + 3x - 2}{x^2 - 1}$, $D(x) = x^2 - 1$ and $N(x) = x^2 + 3x - 2$

To be vertical asymptote D(x) = 0 and N(x) ≠ 0

$$D(x) = 0 \text{ or } x^2 - 1 = 0 \quad \therefore (x - 1)(x + 1) = 0 \quad \therefore x = 1, -1$$

$N(x) \neq 0$ or $N(x) = x^2 + 3x - 2$ $\therefore N(1) = 1 + 3 - 2 = 2 \neq 0$ and $N(-1) = 1 - 3 - 2 = -4 \neq 0$

Since D(1) = 0 and N(1) ≠ 0 so, x = 1 is vertical asymptote.

Horizontal Asymptote

Let y = f(x) be a function. suppose that $\lim_{x \to \infty} f(x) = L$ or $\lim_{x \to -\infty} f(x) = M$

we refer to the each of the line y = L and y = M as a horizontal asymptote of the function f(x).

Let $f(x) = \dfrac{a_m x^m + a_{m-1}x^{m-1} + \cdots \ldots\ldots\ldots\ldots\ldots\ldots\ldots\ldots\ldots\ldots . +a_1x + a_0}{b_n x^n + b_{n-1}x^{n-1} + \cdots \ldots\ldots\ldots\ldots\ldots\ldots\ldots\ldots\ldots\ldots . +b_1x + b_0}$,

$a_m \neq 0,\ b_n \neq 0$ be a rational function.

\# If $m < n$ *then* $y = 0$ *is the horizontal asymptote.*

\# If $m = n$ then $y = \dfrac{a_m}{b_m}$ is the horizontal asymptote.

\# If $m > n$ *then there is no horizontal asymptote.*

Example: − find the horizontal asymptote of $f(x) = \dfrac{3x^3 - 2x^2 + 2}{2x^3 + 3x - 5}$.

Solution: − Here $m = 3,\ n = 3,\ a_m = 3,\ b_n = 2$ $\therefore$ This is the case $m = n$

so, the horizontal asymptote is $\boxed{y = \dfrac{a_m}{b_m} = \dfrac{3}{2}}$ Ans.

Example: − find the horizontal asymptote of $f(x) = \dfrac{2x^4 - x^2 + 3}{5x^5 - 20}$.

Solution: − Here $m = 4,\ n = 5,\ a_m = 2,\ b_n = 5$ $\therefore$ This is the case $m < n$

so, the horizontal asymptote is $\boxed{y = 0}$ Ans.

Example: − find the horizontal asymptote of $f(x) = \dfrac{2x^4 - x^3 - 2}{6x^3 + 2x^2 - 5}$.

Solution: − Here $m = 4,\ n = 3,\ a_m = 2,\ b_n = 6$ $\therefore$ This is the case $m > n$

so, there is no horizontal asymptote. Ans.

Symmetry

(a) If the equation of the curve involves even and only even powers of x, then that curve is symmetrical about y − axis.

e.g. Parabola, $x^2 = 4ay$ or $x^2 - 4ay = 0$

Let $f(x) = x^2 - 4ay$, $f(-x) = x^2 - 4ay$ $\therefore$ $f(x) = f(-x)$

Hence this parabola is symmetrical about y − axis.

(b) If the equation of the curve involves even and only even powers of y, then that curve is symmetrical about x − axis.

e.g. Parabola, $y^2 = 4ax$ or $y^2 - 4ax = 0$

Let $f(y) = y^2 - 4ax$, $f(-y) = y^2 - 4ax$ $\therefore$ $f(y) = f(-y)$

Put $y = -y$ in equation $y^2 = 4ax$ then $y^2 = 4ax$ no any change in the equation $y^2 = 4ax$

Hence then that curve is symmetrical about x − axis.

(c) The curve involves even and only even powers of both x and y then it is symmetrical about both the axes.

e.g. $x^2 + y^2 = 1$ and $\dfrac{x^2}{m^2} + \dfrac{y^2}{n^2} = 1$, Put $x = -x$ and $y = -y$ then $x^2 + y^2 = 1$ and $\dfrac{x^2}{m^2} + \dfrac{y^2}{n^2} = 1$

which is no any change in the curve. hence it is symmetrical about both the axes. (x and y)

(d) If x and y be interchanged and the equation of the curve does not change then it is symmetrical about the line $y = x$. i.e a line through origin making an angle of 45^0 with $x -$ axis.

Point on the curve

Find the point on the curve $x^2 + y^2 + 2x = 3$ ……………… (i)

Put $y = 0$ in equation (i) and find x,

$\therefore\ x^2 + 0 + 2x = 3$ or $x^2 + 2x - 3 = 0$ or $(x-1)(x+3) = 0$

$\therefore\ x = 1, -3 \qquad \therefore$ points are $(1,0)\ \&\ (-3,0)$

Again, Put $x = 0$ in equation (i) and find y, $x^2 + y^2 + 2x = 3$ or $y^2 = 3$ or $y = \pm\sqrt{3}$

$\therefore$ point $= (0, \sqrt{3})\ \&\ (0, -\sqrt{3})$

Points are $(1,0), (-3,0), (0,\sqrt{3})$ and $(0, -\sqrt{3})$.

Draw graph

Tangents

Tangent at origin (0,0) are obtained by equating to zero (0) the lowest degree terms in the equation of the curve $y^2 = 4ax$, $T_{(0,0)} = 4ax = 0$ i.e $x = 0$ or $y -$ axis and curve $x^2 = 4ay$, $T_{(0,0)} = 4ay = 0$ i.e $y = 0$ or $x -$ axis. curve $x^2 + y^2 = 2axy$, $T_{(0,0)} = 2axy = 0$ i.e $x = 0$ and $y = 0$ i.e both the axes are tangents at the origin. Tangent at any other point is given by $\frac{dy}{dx} = -\frac{f_x}{f_y}$ and

slope of the tangent at (h, k), $\left.\frac{dy}{dx}\right|_{(h,k)} = -\frac{f_x}{f_y}$.

Example: $-$ (1) Find the area between the $x -$ axis, the graph of $y = x$ and $x = 2$.

Solution: $-$ Ist Method: $-$ by elementary geometry $-$ Area $\Delta OAB = \frac{OA \times AB}{2}$

$$\text{Area} = \frac{2 \times 2}{2} = 2 \quad \text{Ans.}$$

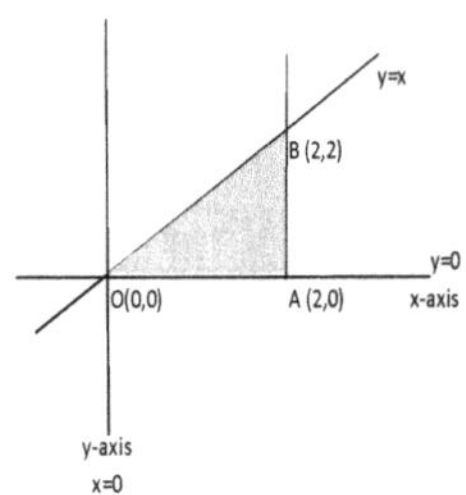

IInd Method: $-$ Area (A) $= \int_0^2 (y_1 - y_2)\,dx = \int_0^2 (x - 0)\,dx = \int_0^2 x\,dx = \left(\frac{x^2}{2}\right)_0^2 = \frac{4}{2} - 0 = 2$ Ans.

Example: $-$ (2) Find the area bounded by $y = 3x, x = 3, x = 5$ and the $x -$ axis.

Solution: $-$ Ist Method: $-$ find the area of ABCD $= AB\left(\frac{AD + BC}{2}\right) = 2\left(\frac{9 + 15}{2}\right) = 24$ Ans.

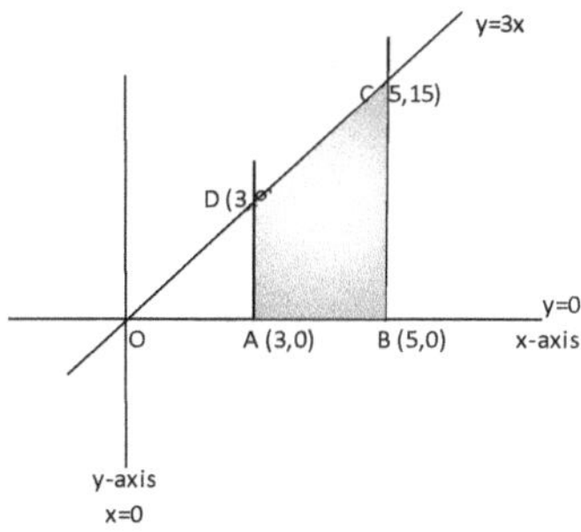

IInd Method: – Area $= \int_3^5 (y_1 - y_2)\,dx = \int_3^5 (3x - 0)\,dx = \int_3^5 3x\,dx = \left(\frac{3x^2}{2}\right)_3^5 = \frac{75}{2} - \frac{27}{2} = \frac{75-27}{2} = \frac{48}{2} = 24$ Ans.

Example: – (3) Find the area under $x = y^2$ from $y = 2$ to $y = 3$ above the x – axis and right sides of y – axis.

Solution: – Area $= \int_2^3 (x_2 - x_1)\,dy = \int_2^3 (y^2 - 0)\,dy = \int_2^3 y^2\,dy = \left(\frac{y^3}{3}\right)_2^3 = \frac{27}{3} - \frac{8}{3} = \frac{27-8}{3} = \frac{19}{3}$ Ans.

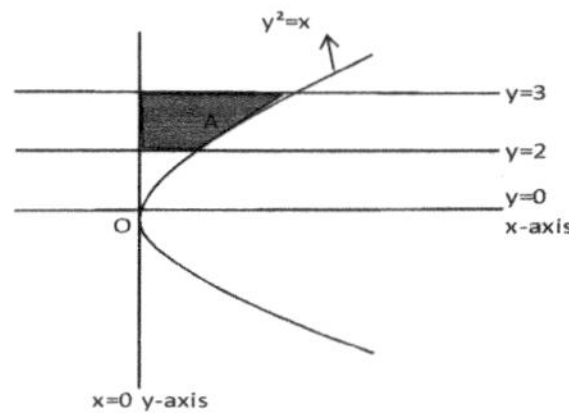

Example: – (4) Find the area between the parabola $y^2 = 4ax$ and $x^2 = 4by$.

Solution: – The two curves meet at point P.

or $y^2 = 4ax$ ………….(i) and $x^2 = 4by$ ……………..(ii)

solving the equation (i) and (ii), we have

From $x^2 = 4by$, $y = \frac{x^2}{4b}$ and put value of y in equation (i), we get

or $y^2 = 4ax$ or $\left(\frac{x^2}{4b}\right)^2 = 4ax$ or $x^4 = 64ab^2x$ or $x^4 - 64ab^2x = 0$ or $x(x^3 - 64ab^2) = 0$

$\therefore\ x = 0$ and $x^3 - 64ab^2 = 0$ or $x^3 = 64ab^2$ $\therefore\ x = 4.a^{\frac{1}{3}}.b^{\frac{2}{3}}$

Required Area $= \int_0^{4.a^{\frac{1}{3}}.b^{\frac{2}{3}}} (y_1 - y_2)\,dx$ ………………….(A)

where y_1 is the ordinate of the upper parabola $y^2 = 4ax$ $\therefore\ y = 2.\sqrt{a}.\sqrt{x}$ or $y_1 = 2.\sqrt{a}.\sqrt{x}$

and y_2 is the ordinate of the lower parabola $x^2 = 4by$ $\therefore\ y = \frac{x^2}{4b}$ or $y_2 = \frac{x^2}{4b}$

Putting for y_1 and y_2 in equation (A), we get

Required Area $= \int_0^{4.a^{\frac{1}{3}}.b^{\frac{2}{3}}} (y_1 - y_2)\,dx = \int_0^{4.a^{\frac{1}{3}}.b^{\frac{2}{3}}} \left(2.\sqrt{a}.\sqrt{x} - \frac{x^2}{4b}\right) dx = \int_0^{4.a^{\frac{1}{3}}.b^{\frac{2}{3}}} (2.\sqrt{a}.\sqrt{x})\,dx - \int_0^{4.a^{\frac{1}{3}}.b^{\frac{2}{3}}} \left(\frac{x^2}{4b}\right) dx$

$= 2\sqrt{a}\int_0^{4.a^{\frac{1}{3}}.b^{\frac{2}{3}}} x^{\frac{1}{2}}\,dx - \frac{1}{4b}\int_0^{4.a^{\frac{1}{3}}.b^{\frac{2}{3}}} x^2\,dx = 2\sqrt{a}.\left(\frac{x^{3/2}}{3/2}\right)_0^{4.a^{\frac{1}{3}}.b^{\frac{2}{3}}} - \frac{1}{4b}.\left(\frac{x^3}{3}\right)_0^{4.a^{\frac{1}{3}}.b^{\frac{2}{3}}}$

$$A = \frac{4\sqrt{a}}{3}\left[\left(4.a^{\frac{1}{3}}.b^{\frac{2}{3}}\right)^{3/2} - 0\right] - \frac{1}{12b}\left[\left(4.a^{\frac{1}{3}}.b^{\frac{2}{3}}\right)^{3} - 0\right] = \frac{4\sqrt{a}}{3}\left[(2)^{2.\frac{3}{2}}.(a)^{\frac{1}{3}\frac{3}{2}}.(b)^{\frac{2}{3}\frac{3}{2}}\right] - \frac{1}{12b}\left[4^3.(a)^{\frac{1}{3}.3}.(b)^{\frac{2}{3}.3}\right] = \frac{4\sqrt{a}}{3}\left[8.a^{\frac{1}{2}}.b\right] - \frac{1}{12b}[64ab^2]$$

$$= \frac{32ab}{3} - \frac{16ab}{3} = \frac{32ab - 16ab}{3} = \frac{16ab}{3} \quad \text{Ans.}$$

Note: – Two parabolas be $y^2 = 4ax$ and $x^2 = 4ay$ then putting $a = b$ the required area is $\frac{16}{3}a^2$ or $\frac{16}{3}b^2$.

Example: – (5) Find the ratio of the area bounded by curves $y^2 = 8x$ and $x^2 = 8y$ is devided by the line $x = 2$.

Solution: – Curves are $y^2 = 8x$ and $x^2 = 8y$, we can write the curves $y^2 = 4.2x$ and $x^2 = 4.2x$

see above question and find the area, $y^2 = 4ax$ and $x^2 = 4ay$ where $a = 2$ and $b = 2$

$$\text{Area (A)} = \frac{16ab}{3} = \frac{16.2.2}{3} = \frac{64}{3}$$

Again, $\text{Area }(A_1) = \int_0^2 (y_1 - y_2)\,dx = \int_0^2 \left(2\sqrt{2}.\sqrt{x} - \frac{x^2}{8}\right) dx = 2\sqrt{2}\int_0^2 x^{\frac{1}{2}}\,dx - \frac{1}{8}\int_0^2 x^2\,dx = 2\sqrt{2}\left(\frac{x^{3/2}}{3/2}\right)_0^2 - \frac{1}{8}\left(\frac{x^3}{3}\right)_0^2 = \frac{4\sqrt{2}}{3}(2)^{3/2} - \frac{1}{24}.2^3$

$$= \frac{4\sqrt{2}.2\sqrt{2}}{3} - \frac{8}{24} = \frac{16}{3} - \frac{1}{3} = \frac{16-1}{3} = \frac{15}{3} = 5$$

or $A_2 = A - A_1 = \frac{64}{3} - 5 = \frac{64-15}{3} = \frac{49}{3}$ Ratio is $\frac{A_1}{A_2} = \frac{5}{\frac{49}{3}} = \frac{15}{49}$ Ans.

Example: – (6) Find the area bounded by the curves $y^2 = 12x$ and $x^2 = 16y$.

Solution: – Curves are $y^2 = 12x$ and $x^2 = 16y$,

we can write the curves $y^2 = 4.3x$ and $x^2 = 4.4y$

or $y^2 = 4ax$ and $x^2 = 4by$ [see question no. –(4)]

compare the curves, $a = 3$ and $b = 4$ then $\text{Area (A)} = \frac{16ab}{3} = \frac{16.3.4}{3} = 64$ Ans.

Example: – (7) Find the area between the parabola $y^2 = 4ax$ and line $y = mx$.

Solution: – The parabola $y^2 = 4ax$ ………………(i) and line $y = mx$ ………………….(ii)

From (i), ⇒ $y^2 = 4ax$ or $(mx)^2 = 4ax$ or $m^2x^2 = 4ax$ or $m^2x^2 - 4ax = 0$

or $x(m^2x - 4a) = 0$ ∴ $x = 0$ or $m^2x - 4a = 0$ ∴ $x = \frac{4a}{m^2}$

The required area $A = \int_0^{\frac{4a}{m^2}} (y_1 - y_2)\,dx$ ………………….(A)

∴ $y^2 = 4ax$ or $y = 2\sqrt{a}\sqrt{x}$ or $y_1 = 2\sqrt{a}\sqrt{x}$ (upper curves)

and $y = mx$ or $y_2 = mx$ (Lower curves)

Put y_1 and y_2 in equation (A), we get ∴ $\text{Area (A)} = \int_0^{\frac{4a}{m^2}} (y_1 - y_2)\,dx = \int_0^{\frac{4a}{m^2}} (2\sqrt{a}\sqrt{x} - mx)\,dx$

$$A = \int_0^{\frac{4a}{m^2}} (2\sqrt{a}\sqrt{x})\,dx + \int_0^{\frac{4a}{m^2}} (mx)\,dx = 2\sqrt{a}\int_0^{\frac{4a}{m^2}} x^{\frac{1}{2}}\,dx + m\int_0^{\frac{4a}{m^2}} x\,dx = 2\sqrt{a}\left(\frac{x^{3/2}}{3/2}\right)_0^{\frac{4a}{m^2}} - m\left(\frac{x^2}{2}\right)_0^{\frac{4a}{m^2}} = \frac{4\sqrt{a}}{3}\left(\frac{4a}{m^2}\right)^{\frac{3}{2}} - \frac{m}{2}\left(\frac{4a}{m^2}\right)^2$$

$$= \frac{4\sqrt{a}}{3m^3}.(4)^{\frac{3}{2}}.a^{\frac{3}{2}} - \frac{16a^2}{2m^3} = \frac{32a\sqrt{a}\sqrt{a}}{3m^3} - \frac{16a^2}{2m^3} = \frac{32a^2}{3m^3} - \frac{8a^2}{m^3} = \frac{32a^2 - 24a^2}{3m^3}$$

$A = \frac{8a^2}{3m^3} = \frac{8}{3}\left(\frac{a^2}{m^3}\right)$ Ans.

Note: – If the curves were $y^2 = 10x$ and $y = 2x$ then $4a = 10$ $\therefore a = \frac{10}{4} = \frac{5}{2}$ and $m = 2$

$$\text{Area (A)} = \frac{8}{3}\left(\frac{a^2}{m^3}\right) = \frac{8}{3}.\frac{\left(\frac{5}{2}\right)^2}{(2)^3} = \frac{8 \times 25}{3 \times 4 \times 8} = \frac{25}{12} \text{ square units} \quad \text{Ans.}$$

Example: – (8) Find the area cut off the parabola $3y = x^2$ by the straight line $2y = x + 3$.

Solution: – The parabola $3y = x^2$ ……………(i) and line $2y = x + 3$ ………………..(ii)

From (ii), $2y = x + 3$ $\therefore$ $y = \frac{x+3}{2}$ Put y in equation (i), we get

$$\therefore 3y = x^2 \text{ or } 3\left(\frac{x+3}{2}\right) = x^2 \text{ or } 3x + 9 = 2x^2 \text{ or } 2x^2 - 3x - 9 = 0 \text{ or } 2x^2 - 6x + 3x - 9 = 0$$

$$\therefore 2x(x-3) + 3(x-3) = 0 \text{ or } (2x+3)(x-3) = 0 \quad \therefore x = -\frac{3}{2}, 3$$

If $x = 3$ then $y = \frac{3+3}{2} = \frac{6}{2} = 3$ and $x = -\frac{3}{2}$ then $y = \frac{-\frac{3}{2}+3}{2} = \frac{-3+6}{4} = \frac{3}{4}$

The Points of intersection are P (3,3) and $Q\left(-\frac{3}{2}, \frac{3}{4}\right)$

$$\text{Area (A)} = \int_{-\frac{3}{2}}^{3}(y_1 - y_2)\,dx = \int_{-\frac{3}{2}}^{3}\left(\frac{x+3}{2} - \frac{x^2}{3}\right)dx = \int_{-\frac{3}{2}}^{3}\left(\frac{3x+9-2x^2}{6}\right)dx = \frac{1}{6}\int_{-\frac{3}{2}}^{3}(3x+9-2x^2)\,dx = \frac{1}{6}\left[3.\frac{x^2}{2} + 9x - 2.\frac{x^3}{3}\right]_{-\frac{3}{2}}^{3}$$

$$= \frac{1}{6}\left\{\left[3.\frac{9}{2} + 9.3 - 2.\frac{27}{3}\right] - \left[3.\frac{\left(-\frac{3}{2}\right)^2}{2} + 9.\left(-\frac{3}{2}\right) - 2.\frac{\left(-\frac{3}{2}\right)^3}{3}\right]\right\}$$

$$A = \frac{1}{6}\left[\frac{27}{2} + 27 - 18\right] - \frac{1}{6}\left[\frac{27}{8} - \frac{27}{2} + \frac{9}{4}\right] = \frac{1}{6}\left[\frac{27+18}{2}\right] - \frac{1}{6}\left[\frac{27-108+18}{8}\right] = \frac{45}{12} + \frac{63}{48} = \frac{135+63}{48} = \frac{198}{48} = \frac{33}{8} \text{ sq. units} \quad \text{Ans.}$$

Example: – (9) Find the area bounded by the curve $x^2 = 2y$ and the straight line $x = 3y - 4$.

Solution: – The curve is $x^2 = 2y$ or $y = \frac{x^2}{2}$ ……………(i)

and line $x = 3y - 4$ or $y = \frac{x+4}{3}$ …………(ii)

From (i), $\Rightarrow$ $x^2 = 2y$ or $x^2 = 2.\left(\frac{x+4}{3}\right)$ or $3x^2 = 2x + 8$ or $3x^2 - 2x - 8 = 0$

$$\therefore x = \frac{-b \pm \sqrt{b^2 - 4ac}}{2a} \quad \text{here } a = 3,\ b = -2 \text{ and } c = -8$$

then $x = \frac{2 \pm \sqrt{4+96}}{6} = \frac{2 \pm 10}{6} = \frac{2+10}{6}$ or $\frac{2-10}{6} = \frac{12}{6}$ or $\frac{-8}{6}$ $\therefore x = 2, -\frac{4}{3}$

If $x = 2$ then $y = \frac{x+4}{3} = \frac{2+4}{3} = 2$ and $x = -\frac{4}{3}$ then $y = \frac{x+4}{3} = \frac{-\frac{4}{3}+4}{3} = \frac{\frac{-4+12}{3}}{3} = \frac{8}{9}$

$$\text{Area (A)} = \int_{-\frac{4}{3}}^{2}(y_1 - y_2)\,dx = \int_{-\frac{4}{3}}^{2}\left(\frac{x+4}{3} - \frac{x^2}{2}\right)dx = \int_{-\frac{4}{3}}^{2}\left(\frac{x+4}{3}\right)dx - \int_{-\frac{4}{3}}^{2}\left(\frac{x^2}{2}\right)dx = \frac{1}{3}\left(\frac{x^2}{2} + 4x\right)_{-\frac{4}{3}}^{2} - \frac{1}{2}\left(\frac{x^3}{3}\right)_{-\frac{4}{3}}^{2}$$

$$= \frac{1}{3}\left[2 + 8 - \frac{8}{9} + \frac{16}{3}\right] - \frac{1}{2}\left[\frac{8}{3} + \frac{64}{81}\right] = \frac{1}{3}\left[\frac{90-8+48}{9}\right] - \frac{1}{2}\left[\frac{216+64}{81}\right] = \frac{130}{27} - \frac{140}{81} = \frac{390-140}{81} = \frac{250}{81} \quad \text{Ans.}$$

Example: – (10) Find the area bounded by the curves $x = y^2$ and $x = \frac{3}{y^2 - 2}$.

Solution: – The points of intersection are given, $x = y^2$ and $x = \frac{3}{y^2 - 2}$

or $y^2 = \frac{3}{y^2 - 2}$ or $y^4 - 2y^2 = 3$ or $y^4 - 2y^2 - 3 = 0$ or $y^2(y^2 - 3) + 1(y^2 - 3) = 0$

or $(y^2 + 1)(y^2 - 3) = 0$ $\therefore$ $(y^2 + 1) = 0$ it is not possible.

or $(y^2 - 3) = 0$ $\therefore$ $y^2 = 3$ $\therefore$ $y = \pm\sqrt{3}$

If $y = \sqrt{3}$ then $x = y^2 = (\sqrt{3})^2 = 3$ and $y = -\sqrt{3}$ then $x = y^2 = (-\sqrt{3})^2 = 3$

The points of intersection $P(3, \sqrt{3})$ and $Q(3, -\sqrt{3})$.

$$\text{Area (A)} = 2\int_0^{\sqrt{3}} (x_1 - x_2)\, dy = 2\int_0^{\sqrt{3}} \left(\frac{3}{y^2 - 2} - y^2\right) dy = 2\int_0^{\sqrt{3}} \frac{3}{y^2 - 2} dy - 2\int_0^{\sqrt{3}} y^2\, dy$$

$$\left[\text{use formula, } \int \frac{dx}{x^2 - a^2} = \frac{1}{2a}\log\left|\frac{x - a}{x + a}\right|\right]$$

$$A = 6.\left[\frac{1}{2.\sqrt{2}}\log\left|\frac{y - \sqrt{2}}{y + \sqrt{2}}\right|\right]_0^{\sqrt{3}} - 2\left[\frac{y^3}{3}\right]_0^{\sqrt{3}} = \frac{3}{\sqrt{2}}\left[\log\left|\frac{\sqrt{3} - \sqrt{2}}{\sqrt{3} + \sqrt{2}}\right| - \log\left|\frac{-\sqrt{2}}{\sqrt{2}}\right|\right] - 2\left[\frac{3\sqrt{3}}{3} - 0\right] = \frac{3}{\sqrt{2}}\log\left|\frac{\sqrt{3} - \sqrt{2}}{\sqrt{3} + \sqrt{2}}\right| - 2\sqrt{3}$$

$$= \frac{3}{\sqrt{2}}\log\left|\frac{\sqrt{3} - \sqrt{2}}{\sqrt{3} + \sqrt{2}} \times \frac{\sqrt{3} - \sqrt{2}}{\sqrt{3} - \sqrt{2}}\right| - 2\sqrt{3} = \frac{3}{\sqrt{2}}\log\left|\frac{(\sqrt{3} - \sqrt{2})^2}{3 - 2}\right| - 2\sqrt{3} = \frac{3}{\sqrt{2}}\log(\sqrt{3} - \sqrt{2})^2 - 2\sqrt{3} = \frac{3.2}{\sqrt{2}}\log(\sqrt{3} - \sqrt{2}) - 2\sqrt{3}$$

$$= 3\sqrt{2}\log(1.732 - 1.414) - 2\sqrt{3} = 3 \times 1.414\log(0.318) - 2 \times 1.732$$

$A = 4.242\log(0.318) - 3.464 = 4.242(-0.4975) - 3.464 = -2.110395 - 3.464 = -1.353605$ $[\log(0.318) = -0.4975]$

Area does not negative so Area (A) = 1.35 sq. units (approx.) Ans.

Example: – (11) The area of the region bounded by the curve $y = 2x + x^2$ and the line $y = 2mx$ equals $\frac{32}{3}$? find the following value of m.

Solution: – The two curves meet at $y = 2x + x^2$ and $y = 2mx$

from, $y = 2x + x^2$ or $2mx = 2x + x^2$ or $x^2 + 2x - 2mx = 0$ or $x^2 + 2x(1 - m) = 0$

or $x[x + 2(1 - m)] = 0$ $\therefore$ $x = 0$ and $x + 2(1 - m) = 0$ $\therefore x = -2(1 - m) = 2m - 2$

If $m > 1$ *then* $2m - 2$ *is positive, Area* $(A) = \int_0^{2m-2} (y_1 - y_2)\, dx = \int_0^{2m-2} (2x + x^2 - 2mx)\, dx$

$$A = \left[2.\frac{x^2}{2} + \frac{x^3}{3} - 2m.\frac{x^2}{2}\right]_0^{2m-2} = \left[x^2 + \frac{x^3}{3} - mx^2\right]_0^{2m-2} = \left[(2m - 2)^2 + \frac{(2m - 2)^3}{3} - m(2m - 2)^2\right]$$

$$A = \frac{32}{3} = \left[\frac{3(2m - 2)^2 + (2m - 2)^3 - 3m(2m - 2)^2}{3}\right] = \frac{(2m - 2)^2}{3}[3 + 2m - 2 - 3m] = \frac{(2m - 2)^2}{3}[1 - m]$$

$$\text{or } 32 = (2m - 2)^2(1 - m) = -[2(m - 1)]^2(m - 1) = -4(m - 1)^3 \quad \therefore -\frac{32}{4} = (m - 1)^3$$

$$\text{or } -8 = (m - 1)^3 \quad \therefore (-2)^3 = (m - 1)^3 \text{ or } m - 1 = -2 \quad \therefore m = -2 + 1 = -1$$

If $m < 1$ *then* $2m - 2$ *is negative, Area* $(A) = \int_{2m-2}^{0} (y_1 - y_2)\, dx = \int_{2m-2}^{0} (2x + x^2 - 2mx)\, dx$

$$A = \frac{32}{3} = \left[2.\frac{x^2}{2} + \frac{x^3}{3} - 2m.\frac{x^2}{2}\right]_{2m-2}^{0} = \left[x^2 + \frac{x^3}{3} - mx^2\right]_{2m-2}^{0} = -\left[(2m-2)^2 + \frac{(2m-2)^3}{3} - m(2m-2)^2\right]$$

$$= -\left[\frac{3(2m-2)^2 + (2m-2)^3 - 3m(2m-2)^2}{3}\right] = -\frac{(2m-2)^2}{3}[3 + 2m - 2 - 3m]$$

or $\frac{32}{3} = \frac{(2m-2)^2}{3}[m-1]$ or $32 = [2(m-1)]^2(m-1) = 4(m-1)^3$ or $\frac{32}{4} = (m-1)^3$

or $8 = (m-1)^3$ or $2^3 = (m-1)^3$ or $m - 1 = 2$ or $m = 3$ $\therefore$ $m = -1,3$ Ans.

Example: – (12) The area bounded by the curve $y = f(x)$and the lines $x = 0, y = 0$

and $x = t$, lies in the interval $\left(\frac{3}{4}, 3\right)$.

Solution: – $-\frac{3}{4} < s < -\frac{1}{2}$ and $\frac{1}{2} < t < \frac{3}{4}$

or $\int_0^{\frac{1}{2}}(4x^3 + 3x^2 + 2x + 1)\,dx < area < \int_0^{\frac{3}{4}}(4x^3 + 3x^2 + 2x + 1)\,dx$

or $[x^4 + x^3 + x^2 + x]_0^{\frac{1}{2}} < area < [x^4 + x^3 + x^2 + x]_0^{\frac{3}{4}}$

or $\frac{1}{16} + \frac{1}{8} + \frac{1}{4} + \frac{1}{2} < area < \frac{81}{256} + \frac{27}{64} + \frac{9}{16} + \frac{3}{4}$ $\therefore$ $\frac{15}{16} < area < \frac{525}{256}$ Ans.

Example: – (13) The area enclosed by the curves $y = \sin x + \cos x$ and $y = |\cos x - \sin x|$ over the interval $\left[0, \frac{\pi}{2}\right]$.

Solution: – Given, $y = \sin x + \cos x$ and $y = |\cos x - \sin x|$

or $y_1 = \sin x + \cos x = \sqrt{2}\left[\sin x \cos\frac{\pi}{4} + \cos x \sin\frac{\pi}{4}\right] = \sqrt{2}\sin\left(x + \frac{\pi}{4}\right)$

similarly, $y_2 = |\cos x - \sin x| = \sqrt{2}\left|\sin\left(\frac{\pi}{4} - x\right)\right|$

$$\Rightarrow \text{Area (A)} = \int_0^{\frac{\pi}{4}}[(\sin x + \cos x) - (\cos x - \sin x)]\,dx + \int_{\frac{\pi}{4}}^{\frac{\pi}{2}}[(\sin x + \cos x) - (\sin x - \cos x)]\,dx$$

$$A = \int_0^{\frac{\pi}{4}} 2\sin x\,dx + \int_{\frac{\pi}{4}}^{\frac{\pi}{2}} 2\cos x\,dx = 2(-\cos x)_0^{\frac{\pi}{4}} + 2(\sin x)_{\frac{\pi}{4}}^{\frac{\pi}{2}} = -2\left(\frac{1}{\sqrt{2}} - 1\right) + 2\left(1 - \frac{1}{\sqrt{2}}\right) = 2 - \frac{2}{\sqrt{2}} + 2 - \frac{2}{\sqrt{2}} = 4 - \frac{2}{\sqrt{2}} - \frac{2}{\sqrt{2}} = 4 - \sqrt{2} - \sqrt{2}$$

$= 4 - 2\sqrt{2}$ Ans.

Example: – (14) Prove that the area between the parabola $x^2 = 4y$, $x^2 = y$ and $y = 4$, $y = 9$ is $\frac{76}{3}$ sq. units.

Solution: – Given, $x^2 = 4y$ and $x^2 = y$ or $x_1 = 2\sqrt{y}$ and $x_2 = \sqrt{y}$

$$\text{Area (A)} = 2\int_4^9 (x_1 - x_2)\,dy = 2\int_4^9 (2\sqrt{y} - \sqrt{y})\,dy = 2\int_4^9 \sqrt{y}\,dy = 2\int_4^9 y^{\frac{1}{2}}\,dy = 2\left[\frac{y^{\frac{1}{2}+1}}{\frac{1}{2}+1}\right]_4^9 = 2\left[\frac{y^{\frac{3}{2}}}{\frac{3}{2}}\right]_4^9 = \frac{4}{3}\left[9^{\frac{3}{2}} - 4^{\frac{3}{2}}\right] = \frac{4}{3}\left[(3^2)^{\frac{3}{2}} - (2^2)^{\frac{3}{2}}\right]$$

$= \frac{4}{3}(3^3 - 2^3) = \frac{4}{3}(27 - 8) = \frac{4}{3}.19 = \frac{76}{3}$ sq. units Proved.

Example: – (15) Find the area bounded by the curve $x = (y + 1)(y - 1)(y - 2)$ lying between the ordinate $y = 0$ and $y = 2$.

Solution: – Given, $x = (y + 1)(y - 1)(y - 2)$

$\therefore x = 0$ then $y = -1,1,2$ and $y = 0, x = 2, y = 2, x = 0$

$x = +ve$ for $y > 2$, $x = -ve\ for\ y < -1$, $x = +ve\ for\ 0 < y < 1$,

$x = -ve$ for $1 < y < 2$, $x = +ve\ for - 1 < y < 0$

$$\text{Area (A)} = \int_0^2 |x|\,dy = \int_0^1 x\,dy + \int_1^2 (-x)\,dy = \int_0^1 (y+1)(y-1)(y-2)\,dy - \int_1^2 (y+1)(y-1)(y-2)\,dy$$
$$= \int_0^1 (y^3 - 2y^2 - y + 2)\,dy - \int_1^2 (y^3 - 2y^2 - y + 2)\,dy$$

$$A = \left[\frac{y^4}{4} - \frac{2}{3}y^3 - \frac{y^2}{2} + 2y\right]_0^1 - \left[\frac{y^4}{4} - \frac{2}{3}y^3 - \frac{y^2}{2} + 2y\right]_1^2 = \left[\frac{1}{4} - \frac{2}{3} - \frac{1}{2} + 2 - 0\right] - \left[\frac{16}{4} - \frac{16}{3} - \frac{4}{2} + 4 - \frac{1}{4} + \frac{2}{3} + \frac{1}{2} - 2\right]$$

$$A = \left[\frac{3 - 8 - 6 + 24}{12}\right] - \left[4 - \frac{16}{3} - 2 + 4 - \frac{1}{4} + \frac{2}{3} + \frac{1}{2} - 2\right] = \frac{13}{12} - \left[4 - \frac{16}{3} - \frac{1}{4} + \frac{2}{3} + \frac{1}{2}\right] = \frac{13}{12} - \left[\frac{48 - 64 - 3 + 8 + 6}{12}\right] = \frac{13}{12} - \frac{62 - 67}{12}$$
$$= \frac{13}{12} + \frac{5}{12} = \frac{18}{12} = \frac{3}{2} = 1\frac{1}{2} \text{ units} \quad \text{Ans.}$$

Example: – (16) Prove that the area bounded by the hyperbola $4x^2 - 9y^2 = a^2$ between the

straight lines $x = 2a$ and $x = 3a$ is $\frac{a}{3}(3\sqrt{35} - 2\sqrt{15}) + \frac{a^2}{6}\log\left(\frac{4 + \sqrt{15}}{6 + \sqrt{35}}\right)$ sq. units.

Solution: – Given, $4x^2 - 9y^2 = a^2$ or $9y^2 = 4x^2 - a^2$ or $y^2 = \frac{4x^2 - a^2}{9}$ or $y = \frac{\sqrt{4x^2 - a^2}}{3}$

$$\text{Area (A)} = \int_{2a}^{3a} \frac{\sqrt{4x^2 - a^2}}{3}\,dx = \frac{1}{3}\int_{2a}^{3a} \sqrt{4x^2 - a^2}\,dx = \frac{1}{3}\int_{2a}^{3a} \sqrt{(2x)^2 - a^2}\,dx = \frac{1}{3}\left[\frac{2x}{2}\sqrt{4x^2 - a^2} - \frac{a^2}{2}\log\left(2x + \sqrt{4x^2 - a^2}\right)\right]_{2a}^{3a}$$
$$= \frac{1}{3}\left[x\sqrt{4x^2 - a^2} - \frac{a^2}{2}\log\left(2x + \sqrt{4x^2 - a^2}\right)\right]_{2a}^{3a}$$

$$\left[\text{use formula, } \int \sqrt{x^2 - a^2}\,dx = \frac{x}{2}\sqrt{x^2 - a^2} - \frac{a^2}{2}\log\left(x + \sqrt{x^2 - a^2}\right)\right]$$

$$A = \frac{1}{3}\left\{3a\sqrt{4(3a)^2 - a^2} - \frac{a^2}{2}\log\left(2.3a + \sqrt{4(3a)^2 - a^2}\right) - \left[2a\sqrt{4(2a)^2 - a^2} - \frac{a^2}{2}\log\left(2.2a + \sqrt{4(2a)^2 - a^2}\right)\right]\right\}$$

$$A = \frac{1}{3}\left[3a\sqrt{36a^2 - a^2} - \frac{a^2}{2}\log\left(6a + \sqrt{36a^2 - a^2}\right) - 2a\sqrt{16a^2 - a^2} + \frac{a^2}{2}\log\left(4a + \sqrt{16a^2 - a^2}\right)\right]$$

$$A = \frac{1}{3}\left[3a\sqrt{35a^2} - \frac{a^2}{2}\log\left(6a + \sqrt{35a^2}\right) - 2a\sqrt{15a^2} + \frac{a^2}{2}\log\left(4a + \sqrt{15a^2}\right)\right]$$

$$A = \frac{1}{3}\left[3\sqrt{35}a - 2\sqrt{15}a + \frac{a^2}{2}\log(4a + \sqrt{15}a) - \frac{a^2}{2}\log(6a + \sqrt{35}a)\right] = \frac{a}{3}(3\sqrt{35} - 2\sqrt{15}) + \frac{a^2}{6}\log\left(\frac{4a + \sqrt{15}a}{6a + \sqrt{35}a}\right)$$

$$A = \frac{a}{3}(3\sqrt{35} - 2\sqrt{15}) + \frac{a^2}{6}\log\left(\frac{4 + \sqrt{15}}{6 + \sqrt{35}}\right) \text{ sq. units} \quad \text{Proved.}$$

Example: – (17) Prove that the area common to the parabolas $y = 3x^2$ and $y = x^2 + 2$ is $\frac{8}{3}$ sq. units.

Solution: – Given parabolas $y = 3x^2 \ldots\ldots\ldots (i)$ and $y = x^2 + 2 \ldots\ldots\ldots\ldots\ldots (ii)$

solve equation (i) and (ii), Put $y = 3x^2$ in equation $y = x^2 + 2$ $\therefore$ $3x^2 = x^2 + 2$ or $2x^2 = 2$

$$\therefore\ x^2 = 1 \quad \therefore\ x = \pm 1$$

$$\text{Area (A)} = \int_{-1}^{1} (y_1 - y_2)\,dx = \int_{-1}^{1} (x^2 + 2 - 3x^2)\,dx = \int_{-1}^{1} (2 - 2x^2)\,dx = \left(2x - \frac{2x^3}{3}\right)_{-1}^{1} = \left(2 - \frac{2}{3}\right) - \left(-2 + \frac{2}{3}\right) = 2 - \frac{2}{3} + 2 - \frac{2}{3} = 4 - \frac{2}{3} - \frac{2}{3}$$
$$= \frac{12 - 2 - 2}{3} = \frac{8}{3} \text{ sq. units} \quad \text{Proved.}$$

Example: – (18) Prove that the area bounded by the parabola $2x = y^2$ and the line $x = 4y$ is $\frac{128}{3}$ sq. units.

Solution: – Given parabola $2x = y^2$ and line $x = 4y$

solve, $2x = y^2$ and $x = 4y$ Put $x = 4y$ in $2x = y^2$ or $2.4y = y^2$

or $y^2 - 8y = 0$ or $y(y-8) = 0$ $\therefore$ $y = 0,8$

$$\text{Area (A)} = \int_0^8 \left(4y - \frac{y^2}{2}\right) dy = \left(\frac{4y^2}{2} - \frac{y^3}{6}\right)_0^8 = \frac{256}{2} - \frac{512}{6} = 128 - \frac{256}{3} = \frac{384-256}{3} = \frac{128}{3} \text{ sq. units. Proved.}$$

(19) Find the area bounded by the curve $y = 3x + x^2$ and the straight line $y = x$.

Solution: – Given curve $y = 3x + x^2$ and line $y = x$ Put $y = x$ in the curve $y = 3x + x^2$ and

find the point, $x = 3x + x^2$ or $x^2 + 3x - x = 0$ or $x^2 + 2x = 0$ or $x(x+2) = 0$ $\therefore$ $x = 0, -2$

$$\text{Area (A)} = \int_{-2}^0 (y_1 - y_2)\, dx = \int_{-2}^0 (x - 3x - x^2)\, dx = \int_{-2}^0 (-2x - x^2)\, dx = \left(\frac{-2x^2}{2}\right)_{-2}^0 - \left(\frac{x^3}{3}\right)_{-2}^0 = (0+4) - \left(0 + \frac{8}{3}\right) = 4 - \frac{8}{3} = \frac{12-8}{3}$$

$$= \frac{4}{3} \text{ sq. units Ans.}$$

(20) Find the area in the plane bounded by the curves $x = y + 1$ and $(x-2)^2 = 2(y-1)$.

Solution: – Given, $x = y + 1$ or $y = x - 1$ ……….(i)

and $(x-2)^2 = 2(y-1)$ or $y = \dfrac{(x-2)^2 + 2}{2}$ …………(ii)

solve the equation (i) & (ii), or $(x-2)^2 = 2(x-1-1)$ or $(x-2)^2 = 2(x-2)$

or $x^2 - 4x + 4 = 2x - 4$ or $x^2 - 6x + 8 = 0$ or $x^2 - 4x - 2x + 8 = 0$ or $x(x-4) - 2(x-4) = 0$

or $(x-2)(x-4) = 0$ $\therefore$ $x = 2,4$

$$\text{Area (A)} = \int_2^4 \left[\frac{(x-2)^2 + 2}{2} - (x-1)\right] dx = \int_2^4 \left[\frac{(x-2)^2 + 2 - 2x + 2}{2}\right] dx$$

$$A = \frac{1}{2}\int_2^4 (x^2 - 4x + 4 - 2x + 4)\, dx = \frac{1}{2}\int_2^4 (x^2 - 6x + 8)\, dx = \frac{1}{2}\left[\left(\frac{x^3}{3}\right)_2^4 - \left(\frac{6x^2}{2}\right)_2^4 + (8x)_2^4\right] = \frac{1}{2}\left[\left(\frac{64}{3} - \frac{8}{3}\right) - 3(16-4) + (32-16)\right]$$

$$= \frac{1}{2}\left[\left(\frac{64-8}{3}\right) - 36 + 16\right] = \frac{1}{2}\left[\frac{56}{3} - 20\right] = \frac{1}{2}\left[\frac{56-60}{3}\right] = \frac{1}{2}\left[\frac{-2}{3}\right] = -\frac{1}{3} \quad \text{(Area is not negative)}$$

Area (A) $= \frac{1}{3}$ sq. units Ans.

(21) Find the area bounded by the parabola $x = y^2 - 6$ and the straight line $x = -y$.

Solution: – Put $x = -y$ in the parabola $x = y^2 - 6$, we get

or $-y = y^2 - 6$ or $y^2 + y - 6 = 0$ or $y^2 + 3y - 2y - 6 = 0$ or $y(y+3) - 2(y+3) = 0$

or $(y-2)(y+3) = 0$ $\therefore$ $y = -3,2$

$$\text{Area (A)} = \int_{-3}^2 (x_1 - x_2)\, dy = \int_{-3}^2 (-y - y^2 + 6)\, dy = \left[-\frac{y^2}{2} - \frac{y^3}{3} + 6y\right]_{-3}^2 = \left[-2 - \frac{8}{3} + 12\right] - \left[-\frac{9}{2} + 9 - 18\right] = \left[10 - \frac{8}{3}\right] - \left[-9 - \frac{9}{2}\right]$$

$$= \left(\frac{30-8}{3}\right) - \left(\frac{-18-9}{2}\right) = \frac{22}{3} + \frac{27}{2} = \frac{44+81}{6} = \frac{125}{6} \text{ sq. units Ans.}$$

(22) The area of the region bounded by the curve $x^2 = y + 1$ and $x + y = 1$.

Solution: – Given, $x^2 = y + 1$ ………(i) and $x + y = 1$ ………(ii) solve equation (i) & (ii), or $x^2 = y + 1$ or $y = x^2 - 1$ and $y = 1 - x$

Put $y = 1 - x$ in equation (i), we get or $x^2 = 1 - x + 1 = 2 - x$ or $x^2 + x - 2 = 0$

or $x^2 + 2x - x - 2 = 0$ or $x(x + 2) - 1(x + 2) = 0$ or $(x - 1)(x + 2) = 0$ $\therefore$ $x = -2, 1$

$$\text{Area (A)} = \int_{-2}^{1} (y_1 - y_2)\,dx = \int_{-2}^{1} [(1 - x) - (x^2 - 1)]\,dx = \int_{-2}^{1} (1 - x - x^2 + 1)\,dx = \int_{-2}^{1} (2 - x - x^2)\,dx = \left[2x - \frac{x^2}{2} - \frac{x^3}{3}\right]_{-2}^{1}$$

$$= \left[2 - \frac{1}{2} - \frac{1}{3}\right] - \left[-4 - \frac{4}{2} + \frac{8}{3}\right] = \left(\frac{12 - 3 - 2}{6}\right) - \left(\frac{-24 - 12 + 16}{6}\right) = \frac{7}{6} + \frac{20}{6} = \frac{7 + 20}{6} = \frac{27}{6} = \frac{9}{2} \text{ sq. units} \quad \text{Ans.}$$

Differential Equation

Definitions: – **(a) Differential equations** – consider the following equations

$$\therefore \frac{d^3y}{dx^3}+2\frac{d^2y}{dx^2}-2\frac{dy}{dx}+y=\cos 2x \ldots\ldots\ldots\ldots\ldots (i)$$

$$\therefore \frac{dy}{dx}=\frac{\sqrt{x}}{\sqrt{1+y^2}} \ldots\ldots\ldots (ii),\quad xy\frac{dy}{dx}=\frac{x^2-1}{1-y} \ldots\ldots\ldots (iii)$$

$$\therefore x\frac{\partial z}{\partial \theta}+y\frac{\partial z}{\partial y}+z=0 \ldots\ldots\ldots (iv) \quad \therefore \frac{\partial^2 y}{\partial t^2}=b^2\frac{\partial^2 y}{\partial x^2} \ldots\ldots\ldots (v)$$

$$\therefore \frac{\left[1+\left(\frac{dy}{dx}\right)^2\right]^{\frac{1}{2}}}{\frac{d^2y}{dx^2}}=\rho \ldots\ldots\ldots\ldots (vi)$$

All the above equation are called differential equations.

(b) ordinary and partial differential equations: –

ordinary differential equations are those which involve only one independent variable.

e.g. $\frac{d^2y}{dx^2}+2\frac{dy}{dx}+y=\sin x \ldots\ldots\ldots\ldots (i)$ and $\frac{dy}{dx}=\frac{x+1}{\sqrt{1+y^2}} \ldots\ldots\ldots\ldots (ii)$

Both of equations (i) & (ii) involve only one independent variable x.

Partial differential equations are those which involve two or more than two independent variable.

e.g. $\frac{\partial^2 y}{\partial t^2}=a^2\frac{\partial^2 y}{\partial x^2}+b\frac{\partial y}{\partial x} \ldots\ldots\ldots\ldots (iii)$ $\therefore x\frac{\partial z}{\partial \theta}+y\frac{\partial z}{\partial y}+z=0 \ldots\ldots\ldots\ldots (iv)$

Both of equations (iii) & (iv), (θ, y) and (t, x) are independent variable respectively.

(c) order and degree of differential equations: – The order of differential equation is defined to be the order of the highest derivative or differential coefficient occurring in it.

Example: – Equation $\frac{dy}{dx}=\frac{\sqrt{x+1}}{\sqrt{1+y^3}}$ and $xy\frac{dy}{dx}=\frac{x^2-1}{y-1}$ are of first order.

Equation $\frac{\left[1-\left(\frac{dy}{dx}\right)^3\right]^{\frac{2}{3}}}{\frac{d^2y}{dx^2}}=\rho$ and $\frac{d^3y}{dx^3}+3\frac{d^2y}{dx^2}-\frac{dy}{dx}+y=\cos 2x$ are of second and third order respectively.

Equation $\frac{\left[1+\left(\frac{dy}{dx}\right)^2\right]^{\frac{3}{2}}}{\frac{d^2y}{dx^2}}=\rho$ or $\left[1+\left(\frac{dy}{dx}\right)^2\right]^3=\rho^2\left(\frac{d^2y}{dx^2}\right)^2$

Hence the equation is of second order and second degree.

Example: – (1) $y=Ae^{-3x}-Be^{3x}$, Eliminating the arbitrary constant A and B.

$$\therefore \frac{dy}{dx}=-3Ae^{-3x}-3Be^{3x} \ldots\ldots\ldots\ldots\ldots (i)$$

Again, Differentiate $\frac{d^2y}{dx^2}=9Ae^{-3x}-9Be^{3x}=9(Ae^{-3x}-Be^{3x})=9y$

$$\therefore \frac{d^2y}{dx^2} - 9y = 0$$ it is differential equation. hence the equation is of second order.

Example: – (2) $y = A\cos 3x - B\sin 3x$ Differentiate, $\frac{dy}{dx} = -3A\sin 3x - 3B\cos 3x$

$$\therefore \frac{d^2y}{dx^2} = -9A\cos 3x + 9B\sin 3x \quad \therefore \frac{d^2y}{dx^2} = -9(A\cos 3x - B\sin 3x) = -9y$$

$$\therefore \frac{d^2y}{dx^2} + 9y = 0,$$ hence the differential equation is of second order.

Example: – (3) (a) Prove that $Ax + By^2 = 1$ is the solution of $y\frac{d^2y}{dx^2} + \left(\frac{dy}{dx}\right)^2 = 0$.

Solution: – Given, $Ax + By^2 = 1 \quad \therefore Ax + By^2 - 1 = 0$

Differentiating, $A + 2By\frac{dy}{dx} = 0$ or $2By\frac{dy}{dx} = -A$ or $y\frac{dy}{dx} = -\frac{A}{2B}$

Again Differentiating, $y\frac{d^2y}{dx^2} + \frac{dy}{dx}.\frac{dy}{dx} = 0 \quad \therefore y\frac{d^2y}{dx^2} + \left(\frac{dy}{dx}\right)^2 = 0$ Proved.

(b) Prove that $By^2 = 1 + Ax^2$ is the solution of $y\frac{d^2y}{dx^2} = \frac{y}{x}.\frac{dy}{dx} - \left(\frac{dy}{dx}\right)^2$.

Solution: – $By^2 = 1 + Ax^2$ Differentiating, $2By\frac{dy}{dx} = 2Ax \quad \therefore \frac{dy}{dx} = \frac{2Ax}{2By} = \frac{Ax}{By}$

Again Differentiating, $\frac{d^2y}{dx^2} = \frac{A}{B}\left[\frac{y.1 - x.\frac{dy}{dx}}{y^2}\right]$ or $By^2\frac{d^2y}{dx^2} = Ay - Ax\frac{dy}{dx}$ .... (A)

Put $\frac{dy}{dx} = \frac{Ax}{By} \quad \therefore By = \frac{Ax}{\frac{dy}{dx}}$ in equation (A), we get

or $\frac{Ax}{\frac{dy}{dx}}.y\frac{d^2y}{dx^2} = Ay - Ax\frac{dy}{dx}$ or $Axy\frac{d^2y}{dx^2} = Ay\frac{dy}{dx} - Ax\left(\frac{dy}{dx}\right)^2$ or $y\frac{d^2y}{dx^2} = \frac{y}{x}.\frac{dy}{dx} - \left(\frac{dy}{dx}\right)^2$ Proved.

Example: – (4)(a) Prove that $y = e^x + \log x$ is the solution of $\frac{d^2y}{dx^2} - \frac{dy}{dx} + \frac{(x+1)}{x^2} = 0$.

Solution: – Given, $y = e^x + \log x \quad \therefore \frac{dy}{dx} = e^x + \frac{1}{x}$

$$\therefore \frac{d^2y}{dx^2} = e^x - \frac{1}{x^2} = \frac{dy}{dx} - \frac{1}{x} - \frac{1}{x^2} = \frac{dy}{dx} - \left(\frac{x+1}{x^2}\right) \quad \therefore \frac{d^2y}{dx^2} - \frac{dy}{dx} + \left(\frac{x+1}{x^2}\right) = 0$$ Proved.

(b) Prove that $ye^x + \log x = 0$ is the solution of $\frac{d^2y}{dx^2} + \frac{dy}{dx} = \frac{(x+1)}{x^2e^x}$ or $x^2e^x\frac{d^2y}{dx^2} + x^2e^x\frac{dy}{dx} = (x+1)$.

Solution: – Given, $ye^x + \log x = 0$ Differentiating, $ye^x + e^x\frac{dy}{dx} + \frac{1}{x} = 0 \quad \therefore e^x\left(y + \frac{dy}{dx}\right) = -\frac{1}{x}$

$$\text{or } y + \frac{dy}{dx} = -\frac{1}{xe^x} \quad \therefore \frac{dy}{dx} = -y - \frac{1}{xe^x}$$

Again Differentiating, $\frac{d^2y}{dx^2} = -\frac{dy}{dx} - \left[\frac{0 - (xe^x + e^x)}{x^2e^{2x}}\right] = -\frac{dy}{dx} + \frac{e^x(x+1)}{x^2e^{2x}} = -\frac{dy}{dx} + \frac{(x+1)}{x^2e^x}$

$$\therefore \frac{d^2y}{dx^2} + \frac{dy}{dx} = \frac{(x+1)}{x^2e^x} \text{ or } x^2e^x\frac{d^2y}{dx^2} + x^2e^x\frac{dy}{dx} = (x+1)$$ Proved.

(5) Find the Order, Degree, Linear and Non – linear of the differential equation: –

(a) $y'' + 3y' + 4y = 0$ or $\frac{d^2y}{dx^2} + 3\frac{dy}{dx} + 4y = 0$

Order = 2, Degree = 1 and Linear differential equation.

(b) $x^2y'' + xy' + 3y = 5x$ or $x^2\frac{d^2y}{dx^2} + x\frac{dy}{dx} + 3y = 5x$ Order = 2, Degree = 1 and Linear.

(c) $(y')^2 + 3xy' + 3y = 0$ Ans: – Order = 1, Degree = 1 and Non – linear.

(d) $\sqrt{1 + 3x^2}\, dx + \sqrt{1 + 3y^2}\, dy = 0 \Rightarrow \frac{dy}{dx} = -\frac{\sqrt{1 + 3x^2}}{\sqrt{1 + 3y^2}}$

Order = 1, Degree = 1 and Non – linear.

(6) Find the differential equation: – (a) $y = Ae^{Px}$ $\therefore \frac{dy}{dx} = APe^{Px} = P(Ae^{Px}) = Py,$

Put $y = Ae^{Px}$ or $\frac{dy}{dx} = Py$ $\therefore \frac{dy}{dx} - Py = 0$ Ans.

(b) $y = A\sin Px - B\cos Px$ Differentiating, $\frac{dy}{dx} = AP\cos Px + BP\sin Px$

Again Differentiating, $\frac{d^2y}{dx^2} = -AP^2\sin Px + BP^2\cos Px = -P^2(A\sin Px - B\cos Px) = -P^2y$

or $\frac{d^2y}{dx^2} = -P^2y$ $\therefore \frac{d^2y}{dx^2} + P^2y = 0$ Ans. [put $y = A\sin Px - B\cos Px$]

(c) $y = A\sin(Px - q)$ Differentiating, $\frac{dy}{dx} = AP\cos(Px - q)$

Again Differentiatin $\frac{d^2y}{dx^2} = -AP^2\sin(Px - q) = -P^2[A\sin(Px - q)] = -P^2y$

$\therefore \frac{d^2y}{dx^2} + P^2y = 0$ Ans. [put $y = A\sin(Px - q)$]

(d) $y = \tan(Px + q)$ Ans: – $y'' + P^2y = 0$ or $\frac{d^2y}{dx^2} + P^2y = 0$ (Do yourself)

(7) Find the differential equation: –

(a) $y = a\sin x + b\cos x$, eliminate the constant a, b from the relation.

Solution: – Given, $y = a\sin x + b\cos x$ Differentiating, $\frac{dy}{dx} = a\cos x - b\sin x$

Again Differentiating, $\frac{d^2y}{dx^2} = -a\sin x - b\cos x = -(a\sin x + b\cos x) = -y$ $\therefore \frac{d^2y}{dx^2} + y = 0$

$\therefore y'' + y = 0$ Ans.

(b) Prove that $y = e^x\log x + x$ is the solution of $x^2\left[\frac{d^2y}{dx^2} - \frac{dy}{dx} + 1\right] = e^x(x - 1)$.

Solution: – Given, $y = e^x\log x + x$ Diff. $\frac{dy}{dx} = e^x.\frac{1}{x} + \log x . e^x + 1 = \frac{e^x}{x} + e^x\log x + 1$

or $\frac{dy}{dx} = \frac{e^x}{x} + e^x\log x + 1$

Again Differentiating, $\frac{d^2y}{dx^2} = \frac{xe^x - e^x}{x^2} + e^x.\frac{1}{x} + \log x . e^x = \frac{e^x(x-1)}{x^2} + \left(\frac{e^x}{x} + e^x \log x\right)$

put $\frac{dy}{dx} = \frac{e^x}{x} + e^x \log x + 1$ or $\frac{e^x}{x} + e^x \log x = \frac{dy}{dx} - 1$ or $\frac{d^2y}{dx^2} = \frac{e^x(x-1)}{x^2} + \frac{dy}{dx} - 1$

$$\text{or } \frac{d^2y}{dx^2} - \frac{dy}{dx} + 1 = \frac{e^x(x-1)}{x^2} \quad \therefore x^2\left[\frac{d^2y}{dx^2} - \frac{dy}{dx} + 1\right] = e^x(x-1) \quad \text{Proved.}$$

(8) (a) Find the differential equation of the family of curves $y = Ae^{2x} + Be^{3x}$.

Solution: – Given, $y = Ae^{2x} + Be^{3x}$ ……………… (i)

Differentiating, $y' = 2Ae^{2x} + 3Be^{3x}$ …………..(ii) and $y'' = 4Ae^{2x} + 9Be^{3x}$ …………….(iii)

Eliminating A and B from the above three equation, we get

$$\text{or } \begin{vmatrix} e^{2x} & e^{3x} & -y \\ 2e^{2x} & 3e^{3x} & -y' \\ 4e^{2x} & 9e^{3x} & -y'' \end{vmatrix} = 0 \text{ or } (-e^{2x}.e^{3x}) \begin{vmatrix} 1 & 1 & y \\ 2 & 3 & y' \\ 4 & 9 & y'' \end{vmatrix} = 0$$

$$\text{or } -e^{2x}.e^{3x}[1(3y'' - 9y') - 1(4y' - 2y'') + y(18 - 12)] = 0 \text{ or } 3y'' - 9y' - 4y' + 2y'' + 6y = 0$$

$$\text{or } 5y'' - 13y' + 6y = 0 \quad \therefore 5\frac{d^2y}{dx^2} - 13\frac{dy}{dx} + 6y = 0 \quad \text{Ans.}$$

(b) Find the differential equation of the family of curves $y = Ae^{3x} + Be^{-5x} + Ce^x$

where A, B and C are arbitrary constant.

Solution: – Do yourself.

Ordinary differential equations of the first order and first degree: –

An ordinary differential equation of the first order and first degree is of the form: –

$$M + N\frac{dy}{dx} = 0 \text{ or } M\,dx + N\,dy = 0$$

where M and N are functions of x and y or constant.

<u>**Some special cases: –**</u>

Case I: – **Variable separable**: – Equations which are capable of being put in the form

$f(x)dx + \phi(y)dy = 0$ will be termed as variable separable.

Hence complete general solution of $f(x)dx + \phi(y)dy = 0$ will be

$$\int f(x)\,dx + \int \phi(y)\,dy = a \text{ (constant)}$$

Example: – Solve, $\frac{dy}{dx} = \frac{1 + x^2}{1 + y^2}$ or $(1 + y^2)dy = (1 + x^2)dx$, Integrating both sides

or $\int (1 + y^2)dy = \int (1 + x^2)dx$ or $y + \frac{y^3}{3} = x + \frac{x^3}{3} + c$ or $\frac{3y + y^3}{3} = \frac{3x + x^3 + 3c}{3}$

or $3y + y^3 = 3x + x^3 + 3c$ or $y^3 - x^3 + 3(y - x) = k$ Ans. (where $3c = k$)

Example: – Solve, $\frac{dy}{dx} = \frac{\sqrt{1 - y^2}}{\sqrt{1 - x^2}}$ or $\frac{dy}{\sqrt{1 - y^2}} = \frac{dx}{\sqrt{1 - x^2}}$

Integrating both of sides, $\int \frac{dy}{\sqrt{1 - y^2}} = \int \frac{dx}{\sqrt{1 - x^2}}$

or $\sin^{-1} y = \sin^{-1} x + c$ or $\sin^{-1} y - \sin^{-1} x = c$ or $\sin^{-1}\left[y\sqrt{1-x^2} - x\sqrt{1-y^2}\right] = c$

use formula, $\displaystyle\int \frac{dx}{\sqrt{a^2 - x^2}} = \sin^{-1}\left(\frac{x}{a}\right)$ and $\sin^{-1} x \pm \sin^{-1} y = \sin^{-1}\left[x\sqrt{1-y^2} \pm y\sqrt{1-x^2}\right]$

or $\sin^{-1}\left[y\sqrt{1-x^2} - x\sqrt{1-y^2}\right] = c$ or $y\sqrt{1-x^2} - x\sqrt{1-y^2} = \sin c = k$ (say)

$$\therefore\ y\sqrt{1-x^2} - x\sqrt{1-y^2} = k \quad \text{Ans.}$$

Remember the following result

(a) $d\left(\frac{x}{y}\right) = \frac{y.dx - x.dy}{y^2}$ (b) $d\left(\frac{y}{x}\right) = \frac{x.dy - y.dx}{x^2}$

(c) $d(x^2 + y^2) = 2(x.dx + y.dy)$ (d) $d(x^2 - y^2) = 2(x.dx - y.dy)$

Case II: – Exact differential equations: – $M\,dx + N\,dy = 0$ where M and N are functions of x and y if $\frac{\partial M}{\partial y} = \frac{\partial N}{\partial x}$ then the equation is exact and its solution is

$$\int M\,dx + \int N\,dy = c$$

where $\int M\,dx \rightarrow y -$ constant and $\int N\,dy \rightarrow$ free form of x

Example: – Solve, $(x^2 + 2xy + 3y^2)\,dx + (y^2 + 6xy + x^2)\,dy = 0$

Solution: – Let $M = x^2 + 2xy + 3y^2$ and $N = y^2 + 6xy + x^2$

or $\frac{\partial M}{\partial y} = 6y + 2x$ and $\frac{\partial N}{\partial x} = 2x + 6y$ since $\frac{\partial M}{\partial y} = \frac{\partial N}{\partial x}$ the equation is exact.

Its solution is $\int (x^2 + 2xy + 3y^2)\,dx + \int N\,dy = c$ where $\int N\,dy \rightarrow$ free form of x

or $\frac{x^3}{3} + x^2y + 3y^2x + \int y^2\,dy = c$ or $\frac{x^3}{3} + x^2y + 3y^2x + \frac{y^3}{3} = c$ or $x^3 + 3x^2y + 9y^2x + y^3 = 3c$

$$\therefore\ x^3 + 3x^2y + 9y^2x + y^3 = k \quad \text{where } k = 3c \quad \text{Ans.}$$

Example: – Solve, $y(1 + e^x)\,dx + (x + e^x)\,dy = 0$

Solution: – Given, $y(1 + e^x)\,dx + (x + e^x)\,dy = 0$ Here $M = y(1 + e^x)$ and $N = x + e^x$

or $\frac{\partial M}{\partial y} = (1 + e^x)$ and $\frac{\partial N}{\partial x} = 1 + e^x$ $\therefore\ \frac{\partial M}{\partial y} = \frac{\partial N}{\partial x}$ the equation is exact.

Its solution is $\int M\,dx + \int N\,dy = c$ or $\int y(1 + e^x)\,dx + \int dy = c$

or $y\int dx + y\int e^x\,dx + \int dy = c$ or $y.x + y.e^x + y = c$ or $y(x + e^x + 1) = c$

$$\therefore\ y(1 + x + e^x) = c \quad \text{Ans.}$$

Case III: – Homogeneous equations: – when $\frac{dy}{dx} = \frac{x+y}{x-y}$ whose N^r and D^r both are homogeneous function of x and y of the same degree then the differential equation is called to be homogeneous equation.

$\frac{dy}{dx} = \frac{f(x,y)}{\phi(x,y)}$ where f and ϕ are both homogeneous function of x and y of the same degree.

Method of solution: – Let $\frac{dy}{dx} = \frac{x+y}{x-y}$ it is homogeneous function.

$$\therefore\ \frac{dy}{dx} = \frac{x\left(1+\frac{y}{x}\right)}{x\left(1-\frac{y}{x}\right)} \qquad \text{Put } y = vx \quad \therefore\ \frac{dy}{dx} = v + x\frac{dv}{dx}$$

Now, substitute the value of $\frac{dy}{dx}$ and y in the given equation

or $\frac{dy}{dx} = \frac{\left(1+\frac{y}{x}\right)}{\left(1-\frac{y}{x}\right)}$ or $v + x\frac{dv}{dx} = \frac{1+v}{1-v}$ or $x\frac{dv}{dx} = \frac{1+v}{1-v} - v = \frac{1+v-v+v^2}{1-v} = \frac{1+v^2}{1-v}$

or $\frac{1-v}{1+v^2}\,dv = \frac{dx}{x}$ or $\frac{1}{1+v^2}\,dv - \frac{v}{1+v^2}\,dv = \frac{dx}{x}$

Integrating, $\int \frac{1}{1+v^2}\,dv - \int \frac{v}{1+v^2}\,dv = \int \frac{dx}{x}$ or $\tan^{-1} v - \frac{1}{2}\log(1+v^2) = \log x + \log c$

or $\tan^{-1}\left(\frac{y}{x}\right) - \log\sqrt{1+\left(\frac{y}{x}\right)^2} = \log x + \log c$ or $\tan^{-1}\left(\frac{y}{x}\right) = \log\left(\sqrt{1+\left(\frac{y}{x}\right)^2}\right) + \log x.c$

or $\tan^{-1}\left(\frac{y}{x}\right) = \log\left(\sqrt{\frac{x^2+y^2}{x^2}}\right) + \log x.c$ or $\tan^{-1}\left(\frac{y}{x}\right) = \log\left\{x.c.\frac{\sqrt{x^2+y^2}}{x}\right\} = \log\left(\sqrt{x^2+y^2}.c\right)$ Ans.

Example: – Solve, $x\frac{dy}{dx} = x\sin\left(\frac{x+y}{x}\right) + y$

Solution: – Given, $x\frac{dy}{dx} = x\sin\left(\frac{x+y}{x}\right) + y$ or $\frac{dy}{dx} = \sin\left[1+\frac{y}{x}\right] + \frac{y}{x}$

Put $v = \frac{y}{x}$ or $y = vx$ $\therefore\ \frac{dy}{dx} = v + x\frac{dv}{dx}$

or $v + x\frac{dv}{dx} = \sin(1+v) + v$ or $x\frac{dv}{dx} = \sin(1+v) + v - v = \sin(1+v)$ or $\frac{dv}{\sin(1+v)} = \frac{dx}{x}$

Integrating, $\int \frac{dv}{\sin(1+v)} = \int \frac{dx}{x}$ or $\int \operatorname{cosec}(1+v)\,dv = \int \frac{dx}{x}$ or $\log\left(\frac{1+v}{2}\right) = \log x + \log c$

or $\log\left(\frac{1+\frac{y}{x}}{2}\right) = \log x.c$ or $\log\left(\frac{x+y}{2x}\right) - \log x.c = 0$ or $\log\left(\frac{x+y}{2x^2c}\right) = 0$ Ans. $\left[\int \operatorname{cosec} x\,dx = \log(\tan {}^x/_2)\right]$

Example: – Solve, $\frac{dx}{dy} = e^{-\frac{x}{2}} + \frac{x}{y}$

Solution: – Given, $\frac{dx}{dy} = e^{-\frac{x}{2}} + \frac{x}{y}$ Put $v = \frac{x}{y}$ or $x = yv$ $\therefore\ \frac{dx}{dy} = v + y\frac{dv}{dy}$

or $v + y\frac{dv}{dy} = e^{-v} + v$ or $y\frac{dv}{dy} = e^{-v}$ or $\frac{dv}{e^{-v}} = \frac{dy}{y}$ or $e^v\,dv = \frac{dy}{y}$

Integrating, $\int e^v\,dv = \int \frac{dy}{y}$ or $e^v = \log y + \log k$ or $e^v = \log(y.k)$ put $v = \frac{x}{y}$

or $e^v = \log(y.k)$ $\therefore$ $e^{\frac{x}{y}} = \log(y.k)$ Ans.

Example: – Solve, $x\frac{dy}{dx} = y + x\,\mathrm{cosec}\left(\frac{y}{x}\right)$

Solution: – Given, $x\frac{dy}{dx} = y + x\,\mathrm{cosec}\left(\frac{y}{x}\right)$ Put $v = \frac{y}{x}$ or $y = vx$ $\therefore$ $\frac{dy}{dx} = v + x\frac{dv}{dx}$

or $x\frac{dy}{dx} = y + x\,\mathrm{cosec}\left(\frac{y}{x}\right)$ or $\frac{dy}{dx} = \frac{y}{x} + \mathrm{cosec}\left(\frac{y}{x}\right)$ or $v + x\frac{dv}{dx} = v + \mathrm{cosec}\,v$ or $x\frac{dv}{dx} = \mathrm{cosec}\,v$

or $\frac{dv}{\mathrm{cosec}\,v} = \frac{dx}{x}$ Integrating, $\int \frac{dv}{\mathrm{cosec}\,v} = \int \frac{dx}{x}$ or $\int \sin v\,dv = \int \frac{dx}{x}$

or $-\cos v = \log x + \log k$ or $-\cos v = \log(x.k)$ Put $v = \frac{y}{x}$ $\therefore$ $\cos\left(\frac{y}{x}\right) + \log(x.k) = 0$ Ans.

Example: – Solve, $(x + y)\frac{dy}{dx} = y$

Solution: – Given, $(x + y)\frac{dy}{dx} = y$ or $\frac{dy}{dx} = \frac{y}{x + y} = \frac{y}{x}.\frac{1}{\left(1 + \frac{y}{x}\right)}$ Put $y = vx$ $\therefore$ $\frac{dy}{dx} = v + x\frac{dv}{dx}$

or $v + x\frac{dv}{dx} = v.\frac{1}{1 + v} = \frac{v}{1 + v}$ or $x\frac{dv}{dx} = \frac{v}{1 + v} - v = \frac{v - v - v^2}{1 + v} = -\frac{v^2}{1 + v}$ or $\frac{1 + v}{v^2}\,dv = -\frac{dx}{x}$

Integrating, $\int \frac{1 + v}{v^2}\,dv = -\int \frac{dx}{x}$ or $\int \left(\frac{1}{v^2} + \frac{v}{v^2}\right)dv = -\int \frac{dx}{x}$ or $\int \frac{1}{v^2}\,dv + \int \frac{dv}{v} = -\int \frac{dx}{x}$

or $-\frac{1}{v} + \log v = -\log x - \log k$ or $-\frac{1}{v} = -[\log x + \log v + \log k]$ or $\frac{1}{v} = \log|x.v.k|$ Put $v = \frac{y}{x}$

or $\frac{1}{\frac{y}{x}} = \log\left|x.\frac{y}{x}.k\right| = \log|y.k|$ $\therefore$ $\frac{x}{y} = \log|y.k|$ $\therefore$ $|y.k| = e^{\frac{x}{y}}$ Ans.

Example: – Solve, $y\frac{dx}{dy} = x + ye^{\frac{x}{y}}$

Solution: – Given, $y\frac{dx}{dy} = x + ye^{\frac{x}{y}}$ or $\frac{dx}{dy} = \frac{y\left(\frac{x}{y} + e^{\frac{x}{y}}\right)}{y}$ Put $v = \frac{x}{y}$ or $x = vy$ $\therefore$ $\frac{dx}{dy} = v + y\frac{dv}{dy}$

or $\frac{dx}{dy} = \frac{y\left(\frac{x}{y} + e^{\frac{x}{y}}\right)}{y} = \frac{x}{y} + e^{\frac{x}{y}}$ or $v + y\frac{dv}{dy} = v + e^v$ or $y\frac{dv}{dy} = e^v$ or $\frac{dv}{e^v} = \frac{dy}{y}$

Integrating, $\int \frac{dv}{e^v} = \int \frac{dy}{y}$ or $\int e^{-v}\,dv = \int \frac{dy}{y}$ or $-e^{-v} = \log y + \log k = \log|y.k|$ put $v = \frac{x}{y}$

or $e^{-v} = -\log|y.k|$ or $e^{-\frac{x}{y}} = -\log|y.k|$ $\therefore$ $e^{-\frac{x}{y}} = \log\left(\frac{1}{|y.k|}\right)$ Ans.

(I). Equations reduciable to homogeneous form: –

Consider the following equations $\frac{dy}{dx} = \frac{ax + by + c}{a'x + b'y + c'}$ ……. (i)

where $\frac{a}{a'} \neq \frac{b}{b'}$ it is not a homogeneous equation.

Method of solution: – Put $x = \alpha + h$ and $y = \beta + k$ where α and β are variables but h and k are constant. then, $\frac{dy}{dx} = \frac{d\beta}{d\alpha}$ and the above equation becomes

from equation (i), $\frac{dy}{dx} = \frac{ax + by + c}{a'x + b'y + c'}$ Put value x, y and $\frac{dy}{dx}$ we have

$$\text{or} \quad \frac{d\beta}{d\alpha} = \frac{a(\alpha + h) + b(\beta + k) + c}{a'(\alpha + h) + b'(\beta + k) + c'} = \frac{a\alpha + b\beta + (ah + bk + c)}{a'\alpha + b'\beta + (a'h + b'k + c')}$$

Now, choose h and k such that $ah + bk + c = 0$ or $a'h + b'k + c' = 0 \ldots\ldots\ldots\ldots\ldots (A)$

Then, $\frac{d\beta}{d\alpha} = \frac{a\alpha + b\beta}{a'\alpha + b'\beta}$ which is a homogeneous equation in α and β.

Put $\beta = v\alpha \quad \therefore \frac{d\beta}{d\alpha} = v + \alpha\frac{dv}{d\alpha}$

$$\text{or} \quad \frac{d\beta}{d\alpha} = \frac{a\alpha + b\beta}{a'\alpha + b'\beta} \quad \text{or} \quad v + \alpha\frac{dv}{d\alpha} = \frac{\alpha\left(a + b.\frac{\beta}{\alpha}\right)}{\alpha\left(a' + b'.\frac{\beta}{\alpha}\right)} = \frac{a + bv}{a' + b'v}$$

$$\text{or} \quad \alpha\frac{dv}{d\alpha} = \frac{a + bv}{a' + b'v} - v = \frac{a + bv - a'v - b'v^2}{a' + b'v} \quad \text{or} \quad \frac{a' + b'v}{a + bv - a'v - b'v^2}\,dv = \frac{d\alpha}{\alpha}$$

In the end put $v = \frac{\beta}{\alpha}$ and $\alpha = x - h$, $\beta = y - k$ where h and k are determined from equation (A).

(II). If $\frac{dy}{dx} = \frac{ax + by + c}{a'x + b'y + c'}$ and $\frac{a}{a'} = \frac{b}{b'} = m$ (say)

Method of solution:– $\frac{dy}{dx} = \frac{ax + by + c}{a'x + b'y + c'} = \frac{ax + by + c}{\frac{1}{m}(ax + by) + c'}$

Put $ax + by = v \quad \therefore a + b\frac{dy}{dx} = \frac{dv}{dx}$

$$\text{or} \quad \frac{dy}{dx} = \frac{1}{b}\left(\frac{dv}{dx} - a\right) \quad \text{or} \quad \frac{dy}{dx} = \frac{ax + by + c}{\frac{1}{m}(ax + by) + c'} \quad \text{or} \quad \frac{1}{b}\left(\frac{dv}{dx} - a\right) = \frac{v + c}{\frac{v}{m} + c'} = \frac{m(v + c)}{v + mc'}$$

Now the variables can be separated.

(III). **Linear Differential Equations**:– A differential equation of the form $\frac{dy}{dx} + Py = Q$

where P and Q are functions of x (not of y)

Similarly, $\frac{dx}{dy} + Px = Q$ where P and Q are functions of y (not of x)

Method of solution:– The equation in the form $\frac{dy}{dx} + Py = Q$

Integrate P with respect to x and make the integrand the power of e to get I. F (Integrating factor I. F).

i.e $I.F = e^{\int P\,dx}$ The required solution as $\boxed{y \times I.F = \int Q.(I.F)\,dx + c}$ formula

Similarly, $\frac{dx}{dy} + Px = Q$ then the required solution as $\boxed{x \times I.F = \int Q(I.F)\,dy + c}$ formula

Example:– Solve, $(x + 3y + 1)\,dy = (2x + y + 3)\,dx$

Solution:– Given, $(x + 3y + 1)\,dy = (2x + y + 3)\,dx$ or $\frac{dy}{dx} = \frac{2x + y + 3}{x + 3y + 1}$

Put $x = \alpha + h$ and $y = \beta + k$ $\therefore \frac{dy}{dx} = \frac{d\beta}{d\alpha}$

or $\frac{d\beta}{d\alpha} = \frac{2(\alpha+h)+(\beta+k)+3}{(\alpha+h)+3(\beta+k)+1} = \frac{2\alpha+\beta+(2h+k+3)}{\alpha+3\beta+(h+3k+1)}$ ………….(A)

Now, choose h and k such that

$\therefore 2h + k + 3 = 0$ ……………..(i) and $h + 3k + 1 = 0$ ………………(ii)

Multiplying, (i) × 1 & (ii) × 2, we have

$\therefore 2h + k + 3 = 0$ ………….(iii) and $2h + 6k + 2 = 0$ ……..…….(iv)

Substracting, $2h + k + 3 - 2h - 6k - 2 = 0$ $\therefore -5k + 1 = 0$ or $5k = 1$ $\therefore k = \frac{1}{5}$

Put value of k in equation (i), we have

or $2h + k + 3 = 0$ or $2h = -3 - \frac{1}{5} = \frac{-15-1}{5} = -\frac{16}{5}$ $\therefore h = -\frac{8}{5}$

From (A), $\frac{d\beta}{d\alpha} = \frac{2\alpha+\beta+(2h+k+3)}{\alpha+3\beta+(h+3k+1)} = \frac{2\alpha+\beta}{\alpha+3\beta} = \frac{\alpha\left(2+\frac{\beta}{\alpha}\right)}{\alpha\left(1+3.\frac{\beta}{\alpha}\right)} = \frac{2+\frac{\beta}{\alpha}}{1+3.\frac{\beta}{\alpha}}$

which is homogeneous equation.

Put $\beta = v\alpha$ $\therefore \frac{d\beta}{d\alpha} = v + \alpha\frac{dv}{d\alpha}$ or $\frac{d\beta}{d\alpha} = \frac{2+\frac{\beta}{\alpha}}{1+3.\frac{\beta}{\alpha}}$ or $v + \alpha\frac{dv}{d\alpha} = \frac{2+v}{1+3v}$ or $\alpha\frac{dv}{d\alpha} = \frac{2+v}{1+3v} - v$

or $\alpha\frac{dv}{d\alpha} = \frac{2+v-v-3v^2}{1+3v} = \frac{2-3v^2}{1+3v}$ or $\frac{1+3v}{2-3v^2}\,dv = \frac{d\alpha}{\alpha}$

Integrating, $\int \frac{1+3v}{2-3v^2}\,dv = \int \frac{d\alpha}{\alpha}$ or $\int \frac{1}{2-3v^2}\,dv + \int \frac{3v}{2-3v^2}\,dv = \int \frac{d\alpha}{\alpha}$ ……………(B)

Let $I_1 = \int \frac{1}{2-3v^2}\,dv = \int \frac{1}{(\sqrt{2})^2 - (\sqrt{3}v)^2}\,dv = \frac{1}{2.\sqrt{2}}\log\left|\frac{\sqrt{2}+\sqrt{3}v}{\sqrt{2}-\sqrt{3}v}\right|$

using formula, $\int \frac{dx}{a^2-x^2} = \frac{1}{2a}\log\left|\frac{a+x}{a-x}\right|$, $x < a$

and $I_2 = \int \frac{3v}{2-3v^2}\,dv$ Let $2 - 3v^2 = z$ or $-6v\,dv = dz$ $\therefore 2.3v\,dv = -dz$ $\therefore 3v\,dv = -\frac{dz}{2}$

$$\therefore I_2 = -\frac{1}{2}\int \frac{dz}{z} = -\frac{1}{2}\log|z| = -\frac{1}{2}\log|2-3v^2|$$

Put value of I_1 and I_2 in equation (B), we get

$$\text{or } \int \frac{1}{2-3v^2}\,dv + \int \frac{3v}{2-3v^2}\,dv = \int \frac{d\alpha}{\alpha} \quad \text{or} \quad \frac{1}{2.\sqrt{2}}\log\left|\frac{\sqrt{2}+\sqrt{3}v}{\sqrt{2}-\sqrt{3}v}\right| - \frac{1}{2}\log|2-3v^2| = \log\alpha + k$$

Put $v = \frac{\beta}{\alpha}$ and $x = \alpha + h$ $\therefore \alpha = x - h$, $y = \beta + k$ $\therefore \beta = y - k$

or $\frac{1}{2.\sqrt{2}}\log\left|\frac{\sqrt{2}+\sqrt{3}.\frac{\beta}{\alpha}}{\sqrt{2}-\sqrt{3}.\frac{\beta}{\alpha}}\right| - \frac{1}{2}\log\left|2 - 3\left(\frac{\beta}{\alpha}\right)^2\right| = \log\alpha + k$

or $\frac{1}{2\sqrt{2}}\log\left|\frac{\sqrt{2}\alpha+\sqrt{3}\beta}{\sqrt{2}\alpha-\sqrt{3}\beta}\right| - \frac{1}{2}\log\left|\frac{2\alpha^2-3\beta^2}{\alpha^2}\right| = \log\alpha + k$

or $\frac{1}{2\sqrt{2}}\log\left|\frac{\sqrt{2}(x-h)+\sqrt{3}(y-k)}{\sqrt{2}(x-h)-\sqrt{3}(y-k)}\right|-\frac{1}{2}\log\left|\frac{2(x-h)^2-3(y-k)^2}{(x-h)^2}\right|=\log(x-h)+k$

Put $\alpha=x-h=x+\frac{8}{5}$ and $\beta=y-k=y-\frac{1}{5}$

or $\frac{1}{2\sqrt{2}}\log\left|\frac{\sqrt{2}\left(x+\frac{8}{5}\right)+\sqrt{3}\left(y-\frac{1}{5}\right)}{\sqrt{2}\left(x+\frac{8}{5}\right)-\sqrt{3}\left(y-\frac{1}{5}\right)}\right|-\frac{1}{2}\log\left|\frac{2\left(x+\frac{8}{5}\right)^2-3\left(y-\frac{1}{5}\right)^2}{\left(x+\frac{8}{5}\right)^2}\right|=\log\left(x+\frac{8}{5}\right)+k$

or $\frac{1}{2\sqrt{2}}\log\left|\frac{\sqrt{2}(5x+8)+\sqrt{3}(5y-1)}{\sqrt{2}(5x+8)-\sqrt{3}(5y-1)}\right|-\frac{1}{2}\log\left|\frac{2(5x+8)^2-3(5y-1)^2}{(5x+8)^2}\right|=\log\left(\frac{5x+8}{5}\right)+k$ Ans.

Example: – Solve, $\frac{dy}{dx}+\frac{y}{x}=\frac{\sin x}{x}$

Solution: – Given, $\frac{dy}{dx}+\frac{y}{x}=\frac{\sin x}{x}$ The equation in the form $\frac{dy}{dx}+Py=Q$

where $P=\frac{1}{x}$ and $Q=\frac{\sin x}{x}$ $\therefore$ $I.F=e^{\int P\,dx}=e^{\int\frac{1}{x}dx}=e^{\log x}=x$

The required equation is – $y\times I.F=\int (I.F).Q\;dx+c$

or $y.x=\int x.\frac{\sin x}{x}\,dx+c=\int \sin x\,dx+c=-\cos x+c$ or $xy=-\cos x+c$ $\therefore$ $xy+\cos x=c$ Ans.

Example: – Solve, $\frac{dx}{dy}-\frac{x}{y}=y\cos y$

Solution: – Given, $\frac{dx}{dy}-\frac{x}{y}=y\cos y$ form of $\frac{dx}{dy}+Px=Q$ where $P=-\frac{1}{y}$ and $Q=y\cos y$

$\therefore$ $I.F=e^{\int P\,dy}=e^{-\int\frac{1}{y}dy}=e^{-\log y}=e^{\log\frac{1}{y}}=\frac{1}{y}$

The required solution is – $x\times I.F=\int I.F\times Q\;dy+c$

or $x.\frac{1}{y}=\int\frac{1}{y}.y\cos y\;dy+c=\int\cos y\;dy+c=\sin y+c$

or $\frac{x}{y}=\sin y+c$ or $x=y\sin y+yc$ or $x-y\sin y=yc$ $\therefore$ $c=\frac{x-y\sin y}{y}$ Ans.

Example: – Solve, $\frac{dy}{dx}=\frac{x+y+1}{x+y+3}$

Solution: – Given, $\frac{dy}{dx}=\frac{x+y+1}{x+y+3}$ here $a=a'$, $b=b'$ $\therefore$ $\frac{dy}{dx}=\frac{(x+y)+1}{(x+y)+3}$

Put $x+y=v$ $\therefore$ $1+\frac{dy}{dx}=\frac{dv}{dx}$ $\therefore$ $\frac{dy}{dx}=\frac{dv}{dx}-1$

or $\frac{dy}{dx}=\frac{(x+y)+1}{(x+y)+3}$ or $\frac{dv}{dx}-1=\frac{v+1}{v+3}$ or $\frac{dv}{dx}=\frac{v+1}{v+3}+1=\frac{v+1+v+3}{v+3}=\frac{2v+4}{v+3}$

or $\frac{v+3}{2v+4}\,dv=dx$

Integrating both of sides, $\int\frac{v+3}{2v+4}\,dv=\int dx$ or $\int\frac{v}{(2v+4)}\,dv+\int\frac{3}{(2v+4)}\,dv=\int dx$

$$\text{or} \quad \int \frac{v}{2(v+2)}\,dv + \int \frac{3}{2(v+2)}\,dv = \int dx \quad \text{or} \quad \frac{1}{2}\int \frac{v}{v+2}\,dv + \frac{3}{2}\int \frac{dv}{v+2} = \int dx$$

$$\text{or} \quad \frac{1}{2}\int \left(1 - \frac{2}{v+2}\right)dv + \frac{3}{2}\int \frac{dv}{v+2} = \int dx \quad \text{or} \quad \frac{1}{2}\int dv - \frac{1}{2}.2\int \frac{dv}{v+2} + \frac{3}{2}\int \frac{dv}{v+2} = \int dx$$

$$\text{or} \quad \frac{1}{2}v - \log|v+2| + \frac{3}{2}\log|v+2| = x + c \quad \text{or} \quad \frac{v - 2\log|v+2| + 3\log|v+2|}{2} = x + c \quad \text{Put } v = x + y$$

$$\text{or} \quad x + y + \log|x+y+2| = 2x + 2c \quad \therefore \ y - x + \log|x+y+2| = k \quad \text{where } k = 2c \qquad \text{Ans.}$$

Solved Example

(1) Solve the following differential equations: – (a) $(x+y)dx - 2x\,dy = 0$

(b) $\frac{dy}{dx} = \frac{yx^3 - xy^3}{x^4}$ (c) $\frac{dy}{dx} = \frac{y}{x} + \cos\left(\frac{y}{x}\right)$ (d) $(x^2 + 2x + 1)dy - (x^2 + 3x + 2)dx = 0$

(e) $dy + \log(x+y)\,dx = \log(2x+y)\,dx$ (f) $e^{\frac{dy}{dx}} = 2^x + 2$

Solution: – (a) $(x+y)dx - 2x\,dy = 0$ or $(x+y)dx = 2x\,dy$ or $2\frac{dy}{dx} = \frac{x+y}{x} = 1 + \frac{y}{x}$ ……..(i)

$$\text{Put } y = vx \quad \therefore \ \frac{dy}{dx} = v + x\frac{dv}{dx} \quad \text{or} \quad y = vx \quad \therefore \ v = \frac{y}{x}$$

Put value of $\frac{dy}{dx}$ in equation (i), we get $2\left(v + x\frac{dv}{dx}\right) = 1 + v$ or $2v + 2x\frac{dv}{dx} = 1 + v$

$$\text{or} \quad 2x\frac{dv}{dx} = 1 + v - 2v \quad \text{or} \quad 2x\frac{dv}{dx} = 1 - v \quad \text{or} \quad \frac{2\,dv}{1-v} = \frac{dx}{x}$$

Integrating, $2\int \frac{dv}{1-v} = \int \frac{dx}{x}$ or $2\log|1-v| = \log|x| + \log k$ or $\log|1-v|^2 = \log|xk|$

$$\text{or} \quad |1-v|^2 = |xk| \quad \text{or} \quad \left|1 - \frac{y}{x}\right|^2 = |xk| \quad \text{or} \quad \left|\frac{x-y}{x}\right|^2 = |xk| \quad \text{or} \quad |x-y|^2 = |x^3k| \qquad \text{Ans.}$$

(b) $\frac{dy}{dx} = \frac{yx^3 - xy^3}{x^4}$ or $\frac{dy}{dx} = \frac{x^4\left[\frac{y}{x} - \left(\frac{y}{x}\right)^3\right]}{x^4}$ or $\frac{dy}{dx} = \frac{y}{x} - \left(\frac{y}{x}\right)^3$ …………(i)

$$\text{Put } v = \frac{y}{x} \quad \text{or} \quad y = vx \quad \therefore \ \frac{dy}{dx} = v + x\frac{dv}{dx}$$

Put value of $\frac{dy}{dx}$ in equation (i), we get $v + x\frac{dv}{dx} = v - v^3$ or $x\frac{dv}{dx} = -v^3$ or $\frac{dv}{v^3} = -\frac{dx}{x}$

Integrating, $\int \frac{dv}{v^3} = -\int \frac{dx}{x}$ or $\int v^{-3}\,dv = -\int \frac{dx}{x}$ or $\frac{v^{-3+1}}{-3+1} = -\log x + \log k$

or $-\frac{1}{2}v^{-2} = -(\log x - \log k)$ or $\frac{1}{2v^2} = \log\left(\frac{x}{k}\right)$ or $1 = 2v^2\log\left(\frac{x}{k}\right)$ or $2\left(\frac{y}{x}\right)^2.\log\left(\frac{x}{k}\right) = 1$

or $\frac{2y^2}{x^2}.\log\left(\frac{x}{k}\right) = 1$ or $2y^2.\log\left(\frac{x}{k}\right) = x^2$ Ans.

(c) $\frac{dy}{dx} = \frac{y}{x} + \cos\left(\frac{y}{x}\right)$ ……………(i) Put $y = vx$ $\therefore$ $\frac{dy}{dx} = v + x\frac{dv}{dx}$

Put value of $\frac{dy}{dx}$ in equation (i), we get $v + x\frac{dv}{dx} = v + \cos v$ or $x\frac{dv}{dx} = \cos v$ or $\frac{dv}{\cos v} = \frac{dx}{x}$

Integrating, $\int \frac{dv}{\cos v} = \int \frac{dx}{x}$ or $\int \sec v\,dv = \int \frac{dx}{x}$ or $\int \frac{\sec v.\tan v}{\tan v}\,dv = \int \frac{dx}{x}$

or $\int \frac{\sec v \,.\tan v}{\sqrt{\sec^2 v - 1}}\, dv = \int \frac{dx}{x}$ Put $\sec v = t$ $\therefore$ $\sec v \,.\tan v\; dv = dt$

or $\int \frac{dt}{\sqrt{t^2-1}} = \int \frac{dx}{x}$ or $\log\left|t + \sqrt{t^2-1}\right| = \log|x| + \log|k|$ using formula, $\int \frac{dx}{\sqrt{x^2-a^2}} = \log\left|x + \sqrt{x^2-a^2}\right|$

or $\log\left|\sec v + \sqrt{\sec^2 v - 1}\right| = \log|xk|$ or $\log\left|\sec\left(\frac{y}{x}\right) + \sqrt{\sec^2\left(\frac{y}{x}\right) - 1}\right| - \log|x| = \log|k|$ Ans.

(d) $(x^2 + 2x + 1)dy - (x^2 + 3x + 2)dx = 0$ $\therefore$ $\frac{dy}{dx} = \frac{x^2+3x+2}{x^2+2x+1} = \frac{(x+1)(x+2)}{(x+1)^2} = \frac{x+2}{x+1}$

or $dy = \frac{x+2}{x+1}\, dx$ Integrating, $\int dy = \int \frac{x+2}{x+1}\, dx$ or $\int dy = \int \left(1 + \frac{1}{x+1}\right) dx$

or $y = x + \log(x+1) + \log c$ $\therefore$ $y - x = \log(x+1) + k$ Ans. $(\log c = k)$

(f) $e^{\frac{dy}{dx}} = 2^x + 2 = 2^{x+1}$ or $\log e^{\frac{dy}{dx}} = \log(2^{x+1}) = \log_e 2^{x+1}$

or $\frac{dy}{dx} = (x+1)\log 2$ or $dy = (x+1)\log 2\; dx$

Intgerating, $\int dy = \int (x+1)\log 2\; dx = \log 2 \int (x+1)\, dx$ or $y = \log 2\left[\frac{x^2}{2} + x\right] + c$

or $y = \frac{\log 2\,(x^2+2x) + 2c}{2}$ or $2y = \log 2\,(x^2 + 2x) + 2c$ $\therefore$ $2y = (x^2+2x)\log 2 + k$ Ans. $(k = 2c)$

(2) (a) Solve, $y^3 \frac{dx}{dy} = x^3 + x^2\sqrt{x^2-y^2}$

Solution: – Given, $y^3 \frac{dx}{dy} = x^3 + x^2\sqrt{x^2-y^2} = x^3 + x^2\sqrt{y^2\left(\frac{x^2}{y^2} - 1\right)} = x^3 + x^2 y\sqrt{\left(\frac{x}{y}\right)^2 - 1}$

or $\frac{dx}{dy} = \left(\frac{x}{y}\right)^3 + \left(\frac{x}{y}\right)^2\sqrt{\left(\frac{x}{y}\right)^2 - 1}$ Put $x = yv$ or $v = \frac{x}{y}$ $\therefore$ $\frac{dx}{dy} = v + y\frac{dv}{dy}$

or $v + y\frac{dv}{dy} = v^3 + v^2\sqrt{v^2-1}$ or $y\frac{dv}{dy} = v^3 + v^2\sqrt{v^2-1} - v = v\left[(v^2-1) + v\sqrt{v^2-1}\right]$

or $y\frac{dv}{dy} = v\sqrt{v^2-1}\left[\sqrt{v^2-1} + v\right]$ or $\frac{dv}{v\sqrt{v^2-1}\left[\sqrt{v^2-1}+v\right]} = \frac{dy}{y}$

or $\frac{\left(\sqrt{v^2-1} - v\right)}{v\sqrt{v^2-1}(v^2 - 1 + v^2)}\, dv = \frac{dy}{y}$ or $\frac{\left(\sqrt{v^2-1} - v\right)}{-v\sqrt{v^2-1}}\, dv = \frac{dy}{y}$

Integrating both of sides, $-\int \frac{\left(\sqrt{v^2-1} - v\right)}{v\sqrt{v^2-1}}\, dv = \int \frac{dy}{y}$ or $\int \frac{\left(v - \sqrt{v^2-1}\right)}{v\sqrt{v^2-1}}\, dv = \int \frac{dy}{y}$

or $\int \frac{dv}{\sqrt{v^2-1}} - \int \frac{dv}{v} = \int \frac{dy}{y}$ or $\log\left|v + \sqrt{v^2-1}\right| - \log v = \log y + \log a$

or $\log\left(\frac{v + \sqrt{v^2-1}}{v}\right) = \log ay$ or $\frac{v + \sqrt{v^2-1}}{v} = ay$ or $ay = \frac{\frac{x}{y} + \sqrt{\left(\frac{x}{y}\right)^2 - 1}}{\frac{x}{y}}$

or $\frac{x + \sqrt{x^2-y^2}}{x} = ay$ using formula, $\int \frac{dx}{\sqrt{x^2-a^2}} = \log\left|x + \sqrt{x^2-a^2}\right|$

or $axy = x + \sqrt{x^2-y^2}$ $\therefore$ $ax = \frac{1}{y}\left(x + \sqrt{x^2-y^2}\right)$ Ans.

(b) Solve, $\frac{dy}{dx} = \frac{x - 2y + 3}{2x - 4y + 5}$

Solution: – Given, $\frac{dy}{dx} = \frac{x - 2y + 3}{2x - 4y + 5} = \frac{(x - 2y) + 3}{2(x - 2y) + 5}$ ……..(i)

Put $x - 2y = v$ or $1 - 2\frac{dy}{dx} = \frac{dv}{dx}$ or $\frac{dy}{dx} = \frac{1 - \frac{dv}{dx}}{2}$

From (i), $\Rightarrow$ $\frac{1 - \frac{dv}{dx}}{2} = \frac{v + 3}{2v + 5}$ or $1 - \frac{dv}{dx} = \frac{2v + 6}{2v + 5}$

or $\frac{dv}{dx} = 1 - \frac{2v + 6}{2v + 5} = \frac{2v + 5 - 2v - 6}{2v + 5} = \frac{-1}{2v + 5}$ or $(2v + 5)\,dv = -dx$

Integrating, $\int (2v + 5)\,dv = -\int dx$ or $2.\frac{v^2}{2} + 5v = -x + c$ or $v^2 + 5v = -x + c$

or $(x - 2y)^2 + 5(x - 2y) = -x + c$ or $x^2 - 4xy + 4y^2 + 5x - 10y + x = c$

or $x^2 + 4y^2 - 4xy + 6x - 10y = c$ $\therefore$ $x^2 - 4xy + 6x - 10y + 4y^2 = c$ Ans.

(c) Solve, $\frac{dy}{dx} = \frac{3(x + y) + 1}{x + y + 3}$

Solution: – Given, $\frac{dy}{dx} = \frac{3(x + y) + 1}{x + y + 3}$ ………(i) Put $x + y = v$ $\therefore$ $1 + \frac{dy}{dx} = \frac{dv}{dx}$ $\therefore$ $\frac{dy}{dx} = \frac{dv}{dx} - 1$

Put value $\frac{dy}{dx}$ in equation (i), we get $\Rightarrow$ $\frac{dv}{dx} - 1 = \frac{3v + 1}{v + 3}$

or $\frac{dv}{dx} = \frac{3v + 1}{v + 3} + 1 = \frac{3v + 1 + v + 3}{v + 3} = \frac{4v + 4}{v + 3}$ or $\frac{dv}{dx} = \frac{4(v + 1)}{v + 3}$ or $\frac{v + 3}{4(v + 1)}\,dv = dx$

Integrating, $\int \frac{v + 3}{4(v + 1)}\,dv = \int dx$ or $\frac{1}{4}\int \frac{v + 3}{v + 1}\,dv = \int dx$ or $\frac{1}{4}\int \frac{(v + 1) + 2}{v + 1}\,dv = \int dx$

or $\frac{1}{4}\int \frac{v + 1}{v + 1}\,dv + \frac{1}{4}\int \frac{2}{v + 1}\,dv = \int dx$ or $\frac{1}{4}\int dv + \frac{1}{2}\int \frac{dv}{v + 1} = \int dx$

or $\frac{1}{4}v + \frac{1}{2}\log|v + 1| = x + c$ or $\frac{v + 2\log|v + 1|}{4} = x + c$ or $v + 2\log|v + 1| = 4x + 4c$

Put $v = x + y$ or $x + y + 2\log|x + y + 1| = 4x + 4c$ or $x + y + 2\log|x + y + 1| - 4x = 4c$

or $y - 3x + 2\log|x + y + 1| = 4c$ $\therefore$ $y - 3x + 2\log|x + y + 1| = k$ where $4c = k$ Ans.

(d) solve, $\frac{dy}{dx} = \frac{2x + 3y - 1}{2x + 3y + 3}$ (Do yourself, solve same as above question)

Ans: – $x + 3y - \frac{4}{5}(10x + 15y + 3) + \frac{12}{5}\log|10x + 15y + 3| = c$

(3) (a) Solve, $\cot^{-1}\left[\frac{x}{y} + \cos^2\left(\frac{x}{y}\right)\right] = \theta$

Solution: – Given, $\cot^{-1}\left[\frac{x}{y} + \cos^2\left(\frac{x}{y}\right)\right] = \theta$ or $\cot\theta = \frac{x}{y} + \cos^2\left(\frac{x}{y}\right)$ $\therefore$ $\tan\theta = \frac{dy}{dx}$ $\therefore$ $\cot\theta = \frac{dx}{dy}$

or $\frac{dx}{dy} = \frac{x}{y} + \cos^2\left(\frac{x}{y}\right)$ Put $v = \frac{x}{y}$ or $x = vy$ $\therefore$ $\frac{dx}{dy} = v + y\frac{dv}{dy}$

or $v + y\frac{dv}{dy} = v + \cos^2 v$ or $y\frac{dv}{dy} = \cos^2 v$ or $\frac{dv}{\cos^2 v} = \frac{dy}{y}$

Integrating, $\int \frac{dv}{\cos^2 v} = \int \frac{dy}{y}$ or $\int \sec^2 v\ dv = \int \frac{dy}{y}$ or $\tan v = \log y + \log a = \log ay$ Put $v = \frac{x}{y}$

or $\tan\left(\frac{x}{y}\right) = \log ay$ or $\frac{x}{y} = \tan^{-1}(\log ay)$ $\therefore$ $x = y\tan^{-1}(\log ay)$ Ans.

(b) Solve, $y\,dx = \left[x + y.\frac{f\left(\frac{x}{y}\right)}{f'\left(\frac{x}{y}\right)}\right] dy$

Solution: – (Do yourself.) Given, $y\,dx = \left[x + y.\frac{f\left(\frac{x}{y}\right)}{f'\left(\frac{x}{y}\right)}\right] dy$ Put $v = \frac{x}{y}$ $\therefore$ $f\left(\frac{x}{y}\right) = ky$ Ans.

(c) Solve, $\frac{dx}{dy} = e^{-\frac{x}{y}} + \frac{x}{y}$

Solution: – Given, $\frac{dx}{dy} = e^{\frac{x}{y}} + \frac{x}{y}$ Put $v = \frac{x}{y}$ or $x = yv$ $\therefore$ $\frac{dx}{dy} = v + y\frac{dv}{dy}$

or $v + y\frac{dv}{dy} = e^{-v} + v$ or $y\frac{dv}{dy} = e^{-v}$ or $\frac{dv}{e^{-v}} = \frac{dy}{y}$ or $e^v\,dv = \frac{dy}{y}$

Integrating, $\int e^v\,dv = \int \frac{dy}{y}$ or $e^v = \log y + \log a = \log|ay|$ or $e^{\frac{x}{y}} = \log|ay|$ Ans. $\left(\text{put } v = \frac{x}{y}\right)$

(d) Solve, $xy\frac{dx}{dy} = x^2 + y^2 e^{\left(\frac{x^2}{y^2}\right)}$

Solution: – Given, $xy\frac{dx}{dy} = x^2 + y^2 e^{\left(\frac{x^2}{y^2}\right)}$ or $\frac{dx}{dy} = \frac{x}{y} + \frac{y}{x}e^{\left(\frac{x^2}{y^2}\right)}$

Put $v = \frac{x}{y}$ or $x = yv$ $\therefore$ $\frac{dx}{dy} = v + y\frac{dv}{dy}$ or $v + y\frac{dv}{dy} = v + \frac{e^{v^2}}{v}$ or $y\frac{dv}{dy} = \frac{e^{v^2}}{v}$ or $\frac{v}{e^{v^2}}\,dv = \frac{dy}{y}$

Integrating, $\int \frac{v}{e^{v^2}}\,dv = \int \frac{dy}{y}$ Let $v^2 = t$ $\therefore$ $2v\,dv = dt$ or $v\,dv = \frac{dt}{2}$

or $\frac{1}{2}\int \frac{dt}{e^t} = \int \frac{dy}{y}$ or $\frac{1}{2}\int e^{-t}\,dt = \int \frac{dy}{y}$ or $-\frac{1}{2}e^{-t} = \log y + \log k = \log|ky|$

or $e^{-t} = -2\log|ky| = \log\left(\frac{1}{ky}\right)^2$ Put $t = v^2$ and $v = \frac{x}{y}$ or $e^{-v^2} = \log\left(\frac{1}{ky}\right)^2$

$$\therefore\ e^{-\left(\frac{x}{y}\right)^2} = \log\left(\frac{1}{ky}\right)^2 \quad \text{Ans.}$$

(4) (a) Solve, $\frac{dy}{dx} = \frac{x + y + 1}{x - y + 5}$

Solution: – Given, $\frac{dy}{dx} = \frac{x + y + 1}{x - y + 5}$ Put $x = \alpha + h$, $y = \beta + k$ $\therefore$ $\frac{dy}{dx} = \frac{d\beta}{d\alpha}$

$$\therefore\ \frac{d\beta}{d\alpha} = \frac{\alpha + h + \beta + k + 1}{\alpha + h - \beta - k + 5} = \frac{\alpha + \beta + (h + k + 1)}{\alpha - \beta + (h - k + 5)}$$ choose h and k such that

or $h + k + 1 = 0$ ……..(i) and $h - k + 5 = 0$ $\therefore$ $k = h + 5$ ……..(ii)

Put $k = h + 5$ in equation (i), we get

$$\therefore\ h + h + 5 + 1 = 0 \text{ or } 2h = -6\ \therefore\ h = -3 \text{ from (ii), } k = h + 5 = -3 + 5 = 2$$

$$\text{or} \quad \frac{d\beta}{d\alpha} = \frac{\alpha+\beta}{\alpha-\beta} = \frac{\alpha\left(1+\frac{\beta}{\alpha}\right)}{\alpha\left(1-\frac{\beta}{\alpha}\right)} = \frac{1+\frac{\beta}{\alpha}}{1-\frac{\beta}{\alpha}} \qquad \text{Put } \beta = v\alpha \ \text{ or } v = \frac{\beta}{\alpha} \quad \therefore \ \frac{d\beta}{d\alpha} = v + \alpha\frac{dv}{d\alpha}$$

$$\text{or} \ v + \alpha\frac{dv}{d\alpha} = \frac{1+v}{1-v} \quad \text{or} \ \alpha\frac{dv}{d\alpha} = \frac{1+v}{1-v} - v = \frac{1+v-v+v^2}{1-v} = \frac{1+v^2}{1-v} \quad \text{or} \ \frac{1-v}{1+v^2}dv = \frac{d\alpha}{\alpha}$$

$$\text{Integrating,} \ \int \frac{1-v}{1+v^2}dv = \int \frac{d\alpha}{\alpha} \quad \text{or} \ \int \frac{1}{1+v^2}dv - \int \frac{v}{1+v^2}dv = \int \frac{d\alpha}{\alpha}$$

$$\text{Put IInd integral } 1+v^2 = z \ \therefore \ 2v\,dv = dz \quad \text{or} \ v\,dv = \frac{dz}{2}$$

$$\text{or} \ \int \frac{1}{1+v^2}dv - \frac{1}{2}\int \frac{dz}{z} = \int \frac{d\alpha}{\alpha} \quad \text{or} \ \tan^{-1} v - \frac{1}{2}\log|z| = \log|\alpha| + k$$

$$\text{or} \ \tan^{-1} v - \frac{1}{2}\log|1+v^2| - \log|\alpha| = k \quad \text{or} \ \tan^{-1}\left(\frac{\beta}{\alpha}\right) - \log\left(\sqrt{1+\left(\frac{\beta}{\alpha}\right)^2}\right) - \log|\alpha| = k$$

Put $x = \alpha + h$, $y = \beta + k \ \therefore \ \alpha = x+3$, $\beta = y-2$

$$\text{or} \ \tan^{-1}\left(\frac{y-k}{x-h}\right) = \log\left(\frac{\sqrt{\alpha^2+\beta^2}}{|\alpha|}\right) + \log|\alpha| + k$$

$$\text{or} \ \tan^{-1}\left(\frac{y-2}{x+3}\right) = \log\left(\sqrt{\alpha^2+\beta^2}\right) - \log|\alpha| + \log|\alpha| + k$$

$$\text{or} \ \tan^{-1}\left(\frac{y-2}{x+3}\right) = \log\left(\sqrt{(x-h)^2+(y-k)^2}\right) + k$$

$$\text{or} \ \tan^{-1}\left(\frac{y-2}{x+3}\right) - \log\left(\sqrt{(x+3)^2+(y-2)^2}\right) = k$$

$$\text{or} \ \tan^{-1}\left(\frac{y-2}{x+3}\right) - \log\left(\sqrt{x^2+6x+9+y^2-4y+4}\right) = k$$

$$\text{or} \ \tan^{-1}\left(\frac{y-2}{x+3}\right) - \log\left(\sqrt{x^2+y^2+6x-4y+13}\right) = k$$

$$\therefore \ \tan^{-1}\left(\frac{y-2}{x+3}\right) - \log\left(\sqrt{x^2+y^2+6x-4y+13}\right) = k \quad \text{Ans.}$$

(b) Solve, $\frac{dy}{dx} = \frac{2x-y+1}{x+2y-5}$

Solution:− Given, $\frac{dy}{dx} = \frac{2x-y+1}{x+2y-5}$ Put $x = \alpha + h$, $y = \beta + k$ $\therefore \frac{dy}{dx} = \frac{d\beta}{d\alpha}$

$$\therefore \ \frac{d\beta}{d\alpha} = \frac{2(\alpha+h)-(\beta+k)+1}{(\alpha+h)+2(\beta+k)-5} = \frac{2\alpha-\beta+(2h-k+1)}{\alpha+2\beta+(h+2k-5)}$$

choose h and k such that $\therefore 2h-k+1 = 0$ ……….(i) and $h+2k-5 = 0$ ……………(ii)

solve (i) & (ii) and find h and k, $\therefore 2h-k+1 = 0$ or $k = 2h+1$

put value of k in equation (ii), we get $h+2k-5 = 0$ or $h+2(2h+1) = 5$ or $h+4h+2 = 5$

$$\text{or} \ 5h = 5-2 = 3 \ \therefore h = \frac{3}{5} \quad \text{from (i),} \ 2h-k+1 = 0 \quad \text{or} \ k = 2.\frac{3}{5}+1 = \frac{6+5}{5} = \frac{11}{5}$$

$$\text{or} \ \frac{d\beta}{d\alpha} = \frac{2\alpha-\beta}{\alpha+2\beta} = \frac{\alpha\left(2-\frac{\beta}{\alpha}\right)}{\alpha\left(1+2.\frac{\beta}{\alpha}\right)} = \frac{2-\frac{\beta}{\alpha}}{1+2.\frac{\beta}{\alpha}} \qquad \text{Put } \beta = v\alpha \ \text{ or } v = \frac{\beta}{\alpha} \quad \therefore \ \frac{d\beta}{d\alpha} = v + \alpha\frac{dv}{d\alpha}$$

or $v + \alpha\frac{dv}{d\alpha} = \frac{2-v}{1+2v}$ or $\alpha\frac{dv}{d\alpha} = \frac{2-v}{1+2v} - v = \frac{2-v-v-2v^2}{1+2v} = \frac{2-2v-2v^2}{1+2v}$

or $\frac{1+2v}{2-2v-2v^2}\,dv = \frac{d\alpha}{\alpha}$

Integrating, $\int \frac{1+2v}{2-2v-2v^2}\,dv = \int \frac{d\alpha}{\alpha}$ Let $2-2v-2v^2 = z$ $\therefore -2(1+2v)\,dv = dz$

or $(1+2v)\,dv = -\frac{dz}{2}$ or $-\frac{1}{2}\int\frac{dz}{z} = \int\frac{d\alpha}{\alpha}$ or $-\frac{1}{2}\log|z| = \log|\alpha| + k$ or $\log\left|\frac{1}{z}\right| = 2\log|\alpha| + 2k$

using formula, $\int \frac{f'(x)}{f(x)}\,dx = \log|f(x)|$

or $\log\left|\frac{1}{2-2v-2v^2}\right| = \log|\alpha|^2 + 2k$ or $\log\left|\frac{1}{2-2.\frac{\beta}{\alpha}-2\left(\frac{\beta}{\alpha}\right)^2}\right| = \log|\alpha|^2 + 2k$

or $\log\left|\frac{\alpha^2}{2\alpha^2-2\alpha\beta-2\beta^2}\right| = \log|\alpha|^2 + 2k$

or $\log\left|\frac{(x-h)^2}{2(x-h)^2-2(x-h)(y-k)-2(y-k)^2}\right| = \log|x-h|^2 + 2k$

or $\log(x-h)^2 - \log[2(x-h)^2 - 2(x-h)(y-k) - 2(y-k)^2] = \log|x-h|^2 + 2k$

Put $h = \frac{3}{5}$ and $k = \frac{11}{5}$

or $\log\left(x-\frac{3}{5}\right)^2 - \log\left[2\left(x-\frac{3}{5}\right)^2 - 2\left(x-\frac{3}{5}\right)\left(y-\frac{11}{5}\right) - 2\left(y-\frac{11}{5}\right)^2\right] = \log\left(x-\frac{3}{5}\right)^2 + 2k$

or $-\log\left[\frac{2(5x-3)^2 - 2(5x-3)(5y-11) - 2(5y-11)^2}{25}\right] = 2k$

$$\therefore -\log\left\{\frac{2}{25}[(5x-3)^2 - (5x-3)(5y-11) - (5y-11)^2]\right\} = 2k \quad \text{Ans.}$$

(5) (a) Solve, $\frac{dy}{dx} + \frac{y}{x} = e^x$

Solution: – Given, $\frac{dy}{dx} + \frac{y}{x} = e^x$ form of $\frac{dy}{dx} + Py = Q$ here $P = \frac{1}{x}$ and $Q = e^x$

$I.F = e^{\int P\,dx} = e^{\int\frac{1}{x}dx} = e^{\log x} = x$

The required solution is $y \times I.F = \int I.F \times Q\,dx$

or $y.x = \int xe^x\,dx = x\int e^x\,dx - \int\left[\frac{d(x)}{dx}.\int e^x\,dx\right]dx$

or $xy = xe^x - \int e^x\,dx$ or $xy = xe^x - e^x + c$ $\boxed{\therefore\ xy = e^x(x-1) + c}$ Ans.

(b) Solve, $\frac{1}{\sin x}.\frac{dy}{dx} + y\cos x = \cos x$

Solution: – Given, $\frac{1}{\sin x}.\frac{dy}{dx} + y\cos x = \cos x$ or $\frac{dy}{dx} + y\sin x\cos x = \sin x\cos x$

Here $P = \sin x\cos x$ and $Q = \sin x\cos x$ $\therefore$ $I.F = e^{\int P\,dx} = e^{\int \sin x\cos x\,dx} = e^{\int z\,dz} = e^{\frac{z^2}{2}} = e^{\frac{\sin^2 x}{2}}$

The required solution is $y \times I.F = \int I.F \times Q\ dx$ or $y.e^{\frac{\sin^2 x}{2}} = \int e^{\frac{\sin^2 x}{2}}.\sin x \cos x\ dx$

Put $\frac{\sin^2 x}{2} = t$ or $\sin^2 x = 2t$ $\therefore$ $2\sin x \cos x\ dx = 2\ dt$ or $\sin x \cos x\ dx = dt$

or $y.e^{\frac{\sin^2 x}{2}} = \int e^t\ dt$ or $y.e^{\frac{\sin^2 x}{2}} = e^t + c$ or $y.e^{\frac{\sin^2 x}{2}} = e^{\frac{\sin^2 x}{2}} + c$ or $y = 1 + ce^{-\frac{\sin^2 x}{2}}$

$\therefore$ $y = 1 + k$ Ans. where $k = ce^{-\frac{\sin^2 x}{2}}$

(c) Solve, $\frac{dx}{dy} - \frac{xy}{1+y^2} = y + y^3$

Solution:– Given, $\frac{dx}{dy} - \frac{xy}{1+y^2} = y + y^3$ or $\frac{dx}{dy} - \frac{y}{1+y^2}x = y + y^3$

Here $P = -\frac{y}{1+y^2}$ and $Q = y + y^3$ $\therefore$ $I.F = e^{\int P\,dy} = e^{-\int\frac{y}{1+y^2}dy}$

Put $1 + y^2 = z$ $\therefore$ $2y\ dy = dz$ or $y\ dy = \frac{dz}{2}$

$$\therefore\ I.F = e^{-\frac{1}{2}\int\frac{dz}{z}} = e^{-\frac{1}{2}\log z} = e^{\log\left(\frac{1}{\sqrt{z}}\right)} = e^{\log\left(\frac{1}{\sqrt{1+y^2}}\right)} = \frac{1}{\sqrt{1+y^2}}$$

The required solution is – $x \times I.F = \int I.F \times Q\ dy$

or $x.\frac{1}{\sqrt{1+y^2}} = \int \frac{1}{\sqrt{1+y^2}}.(y + y^3)\ dy = \int \frac{y(1+y^2)}{\sqrt{1+y^2}}\ dy$ or $\frac{x}{\sqrt{1+y^2}} = \int y\sqrt{1+y^2}\ dy$

Put $1 + y^2 = t^2$ or $y = \sqrt{t^2 - 1}$ $\therefore$ $2y\ dy = 2t\ dt$ or $y\ dy = t\ dt$

or $\frac{x}{\sqrt{1+y^2}} = \int t^2\ dt$ or $\frac{x}{\sqrt{1+y^2}} = \frac{t^3}{3} + c$

or $\frac{x}{\sqrt{1+y^2}} = \frac{(1+y^2)\sqrt{1+y^2}}{3} + c = \frac{(1+y^2)\sqrt{1+y^2} + 3c}{3}$

or $3x = (1+y^2)\sqrt{1+y^2}.\sqrt{1+y^2} + 3c\sqrt{1+y^2}$ $\therefore$ $3x = (1+y^2)^2 + 3c\sqrt{1+y^2}$ Ans.

(6) (a) Solve, $\frac{dy}{dx} + y\cot x = y^2\cos^3 x$

Solution:– Given, $\frac{dy}{dx} + y\cot x = y^2\cos^3 x$ or $\frac{1}{y^2}\frac{dy}{dx} + \frac{\cot x}{y} = \cos^3 x$ Divide by y^2

Let $\frac{1}{y} = z$ $\therefore$ $-\frac{1}{y^2}\frac{dy}{dx} = \frac{dz}{dx}$ or $\frac{1}{y^2}\frac{dy}{dx} = -\frac{dz}{dx}$

Now, the original equation is $-\frac{dz}{dx} + z\cot x = \cos^3 x$ or $\frac{dz}{dx} - z\cot x = \cos^3 x$

Here $P = -\cot x$ and $Q = \cos^3 x$

$$\therefore\ I.F = e^{\int P\,dx} = e^{-\int\cot x\,dx} = e^{-\log(\sin x)} = e^{\log(\sin x)^{-1}} = (\sin x)^{-1} = \frac{1}{\sin x}$$

The required solution is – $z \times I.F = \int (I.F \times Q)\ dx$

or $z.\frac{1}{\sin x} = \int \frac{1}{\sin x}.\cos^3 x\, dx = \int \frac{\cos^2 x.\cos x}{\sin x} dx$ or $\frac{z}{\sin x} = \int \frac{(1-\sin^2 x).\cos x}{\sin x} dx$

Put $\sin x = t$ $\therefore$ $\cos x\, dx = dt$ also put $z = \frac{1}{y}$

or $\frac{1}{y\sin x} = \int \frac{1-t^2}{t} dt = \int \frac{dt}{t} - \int \frac{t^2}{t} dt = \log t - \int t\, dt = \log t - \frac{t^2}{2} + c = \log(\sin x) - \frac{\sin^2 x}{2} + c$

or $\frac{1}{y\sin x} = \frac{2\log(\sin x) - \sin^2 x + 2c}{2}$ or $2 = y\sin x\,[2\log(\sin x) - \sin^2 x + 2c]$ Ans.

(b) Solve, $\frac{dy}{dx}(x^2y + xy) = \frac{1}{y}$

Solution: – Given, $\frac{dy}{dx}(x^2y + xy) = \frac{1}{y}$ or $\frac{dy}{dx} = \frac{1}{y(x^2y + xy)}$ or $\frac{dx}{dy} = y(x^2y + xy)$

or $\frac{dx}{dy} = x^2y^2 + xy^2$ or $\frac{dx}{dy} - xy^2 = x^2y^2$ or $\frac{1}{x^2}\frac{dx}{dy} - \frac{y^2}{x} = y^2$

Put $\frac{1}{x} = z$ or $-\frac{1}{x^2}\frac{dx}{dy} = \frac{dz}{dy}$ or $\frac{1}{x^2}\frac{dx}{dy} = -\frac{dz}{dy}$

Now, the original equation is $-\frac{dz}{dy} - zy^2 = y^2$ or $\frac{dz}{dy} + zy^2 = -y^2$

$\therefore$ I.F $= e^{\int P\, dy} = e^{\int y^2\, dy} = e^{\frac{y^3}{3}}$

The required solution is – $z.e^{\frac{y^3}{3}} = -\int e^{\frac{y^3}{3}}.y^2\, dy$

Let $\frac{y^3}{3} = t$ or $3y^2dy = 3dt$ or $y^2dy = dt$

or $\frac{e^{\frac{y^3}{3}}}{x} = -\int e^t\, dt = -e^t - c$ or $e^{\frac{y^3}{3}} = -xe^{\frac{y^3}{3}} - cx$ or $1 = -x - cxe^{\frac{-y^3}{3}}$

or $x + 1 = -cxe^{\frac{-y^3}{3}}$ Ans.

IInd Method: – $\frac{dz}{dy} + zy^2 = -y^2$ or $\frac{dz}{dy} = -y^2 - zy^2 = -y^2(z+1)$ or $\frac{dz}{z+1} = -y^2dy$

Integrating, $\int \frac{dz}{z+1} = -\int y^2\, dy$ or $\log|z+1| = -\frac{y^3}{3} - c$ or $\log\left(\frac{1+x}{x}\right) = -\left(\frac{y^3}{3} + c\right)$ Ans.

(c) Solve, $(xy^2 + xy)dx = dy$

Solution: – Given, $(xy^2 + xy)dx = dy$ or $\frac{dy}{dx} = xy^2 + xy$ or $\frac{dy}{dx} - xy = xy^2$ or $\frac{1}{y^2}\frac{dy}{dx} - \frac{x}{y} = x$

Let $\frac{1}{y} = z$ $\therefore$ $-\frac{1}{y^2}\frac{dy}{dx} = \frac{dz}{dx}$ or $\frac{1}{y^2}\frac{dy}{dx} = -\frac{dz}{dx}$

Now, the original equation is $-\frac{dz}{dx} - zx = x$ or $\frac{dz}{dx} + zx = -x$ $\therefore$ I.F $= e^{\int x\, dx} = e^{\frac{x^2}{2}}$

The required solution is – $z \times \text{I.F} = \int (\text{I.F} \times Q)\, dx$ or $z.e^{\frac{x^2}{2}} = -\int e^{\frac{x^2}{2}}.x\, dx$

Put $\frac{x^2}{2} = t$ or $x\, dx = dt$ or $z.e^{\frac{x^2}{2}} = -\int e^t\, dt = -e^t - c$ or $\frac{e^{\frac{x^2}{2}}}{y} = -e^{\frac{x^2}{2}} - c$

$$\text{or } 1=-\left(y+cye^{-\frac{x^2}{2}}\right) \text{ or } 1+y=-cye^{-\frac{x^2}{2}} \quad \text{Ans.}$$

(d) Solve, $\frac{dy}{dx}=xy+x^5y^4$

Solution: – Given, $\frac{dy}{dx}=xy+x^5y^4$ or $\frac{dy}{dx}-xy=x^5y^4$ Divide by y^4

or $\frac{1}{y^4}\frac{dy}{dx}-\frac{x}{y^3}=x^5$ Let $\frac{1}{y^3}=z$ $\therefore -\frac{3y^2}{y^6}\frac{dy}{dx}=\frac{dz}{dx}$ or $\frac{1}{y^4}\frac{dy}{dx}=-\frac{1}{3}.\frac{dz}{dx}$

The original equation is $-\frac{1}{3}.\frac{dz}{dx}-zx=x^5$ or $\frac{dz}{dx}+3xz=-3x^5$ $\therefore$ I. F $=e^{\int 3x\,dx}=e^{\frac{3x^2}{2}}$

The required solution is – $z\times \text{I.F}=\int(\text{I.F}\times Q)\,dx$

or $z.e^{\frac{3x^2}{2}}=-3\int e^{\frac{3x^2}{2}}.x^5\,dx=-3\int e^{\frac{3x^2}{2}}.x^4.x\,dx$

Let $\frac{3x^2}{2}=t$ or $x^2=\frac{2t}{3}$ or $6x\,dx=2\,dt$ $\therefore$ $3x\,dx=dt$

or $\frac{e^{\frac{3x^2}{2}}}{y^3}=-\int e^t.\left(\frac{2t}{3}\right)^2 dt$ or $\frac{e^{\frac{3x^2}{2}}}{y^3}=-\frac{4}{9}\int t^2.e^t\,dt$ (use integration by part formula)

or $\frac{e^{\frac{3x^2}{2}}}{y^3}=-\frac{4}{9}.e^t[t^2-2t+2]-c$ or $\frac{e^{\frac{3x^2}{2}}}{y^3}=-\frac{4}{9}.e^{\frac{3x^2}{2}}\left[\left(\frac{3x^2}{2}\right)^2-2\left(\frac{3x^2}{2}\right)+2\right]-c$ Ans.

(8) (a) Solve, $\frac{dy}{dx}+\frac{y}{x}=\frac{\sqrt{y}}{x\sqrt{x}}$

Solution: – Given, $\frac{dy}{dx}+\frac{y}{x}=\frac{\sqrt{y}}{x\sqrt{x}}$ Divide by $\sqrt{y}$, $\therefore$ $\frac{1}{\sqrt{y}}\frac{dy}{dx}+\frac{y}{x\sqrt{y}}=\frac{1}{x\sqrt{x}}$

or $\frac{1}{\sqrt{y}}\frac{dy}{dx}+\frac{\sqrt{y}}{x}=\frac{1}{x\sqrt{x}}$ Let $\sqrt{y}=z$ $\therefore$ $\frac{1}{2\sqrt{y}}\frac{dy}{dx}=\frac{dz}{dx}$ or $\frac{1}{\sqrt{y}}\frac{dy}{dx}=2\frac{dz}{dx}$

The original equation is $2\frac{dz}{dx}+\frac{z}{x}=\frac{1}{x\sqrt{x}}$ or $\frac{dz}{dx}+\frac{z}{2x}=\frac{1}{2x\sqrt{x}}$

$\therefore$ I. F $=e^{\int\frac{dx}{2x}}=e^{\frac{1}{2}\int\frac{dx}{x}}=e^{\frac{1}{2}\log x}=e^{\log\sqrt{x}}=\sqrt{x}$

The required solution is – $z\times \text{I.F}=\int(\text{I.F}\times Q)\,dx$

or $z.\sqrt{x}=\int\sqrt{x}.\frac{1}{2x\sqrt{x}}dx=\frac{1}{2}\int\frac{dx}{x}=\frac{1}{2}\log x+\log k$

or $\sqrt{y}.\sqrt{x}=\log\sqrt{x}+\log k=\log(\sqrt{x}.k)$ or $\sqrt{xy}=\log(\sqrt{x}.k)$ Ans.

(b) Solve, $\frac{dy}{dx}+\frac{1}{1+x}=\frac{e^{-y}}{x}$

Solution: – Given, $\frac{dy}{dx}+\frac{1}{1+x}=\frac{e^{-y}}{x}$ Divide by e^{-y}, we get

$\therefore$ $\frac{1}{e^{-y}}.\frac{dy}{dx}+\frac{1}{e^{-y}(1+x)}=\frac{1}{x}$ or $e^y\frac{dy}{dx}+\frac{e^y}{1+x}=\frac{1}{x}$

Let $e^y=z$ Differentiate, $e^y\frac{dy}{dx}=\frac{dz}{dx}$

The original equation is $\frac{dz}{dx}+\frac{z}{1+x}=\frac{1}{x}$ $\quad\therefore$ $I.F = e^{\int P\,dx} = e^{\int\frac{dx}{1+x}} = e^{\log(1+x)} = 1+x$

The required solution is $z\times I.F=\int (I.F\times Q)\,dx$ or $z.(1+x)=\int\frac{1+x}{x}\,dx=\int\frac{dx}{x}+\int dx$

or $e^{y}(1+x)=\log x+x+k$ or $-x-\log x+e^{y}(1+x)=k$ Ans.

(c) Solve, $x^2\frac{dy}{dx}-xy=y^3\cos x$

Solution: – Given, $x^2\frac{dy}{dx}-xy=y^3\cos x$ Divid by x^2y^3, we get $\frac{1}{y^3}\frac{dy}{dx}-\frac{1}{xy^2}=\frac{\cos x}{x^2}$

Let $\frac{1}{y^2}=z$ or $-\frac{2}{y^3}\frac{dy}{dx}=\frac{dz}{dx}$ or $\frac{1}{y^3}\frac{dy}{dx}=-\frac{1}{2}.\frac{dz}{dx}$

The original equation is $-\frac{1}{2}.\frac{dz}{dx}-\frac{z}{x}=\frac{\cos x}{x^2}$ or $\frac{dz}{dx}+\frac{2z}{x}=-\frac{\cos x}{x^2}$

$\therefore$ $I.F=e^{2\int\frac{dx}{x}}=e^{2\log x}=e^{\log x^2}=x^2$

The required solution is $z\times I.F=\int (I.F\times Q)\,dx$

or $z.x^2=-\int\frac{\cos x}{x^2}.x^2\,dx=-\int\cos x\,dx=-\sin x-c$

$\therefore$ $\frac{x^2}{y}=-\sin x-c$ or $x^2=-y\sin x-cy$ Ans.

(d) Solve, $\frac{1}{y}.\frac{dy}{dx}+\frac{1}{1+x}=\frac{x}{y^3}$

Solution: – Given, $\frac{1}{y}.\frac{dy}{dx}+\frac{1}{1+x}=\frac{x}{y^3}$ Multiplying by y^3, we get $y^2\frac{dy}{dx}+\frac{y^3}{1+x}=x$

Let $y^3=z$ Diff. $3y^2\frac{dy}{dx}=\frac{dz}{dx}$ $\therefore$ $y^2\frac{dy}{dx}=\frac{1}{3}.\frac{dz}{dx}$

The original equation is $\frac{1}{3}.\frac{dz}{dx}+\frac{z}{1+x}=x$ or $\frac{dz}{dx}+\frac{3z}{1+x}=3x$

$\therefore$ $I.F=e^{\int\frac{3}{1+x}dx}=e^{3\log(1+x)}=e^{\log(1+x)^3}=(1+x)^3$

The required solution is $z\times I.F=\int (I.F\times Q)\,dx$ or $z(1+x)^3=\int x.(1+x)^3\,dx$

Put $1+x=t$ $\therefore$ $dx=dt$

or $(1+x)^3.y^3=\int (t-1).t^3\,dt=\int t^4\,dt-\int t^3\,dt=\frac{t^5}{5}-\frac{t^4}{4}+c=\frac{(1+x)^5}{5}-\frac{(1+x)^4}{4}+c$

or $y^3=\frac{(1+x)^2}{5}-\frac{1+x}{4}+\frac{c}{(1+x)^3}$ Ans.

Exercise – A11

(1) Find the differential equation: –

(a) $y=(c_1+c_2)e^x+\cos(x+c_3)$ where c_1, c_2 and c_3 are constant.

(b) $y=c_1\sin(x+c_2)-c_3e^{x+c_4}$ where c_1, c_2, c_3 and c_4 are constant.

(c) $xy = Ae^{3x} + Be^{-3x}$ where A and B are constant.

(2) (a) Prove that the differential equation of the family of parabola

$$y^2 = 4ax \text{ is } 2x\frac{d^2y}{dx^2} + \frac{dy}{dx} = 0 \text{ or } x\frac{d^2y}{dx^2} - y = 0.$$

(b) Prove that the differential equation of the family of parabola

$$x^2 = 4ay \text{ is } x\frac{dy}{dx} - 2y = 0 \text{ or } x^2\frac{d^2y}{dx^2} - 2y = 0.$$

(c) Prove that the differential equation of the circle $x^2 + y^2 = a^2$ is $y\frac{dy}{dx} + x = 0.$

(3) (a) Eliminate the constants a, b from the relation $y = a\cos x + b\sin x + x\cos x.$

(b) Eliminate the constants c_1, c_2 from the relation $y = c_1e^x\sin x + c_2e^x\cos x.$

(4) (a) Prove that $y = ax^2 + bx + c$ is a solution of $x^2\frac{d^2y}{dx^2} - 2x\frac{dy}{dx} + 2y = 0.$

(b) Prove that $u = \frac{A}{v} + Bv$ is a solution of $v^2\frac{d^2u}{dv^2} + v\frac{du}{dv} - u = 0.$

(c) Find the differential equation corresponding to the family of curves $xy = (x + k)^2$

where k is an arbitrary constant.

(5) (a) Find the differential equation of the family of the curves $y = Ae^x + Be^{3x}.$

(b) Find the differential equation of the curves $y = ae^x + be^{-2x} + ce^{3x}$ where a, b and c are arbitrary constant.

(c) Find the differential equation of the curves $y = Ae^{5x} + Be^{-7x}.$

(6) Find the differential equation of the family of the curves

(a) $x^2 + y^2 + 2ax + c = 0$ (b) $x^2 + y^2 + 2by + c = 0$ (c) $\frac{x^2}{a^2 + m} + \frac{y^2}{b^2} = 1$ where m is parameter.

(7) (a) $(1 + x^2)^2\frac{d^2y}{dx^2} + 2x(1 + x^2)\frac{dy}{dx} - 3y = 0$ it being given that $x = \tan\theta.$

(b) $(1 - x^2)\frac{d^2y}{dx^2} - x\frac{dy}{dx} + 2y = 0$ it being given that $x = \sin\theta.$

(c) $(1 - e^{2x})\frac{d^2y}{dx^2} - \frac{e^{2x}}{\sqrt{1 - e^{2x}}}.\frac{dy}{dx} = 0$ it being given that $x = \log(\sin\theta).$

(8) (a) $x[1 - (\log x)^2].\frac{d^2y}{dx^2} + [1 + \log x - (\log x)^2].\frac{dy}{dx} - 3y = 0$ it being given that $x = e^{\cos\theta}.$

(b) $\sin x.\frac{d^2y}{dx^2} - \cos x.\frac{dy}{dx} + 2y\sin^3 x = 3\sin^5 x$ it being given that $\theta = \cos x.$

(c) If $x = \cos\theta,\ y = \cos k\theta$ then show that $(1 - x^2)\frac{d^2y}{dx^2} - x\frac{dy}{dx} + k^2y = 0.$

Answer

(1) (a) $\frac{d^4y}{dx^4} = y$ or $y^{iv} - y = 0$ or $\frac{d^4y}{dx^4} - y = 0$ (b) $\frac{d^4y}{dx^4} = y$ or $\frac{d^4y}{dx^4} - y = 0$

(c) $xy'' + 2y' = 9xy$ or $x\frac{d^2y}{dx^2} + 2\frac{dy}{dx} = 9xy$

(2) (a) $y^2 = 4ax$ Differentiating, $\therefore\ 2y\frac{dy}{dx} = 4a$ or $y\frac{dy}{dx} = 2a$ $\therefore\ \frac{dy}{dx} = \frac{2a}{y}$ or $\frac{dy}{dx} = \frac{2ay}{y^2} = \frac{2ay}{4ax} = \frac{y}{2x}$

or $2x\frac{dy}{dx} = y$ $\therefore\ 2x\frac{dy}{dx} - y = 0$ proved.

Again Differentiating, $y\frac{d^2y}{dx^2} + \frac{dy}{dx}.\frac{dy}{dx} = 0$ or $y\frac{d^2y}{dx^2} + \frac{2a}{y}.\frac{dy}{dx} = 0$

or $\frac{d^2y}{dx^2} + \frac{2a}{y^2}.\frac{dy}{dx} = 0$ or $\frac{d^2y}{dx^2} + \frac{2a}{4ax}.\frac{dy}{dx} = 0$ or $\frac{d^2y}{dx^2} + \frac{1}{2x}.\frac{dy}{dx} = 0$ or $2x\frac{d^2y}{dx^2} + \frac{dy}{dx} = 0$ Proved.

(b) and (c) Do yourself. (see above question)

(3) (a) $\frac{d^2y}{dx^2} + y = -2\sin x$ (b) $\frac{d^2y}{dx^2} - 2\frac{dy}{dx} + 2y = 0$

(4) (a) $y = ax^2 + bx + c$ Diff. $\frac{dy}{dx} = 2ax + b$ Again Diff. $\frac{d^2y}{dx^2} = 2a$

or $x^2\frac{d^2y}{dx^2} - 2x\frac{dy}{dx} + 2y = 0$ or $x^2.2a - 2x(2ax + b) + 2(ax^2 + bx + c) = 0$

or $2ax^2 - 4ax^2 - 2bx + 2ax^2 + 2bx + 2c = 0$ or $2c = 0$ $\therefore\ c = 0$ Ans.

(b) Do yourself.

(c) Do yourself, $\left(x\frac{dy}{dx} + y\right)^2 = 4xy$ or $x^2\left(\frac{dy}{dx}\right)^2 + 2xy\frac{dy}{dx} + y^2 = 4xy$ Ans.

(5) (a) $2\frac{d^2y}{dx^2} - 5\frac{dy}{dx} + 3y = 0$ (b) $5\frac{d^3y}{dx^3} - 18\frac{d^2y}{dx^2} + 19\frac{dy}{dx} - 6y = 0$ (c) $6\frac{d^2y}{dx^2} - 37\frac{dy}{dx} + 35y = 0$

(6) (a) $y\frac{d^2y}{dx^2} + \left(\frac{dy}{dx}\right)^2 + 1 = 0$ (b) $x\frac{d^2y}{dx^2} = \frac{dy}{dx}\left[\left(\frac{dy}{dx}\right)^2 + 1\right]$ (c) $xy\frac{dy}{dx} - y^2 + b^2 = 0$

(7) (a) $\frac{d^2y}{d\theta^2} - 3y = 0$ (b) $\frac{d^2y}{d\theta^2} + 2y = 0$ (c) $\sin\theta\frac{d^2y}{d\theta^2} + \cos\theta\frac{dy}{d\theta} + y = 0$

(8) (a) $\frac{1}{x}.\frac{d^2y}{d\theta^2} - 3y = 0$ (b) $\frac{d^2y}{d\theta^2} + 2y = 3(1 - \theta^2)$

Exercise – A12

(1) solve the following differential equation: – (a) $(2 + x)\,dy - (3 - y)\,dx = 0$

(b) $(e^y + 1)x\,dx = e^y(x + 1)\,dy$ (c) $\cot x\,dy + \cot y\,dx = 0$ (d) $\text{cosec}^2 x.\cot y\,dx = \text{cosec}^2 y.\cot x\,dy$

(2) (a) $x\sin^2 y\,dx - y\sin^2 x\,dy = 0$ (b) $\frac{dy}{dx} = \cos(5x + 3y)$ (c) $\cos(x + y)\,dy = \sin(x + y)\,dx$

(d) $e^x(y + 1)dy = e^y\,dx$

(3) (a) $y\,dx + x\,dy = x\,dx$ (b) $x\,dy - y\,dx = xy\,dx$ (c) $\frac{dx}{dy} - \sqrt{\frac{1 - x^2}{1 - y^2}} = 0$ (d) $\log\left(\frac{dy}{dx}\right) = x - y$

(4) (a) $\frac{dy}{dx} - \frac{xy - x}{xy - y} = 0$ (b) $\frac{dy}{dx} + \frac{e^x\tan y}{(e^x + 1)\sec^2 y} = 0$ (c) $\frac{y}{1 - x}.\frac{dy}{dx} = \frac{1 - y^2}{1 - x^2}$ (d) $\sin^2 x\frac{dy}{dx} = \sin 2x.\cos^2 y$

(5) (a) $\frac{dy}{dx} = (x + y + 1)^2$ (b) $\frac{dy}{dx} = (4x + y)^2$ (c) $\frac{dy}{dx} - 1 = e^{x-y}$ (d) $\frac{dy}{dx} + \sin(x + y) = \sin(x - y)$

(6) (a) $\sin^{-1}\left(\frac{dy}{dx}\right) = 3x + y$ (b) $\frac{dy}{dx} = \cot(y - x) + 1$ (c) $\frac{dy}{dx} - 1 = x\sec(x - y)$ (d) $(3x + y - 1)^2.\frac{dy}{dx} = a^2$

(7) (a) $(x - y + 1)\,dy = (3x - 3y + 5)\,dx$ (b) $(2x - y - 3)\,dy + (4x - 2y + 5)\,dx = 0$

(c) $\cot y . \frac{dy}{dx} = \cos(x + y) + \cos(x - y)$ (d) $x \operatorname{cosec}^2 y\,dy = (x + 1)\cot y\,dx$

(8) (a) $(x + y)^2 . \frac{dy}{dx} = 4$ (b) $(y + 5)\cot x . \frac{dy}{dx} = -y \operatorname{cosec}^2 x$ (c) $\log\left(\frac{dy}{dx}\right) = 3x + 4y$ (d) $y - (1 + x)\frac{dy}{dx} = (1 + x^2)\frac{dy}{dx}$

(9) (a) $2xb^2 + 2ya^2 . \frac{dy}{dx} = 2x$ (b) $\frac{dy}{dx} + x^2 e^{x+by} = e^{ax+by}$ (c) $\frac{dy}{dx} = \sin(10x + 6y) . \cos(10x + 6y)$ (d) $\frac{dy}{dx} = \frac{x + 1}{y}$

(10) (a) $\frac{dy}{dx} + 1 = x \sin(x + y)$ (b) $x \operatorname{cosec}^2 y + (x + 3)\cot y . \frac{dx}{dy} = 0$ (c) $x - y . \frac{dx}{dy} = x^2 - \frac{dx}{dy}$

(d) $y \cos y . \frac{dy}{dx} = x(2 \log x - 1)$

(11) (a) $y\,dx - x dy = \sqrt{x^2 + y^2}\,dy$ (b) $y\,dx - x\,dy = \sqrt{x^2 - y^2}\,dy$ (c) $x\,dy + y\,dx = xy\,dx$ (d) $y\,dx - x\,dy = \frac{x}{y}\,dy$

(12) (a) $\frac{y - x . \frac{dy}{dx}}{\sqrt{x^2 + y^2}} = \frac{x^2 + y^2}{x}$ (b) $- x\,dy - y\,dx = (x^2y^2 + xy)dx$ (c) $dx + \frac{(x + 1)}{y}dy = y(x + 1)\,dx$

(d) $(x + y)\frac{dy}{dx} = \frac{y}{x} + (x - y)\frac{dy}{dx}$

(13) (a) $\frac{\cos y}{\sin x} . \frac{dy}{dx} = \frac{\sqrt{1 + \sin^2 y}}{\sqrt{1 + \cos^2 x}}$ (b) $e^{y+1} . x\frac{dy}{dx} = \frac{x + 1}{y + 1}$ (c) $(x - 1)y\,dx = (y + 2)x\,dy$

(d) $(y + 1) . \cos(y + 1)\,dy = (x + 2) . \sin(x + 2)\,dx$ (e) $e^{y-x} . \frac{dy}{dx} = \frac{e^y + 1}{e^x + 1}$

Answer

(1) (a) $(2 + x)(3 - y) = \frac{1}{k}$ (b) $(x + 1)(e^y + 1) = ke^x$ (c) $\cos x \cos y = \frac{1}{c}$ or $c(\cos x \cos y) = 1$

(d) $\tan y - \tan x = c$

(2) (a) $\log\left(\frac{\sin x}{\sin y}\right) = x \cot x - y \cot y - c$ or $\sin x \sin y . \log\left(\frac{\sin x}{\sin y}\right) = x \cos x \sin y - y \cos y \sin x + k$

(b) $\frac{1}{2}\tan^{-1}\left[\frac{\sqrt{7}}{2}\tan(5x + 3y)\right] + c$ (c) $e^{x+c} . \sqrt{2 + 2\tan(x + y) + \tan^2(x + y)} = 1 + \tan(x + y)$

(d) $e^{x-y}(y - 2) = 1 + k$

(3) (a) $x^2 - 2xy + 2c = 0$ (b) $k = 1 - x \log x$ or $x \log x = 1 - k$ (c) $\sin^{-1} y - \sin^{-1} x = c$ (d) $e^y - e^x = c$

(4) (a) $y - x + \log\left(\frac{y - 1}{x - 1}\right) = c$ or $\frac{y - 1}{x - 1} = e^{x-y+c}$ or $y = (x - 1) . e^{x-y+c} + 1$

(b) $\sin y\,(e^x + 1) = k \cos y$ [where $k = \log c$] (c) $(1 + x)^2(1 - y^2) = \frac{1}{c^2}$ (d) $c . \sin^2 x = e^{\tan y}$ or $\frac{e^{\tan y}}{\sin^2 x} = c$

(5) (a) $\tan^{-1}(x + y + 1) = x + c$ or $x + y + 1 = \tan(x + c)$ or $x + y = \tan(x + c) - 1$

(b) $\tan^{-1}\left(\frac{4x + y}{2}\right) = 2(x + c)$ (c) $(x + c)e^{x-y} = 1$ (d) $\log(\sec x + \tan x) = -2 \sin x + c$

(6) (a) $\frac{\sqrt{6}}{2}\tan^{-1}\left[\frac{3\tan\left(\frac{3x + y}{2}\right) + 1}{2\sqrt{2}}\right] = x + c$ (b) $\cos(y - x)\,e^{x+c} = 1$

(c) $2\sin(x-y)+x^2=2c$ (d) $y-3c=1+a\tan^{-1}\left[\frac{\sqrt{3}}{a}(3x+y-1)\right]$

(7) (a) $3x-y-\log(x-y+2)=c$ (b) $x-y-8\log(8x-4y-1)=c$

(c) $\log\left(\tan\frac{y}{2}\right)=2\sin x+c$ (d) $\log(x\cot y)=-(x+c)$ or $x\cot y=e^{-(x+c)}$

(8) (a) $y-2\tan^{-1}\left(\frac{x+y}{2}\right)=c$ (b) $y^5.e^y=k\cot x$ or $y^5=\cot x.e^{k-y}$ where $k=\log c$

(c) $-\frac{1}{4}e^{-4y}=\frac{1}{3}e^{3x}+c$ or $3e^{-4y}+4e^{3x}=k$ where $k=4c$ (d) $\log y=\frac{2}{\sqrt{7}}\tan^{-1}\left(\frac{2x+1}{\sqrt{7}}\right)+c$

(9) (a) $a^2y^2-(1-b^2)x^2=c$ (b) $\frac{e^{-by}}{b}=e^x(x^2-2x+2)-\frac{e^{ax}}{a}-c$

(c) $\frac{2\sqrt{5}}{\sqrt{91}}\tan^{-1}\left[\frac{10\tan\left(\frac{10x+6y}{2}\right)+3}{\sqrt{91}}\right]=x+c$ (d) $y^2-x^2-2x=2c$ or $y^2-x^2-2x=1+2c$

(10) (a) $\log\left[\tan\left(\frac{x+y}{2}\right)\right]=\frac{x^2}{2}+c$ (b) $\log\left(\frac{\cot y}{x^3}\right)=x+c$

(c) $x(1-y)=(x-1)c$ or $x-xy=(x-1)c$ (d) $y\sin y-\cos y=x^2(\log x-1)+c$

(11) (a) $x+\sqrt{x^2+y^2}=y^2k$ or $x=y\sin(\log ky)$ (b) $x+\sqrt{x^2-y^2}=y^2k$ or $x=y\cos(\log ky)$

(c) $\log(xy)=x+k$ or $xy=e^{x+k}$ (d) $\log\left(\frac{x}{y}\right)+\frac{1}{y}=c$

(12) (a) $x\sqrt{x^2+y^2}+y=k$ where $k=-c\sqrt{x^2+y^2}$ (b) $-x+\log\left(1+\frac{1}{xy}\right)=c$

(c) $\log(x+1)+\frac{1}{y(x+1)}=-k$ where $k=\log c$ (d) $y=\log\sqrt{x}+\log k$ or $y=\log(k\sqrt{x})$

(13) (a) $\left(\sin y+\sqrt{1+\sin^2 y}\right)\left(\cos x+\sqrt{1+\cos^2 x}\right)=k$ (b) $ye^{y+1}-x-\log x=k$

(c) $y-x+\log xy^2=k$ (d) $t\sin t+\cos t+z\cos z-\sin z=k$ where $t=y+1, z=x+2$

(e) $e^y+1=(e^x+1)e^k$

Exercise – A13

Solve the following differential equations: –

(1) (a) $(x-y)\frac{dy}{dx}=y$ (b) $(x^3+y^3)\frac{dy}{dx}=x^2y$ (c) $(x+y)\frac{dy}{dx}=y$ (d) $\frac{dy}{dx}=\frac{y}{x}+\frac{1}{\log\left(1+\frac{y}{x}\right)}$

(2) (a) $(x-y)\frac{dy}{dx}=x+y$ (b) $\frac{dy}{dx}=\frac{3y-x}{3x-y}$ (c) $(y^2-x^2)\,dx=xy\,dy$ (d) $xy^2\,dy=(x^3+y^3)\,dx$

(3) (a) $(x^2+y^2)\,dy=\frac{y^3}{x}\,dx$ (b) $x+y\frac{dy}{dx}=\sqrt{x^2+y^2}$ (c) $(xy+x^2)\frac{dx}{dy}=(y^2+xy)$ (d) $xy+y^2\frac{dx}{dy}=x^2+x^2\frac{dx}{dy}$

(4) (a) $xy\frac{dy}{dx}=y^2+x^2\sec\left(\frac{y}{x}\right)$ (b) $\frac{y}{x}.\frac{dy}{dx}=\left(\frac{y}{x}\right)^2+\left(\frac{x^2+y^2}{x^2}\right)$ (c) $x\sqrt{1+\frac{x}{y}}.\frac{dy}{dx}=\sqrt{1+\frac{y}{x}}+y\sqrt{1+\frac{x}{y}}$

(d) $(x^2+y^2)\frac{dy}{dx}=xy$ (e) $\frac{dy}{dx}=\frac{x-y}{x-y+1}$

(5) (a) $y\frac{dx}{dy} = x - y\sin^2\left(\frac{x}{y}\right)$ (b) $3xy^2\frac{dy}{dx} = x^3 + y^3$ (c) $2x\frac{dy}{dx} = \log_y y - \log_x y + 2y$ (d) $\frac{dy}{dx} = \frac{y(x-y)}{x^2}$

(6) (a) $2x\frac{dy}{dx} = 2y - \sqrt{x^2 - y^2}$ (b) $y\frac{dx}{dy} = x - \sqrt{x^2 + y^2}$ (c) $x\,dy = (\sqrt{xy} + y)\,dx$ (d) $\left(y + \sqrt{x^2 + xy + y^2}\right)dx = x\,dy$

(7) (a) $y^2\,dx = (x^2 + xy + y^2)\,dy$ (b) $x^2y\frac{dy}{dx} = x(y^2 + xy + x^2)$ (c) $(x^2 + 2xy + y^2)\frac{dy}{dx} = x^2 - y^2$

(8) (a) $\frac{dy}{dx} - \frac{y}{x} = \cos^2\left(\frac{y}{x}\right)$ (b) $\frac{y}{x}.\frac{dy}{dx} = \frac{x^2 + y^2}{x^2} + \cos\left(\frac{2y^2}{x^2}\right)$ (c) $\frac{dy}{dx} + \frac{y^2}{x^2} = 1 + \frac{y}{x}$

(d) $\cos\left(\frac{y}{x}\right)\frac{dy}{dx} = \frac{y}{x}\cos\left(\frac{y}{x}\right) + \sin\left(\frac{y}{x}\right)$

(9) (a) $\frac{y}{x}\cot\left(\frac{y}{x}\right) + \frac{y^2}{x^2} = \frac{y}{x}.\frac{dy}{dx}$ (b) $y^2\frac{dx}{dy} = \frac{y(x+y)}{3}$ (c) $x\frac{dy}{dx} + y\log x = y\log y$ (d) $(x + 2y)\,dy - y\,dx = x\,dx$

(10) (a) $\frac{dy}{dx} = \frac{3x + 5y + 1}{x + y + 2}$ (b) $\frac{dy}{dx} = \frac{2x + y + 1}{x + 2y + 3}$ (c) $\frac{dy}{dx} = \frac{3y - x + 5}{x - y + 7}$ (d) $\frac{dy}{dx} = \frac{x - 4y + 1}{4x - 3y + 5}$

(11) (a) $\frac{dx}{dy} = \frac{x - y + 1}{x + y + 3}$ (b) $\frac{dx}{dy} = \frac{x + 2y + 1}{x + y}$ (c) $\frac{dy}{dx} = \frac{10x + 3y + 15}{5x + 6y + 30}$ (d) $\frac{dy}{dx} = \frac{x - y}{2y - 3x + 5}$

(12) (a) $\frac{dy}{dx} = \frac{7x + 5y + 14}{5x - 3y + 10}$ (b) $\frac{dy}{dx} = \frac{10x + 7y + 20}{9x + 5y + 30}$

Answer

(1) (a) $xc = e^{-\frac{x}{y}}$ or $xe^{\frac{x}{y}} = \frac{1}{c}$ (b) $\frac{-x^3 + 2y^3}{2xy^2} = \log x + k$ (c) $x = ce^{\frac{x}{y}}$ (d) $(x + y)\left[\log\left(\frac{x+y}{x}\right) - 1\right] = x\log x + k$

(2) (a) Hint:– $\tan^{-1}\left(\frac{y}{x}\right) - \frac{1}{2}\log\left(\frac{x^2 + y^2}{x^2}\right) = \log x + \log c$ or $2\tan^{-1}\left(\frac{y}{x}\right) - \log\left(\frac{x^2 + y^2}{x^2}\right) = 2\log x + 2\log c$

or $2\tan^{-1}\left(\frac{y}{x}\right) = \log\left(\frac{x^2 + y^2}{x^2}\right) + \log x^2 + k$ or $2\tan^{-1}\left(\frac{y}{x}\right) = \log\left(\frac{x^2 + y^2}{x^2}.x^2\right) + k$

or $2\tan^{-1}\left(\frac{y}{x}\right) - \log(x^2 + y^2) = k$ Ans. where $k = 2\log c$

(b) $\log\left(\frac{x}{x+y}\right) = x + c$ (c) $\frac{y^2}{2x^2} + \log x = k$ (where $k = \log c$) (d) $\frac{y^3}{3x^3} - \log x = k$

(3) (a) $2x^2.\log y + y^2 = 2x^2k$ where $k = \log c$ (b) $\log\left(\frac{x - \sqrt{x^2 + y^2}}{x^2}\right) = k$

(c) $\frac{x^2}{2y^2} + \log\left(\frac{x}{y}\right) - \log y = k$ (d) $x^2 + 2xy = e^{2k}$ where $k = \log c$

(4) (a) $y\sin\left(\frac{y}{x}\right) + x\cos\left(\frac{y}{x}\right) = x\log x + xk$ where $k = \log c$

(b) $\sqrt{x^2 + y^2} = x^2k$ or $x^2 + y^2 = x^4k^2$ (c) $2\sqrt{\frac{y}{x}} - \log x = k$

(d) $\log x - \frac{x^2}{2y^2} = k$ (e) $x^2 + y^2 - 2y(x + 1) = k$

(5) (a) $ye^{-\cot\left(\frac{x}{y}\right)} = k$ (b) $x^3 - 2y^3 = \frac{x}{k^2}$ (c) $y = xe^{\sqrt{x}.k}$ or $\log\left(\frac{y}{x}\right) = k\sqrt{x}$ (d) $k = xe^{-\frac{x}{y}}$

(6) (a) $2\sin^{-1}\left(\frac{y}{x}\right) + \log x = k$ (b) $x + \sqrt{x^2 + y^2} = e^k$ where $k = \log c$

(c) $2\sqrt{y} = \sqrt{x}(\log x + k)$ (d) $2y + x + 2\sqrt{x^2 + xy + y^2} = 2x^2k$

(7) (a) $\tan^{-1}\left(\frac{x}{y}\right) = \log y + k$ or $x = y\tan(\log ky)$ (b) $y - x\log(x + y) = k$ (c) $\log\left(\frac{1}{\sqrt{x^2 - 2xy - y^2}}\right) = k$

(8) (a) $\tan\left(\frac{y}{x}\right) - \log x = k$ (b) $\frac{1}{4}\tan\left(\frac{y^2}{x^2}\right) = \log(k.x)$ (c) $\sin^{-1}\left(\frac{y}{x}\right) - \log x = k$ (d) $\sin\left(\frac{y}{x}\right) = kx$

(9) (a) $\sec\left(\frac{y}{x}\right) = kx$ (b) $-\frac{3}{2}\log(y - 2x) - \frac{1}{2}\log y = k$ (c) $y = xe^{1+xc}$

(d) $\frac{1}{2}\log\left(\frac{x + \sqrt{2}y}{x - \sqrt{2}y}\right) - \frac{1}{4}\log\left(\frac{x^2 - 2y^2}{x^2}\right) = \log(xk)$

(10) (a) $-\frac{3}{2\sqrt{7}}\log\left|\frac{\sqrt{7} + (v - 2)}{\sqrt{7} - (v - 2)}\right| - \frac{1}{2}\log|3 + 4v - v^2| = \log|xk|$

Put $v = \frac{\beta}{\alpha}$ and $\alpha = x - h,\ \beta = y - k$ value of $k = \frac{5}{2}, h = -\frac{9}{2}$

(b) $\frac{1}{4}\log\left|\frac{3x + 3y + 4}{3x - 3y - 6}\right| - \frac{1}{2}\log\frac{2}{3}|3x^2 - 3y^2 - 2x - 10y - 8| = k$ where $k = \log c$

(c) $\frac{1}{\sqrt{2}}\log\left|\frac{(x + y + 19) - \sqrt{2}(x + 13)}{(x + y + 19) + \sqrt{2}(x + 13)}\right| - \frac{1}{2}\log|y^2 - x^2 + 2xy + 38y - 14x + 23| = k$ where $k = \log c$

(d) $-\frac{1}{2}\log\left|\frac{3(13y + 1)^2 - 8(13y + 1)(13x + 17) + (13x + 17)^2}{169}\right| = k$ where $k = \log c$

(11) (a) $-\frac{1}{2}\log|x^2 + y^2 + 4x + 2y + 5| - \tan^{-1}\left(\frac{x + 2}{y + 1}\right) = k$ where $k = \log c$

(b) $\frac{1}{2\sqrt{2}}\log\left|\frac{\sqrt{2}(y + 1) + (x - 1)}{\sqrt{2}(y + 1) - (x - 1)}\right| - \frac{1}{2}\log|2y^2 - x^2 + 4y + 2x + 1| = k$ where $k = \log c$

(c) $-\frac{1}{2}\log|3y^2 - 5x^2 + xy + 30y + 5x + 75| - \frac{2}{\sqrt{183}}\log\left|\frac{12\sqrt{3}(y + 5) + x(2\sqrt{3} - \sqrt{61})}{12\sqrt{3}(y + 5) + x(2\sqrt{3} + \sqrt{61})}\right| = k$

(d) $\frac{1}{\sqrt{3}}\log\left|\frac{2y - x(1 + \sqrt{3})}{2y - x(1 - \sqrt{3}) - 10}\right| - \frac{1}{2}\log|x^2 - 2y^2 + 2xy - 20x + 10y + 25| = k$ where $k = \log c$

(12) (a) $\frac{5}{\sqrt{21}}\tan^{-1}\left[\frac{\sqrt{3}y}{\sqrt{7}(x + 2)}\right] - \frac{1}{2}\log|7x^2 + 3y^2 + 28x + 28| = k$ where $k = \log c$

(b) $-\frac{1}{2}\log|5\beta^2 + 2\alpha\beta - 10\alpha^2| - \frac{4\sqrt{5}}{5\sqrt{51}}\log\left|\frac{\sqrt{5}(5\beta + \alpha) - 5\sqrt{51}\alpha}{\sqrt{5}(5\beta + \alpha) + 5\sqrt{51}\alpha}\right| = k$ where $k = \log c$

Put $\alpha = x - h,\ \beta = y - k$ and $h = -\frac{110}{13},\ k = \frac{120}{13}$

Exercise – A15

Solve the following differential equations: –

(1) (a) $\frac{dy}{dx} + \frac{y}{x} = x^2$ (b) $\frac{dy}{dx} = 1 + \frac{y}{x}$ (c) $\frac{dy}{dx} + \frac{y}{x} = \log x$ (d) $\frac{dy}{dx} = (e^x + 1) + y$

(2) (a) $\frac{dy}{dx} + y\sin x = \sin x\cos x$ (b) $\frac{dx}{dy} + \frac{xy}{\sqrt{1 + y^2}} = y + y^3$ (c) $\frac{dy}{dx} = xe^{2x} - y$ (d) $\frac{dy}{dx} + \frac{y}{x} = \sqrt{1 - x^2}$

(3) (a) $\frac{dx}{dy} - \frac{2x}{y} = \log y$ (b) $(x + 1)\,dy - (y + 1)\,dx$ (c) $\frac{dy}{dx} = \frac{y}{1 + x} + \frac{x}{\sqrt{1 + x}}$ (d) $\frac{dy}{dx} + y\tan x = \frac{\sin x}{\cos^2 x}$

(4) (a) $\frac{dy}{dx}+\frac{y}{x}\log y = y(\log y)^2$ (b) $x\frac{dy}{dx}+\frac{xy}{1+x}=\sqrt{1-x}$ (c) $(1+y)\frac{dy}{dx}+\frac{y}{x}=e^{x-y}$ (d) $\frac{dy}{dx}+\frac{y}{x\log x}=\frac{\log x}{x}$

(5) (a) $\frac{dy}{dx}-y\tan x = 2x\sec x$ (b) $xy\frac{dy}{dx}+\frac{xy^2}{\tan x}=3y\,\text{cosec}\,x$ (c) $\tan x\frac{dy}{dx}-y\sin x=\sin x$

(d) $(1+\log x)\frac{dy}{dx}+\frac{y}{x}=e^x$

(6) (a) $\frac{dy}{dx}+\frac{y(\cos^2 x-\sin^2 x)}{\sin x\cos x}=e^{\cos 2x}$ (b) $\cos x\frac{dy}{dx}+\frac{y}{\sin x}=\frac{1}{\cos^3 x}$ (c) $\frac{dy}{dx}+\frac{y}{\cot x}=e^{\tan x}.\sec x$

(d) $xdy+\frac{y(x+1)}{x}dx=\log x\,dx-\log x\,dy$ (e) $\frac{1}{\cos^2 y}dy+\frac{\tan y}{x}dx=\cos x\,dx$ (f) $\frac{1}{\sqrt{1-y^2}}dy+\frac{\sin^{-1}y}{x}dx=\frac{dx}{x}$

(7) (a) $\sin y\cos y\frac{dy}{dx}+\frac{\sin^2 y}{x}=\frac{1}{x^3-1}$ (b) $\frac{dy}{dx}+\frac{xy}{1-x^2}=\frac{2x}{\sqrt{1-x^2}}$ (c) $2\frac{dy}{dx}+\frac{y}{x}=\frac{y^2}{x}$

(d) $\frac{dy}{dx}-\frac{y}{x}=\frac{y^2}{x^2}$ (e) $2\frac{dy}{dx}-\frac{3y}{x+1}=\frac{x^2}{y}$ (f) $\frac{1}{\sin^2 y}.\frac{dy}{dx}+\frac{x\cot y}{1+x^2}=1+x$

(g) $y\frac{dy}{dx}+\frac{y^2}{x}=\sqrt{1+x}$ (h) $\frac{1}{x}.\frac{dx}{dy}+\frac{\log x}{1+y}=\frac{1}{y}$

(8) (a) $\frac{dy}{dx}+\frac{y\cos x}{\sqrt{1+\sin x}}=\sin x\cos x$

Answer

(1) (a) $4xy=x^4+4c$ (b) $y=x\log x+xc$ (c) $y=\frac{x}{2}\left[\log x-\frac{1}{2}\right]+k$ (d) $y=xe^x-1+ce^x$

(2) (a) $y-\cos x-1=ce^{\cos x}$ (b) $x=t^3-3t^2+6t-6+\frac{c}{e^t}$ $\left(\text{Put } t=\sqrt{1+y^2}\right)$

(c) $ye^x=\frac{xe^{3x}}{3}-\frac{e^{3x}}{9}+c$ (d) $xy=-\frac{(1-x^2)^{\frac{3}{2}}}{3}+c$

(3) (a) $x+y(\log y+1)=cy^2$ (b) $y+1=c(x+1)$

(c) $2\left[(1+x)^{\frac{3}{2}}+(1+x)^{\frac{1}{2}}\right]+c(1+x)$ (d) $2y-\sec x=2c\cos x$

(4) (a) $x\log y.\log\left(\frac{c}{x}\right)=1$ (b) Do yourself (c) $xye^y=e^x(x-1)+c$ (d) $3y=(\log x)^2+k$ where $k=\frac{3c}{\log x}$

(5) (a) $y\cos x=x^2+c$ (b) $y\sin x=3\log x+k$ (c) $y+1=ce^{\sin x}$ (d) $y(1+\log x)=e^x+c$

(6) (a) $y\sin 2x+\frac{1}{2}e^{\cos 2x}=c$ (b) $y\tan x=\frac{1}{4\cos^4 x}+c$ (c) $y=\cos x.e^{\tan x}+c$

(d) $y(x+\log x)=x(\log x-1)+k$ (e) $x\tan y=x\sin x+\cos x+c$ (f) $\sin^{-1}y=1+k$ where $k=\frac{c}{x}$

(7) (a) $x^2\sin^2 y=\frac{2}{3}\log|x^3-1|+k$ (b) $y=\sqrt{1-x^2}\log\left|\frac{1}{1-x^2}\right|+k$ where $k=\sqrt{1-x^2}\log|c|$

(c) solution: – $2\frac{dy}{dx}+\frac{y}{x}=\frac{y^2}{x}$ or $\frac{2}{y^2}\frac{dy}{dx}+\frac{1}{xy}=\frac{1}{x}$ Let $\frac{1}{y}=z$ or $-\frac{1}{y^2}\frac{dy}{dx}=\frac{dz}{dx}$

$$\therefore\ \frac{1}{y^2}\frac{dy}{dx}=-\frac{dz}{dx} \text{ or } -2\frac{dz}{dx}+\frac{z}{x}=\frac{1}{x} \text{ or } 2\frac{dz}{dx}-\frac{z}{x}=-\frac{1}{x} \text{ or } \frac{dz}{dx}-\frac{z}{2x}=-\frac{1}{2x}$$

$$\therefore\ \text{I.F}=e^{-\int\frac{dx}{2x}}=e^{-\frac{1}{2}\log x}=e^{\log(x)^{-\frac{1}{2}}}=x^{-\frac{1}{2}}=\frac{1}{\sqrt{x}}$$

The required solution is – $y.\frac{1}{\sqrt{x}} = -\int \frac{1}{2x}.\frac{1}{\sqrt{x}}dx = -\frac{1}{2}\int \frac{dx}{x^{\frac{3}{2}}} = -\frac{1}{2}\int x^{-\frac{3}{2}}dx = -\frac{1}{2}.\frac{x^{-\frac{3}{2}+1}}{-\frac{3}{2}+1}$

or $y.\frac{1}{\sqrt{x}} = x^{-\frac{1}{2}} + c$ or $y.\frac{1}{\sqrt{x}} = \frac{1}{\sqrt{x}} + c$ $\therefore \frac{y}{\sqrt{x}} = \frac{1}{\sqrt{x}} + c$ $\therefore y = 1 + c\sqrt{x}$ Ans.

(d) $x = y\log\left(\frac{c}{x}\right)$ Ans. (Do yourself, same as above question)

(e) $3y^2 = 3(x+1)^3\log(x+1) + 6(x+1)^2 - \frac{3}{2}(x+1) + c$ Ans. (Do yourself)

(f) $1 + x^2 + \cot y = -\sqrt{1+x^2}\log\left|x + \sqrt{1+x^2}\right| + k$ (Do yourself)

(g) $x^2y^2 = \frac{2}{7}.(1+x)^3\sqrt{1+x} - \frac{4}{5}.(1+x)^2\sqrt{1+x} + \frac{2}{3}.(1+x)\sqrt{1+x} + k$ (h) $y(\log x - 1) + \log\left(\frac{x}{y}\right) = k$

(8) (a) $y = (1+\sin x)^{\frac{3}{2}} + \frac{1}{2}(1+\sin x)^{\frac{1}{2}} - \frac{3}{2}(1+\sin x) + k$ where $k = ce^{-2\sqrt{1+\sin x}}$

www.ingramcontent.com/pod-product-compliance
Lightning Source LLC
Chambersburg PA
CBHW080803180526
45168CB00006B/2311
* 9 7 8 1 5 1 7 1 7 1 6 0 5 *